ADVANCES IN PROTEIN KINASES

Edited by **Gabriela Da Silva Xavier**

Advances in Protein Kinases
Edited by Gabriela Da Silva Xavier

Published by InTech
Janeza Trdine 9, 51000 Rijeka, Croatia

Publishing Process Manager Romina Skomersic

Technical Editor Teodora Smiljanic

Cover Designer InTech Design Team

Additional hard copies can be obtained from orders@intechopen.com

Advances in Protein Kinases, Edited by Gabriela Da Silva Xavier
p. cm.
ISBN 978-953-51-0633-3

Meet the editor

My research interest is in the fuel sensing protein kinases - AMP-activated protein kinase (AMPK) and PAS-domain containing protein kinase (PASK)- and their roles in pancreatic endocrine cell function and development. I completed my PhD in Biochemistry at the University of Bristol in 2001. After two postdoctoral positions at the University of Bristol, I moved to Imperial College in 2006 where I was appointed lecturer in 2008 and have been involved in collaborative efforts to elucidate how the Type 2 Diabetes genes identified by Genome Wide Association Studies - TCF7L2, SLC30A8, WFS1, and HHEX - may affect beta-cell function and development. My work has been funded by the Juvenile Diabetes Research Foundation and European Foundation for the Study of Diabetes.

Contents

Preface

Proteins are the work horses of the cell. As regulators of protein function, protein kinases are involved in the control of cellular functions via intricate signalling pathways, allowing for fine tuning of physiological functions. This book is a collaborative effort, with contribution from experts in their respective fields, reflecting the spirit of collaboration - across disciplines and borders - that exists in modern science. Here, we review the existing literature and, on occasions, provide novel data on the function of protein kinases in various systems. We also discuss the implications of these findings in the context of disease, treatment, and drug development.

Gabriela Da Silva Xavier
Imperial College London, Section of Cell Biology,
Division of Diabetes, Endocrinology and Metabolism,
London,
UK

1

Cross-Regulation of JAK-STAT Signaling: Implications for Approaches to Combat Chronic Inflammatory Diseases and Cancers

Claire Rutherford, Hayley D. Woolson and Timothy M. Palmer
Institute of Cardiovascular and Medical Sciences,
College of Medical, Veterinary and Life Sciences,
University of Glasgow,
Scotland,
U.K.

1. Introduction

The Janus kinase–signal transducer and activator of transcription (JAK–STAT) pathway is utilized by a range of cytokines (interferons, IL-2 and IL-6 amongst others) that control survival, proliferation and differentiation responses in diverse cell types. The realisation that unregulated activation of this pathway is a key driver of not only chronic inflammatory diseases such as rheumatoid arthritis, colitis and psoriasis, but also many cancers has identified its components as targets for therapeutic intervention by small molecule inhibitors and biologicals. In this article, we will discuss how an increased understanding of JAK-STAT pathway architecture, the basis for its dysfunction in pathological states, and its regulation by other intracellular signaling pathways are illuminating multiple strategies to manipulate this pathway in several disease arenas.

2. Basic architecture of the JAK-STAT pathway

2.1 Janus Kinases (JAKs)

JAKs encompass a family of four of cytoplasmic tyrosine kinases (JAK1–JAK3, TYK2) that function as essential signaling components immediately downstream of receptors for many haematopoietic cytokines, such as granulocyte-macrophage colony-stimulating factor (GM-CSF), erythropoietin (Epo), interferons (IFNs), interleukins (e.g. IL-2, IL-6) as well as growth hormone and leptin. While JAK1, JAK2 and TYK2 are widely expressed, JAK3 expression is limited to haematopoietic cells where it is used by the receptors for a selected group of cytokines that are critical in T cell, B cell and natural killer cell development (Ghoreschi et al., 2009). Importantly, functional deficiencies in JAK3 have been shown to account for autosomal recessive "severe combined immunodeficiency" (SCID) syndrome (O'Shea et al., 2004). However, despite these differences, JAKs are thought to function downstream of individual cytokine receptors in a similar manner.

Structurally JAKs comprise seven conserved domains (JAK homology (JH) domains 1-7) numbered from the carboxyl to the amino terminus (Fig. 1). The hallmark of the JAK family is the presence of JH1, which comprises a functional tyrosine kinase domain, and JH2, which was originally thought to be a catalytically inactive pseudokinase domain. However, it has recently been demonstrated that the JAK2 JH2 has dual-specificity protein kinase activity that phosphorylates Ser523 and Tyr570, which are critical negative regulatory sites, although the specific kinase activity of JH2 is approximately ten-fold less than that of JH1 (Ungureanu et al., 2011). It had long been established that JH2 has a negative regulatory function, as deletion of this domain has been shown to increase JAK2 and JAK3 phosphorylation and downstream activation of STATs. This is achieved *via* an intramolecular interaction between JH1 and JH2, which effectively suppresses basal kinase activity. Upon ligand binding, conformational changes relieve this interaction allowing activation of JH1 by phosphorylation of two activation loop Tyr residues (Tyr 1021/1022 in JAK1, Tyr 1007/1008 in JAK2, Tyr 980/981 in JAK3, Tyr 1054/1055 in TYK2) (Saharinen et al., 2000; Saharinen & Silvennoinen, 2002). Abrogation of JH2's dual specificity kinase activity was found to be sufficient to increase basal JAK2 activity and downstream signaling to STAT1, suggesting that this is a key element of the JH2 domain's suppressive effect on JH1. As described later in this chapter, this provides an essential "braking" function as mutations that disrupt the JH2 domain's suppression of JH1 activity are found in several haematological disorders.

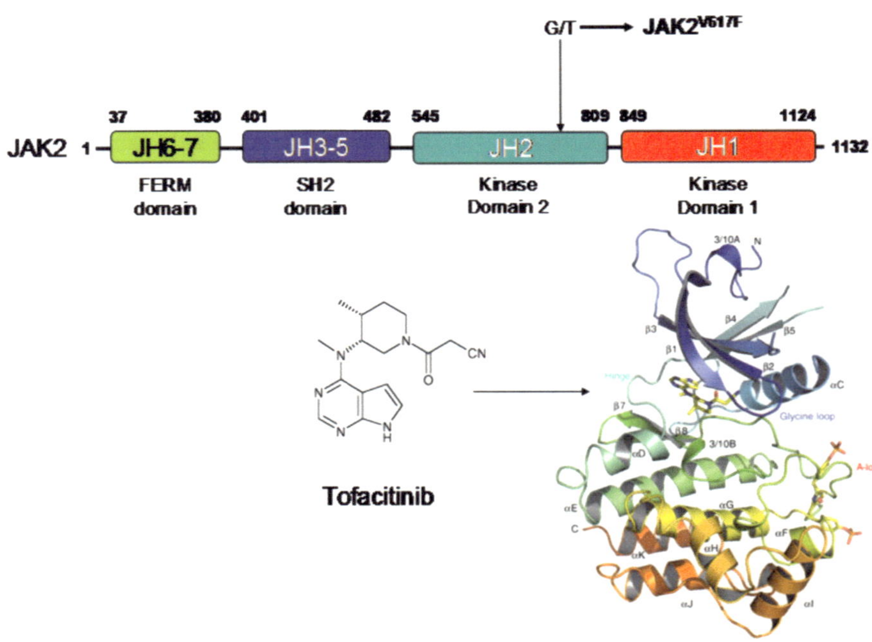

Fig. 1. Domain structure of human JAK2. The crystal structure of the JAK2 JH1 domain bound to JAK inhibitor tofacitinib is reproduced from Williams et al. (2009) with permission.

JH3-5 comprise an SH2 domain typically indicative of PTyr-dependent interactions with other signaling components but surprisingly partners for this domain have yet to be identified for any of the JAK family (Ingley & Klinken, 2006). Lastly, JH6-JH7 constitute a

Band 4.1, ezrin, radaxin, moesin (FERM) domain implicated in interactions with cytokine receptors required for cell surface expression (Huang et al., 2001) and also the JH1 domain to regulate its kinase activity (Zhou et al., 2001).

JAKs have typically been thought to be constitutively associated with dimeric cytokine receptor complexes at the plasma membrane, although a role for nuclear-localised JAKs in controlling the phosphorylation of histone H3 has emerged that may be particularly important in haematological malgnancies and embryonic stem cell renewal (Dawson et al., 2009; Griffiths et al., 2011). However, following cytokine binding, a conformational change within the dimeric receptor complex allows activation of receptor-bound JAKs due to transphosphorylation of JH1 activation loop Tyr residues. The structural basis of the JAK-receptor interaction and the mechanism by which the receptor re-orientates to receive the phosphorylation are currently unclear and the subject of intense research (Lupardus et al., 2011). Nevertheless, after receptors have been phosphorylated at specific Tyr residues, SH2 domain-containing proteins are recruited to activate downstream signaling. The major intracellular mediators commonly activated by multiple cytokine receptors are the signal transducer and activator of transcription (STAT) proteins.

2.2 Signal Transducers and Activators Of Transcription (STATs)

The STAT family of transcription factors were first described as interferon (IFN)-inducible transcription factors (Fu, 1992; reviewed by Darnell *et al.*, 1994;). The STAT family comprises seven mammalian members (STAT1, STAT2, STAT3, STAT4, STAT5a, STAT5b and STAT6). However alternative splicing of STAT1, 3, 4, 5a and 5b primary transcripts yields isoforms with truncated C-terminal domains; for example the STAT3β isoform lacks the 55 C-terminal amino acids of STAT3α but acquires a unique 7 amino acid sequence (Dewilde et al., 2008). While the β isoforms have been reported to act as dominant negative regulators of transcription, it is now apparent that these isoforms can differ in their transcriptional activities, for example isoform-specific deletions of STAT3β and STAT3α have shown that STAT3β activates a distinct subset of STAT3 genes in response to IL-6 and may not act as a dominant negative *in vivo* (Maritano et al., 2004; Dewilde et al., 2008). Exactly how the discrete functional differences observed between STAT3α and STAT3β relate to their distinct subcellular trafficking kinetics remains to be determined (Huang et al., 2007).

STATs are activated by a wide range of cytokines and other growth factors. Depending on the cell type involved, a range of sometimes overlapping ligands can activate STAT1, STAT3, STAT5a and STAT5b, whereas only a few cytokines are capable of activating STAT2, STAT4 and STAT6 (Lim & Cao, 2006). For example, in addition to multiple IL-6 family members, STAT3 can also be activated by G-protein coupled receptors for angiotensin II and thrombin, as well as the VEGFR2 receptor tyrosine kinase (Madamanchi et al., 2001; Pelletier et al., 2003; Bartoli et al., 2000). Functionally, STAT3 has been implicated in cell proliferation, survival and anti-apoptotic responses in multiple cell types (Yu et al., 2009). Conversely, STAT6 is predominantly activated by IL-4 and is involved in the differentiation of $CD4^+$ T cells into T helper 2 (T_h2) cells (Hebenstreit et al., 2006).

Structurally, STATs comprise several distinct functional domains; these include an N-terminal domain, a coiled-coil domain, a DNA binding domain, a linker domain, an SH2 domain and a C-terminal transactivation domain. The N-terminal domain is involved in

STAT dimerisation and tetramerisation, and for some STATs the recruitment of regulatory proteins such as STAT E3 ubiquitin ligase SLIM (Tanaka et al., 2005). The coiled-coil domain is also implicated in protein-protein interactions; for example, the transcription factor c-Jun has been demonstrated to interact with this domain in STAT3, facilitating co-operation required for maximal IL-6-dependent acute-phase response gene activation driven by the 2-macroglobulin enhancer (Zhang et al., 1999b). This domain may also be involved in ensuring high affinity binding to cytokine receptors, as mutations within it impair recruitment of STAT3 to Tyr-phosphorylated gp130 (Zhang et al., 2000). The DNA-binding domain is highly conserved amongst the STAT family and as well as binding DNA, it also controls STAT translocation from the cytoplasm to the nucleus. It has been proposed to achieve this by maintaining the necessary conformation for binding of importins on the nuclear membrane (Ma & Cao, 2006). Once in the nucleus, STAT dimers can bind DNA motifs known as GAS (γ activated sequence) elements ($TTN^{5\text{-}6}AA$, N=any nucleotide) except in the case of the IFNα/β response, where complexes formed between STAT1, STAT2 and IRF9 (interferon regulatory factor 9) bind to the IFNα/β-response element (ISRE) $AGTTN_3TTTC$ (O'Shea et al., 2002). The linker domain has been implicated in transcriptional activation, since point mutations within this region of STAT1 have been found to abolish transcriptional responses to IFNγ (Yang *et al.*, 1999). Additionally, this domain also participates in distinct protein-protein interactions, as demonstrated by the interaction of STAT3 with "genes associated with retinoid–IFN-induced mortality-19" (GRIM-19), which blocks STAT3-mediated transcriptional activity (Kalakonda et al., 2007).

The most conserved domain within the STAT family is the SH2 domain. This domain is essential for binding to activated receptors and is also responsible for cytokine-triggered dimerisation *via* specific phospho-Tyr residues (Shuai et al., 1994; Hemmann et al., 1996). Different receptor motifs determine which STATs are recruited; for example, STAT3 will bind to pYXXQ (Stahl et al., 1995) while STAT1 will only bind to pYXPQ (Gerhartz et al., 1996). This difference has been shown to be due to the SH2 domain through the creation of a chimaeric STAT3 molecule in which the SH2 domain of STAT3 was substituted with a STAT1 SH2 domain, resulting in a molecule that showed the receptor motif binding preference of STAT1 (Hemmann et al., 1996). On recruitment to an activated cytokine receptor, STATs are then phosphorylated by JAKs on a single conserved tyrosine residue at the carboxyl end of the SH2 domain (e.g. Tyr701 in STAT1 (Shuai et al., 1994) and Tyr705 in STAT3 (Kaptein et al., 1996)). This enables them to form dimers through an interaction of the P-Tyr on one STAT with the SH2 domain of another.

Lastly, the C-terminal transactivation domain mediates protein-protein interactions necessary for optimal transcriptional activation; these include interactions with the transcriptional co-activators "cAMP response element binding protein" (CREB)-binding protein (CBP) and p300 (Paulson et al., 1999). In the cases of STAT1 and STAT3, optimal interaction with p300/CBP requires phosphorylation of Ser727 by any of several Ser/Thr kinases, including extracellular signal-regulated kinases 1 and 2 (ERK1,2) (Schuringa et al., 2001; Heinrich et al., 2003).

To demonstrate how these proteins function in context, we will describe signaling pathway activation in response to the activation of two key cytokine receptor systems implicated in several haematological and non-haematological conditions: the IL-6 and IFN signaling complexes.

2.3 Activation of the JAK-STAT and ERK1,2 pathways by IL-6

The IL-6 receptor is composed of two different subunits, an 80 kDa IL-6-binding protein (IL-6Rα) and a 130 kDa signal-transducing subunit (gp130), which is shared by all IL-6-family cytokines, which include IL-11, IL-27, oncostatin M, leukaemia inhibitory factor (LIF), cardiotrophin-1 (CT-1) and ciliary neurotrophic factor (CNTF) amongst others (Heinrich et al., 2003). The gp130 subunit is ubiquitously expressed while IL-6Rα expression is restricted to hepatocytes, monocytes, neutrophils and some B and T cell subsets. However, IL-6 can also bind to a soluble form of the receptor (sIL- 6Rα) which is either shed from cell membranes by limited proteolytic cleavage of membrane-bound IL-6Rα by the metalloproteinases ADAM10 and ADAM17 or created by alternative splicing of IL-6Rα primary transcripts (Kallen, 2002). The resulting sIL-6Rα/IL-6 so-called "trans-signaling" complex can associate with gp130 on cells that do not express the membrane-bound IL-6R thereby widening the spectrum of IL-6-responsive cells. For example, vascular endothelial cells which express only gp130 are rendered responsive to IL-6 due to the shedding of sIL-6Rα from activated neutrophils (Marin et al., 2001).

Binding of the trans-signaling complex to gp130 or the interaction of IL-6 with IL-6Rα triggers dimerisation of the gp130 subunits and the formation of an active receptor signaling complex (Murakami et al., 1993; Fig. 3). JAK1, JAK2 and TYK2 have each been shown to be activated upon receptor stimulation (Stahl *et al.*, 1994; Narazaki et al., 1994) and can phosphorylate gp130 on Tyr's 683, 759, 767, 814, 905 and 915 (Stahl et al., 1994; Stahl et al., 1995; Gerhartz et al., 1996). (Fig. 2) Studies using cell lines lacking either JAK1, JAK2 or TYK2 have revealed that whereas JAK2 and TYK2 may be functionally interchangeable, effective downstream signaling absolutely depends on the presence of JAK1 (Guschin et al., 1995). Thus, phosphorylation of gp130 was demonstrated to be significantly reduced in the absence of JAK1 but was unimpaired in the absence of either JAK2 or TYK2. The membrane-proximal regions of gp130 are predominantly responsible for binding JAK1. These regions contain conserved box1 and box2 motifs which are both required for efficient JAK binding. Either deletions or mutations to box1 result in the impaired binding of JAKs to gp130 (Haan et al., 2000) while deletion of box2 only leads to JAK association when the kinase is over-expressed (Tanner et al., 1995), suggesting that box2 increases the affinity of JAK binding. In addition, an interbox1-2 region on gp130 is also involved in JAK binding and again, in studies where this region has been mutated, defective JAK signaling has been observed (Haan et al., 2000).

All IL-6 type cytokines are capable of activating STAT1 and STAT3 *via* gp130. However, STAT3 activation has been observed to a greater extent than STAT1 activation (Heinrich et al., 2003). STAT recruitment to activated IL-6 type receptors is mediated by the STAT SH2 domain and requires the phosphorylation of specific Tyr residues. In particular, STAT3 binds four phospho pYXXQ motifs of gp130 (Y^{767}RHQ, Y^{814}FKQ, Y^{905}LPQ and Y^{915}MPQ) (Stahl et al., 1995), whereas STAT1 is more restricted and binds two pYXPQ motifs in gp130 (Y^{905}LPQ and Y^{915}MPQ) (Gerhartz et al., 1996) (Fig. 2). Once recruited, STATs are phosphorylated by JAKs on a single Tyr residue (Tyr701 in STAT1 and Tyr705 in STAT3 (Kaptein *et al.*, 1996; Shuai *et al.*, 1993) (Fig. 2). In addition, as discussed earlier, both STAT1 and STAT3 can be phosphorylated by ERK1,2 on Ser727 to enhance p300/CBP recruitment for maximal transcriptional activity (Schuringa et al., 2001; Heinrich et al., 2003). Following phosphorylation, activated STATs form homo- and/or hetero-dimer complexes, consisting of STAT1-STAT1, STAT1-STAT3 or STAT3-STAT3 dimers, which translocate to the nucleus to bind response elements of IL-6 inducible genes. STATs bind to essentially two types of response elements; interferon stimulated response elements (ISREs) and gamma-activated sites (GAS). The ISRE appears to be restricted to IFN signaling (Fu et al., 1990), whereas the GAS, including *sis*-inducible element (SIE), acute phase response element (APRE) and other GAS-like sequences are present in promoters of genes such as *c-fos* and the acute phase proteins that are well-defined STAT targets (Wegenka et al., 1993). Other target genes downstream of STAT3 include cell cycle regulators such as cyclin D1, and anti-apoptotic genes such as survivin and Bcl-X_L (Alvarez & Frank., 2004).

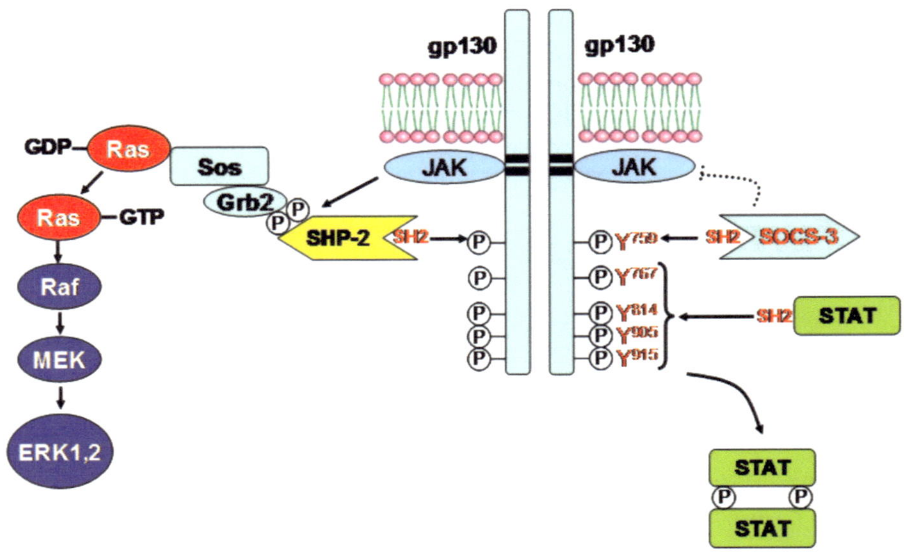

Fig. 2. JAK-mediated activation of ERK1,2 and STATs *via* gp130

Finally, it should be emphasized that SHP2 (SH2-domain-containing cytoplasmic protein tyrosine phosphatase), a ubiquitously expressed and highly conserved PTP, is also recruited to the activated IL-6 receptor. SHP2 comprises two N-terminal SH2 domains and a C-terminal PTP domain. It is recruited to pTyr759 on gp130 following receptor activation and is subsequently phosphorylated by JAKs. The phosphorylation of SHP2 provides SH2 domain docking sites for the adapter protein Grb2 (growth factor receptor binding protein 2), which is constitutively associated with the Ras guanine nucleotide exchange factor Son of sevenless (Sos) (Fig. 2). It has been proposed that the C-terminal domain residues Tyr542 and Tyr580 within SHP2 interact with the Grb2-Sos complex (Heinrich et al., 2003). Sos recruitment to the receptor complex allows for the activation of Ras at the plasma membrane, which in turn leads to the activation of the Ras-Raf-MEK-ERK1,2 cascade. ERK1,2 activation results in the preferential phosphorylation of substrates with the consensus sequence PXS/TP and more than 150 substrates have been identified to date (Yoon & Seger, 2006).

2.4 Activation of the JAK-STAT pathway by interferons (IFNs)

The IFNs comprise three families of secreted proteins that work in both an autocrine and paracrine manner to trigger intracellular signaling networks designed to combat viral infection. Type I IFNs in humans are derived from seventeen genes clustered within chromosome 9. In humans, type I IFNs comprise thirteen IFNα's and a single version of each of IFNβ, ω, ε and κ (Borden et al., 2007). Interestingly, expression of IFN-κ and IFN-ε seems tissue specific, but all cells have the potential to produce other IFNs although the spectrum of IFNs actually induced appears to vary in a stimulus-specific manner (LaFleur et al., 2001; Hardy et al., 2004). IFNγ is designated as a distinct type II IFN because of its distant sequence homology with the type I IFNs, as well as its production by natural killer (NK) or activated T cells. The type III IFN family comprises three IFN-λ subtypes which are co-produced with IFN-β, and which activate the same main signaling pathways as type I IFNs but *via* different receptor complexes (Borden et al., 2007).

2.4.1 IFNγ signaling

The functional IFNγ receptor (IFNGR) comprises two 90 kDa IFNGR1 and two 62 kDa IFNGR2 chains. IFNGR1 is involved in ligand binding and signal transduction while IFNGR2 has a limited role in binding but is essential for downstream signaling (Stark et al., 1998). Originally, these subunits were not thought to interact in unstimulated cells but advances in imaging techniques in living cells have shown that the receptor is pre-assembled and ligand binding results in a conformational change to allow signaling to occur (Krause et al., 2002). IFNGR1 and IFNGR2 have binding motifs for JAK1 and JAK2 respectively. JAK1 binds to the membrane proximal sequence, L^{266}PKS on IFNGR1 (Kaplan et al., 1996) while JAK2 binds a Pro-rich noncontiguous motif of P^{263}PSIP followed by I^{270}EEYL on IFNGR2 (Bach et al., 1996).

Upon receptor activation, JAK2 is autophosphorylated, enabling it to trans-phosphorylate JAK1 (Briscoe et al., 1996). The activated JAKs then phosphorylate each IFNGR1 chain on Tyr440 within the sequence Y^{440}DKPH, which creates a pair of docking sites for STAT1 SH2 domains. STAT1 is then phosphorylated on Tyr701, dissociates from the receptor, forms homodimers and translocates to the nucleus. Activation of the phosphatidylinositol 3-kinase (PI3K) pathway also appears to play a role in IFNγ-induced STAT1-mediated transcriptional regulation, as inhibition of either PI3K or one of its downstream effectors, protein kinase C δ (PKCδ), blocks STAT1 phosphorylation on Ser727, thus reducing its transcriptional activity. Therefore, since IFNγ has been shown to activate PKCδ in a PI3K dependent manner, it has been suggested that PKCδ is an IFNγ-regulated Ser 727 kinase for STAT1 (Nguyen et al., 2001; Deb *et al.*, 2003), although a role for PKC-activated p38 MAP kinase has recently emerged (Borden et al., 2007).

2.4.2 IFNα/β signaling

Type I IFNs belong to the cytokine group that display similar secondary structures of a five α-helix bundle stabilized by two disulphide bonds (Borden et al., 2007). Binding of type I IFNs to their cognate receptor complex, which comprises one chain each of IFNAR1 and IFNAR2c, activates TYK2 and JAK1 constitutively associated on the respective chains. Interestingly, the intracellular domain of IFNAR2c plays the key role in the recruitment and docking of STATs while deletion of much of the cytoplasmic domain of IFNAR1 is without effect on signaling. The presence of a single Tyr residue within the IFNAR2c cytoplasmic domain (either Tyr337 or Tyr512) is required for a full IFN response from this receptor (Borden et al., 2007; van Boxel-Dezaire & Stark, 2007).

Activated TYK2 then phosphorylates receptor-bound STAT2, an essential component of type I IFN signaling, on Tyr690 allowing it to interact with STAT1, which is then phosphorylated on Tyr701. STAT1 and STAT2 form a heterodimeric complex which dissociates from the receptor and translocates to the nucleus. It is now apparent that type I IFNs can activate all STAT members, but the transcription factor complex unique to type I IFNs is ISGF3 (IFN-stimulated gene factor 3), a trimeric complex of STAT1, STAT2 and a 48kDa DNA-binding protein IRF9 (IFN regulatory factor 9). Upon nuclear import, the STAT1 and IRF9 components of the complex specifically bind ISREs within target gene promoters. STAT2 does not contribute to the DNA binding activity of the ISGF3 complex, instead contributing a powerful C-terminal transactivation domain responsible for recruitment of p300/CBP co-activators (Borden et al., 2007).

3. Negative regulation of cytokine signaling

Cytokine signaling is typically transient, suggesting the involvement of negative regulatory steps aimed at terminating or resolving responses. Indeed, controlling these responses is crucial for avoiding detrimental inflammatory outcomes, including the development of diseases such as rheumatoid arthrisis (RA), Crohn's disease and Castleman's disease. There are many mechanisms with which to negatively control cytokine signaling but for the purposes of this chapter we will focus on soluble ligand traps, suppressors of cytokine signaling (SOCS) proteins and PTPs.

3.1 Soluble ligand traps

As previously mentioned, the IL-6 receptor is composed of two different subunits, an 80 kDa IL-6Rα and a 130 kDa signal-transducing gp130 subunit. A soluble form of the signal transducer protein gp130 (sgp130) was also detected in the circulation at relatively high concentrations (100–400 ng/ml in human plasma) (Narazaki et al., 1993; Chalaris et al., 2011). Sgp130 is produced by alternative splicing and has been shown to bind the sIL-6R/IL-6 complex in the circulation, thus acting as a selective inhibitor of IL-6 mediated trans-signaling, as classic signaling *via* membrane-localised IL-6R is not affected (Muller-Newen et al., 1998). Moreover, sgp130 appears to be specific for the IL-6/sIL-6R complex since signaling from gp130 in response to other IL-6-type cytokines such as LIF and OSM were only inhibited at 100–1000-fold higher concentrations (Chalaris et al., 2011).

As discussed later, the ability of either endogenous or genetically engineered soluble cytokine ligand traps to block binding to and activation of signaling from endogenous cytokine receptors has emerged as a very useful strategy to turn off excessive cytokine signaling associated with inflammatory and autoimmune diseases (Jones et al., 2011). In the case of IL-6, they have also increased our understanding of the contributions of classical *versus* trans-signaling in specific biological processes (Scheller et al., 2011).

3.2 Suppressor of Cytokine Signaling (SOCS) proteins

There are eight members of the suppressor of cytokine signaling (SOCS) family of proteins; CIS (cytokine-inducible SH2 domain-containing protein) and SOCS1 through to SOCS7 (Fig.

3). SOCS1 was the first member to be discovered in 1997 by three independent groups (Endo et al., 1997; Naka et al., 1997; Starr et al., 1997). Using the predicted amino acid sequence of SOCS1 as a probe, database searches identified 20 proteins with shared sequence homology within the C-terminal SOCS box region. Based on the presence of a central SH2 domain, the SOCS proteins were subdivided into a group of their own. The remaining proteins were divided into the following groups; WD-40-repeat proteins with a SOCS box (WSB proteins), ankyrin repeat proteins with a SOCS box (ASB proteins), sprouty (SPRY) domain-containing SOCS box proteins (SSB proteins) and GTPase domain-containing proteins (RAR and RAR-like proteins) (Krebs & Hilton, 2001). In addition to a central SH2 domain, all members of the SOCS family contain an amino-terminal region of variable length (50-380 amino acids) and a conserved 40 amino acid carboxyl terminal "SOCS box" (Alexander, 2002; Yoshimura et al., 2007). Analysis of the primary amino acid sequences of all SOCS members has revealed paired associations according to sequence similarity. Thus, CIS/SOCS2, SOCS1/SOCS3, SOCS4/SOCS5 and SOSC6/SOCS7 form related pairs. CIS and SOCS1-3 are the best characterised members of the family, while the remainder are poorly understood in comparison. Since both SOCS1 and SOCS3 are well studied, homology-paired and have been shown to potently inhibit cytokine signaling, focus will be placed on these SOCS members in particular.

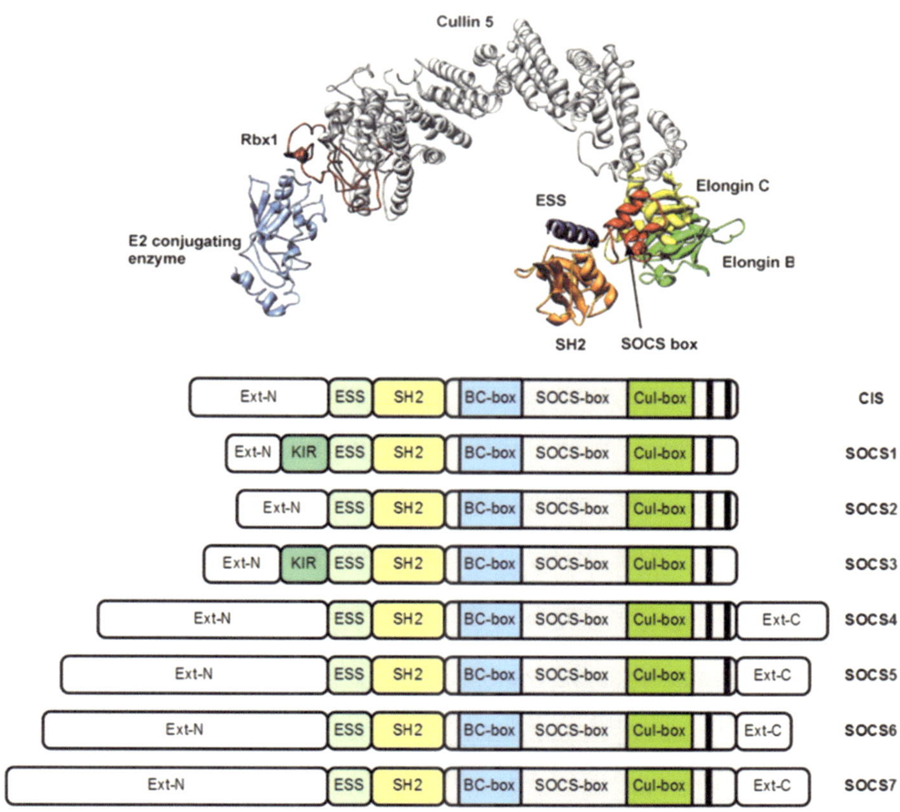

Fig. 3. Elongin-Cullin-SOCS-box (ECS) family of E3 ubiquitin ligases. Structure taken from Piessevaux et al. (2008) with permission.

3.2.1 SOCS proteins as inhibitors of cytokine signaling

SOCS proteins function as classical negative feedback inhibitors of cytokine signaling, since most SOCS proteins are themselves induced by cytokines. Cytokines shown to induce SOCS include the gp130 signaling cytokines, IL-2, IL-3, IL-4, IL-10, IFNs, G-CSF and leptin. (Alexander, 2002; Yoshimura et al., 2007). Other inducers include Toll-like receptor (TLR) agonists (e.g. LPS, CpG-DNA), growth hormone (GH), prolactin and cyclic AMP (cAMP) (Dalpke et al., 2001; Gasperini et al., 2002; Lang et al., 2003; Sands et al., 2006). The SOCS proteins can inhibit signaling by multiple mechanisms according to the SOCS member and signaling pathway involved. SOCS proteins can bind specific PTyr residues *via* their SH2 domain. Thus, SOCS3 binds to PTyr759 (PTyr757 in mouse) on gp130 (Nicholson et al., 2000) and physically occupies the same sites as other SH2 domain-containing signaling components such as SHP2, thereby competing with and blocking activation of other signaling pathways (De Souza et al., 2002; Heinrich et al., 2003) (Fig. 2). *In vitro* studies have shown that the phospho-peptide binding specificity of SOCS3 is very similar to that of SHP2, with optimal SOCS3 and SHP2 phospho-peptide ligands containing overlapping consensus sequences (De Souza et al., 2002). The same group also demonstrated that SOCS3 binds to the gp130 receptor with a much higher affinity than the leptin receptor ObRb. However, the finding that SOCS3 can bind two sites on the leptin receptor versus one site on gp130 may compensate for the low affinity each ObRb site exhibits (De Souza et al., 2002).

The kinase inhibitory region (KIR) of SOCS1 and SOCS3, located upstream of the SH2 domain, is capable of interacting with the substrate binding site of the JH1 domain of JAKs, acting as a pseudosubstrate and thus inhibiting catalytic activity and downstream signaling from the associated receptor (Sasaki et al., 1999; Yasukawa et al., 1999). Specifically, Tyr31 of SOCS3 and Tyr65 of SOCS1 have been identified as the critical residues responsible for the pseudosubstrate inhibition of JAK2 (Bergamin et al., 2006). Interestingly, structural data relating to this interaction has revealed that it is implausible for Tyr31 or Tyr65 to reach the active kinase domain of JAK2 if bound *via* the SH2 domain i.e. in *cis* (Bergamin et al., 2006). This does not rule out the possibility that the SOCS proteins could bind to one JAK molecule *via* their SH2 domain and inhibit another JAK *via* pseudosubstrate inhibition i.e. in *trans,* or the possibility that binding of the SOCS SH2 domain to the specific PTyr residues positions the KIR for binding to the kinase domain of associated JAK2. The latter possibility appears to be the more likely scenario, since the crystal structure of the SOCS3/gp130 complex and various structural data favour the physiological target of SOCS3 SH2 domain to be pTyr757/759 of mouse/human gp130 and not the activation loop of JAK2 (Bergamin et al., 2006).

The SOCS box present within all SOCS members can recruit elongins B and C, which together with cullin 5 and RING-box 2 (Rbx2) form an E3 ubiquitin-ligase complex (Fig. 3). This complex associates with enzymes E1, a ubiquitin-activating enzyme and E2, a ubiquitin-conjugating enzyme, to mediate Lys48 polyubiquitylation and subsequent proteasomal degradation of signaling components bound to the SOCS proteins *via* their SH2 domains (Kamura et al., 2004; Ungureanu et al., 2002; Zhang et al., 1999a) (Fig. 3). A possible ubiquitination site, Lys-6 is also present at the N-terminus of SOCS3, and an N-terminally truncated isoform of SOCS3 lacking this site is significantly more stable than wild-type SOCS3, suggesting that polyubiquitylation of Lys6 plays an important role in regulating turnover. Moreover, this demonstrates one level by which SOCS3 expression can be regulated post-translationally (Sasaki et al., 2003).

SOCS proteins can theorectically target the whole receptor-cytokine complex including the JAKs, plus the SOCS proteins themselves, for proteasomal degradation. Therefore, it raises the question of how SOCS proteins selectively block JAK signaling at specific receptors. This could possibly be explained by the realization that SOCS SH2 domains preferentially bind specific PTyr motifs on activated receptors rather than JAKs, thereby causing degradation of associated JAKs as well as the receptor-cytokine complex, and also achieving specificty at the receptor level. Indeed, it has been shown that mutation of the SOCS3 binding site Tyr757 to non-phosphorylatable Phe on murine gp130 is sufficient to cause enhanced IL-6-inducible gene expression (Anhuf et al., 2000). Furthermore, bone marrow-derived macrophages (BMDMs) isolated from mice with a Tyr757Phe mutation in gp130 switch the IL-6 mediated response to an "IL-10-like" *anti*-inflammatory response, in terms of inhibiting LPS-induced induction of pro-inflammatory cytokines (El Kasmi et al., 2006). Previous studies have linked the absence of SOCS3 with the establishment of the anti-inflammatory response following IL-6 treatment (Yasukawa et al., 2003), so the data suggests that mutation of only the specific PTyr that binds SOCS3 is sufficient to cause cytokine receptors to become refractory to SOCS inhibition, despite the presence of JAKs. This is in contrast to SOCS1, for which it has been shown that the phenotype of SOCS1-deficient mice can only be partially rescued in mice with SOCS1 lacking the SOCS box, but retaining the SH2 domain. This shows that both the SOCS box and SH2 domain are required for the inhibitory effects on IFN-signaling (Zhang *et al.*, 2001). Mice in which endogenous SOCS3 has been replaced with a truncated SOCS3 mutant lacking the SOCS box have been shown to be viable but exhibit hyper-responsivenenss to G-CSF, suggesting that linkage of SOCS3 to the ubiquitin machinery is important in restraining G-CSF signaling in vivo (Boyle et al., 2007), possibly through controlling the ubiquitination-mediated routing of the G-CSF receptor to lysosomes (Irandoust et al., 2007).

In contrast to findings of SOCS interaction with elongins B and C leading to proteasomal degradation, some studies have found that interaction with the elongin BC complex can stabilise both SOCS3 (Haan et al., 2003) and SOCS1 (Kamura et al., 1998). Haan et al. (2003) showed that Tyr phosphorylation of SOCS3 disrupted interaction with elongins, which accelerated SOCS3 degradation. This may suggest that Tyr phosphorylation of SOCS3 is a prerequisite to its subsequent proteasomal degradation. Indeed, Haan *et al.* (2003) proposed that the elongin BC interaction with SOCS3 may function to associate SOCS3 with a latent ubiquitination complex that only becomes active when SOCS3 is phosphorylated. SOCS3 phosphorylation on Tyr204 and/or Tyr221 causes the dissociation of elongin C and the bringing together of the ubiquitination machinery into close proximity with SOCS3, subsequently triggering its degradation (Haan et al., 2003).

3.2.2 SOCS proteins as regulators of other Tyr phosphorylation-dependent pathways

In addition to the involvement of SOCS proteins in cytokine signaling, SOCS1 and SOCS3 have been shown to bind receptors for EGF and FGF receptors and affect downstream signaling events both positively and negatively (Ben-Zvi et al., 2006; Xia et al., 2002). With regards to EGF signaling, SOCS1 and 3 have been shown to facilitate EGFR proteasomal degradation in HEK293 cells (Xia et al., 2002), while SOCS1 has been shown to inhibit STAT1 phosphoryation and elevate ERK1,2 phosphorylation in response to FGF treatment of rat chondrosarcoma (RCS) cells (Ben-Zvi et al., 2006). Furthermore, SOCS1 and SOCS3 have both been shown to associate with insulin receptor substrates 1 (IRS1) and IRS2 following insulin stimulation, and interact with the elongin BC ubiquitin-ligase complex to promote their polyubiquitylnation and subsequent degradation (Kawazoe et al., 2001; Rui et al., 2002).

As mentioned earlier, SOCS3 offers another level of regulation by being able to become phosphorylated on Tyr204 and Tyr221 within the SOCS box in response to IL-2, Epo and other stimuli (Cacalano et al., 2001; Haan et al., 2003). Using a murine B cell line, it was found that WT SOCS3 could inhibit IL-2-mediated STAT5 phosphorylation, but maintain IL-2-mediated ERK1,2 phosphorylation, whereas a Tyr204/221Phe double mutant SOCS3 still inhibited STAT5 phosphorylation, but in contrast to the WT, induced the abolition of ERK1,2 phosphorylation, suggesting a phosphorylation-dependent maintenance of ERK signaling. (Cacalano et al., 2001). The inhibitory effect of the mutant was also observed following Epo and PDGF treatment. Thus, phosphorylation of Tyr204 and Tyr221 of SOCS3 following growth factor stimulation has been proposed to trigger PTyr221 interaction with the SH2 domain of RasGAP, thereby sustaining GTP accumulation on Ras and subsequent activation of ERK1,2. It is well established that the duration of ERK1,2 signaling is important for determining biological outcome; for example sustained activation of ERK has been shown to be required for the control of G1 progression by regulating cyclin D1 activation in some systems (Weber et al., 1997). SOCS3 therefore appears to have both pathway-specific and receptor-specific effects, and can positively regulate activation of specific signaling pathways, adding further complexities to its actions.

An additional level of complexity demonstrated by SOCS proteins is their ability to interact with other SOCS family members (Piessevaux *et al.*, 2006; Tannahill *et al.*, 2005). For example, although SOCS2 plays a major role in the negative regulation of GH signaling (Greenhalgh *et al.*, 2002), it has also been shown to enhance GH signaling. This is believed to be caused by the binding of SOCS2 to other SOCS members and modulating their activity *via* the elongin BC complex, with subsequent proteasomal degradation (Piessevaux et al., 2006; Tannahill et al., 2005). This SOCS2-mediated inhibitory effect on other SOCS members has been observed on SOCS1- and SOCS3-dependent inhibition of GH signaling, thus potentiating it (Piessevaux et al., 2006). SOCS2 has also been shown to enhance IL-2 and IL-3 signaling (Tannahill et al., 2005) by accelerating proteasome-dependent degradation of SOCS3. Similar effects again have been shown on signaling *via* the IFN type 1 and leptin receptors (Piessevaux et al., 2006). These observations imply that SOCS2 is counteracting the effects of other SOCS proteins, rather like a secondary negative feedback mechanism, to limit the effects of excessive levels of SOCS proteins. This assumption is supported by the findings that SOCS2 induction is usually initiated after a significant delay following cytokine stimulation and is prolonged, whereas SOCS1 and SOCS3 expression is typically rapid and transient (Adams et al., 1998). Although poorly understood, SOCS6 and SOCS7 have also been shown to bind other SOCS members and similar effects to SOCS2 have been observed for SOCS6 (Piessevaux et al., 2006). Again, this data suggests that SOCS proteins can act as positive and negative regulators of signaling pathways and could explain some reported anomalies, such as the enhanced insulin signaling observed in transgenic mice overexpressing SOCS6 (Li et al., 2004) or the gigantism observed in transgenic mice overexpressing SOCS2 (Greenhalgh et al., 2002).

3.2.3 Functional roles of SOCS proteins

The functions of SOCS proteins *in vivo* have largely been elucidated by the generation of mice engineered to lack particular SOCS genes. These studies have greatly enhanced our understanding of their roles particularly with regards to the immune response, and have also identified key definitive roles of individual SOCS members, such as the non-redundant role SOCS1 appears to play in IFN signaling (Alexander et al., 1999). However, this is not always the case and knock-out models can encounter problems. Due to placental insufficiency, SOCS3-null mice die at mid-gestation (Roberts et al., 2001; Takahashi et al., 2003) and to overcome this, other ways of investigating SOCS3 deficiency have been explored. A genetic cross study conducted by Robb et al. (2005) showed that mice on a double LIF/SOCS3-null background were rescued from embryonic lethality due to placental failure, and the mice appeared normal at birth (Robb et al., 2005). It is believed that the deletion of SOCS3 leads to dysregulated LIF signaling, which alters trophoblast differentiation and causes placental defects (Boyle & Robb, 2008). In support of this is the finding that the number of trophoblast giant cells are reduced in LIFR-null mice, compared with an abnormally high number of trophoblast giant cells in SOCS3-null mice (Takahashi *et al.*, 2003). Although embryonic lethality is rescued, a high neonatal mortality rate is observed in SOCS3$^{-/-}$LIF$^{-/-}$ null mice and adult animals develop a fatal inflammatory disease very similar to that seen in mice with a conditional deletion of SOCS3 in haematopoietic cells (Croker *et al.*, 2003). On the other hand LIF$^{-/-}$ mice have a normal lifespan and do not exhibit any major haematopoietic abnormalities, suggesting that SOCS3 plays a vital role in the negative regulation of the inflammatory response.

Another way to overcome SOCS3 embryonic lethality is the generation of conditional knock-outs using the inducible Cre recombinase (Cre)-loxP system. In this way, the modified target gene can be ablated in adulthood, thus avoiding the placental insufficiency observed with constitutive global SOCS3 knockouts, and the ablation of the gene can be targeted to any tissue at a defined time. This is a powerful tool for the examination of genes that appear to be crucial during embryonic development but may also play important roles in particular adult tissues. Using these systems, the SOCS3 gene has been specifically deleted in the liver and in macrophages. The absence of SOCS3 results in sustained STAT3 and STAT1 activation following IL-6 treatment, but normal activation of STAT1 in response to IFNγ and normal activation of STAT3 in response to IL-10 (Croker et al., 2003; Lang et al., 2003). SOCS3 deficiency was also shown to trigger up-regulation of several IFN-responsive genes following IL-6 treatment, which is not normally observed upon IL-6 stimulation of cells with functional SOCS3 alleles. This suggested that in the absence of SOCS3, sustained STAT1 activation provokes a dominant IFN-like gene expression response. Furthermore, a mutation in gp130 (Tyr757Phe) in mice, which impedes SOCS3 and/or SHP2 recruitment, was shown to result in a phenotype displaying characteristics of RA, a condition already well established to be associated with deregulation of IL-6 signaling (Atsumi et al., 2002). Collectively, these studies demonstrate that SOCS3 is the main physiological regulator of IL-6 signaling and that SOCS3 can regulate the "identity" of the cytokine response as well as the duration of the signal (Croker et al., 2003; Lang et al., 2003).

Interestingly, in the absence of SOCS3 in mouse macrophages, IL-6 has been shown to induce an "IL-10-like" anti-inflammatory response, as demonstrated by a reduction in LPS-induced production of TNFα and IL-12 (Yasukawa et al., 2003). This is an interesting observation because until this point there was no obvious explanation as to why these two cytokines should have such diverse effects. Both cytokines use identical JAK-STAT members and yet have very distinct gene expression patterns (Murray, 2007; O'Shea & Murray, 2008). IL-10 has been shown to be anti-inflammatory in macrophages and dendritic cells, activating a different set of genes from IL-6, but both cytokines also activate a common pool of genes, including SOCS3 (Murray, 2007). Yasukawa *et al.* (2003) proposed that the difference in gene expression may be due to the intensity of the STAT3 signal. However, Murray (2007) has identified flaws in this concept; for example the strength of the signal does not account for the commonality of genes activated by the two cytokines. One obvious difference between the two cytokines is the involvement of SOCS3 as an inhibitory regulator of IL-6 but not IL-10 signaling. Studies have shown that if modified receptors are used, which are either naturally insensitive to SOCS3 (e.g. IL-22R) or engineered to be insensitive (e.g. IL-6, leptin receptors) but still activate STAT3, an anti-inflammatory response is triggered (El Kasmi et al., 2006). Thus, based on SOCS3 involvement, a hypothesis has been proposed describing the activation of a generic pool of STAT3 by the IL-10R, which is not subjected to any inhibition by SOCS3. The IL-6R, on the other hand may activate a different pool of STAT3 which can be specifically inhibited by SOCS3, possibly *via* post-translational modification by kinases, phosphatases, methylases or other regulators. These distinct STAT3 pools may therefore go on to activate different sets of genes. However, this is just one idea and ultimately ChIP-Seq experiments will be necessary to identify any differences in the genomic locations to which STAT3 can be recruited following stimulation with either IL-10 or IL-6.

With regards to leptin signaling, mice with a neural-specific deletion of SOCS3 have been generated using the Cre-loxP system. Similar to the observations for IL-6, SOCS3 deletion resulted in prolonged activation of STAT3 in response to leptin. Moreover, SOCS3-deficient mice exhibited a greater weight loss compared to their wild-type littermates. These knock-out mice were also resistant to high fat diet-induced weight gain and hyperleptinaemia, and retained insulin sensitivity. This study showed that SOCS3 is a key regulator of leptin signaling and hence plays an important role in diet-induced leptin and insulin resistance (Mori et al., 2004). A number of studies support this link between SOCS3 and leptin resistance, whereby leptin-mediated induction of SOCS3 has been associated with the attenuation of ObRb signaling (Bjorbaek et al., 1998). Chronic stimulation of ObRb has been shown to result in the desensitisation of ObRb signaling, whereby the receptor becomes refractory to re-stimulation. Mutation of the STAT3 binding site on ObRb (Tyr1138Phe), which mediates STAT3-induced SOCS3 induction, alleviates this feedback inhibition. Moreover, RNAi-mediated knock-down of SHP2 had no effect on the attenuation of ObRb signaling, suggesting a role for SOCS3 in the feedback inhibition of ObRb signaling and not SHP2 (Dunn *et al.*, 2005). Consistent with a role for SOCS3 as a central regulator of leptin responsiveness, it has been shown recently that the ability of intracellular cAMP sensor Epac1 (exchange protein directly activated by cAMP-1) to trigger the induction of SOCS3 (Sands et al., 2006) blocks multiple signaling pathways downstream of the leptin receptor ObRb, thus suppressing leptin function in hypothalamic pro-opiomelanocortin neurons (Fukuda et al., 2011).

3.3 Protein Tyrosine Phosphatases (PTPs)

Protein phosphatases reverse the effects of protein kinases by catalysing the removal of phosphoryl groups to initiate, sustain or terminate signals (Andersen et al., 2001). Protein tyrosine phosphatases (PTPs) comprise a large family of these proteins and are distinguished by a unique signature motif. Residues in this motif form the phosphate-binding loop and two residues (a Cys and Arg) are critical for the catalytic activity of PTPs (Andersen et al., 2001; Tiganis & Bennett, 2007). PTPs can be grouped into two general families; (1) the Tyr-specific PTPs, which can dephosphorylate substrate proteins on Tyr; these can be further sub-divided into transmembrane receptor-like PTPs and non-transmembrane PTPs, and (2) the dual-specificity phosphatases (DUSPs), which can dephosphorylate protein substrates on Tyr, Ser and Thr (Tiganis & Bennett, 2007). Our understanding of PTPs is greatly lagging behind that of PTKs, partly due to the discovery of PTKs a decade before PTPs. However like most kinases, PTPs exhibit a high degree of specificity towards their substrates. This is achieved by the PTP catalytic domain, which recognises phosphorylated Tyr residues and the flanking amino acids within the substrate, and the non-catalytic N- and C- terminal domains, which target the PTP to particular intracellular compartments for substrate recognition (Andersen et al., 2001). This is exemplified by the DUSPs, which dephosphorylate MAPKs on Tyr and Thr residues. Of the 10 members that make up this family, some can specifically target one class of MAPK (e.g. DUSP6/MKP-3 which dephosphorylates ERK1,2) while others can target more than one MAPK class (e.g. DUSP1/MKP-1 which can dephosphorylate ERK, JNK and p38 MAP kinases) (Owens & Keyse, 2007).

Given their importance in cytokine signaling, we will describe three PTPs in detail: PTP1B, TC-PTP and SHP2.

3.3.1 PTP1B

The prototypical PTP is PTP1B, which was first identified in 1988 (Charbonneau *et al.*, 1988). It has been shown to have numerous substrates, but the most extensively studied of these include the insulin receptor (IR) and JAK2 (Tiganis & Bennett, 2007). Much of this information has come from analysis of PTP1B-null mice, which exhibit enhanced insulin sensitivity resulting from increased insulin-stimulated Tyr phosphorylation of the insulin receptor in muscle and liver. Furthermore, these mice are resistant to diet-induced obesity (Elchebly et al., 1999). Additional studies have since revealed the involvement of leptin signaling in the above phenotype and have demonstrated PTP1B inhibition of leptin signaling *via* dephosphorylation of JAK2 (Cheng et al., 2002).

3.3.2 T cell PTP (TC-PTP)

T-cell-specific protein tyrosine phosphatase (TC-PTP), which as the name suggests was originally cloned from a peripheral T cell cDNA library, is a ubiquitously expressed PTP. The primary transcript is processed into two splice variants that encode TC45 and TC48 isoforms. The resulting differences in primary sequence at the C-terminal domains of TC45 and TC48 are responsible for their distinct intracellular localization patterns (Lorenzen et al., 1995). Specifically, TC45 is localized to the nucleus due to a R^{378}KRK sequence, while TC48 is localized to the endoplasmic reticulum by its unique C-terminal 19 amino acids (Lorenzen et al., 1995). The nuclear TC45 isoform has several proposed substrates, including the IR and EGFR (Tiganis & Bennett, 2007), JAK1 and JAK3 (Simoncic et al., 2002), and also STAT1 and STAT3 (ten Hoeve et al., 2002; Yamamoto et al., 2002). Of interest, Yamamoto et al. (2002) have demonstrated TC45-mediated suppression of STAT3 activation in response to IL-6 in

293T cells, implicating the nuclear isoform of TC-PTP as an important negative regulator of IL-6 signaling. This is supported by studies of TC-PTP-deficient mice, which display a more complex phenotype as compared to mice deficient in the closely related PTP1B. Specifically, TC-PTP-null mice are viable but exhibit profound defects, resulting in splenomegaly, lymphadenopathy and thymic atrophy. As a result, the mice typically die at 3-5 weeks. Homozygous mice display defects in bone marrow, B cell lymphopoiesis, erythropoiesis, and impaired T and B cell functions. Taken together, the abnormalities displayed in TC-PTP-deficient mice strongly suggest a crucial role in haematopoiesis and immunity (You-Ten *et al.*, 1997). However, targeted deletion of TC-PTP in defined cell types will undoubtedly reveal previously unknown aspects of its function. For example, a recent study has shown mice lacking TC-PTP in neurons are hypersensitive to leptin and are resistant to diet-induced weight gain compared with WT animals (Loh et al., 2011). Animals in which both PTP-1B and TC-PTP were deleted showed additive effects compared with the single knockouts, suggesting that both PTPs regulate ObRb signaling. However, it was found that TC-PTP levels are elevated in hypothalamus extracts from obese mice, raising the possibility that this may contribute relatively more to the leptin resistance observed in obesity (Loh et al, 2011; Myers et al., 2008).

3.3.3 SH2 domain-containing PTP 2 (SHP2)

The SH2-domain containing PTPs (SHPs) are a subfamily of non-transmembrane PTPs comprising two vertebrate members, SHP1 and SHP2. While SHP1 expression is restricted to cells of the haematopoietic system, SHP2 is ubiquitously expressed. Both proteins contain two N-terminal SH2 domains (N-SH2 and C-SH2) and a C-terminal catalytic PTP domain. As such, both proteins have the unique ability to function as PTPs, dephosphorylating signaling components and thus *inhibiting* signal transduction, whilst also serving as adapter molecules *via* their SH2 domains to recruit further adapter molecules and *activate* downstream signaling (Heinrich et al., 2003; Neel et al., 2003). Both SHP1 and SHP2 appear to have non-redundant roles since homozygous deletion of either gene in mice results in death at 2-3 weeks due to severe inflammation. The SHP1-null phenotype is termed "motheaten" because of the patchy hair loss caused by sterile dermal abscesses (Neel et al., 2003) while SHP2 deletion produces embryonic lethality due to defective gastrulation or mesodermal differentiation (Neel et al., 2003). Thus, one SHP does not appear to compensate for the other in these phenotypes. The functional differences between SHP1 and SHP2 appear to be due to the differences in SH2 domain-mediated protein interactions, as well as differences in the PTP domains of both proteins (Salmond & Alexander, 2006).

The finding that ERK1,2 activation by IL-6 is inhibited in either knock-in mice expressing a Tyr759Phe-mutated gp130 (gp130$^{F759/F759}$) or cells transfected with a mutated Tyr759Phe gp130 construct, establishes a positive regulatory role of SHP2 on the ERK1,2 pathway (Kim & Baumann, 1999; Ohtani et al., 2000). Indeed, studies on growth factor receptor signaling, including EGF, insulin and platelet-derived growth factor (PDGF) signaling have each demonstrated a positive regulatory role of SHP2 on ERK1,2 signaling and gene expression (e.g. Yamauchi et al., 1995). In contrast to these positive effects, gp130$^{F759/F759}$ mice and cells transfected with a Tyr759Phe-mutated gp130 demonstrate impaired SHP2 activation, prolonged STAT3 and STAT1 activation and enhanced acute-phase protein gene induction, suggesting a negative role of SHP2 on the STAT pathway (Ohtani et al., 2000; Schaper et al.,

1998). In support of this negative regulatory role, the gp130$^{F759/F759}$ mouse displays splenomegaly, lymphadenopathy and an enhanced acute phase reaction (Ohtani et al., 2000). However, these results are complicated by the findings that SOCS3 binds to the same PTyr site on gp130 as SHP2 (De Souza et al., 2002) and could therefore potentially contribute towards these negative effects. To address this issue, studies have employed catalytically inactive dominant-negative SHP2 mutants in gp130 signaling. Expression of these mutants result in increased gp130, JAK and STAT3 phosphorylation as well as elevated gene induction (Lehmann et al., 2003; Symes et al., 1997), thus confirming the involvement of SHP2 in the negative regulation of the STAT pathway activated by gp130.

These negative effects of SHP2 on gp130 signaling may be receptor-specific as they do not appear to be observed for signaling from the leptin receptor ObRb, since mutation of SHP2 binding site Tyr985 of ObRb has no effect on STAT activation (Li & Friedman, 1999). Also, using dominant negative SHP2 strategies in COS-1 cells, SHP2 was shown not to have an effect on STAT3 phosphorylation or STAT3-mediated gene transcription from the SOCS3 promoter (Bjorbaek et al., 2001). However, these studies demonstrated no effect of SHP2 on STAT activation and STAT-dependent promoter activity following 15 minutes and 6 hours of leptin treatment respectively. Further research by this group has demonstrated that following 24 hours of leptin treatment, STAT-mediated transcription was enhanced in cells expressing mutated Tyr985Phe ObRb (Bjorbaek et al., 2001) and suggested that the induction of SOCS3 by leptin over prolonged leptin treatment could account for the enhanced STAT3 response, implicating SOCS3 involvement and not SHP2. In addition, over these extended periods of leptin treatment, SHP2 could possibly act as an indirect positive regulator of the STAT pathway, preventing SOCS3 binding to the ObRb at Tyr985 (Bjorbaek et al., 2001).

Thus, in general, a positive role of SHP2 in cytokine-induced ERK1,2 activation and a negative or positive role of SHP2 in cytokine-induced STAT activation have been proposed. How SHP2 mediates this positive effect on ERK activation and the contribution of SHP2 to the negative regulation of STAT signaling in relation to SOCS3 are still areas under investigation. To address the first point, SHP2 appears to exert a positive effect on ERK signaling by acting as an adapter protein, wherein SHP2 becomes recruited to PTyr residues on activated cytokine receptors and following activation, associates with the adapter protein Grb2, which is bound to the Ras GDP-GTP exchange factor Sos. The membrane-localised Grb2/Sos complex can then go on to activate the Ras-Raf-MAK-ERK pathway (Li et al., 1994). With regards to gp130 signaling, it has been proposed that the SHP2-Grb2/Sos mode of ERK activation is adopted (Heinrich et al., 2003). However, in addition to this model, Grb2-associated binder-1 (Gab1) has also been shown to become tyrosine phosphorylated to associate with SHP2 and phosphatidylinositol 3-kinase (PI3K) to activate ERK in response to IL-6 (Takahashi-Tezuka *et al.*, 1998). Alternatively, SHP2 may act as a PTP, thus dephosphorylating and inactivating specific substrates that normally suppress ERK1,2 activation. For example, SHP2 has been shown to dephosphorylate tyrosine residues on EGFR required for RasGAP recruitment to the receptor, thereby inhibiting phosphorylation-dependent translocation of RasGAP to the plasma membrane and thereby maintain Ras-GTP levels to sustain ERK1,2 activation (Agazie & Hayman, 2003).

Finally a model has been proposed whereby the binding of SHP2 to PTyr residues has been shown to activate its PTP activity *via* either of two mechanisms (Barford & Neel, 1998); (1) the SH2 domains can bind to PTyr motifs on activated receptors such as pTyr759 on gp130, which leads to unfolding of the protein and subsequent activation of phosphatase activity, or (2) the SH2 domains can bind to PTyr542 and 580 on the C-terminal tail of SHP2 itself and cause conformational changes leading to activation. If PTyr binding does not occur, SHP2 remains in an inactive state whereby the N terminal SH2 (N-SH2) domain sterically hinders access of phosphorylated substrates to the PTP domain, as demonstrated by its crystal structure (Hof et al., 1998). Therefore, the N-SH2 domain of SHP2 can either bind and inhibit the phosphatase, or bind to PTyr residues on substrates to activate the enzyme. Interestingly, in the human autosomal dominant disorder Noonan Syndrome, approximately 50% of all cases are caused by mutations in the SHP2 gene *PTPN11*, and specifically in portions of the amino N-SH2 domain. These mutations lock SHP2 in its active conformation and subsequently cause excessive SHP2 activity (Tartaglia et al., 2001). Noonan Syndrome is characterised by short stature, cardiac defects, facial dysmorphia and an increased risk of developing leukaemia (Salmond & Alexander, 2006). Interestingly this syndrome displays aberrant hyperactivation of the ERK1,2 pathway (Bentires-Alj et al., 2006). Thus, Noonan Syndrome demonstrates an involvement of SHP2 in regulating ERK activation in human disease although further research is required to better understand the exact mechanisms involved.

Likewise, the mechanisms by which SHP2 negatively or positively regulates STAT activation are not well understood and the relative contribution of both SHP2 and SOCS on gp130 or ObRb signaling is still unclear. There are reports showing that SHP2 can act directly as a STAT phosphatase (Chen et al., 2003) and STAT3/SHP2 complexes have been detected in cells (Gunaje & Bhat, 2001). A large body of evidence supporting a negative role of SHP2 in gp130-mediated STAT activation exists, whereas conversely, SHP2 has been shown to have no effects on ObRb-mediated STAT3 phosphorylation (Bjorbaek et al., 2001). Therefore the possibility exists that SHP2 acts as an indirect positive regulator of STAT3 by impeding SOCS3 recruitment at PTyr985 (Bjorbaek et al., 2001).

4. Cross-regulation of signaling pathways via SOCS proteins

Cells are typically exposed to a changing milieu of cytokines and other extracellular regulators. Therefore to produce an appropriate response, effective crosstalk between distinct signaling pathways is essential. In terms of cytokine receptor signaling, the levels at which such crosstalk can occur are now beginning to be revealed, and here we will discuss two recently characterized examples of this type of regulation.

4.1 Regulation of SOCS3 expression by cAMP/Epac1

Many studies have shown that prototypical second messenger cAMP exerts profound anti-inflammatory and immunosuppressive effects on many target cell types, including vascular endothelial cells, neutrophils, monocytes, CD4 T cells and regulatory T cells (Mosenden & Tasken, 2011). Many of these effects are mediated *via* defined gene transcription programmes, which include the up-regulation of protective molecules, such as the immunosuppressive cytokine IL-10, while down-regulating pro-inflammatory molecules such as TNFα and IL-1 (e.g. Wall et al., 2009). A large body of research has identified members of the cAMP-response element binding (CREB) protein family as the principal mediators of changes in gene expression in response to cAMP following their phosphorylation by PKA. However, observations of cAMP-mediated induction of specific genes occurring *via* PKA-independent mechanisms have demonstrated that other mechanisms must also exist. For example, we have demonstrated that cAMP elevation is capable of inducing the SOCS3 gene *via* a PKA-independent mechanism that requires Epac1 (Sands et al., 2006; Woolson et al., 2009). Epac1 and Epac2 were originally identified as guanine nucleotide exchange factors (GEFs) for the Rap 1 and Rap2 small GTPases. Each has several important structural features, including a cyclic nucleotide binding domain (Epac2 has two such domains), a DEP (Dishevelled, Egl, Pleckstrin) domain, a REM (Ras exchanger motif) domain and a Cdc25 homology domain that exhibits GEF activity (Borland et al., 2009b). Identification of SOCS3 as an Epac-regulated target gene revealed a new mechanism by which cAMP could inhibit signaling from specific SOCS3-targeted cytokine receptors, such as the leptin receptor ObRb and gp130 (Sands et al., 2006; Fukuda et al., 2011). The mechanism of SOCS3 induction is mimicked by expression of a constitutively active Val12 mutant of Rap1a and is blocked by either siRNA-mediated depletion of Rap1 or overexpression of RapGAPII, suggesting that the ability of Epac1 to stimulate the accumulation of GTP-bound active Rap1 is essential (Yarwood et al., 2008). Rap1 then activates phospholiase C ε to generate intracellular *sn* 1,2-diacylglycerol (DAG) and mobilize intracellular calcium, resulting in the activation of specific protein kinase C isoforms (Borland et al., 2009a), although exactly how PKC regulates SOCS3 transcription is unclear.

We have also identified the transcription factor(s) targeted by Epac1, demonstrating that Epac1 activation increases the binding of the CCAAT/enhancer binding protein β (C/EBPβ) to a region within the human SOCS-3 promoter enriched in C/EBP-responsive elements in vascular endothelial cells *in situ*. In addition, overexpression of specific C/EBP isoforms in endothelial cells can potentiate Epac-mediated induction of SOCS-3, while selective knockdown of either C/EBPβ or C/EBPδ abolishes induction. Similar experiments performed on C/EBP knockout murine embryonic fibroblasts have confirmed these findings. Interestingly, SOCS3 induction in response to cAMP requires activation of the ERK1,2 pathway, which triggers phosphorylation of C/EBPβ on Thr225. The importance of this event is demonstrated by the ability of Thr225Ala-mutated C/EBPβ to function as a dominant-negative inhibitor of SOCS3 induction (Borland et al., 2009a). However, siRNA-mediated depletion of Epac1 fails to diminish cAMP-stimulated ERK1,2 phosphorylation, which occurs *via* a PKA-independent mechanism, suggesting that an additional PKA-independent pathway is involved in C/EBP-mediated SOCS3 induction (Woolson et al., 2009).

While a role for specific transcription factors has yet to be addressed, characterisation of the induction of the pro-glucagon and AQP2 genes in response to cAMP has revealed some similarities to the induction of SOCS3. In each case, induction by cAMP-elevating agents is resistant to inhibition of PKA and can be mimicked by selective activation of Epac by the Epac-selective activator 8CPT-2′-O-Me-cAMP. There also appears to be a requirement for activation of the ERK1,2 pathway by cAMP (Lotfi et al., 2006; Umenishi et al., 2006). These similarities would suggest that the PKA-independent mechanisms by which the pro-glucagon, AQP2 and SOCS3 genes are induced in response to cAMP may be related and presumably apply to other yet to be identified Epac-inducible genes.

4.2 Regulation of multiple steps within the NF-κB pathway by SOCS proteins

Given that SOCS family members each contain an SH2 domain, it is not surprising that their influence on cellular processes have been found to extend beyond classical JAK-STAT signaling processes. These include the inhibition of insulin signaling due to SOCS3-mediated ubiquitylation and degradation of IRS1 and/or IRS2 (Kawazoe et al., 2001; Rui et al., 2002), and the regulation of FAK1 (Liu et al., 2003; Stevenson et al., 2010). Importantly, many studies have also revealed that SOCS proteins are able to regulate the NF-κB pathway at several levels, providing a key mechanism by which JAK-STAT-mobilising cytokines can control pro-inflammatory responses.

Members of the TLR family of receptors and the receptor for IL-1 trigger an inflammatory response via a conserved mechanism. The cytosolic region of these receptors contains a Toll/IL-1 receptor (TIR) domain that interacts with a panel of adaptor molecules to activate defined intracellular signaling cascades, including p38 and c-Jun N-terminal kinase MAP kinase pathways and the NF-κB pathway (O'Neill, 2009). While several TIR domain-interacting proteins have been identified, the best characterized are MyD88 (myeloid differentiation primary response gene 88), Mal (MyD88 adaptor-like), TRIF (TIR domain-containing domain that induced IFNβ) and TRAM (TRIF-related adaptor molecule). All TLRs activate the NF-κB pathway but different receptors utilize specific subsets of adaptors to achieve this; for example, both TRAM and TRIF have been shown to be required for activation of IRF3/7 and NF-κB by TLR4, while the absence of Mal abolishes TLR2- and TLR4-stimulated pro-inflammatory cytokine production (O'Neill, 2009; Brikos & O'Neill, 2008). The requirement for Mal is thought to be due to an interaction with TRAF6 that promotes phosphorylation of the RelA subunit of NF-κB on Ser536, which is required for binding of transcriptional co-activators, rather than any effect on transcription factor translocation to the nucleus from the cytosol (Verstak et al., 2009). The ability of SOCS1 to regulate TLR-mediated responses has been reported by multiple investigators, but no one unifying mechanism has emerged. One study has demonstrated that activation of either TLR2 or TLR4 with selective agonists can trigger the proteasomal degradation of Mal but not MyD88 (Mansell et al., 2006). Further characterization of this phenomenon revealed that this process required the Tyr phosphorylation of Mal by Bruton's tyrosine kinase (BTK), which is also activated by TLRs. Upon SOCS1 induction in response to TLR activation, the SOCS1 SH2 domain is thought to bind phosphorylated Mal. Once bound, SOCS1 functions as an E3 ubiquitin ligase to catalyse the Lys48 polyubiquitylation of Mal, thereby targeting it for degradation by the proteasome, blocking RelA phosphorylation and turning off NF-κB-mediated gene transcription (Mansell et al., 2006) (Fig. 4). However it should be noted that others have found that SOCS3 plays an important role in the suppression of IL-1 signaling by binding and inhibiting TAK1, a kinase which is required for IL-1 receptor-mediated initiation of MAP kinase and NF-κB pathways (Frobøse et al., 2006). The significance of this interaction for SOCS-mediated suppression of TLR signaling has yet to be investigated.

Several recent studies have also shown that SOCS1-mediated negative regulation of NF-κB is not restricted to activation by TIR domain-containing receptors due to its incorporation into a multimeric complex by COMMD1 (copper metabolism gene MURR1 domain-containing protein 1) (Maine & Burstein, 2007). The defining ~70 residue domain conserved within all 10 members of the COMMD family is located at their C-termini, and mutation of conserved residues within this domain has been shown to ablate inhibition of NF-κB. Interestingly, COMMD1 appears to associate with the RelA component of NF-κB complexes on target gene promoters, i.e. after translocation of NF-κB from the cytoplasm to the nucleus (Maine et al., 2007; Geng et al., 2009). Current research suggests that RelA-bound COMMD1 then recruits a SOCS1-elongin B/C-Cul2-Rbx1 E3 ubiquitin ligase complex which promotes the polyubiquitylation and proteasomal degradation of RelA, thus turning off NF-κB-mediated target gene transcription (Fig. 4). Accordingly, COMMD1 deficiency results in a sustained nuclear accumulation of RelA without affecting nuclear entry of NF κB (Maine et al., 2007; Geng et al., 2009).

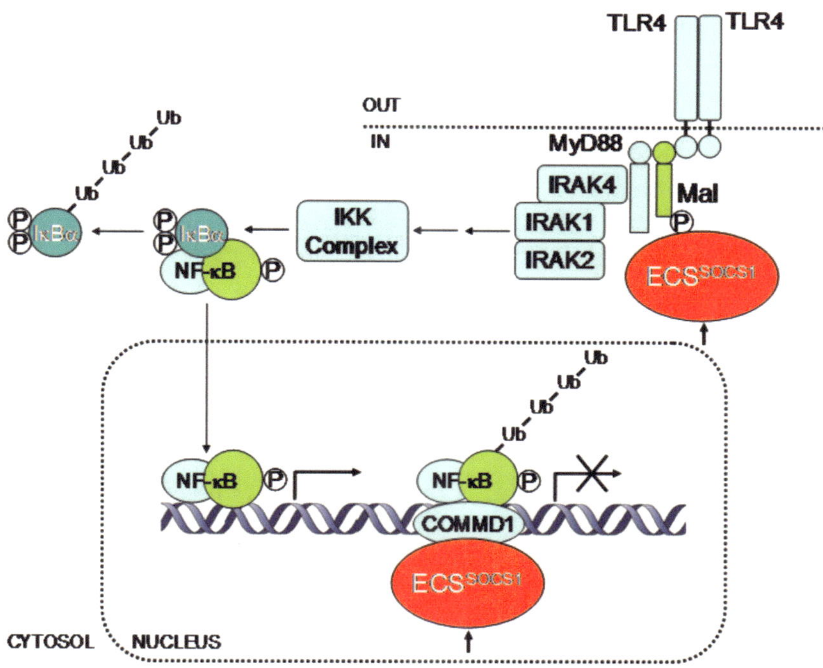

Fig. 4. Control of the NF-κB pathway at multiple levels by a elongin-cullin-SOCS1 (ECS^{SOCS1}) ubiquitin E3 ligase complex.

5. Implications for therapeutic strategies to treat chronic inflammatory diseases and cancers

Over the last decade, the importance of chronic inflammation to the pathophysiology of non-infectious diseases, including cancers, cardiovascular disease and diabetes mellitus, has been established (O'Shea & Murray, 2008). In parallel with these developments, genome wide association studies have identified polymorphisms in cytokine receptors and their associated signaling molecules associated with autoimmune diseases such as type 1 diabetes and inflammatory bowel syndrome (e.g. Rossin et al., 2011), while new therapies have been approved deriving from our understanding of how pro-inflammatory signaling drives pathology (e.g. anti-TNF and IL-6 antibodies for RA and other indications) and several more are in development (e.g. JAK inhibitors) (Taylor & Feldmann, 2009; Ding et al., 2009; Quintás-Cardama et al., 2011).

5.1 Polycythaemia Vera (PV) and other myoproliferative neoplasms

Myoproliferative neoplasms arise from either haematopoietic stem or progenitor cells, and are characterized by the proliferation of terminally differentiated myeloid cells (Quintás-

Cardama et al., 2011). Along with essential thrombocythaemia (ET) and primary myelofibrosis (PMF), polycythaemia vera (PV) represents one of the most studied classes of myoproliferative neoplasms, with patients displaying hyperactivation and proliferation of haematopoietic stem cells in bone marrow, resulting in increased production of platelets, white and red blood cells (Quintás-Cardama et al., 2011). The importance of JAK2 in haematopoiesis had already been established from studies demonstrating that homozygous JAK2 deletion produces embryonic lethality due to insufficient erythropoeisis. This arises due to the pivotal role of JAK2 activation downstream of receptors for cytokines that drive this process, including Epo, thrombopoietin (Tpo), G-CSF, and GM-CSF (Rane & Reddy, 2002). It was subsequently found that approximately 95% of patients presenting with PV have a G-to-T substitution at position 1849 of the JAK2 gene that results in a Val617Phe dominant gain-of-function mutation within the JH2 domain (*JAK2^{V617F}*) (Baxter et al., 2005; James et al., 2005; Kralovics et al., 2005; Levine et al., 2005) (Fig. 1). As mentioned previously, a recent study demonstrating that the JH2 domain functions as a dual specificity kinase module has shown that the ability of the JH2 domain to suppress JAK2 Tyr kinase activity is disrupted by the Val617Phe mutation due to a reduced JH2-mediated regulatory phosphorylation of Ser523 and Tyr570 (Ungureanu et al., 2011). Additional studies examining those PV patients who are *JAK2^{V617F}*-negative have found that the phenotype is associated with additional mutations within exon 12 of the JAK2 gene. Thus, essentially all cases of PV appear to be driven by somatic activating mutations within the JAK2 gene.

Not surprisingly, expression of Val617PheJAK2 activates signaling pathways that drive proliferation and resistance to apoptosis in a cytokine-independent manner. STAT3 and STAT5, as well as the ERK1,2 and PI3K pathways, are the main effectors of JAK2 activation responsible for the PV phenotype. However it has also recently been proposed that Val617PheJAK2 is resistant to the suppressive effects of SOCS induction. In fact, SOCS3 appears to specifically enhance downstream activation of STAT5 in Epo receptor- and Val617PheJAK2-expressing cells (Hookham et al., 2007). The mutated JAK2 was also found to enhance the Tyr phosphorylation of SOCS3, and this has been proposed as the mechanism by which SOCS3 is rendered incapable of blocking signaling from the Epo receptor and activated Val617PheJAK2. Conversely, others have found that both SOCS1 and SOCS3 are still capable of binding and regulating Val617PheJAK2 expression levels by targeting it for degradation: indeed the mutated constitutively active JAK2 was found to have a reduced stability compared with WT JAK2 (Haan et al., 2009). Furthermore, the authors found that Val617PheJAK2 expression levels have to exceed a minimum threshold before cytokine-independent constitutive activation of downstream signaling is observed, and they propose that such a scenario may explain progression to homozygocity in myoproliferative neoplasms such as PV (Haan et al., 2009). Perhaps related to this observation, the expression ratio of mutated to wild-type JAK2 seems to modulate the phenotypic manifestations of myoproliferative neoplasms, with high ratios favouring the development of a PV-like phenotype and low ratios inducing an ET-like phenotype (Kralovics et al., 2005). In this context, the hypermethylation-mediated silencing of SOCS gene induction seen in idiopathic myelofibrosis (Fourouclas et al., 2008) would be predicted to play an important role in determining pathological severity arising from somatic activating JAK2 mutations.

Given their role in disease and their ability to recapitulate disease phenotypes in mouse models, mutated JAK2 proteins represent attractive targets for drugs to treat myoplastic neoplasms. Incyte's ruxolitinib, an orally available non-selective JAK1/JAK2 inhibitor and one of the first of these agents to be developed, has been shown to improve symptoms such as itching and reduce spleen size in PMF patients but without producing observable remission (Verstovsek et al. 2010; Tefferi et al., 2011). Remarkably, these effects occur regardless of whether the *JAK2*^{V617F} allele is present. Therefore, determining exactly how this class of drugs can exert its effects is necessary if clinical efficacy is to be improved. It has been suggested that the lack of selectivity of ruxolitinib may be responsible for its limited efficacy in PMF, but comparable studies with more selective inhibitors such as LY3009104 (Lilly, currently in phase II trials for RA), which is approximately 40-fold more selective for JAK2 *versus* JAK1, will be required to address this issue. Tofacitinib (Pfizer, currently in phase III trials for multiple immune/inflammatory disorders including psoriasis and RA) was initially developed as part of a strategy to develop JAK3-selective inhibitors but upon detailed characterization was also shown to have significant activity against JAK1 and JAK2. Encouragingly, JAK inhibitors such as tofacitinib have shown efficacy in several models of *JAK2*^{V617F}-mediated proliferation (Manshouri et al., 2008) and therefore may prove useful if ultimately approved for their primary indications.

However, as with almost all kinase inhibitors currently used for treatment of hyperproliferative disorders such as chronic myelogenous leukaemia (Diamond & Melo, 2011), there is the strong possibility of drug resistance developing following chronic drug administration. Thus, new drug-resistant activating JAK2 point mutations will likely emerge and alternative signaling pathways able to drive proliferation and resistance to apoptosis may be up-regulated to bypass drug-mediated inhibition of JAK2 (Deshpande et al., 2011). As a result, there will undoubtedly be a need to develop additional therapeutics for myeloproliferative disorders to account for these possibilities once first generation JAK inhibitor drugs have been approved.

5.2 RA and other autoimmune disorders

RA is a chronic, systemic autoimmune/inflammatory disease that predominantly results in the deterioration of synovial joints, although other tissues may be affected. The disease process often typically leads to the destruction of articular cartilage and ankylosis of the joints, resulting in discomfort and pain for affected individuals. Key cytokines involved in driving pathogenesis include TNFα, IL-1 and IL-6, which exert their effects on immune cells and synovial fibroblasts (Feldman et al., 1996). Up until approximately ten years ago, the treatment options for RA were limited to disease-modifying anti-rheumatic drugs (DMARDs) such as methotrexate, sulphasalazine and leflunomide. While DMARDs are effective for the majority of patients, a significant proportion display adverse reactions to chronic administration, including renal and hepatotoxic effects. This has led to the development of several biological drugs designed to block cytokine receptor activation by binding and sequestering free cytokine to prevent engagement with the cognate receptor, and each is typically administered either as part of a combination therapeutic regimen with one or more DMARDs or as monotherapy in patients not able to tolerate DMARD therapy or for whom it has proven ineffective. They include TNFα blockers (e.g. etanercept, infliximab and adalimumab, the first fully human antibody drug to be FDA approved) and IL-1 blockers (e.g. anakinra) (Taylor & Feldmann, 2009; Gabay et al., 2010).

One of the more recently approved biological therapies is the humanized antibody tocilizumab, which blocks IL-6 signaling by binding to the IL-6R to prevent receptor activation (Ding et al., 2009). The development of IL-6 blockers came from a wealth of research implicating this pleiotropic cytokine in the progression of RA and other autoimmune conditions at several levels. Seminal studies demonstrated that patients with RA, but not osteoarthritis, had elevated levels of synovial IL-6, thereby linking the cytokine to an autoimmune syndrome (Hirano et al., 1988). Studies of knock-in mice expressing a Tyr759Phe-mutated gp130 resistant to SOCS3-mediated inhibition and incapable of activating ERK1,2 demonstrated that these animals spontaneously developed an RA-like autoimmune disease, implicating IL-6 activation of STAT3 as a key mediator of the pathology (Atsumi et al., 2002). Additionally, IL-6 *via* activation of STAT3 can induce the expansion of Th17 cells, which are intimately involved in the development of several autoimmune diseases including RA. In addition, IL-17A can trigger the induction of IL-6 in synovial fibroblasts, thereby creating a positive feedback loop that sustains the chronic pro-inflammatory state (Ogura et al., 2008). IL-6 has also been found to suppress generation of Treg cells, which restrain effector T cell responses (Hirano, 2010). The balance between these two subsets is critical for maintaining homeostasis, and so the ability of IL-6 to drive the balance in favour of Th17 cell dominance is likely to be a be a major cause of pathology in RA and other disorders. Interestingly, it was also found that Tyr759Phe-mutated gp130 must be present in non-haematopoietic cells, and that in response to IL-6 stimulation non-haematopoietic cells expressing Tyr759Phe-mutated gp130 show enhanced production of IL-7, leading to the activation and proliferation of $CD4^+$ T cells (Sawa et al., 2006). These observations demonstrate that IL-6-regulated immune and non-immune cell interactions are critical in the development of RA, and suggest that such interactions may play a similar etiological role in other autoimmune diseases in which IL-6 plays an integral role such as psoriasis, uveitis and inflammatory bowel disease.

5.3 STAT3 activation and tumour development

Mutational activation of proto-oncogenes, coupled with the mutational inactivation of tumor suppressors, are critical events in pathogenesis of cancers (Hanahan & Weinberg, 2011). However several genes that play pivotal roles in the development of malignancies are not activated by mutation. These include the STATs (particularly STAT3) and NF-κB transcription factors, which are activated by extracellular stimuli generated within the tumor microenvironment (Grivennikov & Karin, 2010). As described earlier in this chapter, STATs are typically activated transiently due to the existence of multiple negative feedback mechanisms that act at several levels to terminate signaling, and the same is true for activation of both the canonicical and non-canonical NF-κB pathways (Liu & Chen, 2011). Thus, activated nuclear STAT3 is found in many cancers, including breast, colon and prostate among others (Grivennikov et al., 2009; Marotta et al. 2011; Hedvat et al., 2009). Hyperactivation of STAT3 in infiltrating immune cells and stromal cells is due to the presence of multiple STAT3-activating stimuli, including IL-6 and other cytokine capable of utilizing gp130, IL-10 family members and also receptor tyrosine kinases for VEGF and EGF. Importantly, several of the STAT3-inducible gene products generated by these stimuli, including IL-6 itself, IL-10, IL-23, sphingosine-1-phosphate receptor $S1P_1$, and angiogenic growth factor VEGF, perpetuate a positive feedback loop that promotes sustained pathogenic STAT3 activation within the tumour (Yu et al., 2009; Lee et al., 2010). In conditions such as hepatocellular carcinoma, malignant melanoma, prostate cancer and others, this STAT3 activation loop can be amplified further due to methylation of the CpG island within the SOCS3 gene promoter, thereby blocking its induction (Niwa et al., 2005; Fourouclas et al., 2008; Pierconti et al., 2011). Indeed, SOCS3 hypermethylation has been shown to be associated with prostate cancer tumours with a high Gleason score, suggesting a causative role in pathogenesis (Pierconti et al., 2011).

Importantly, while STAT3 and NF-κB trigger distinct gene expression programmes in various cell types, many gene products are induced by both, including IL-6, VEGF, chemokines CCL2 and CXCL2, anti-apoptotic proteins Bcl-X_L and Bcl-2, and SOCS3 (Yu et al., 2009). The different mechanisms by which STAT3 and NF-κB (predominantly the RelA/p50 heterodimer) interact to enhance target gene transcription have been shown to be context-dependent but typically result in enhanced target gene induction. For example, Tyr705-phosphorylated STAT3 has been shown to bind promoter-bound RelA to facilitate the recruitment of the transcriptional co-activator p300, thereby triggering RelA acetylation. As a result, interaction with IκBα is inhibited, resulting in reduced export of RelA/p50/IκBα complexes and thereby prolonging its nuclear localization and potentially its transcriptional capacity even if upstream IKK-mediated signaling is only weakly activated (Lee et al., 2009). Other groups have identified a different mechanism whereby non-phosphorylated STAT3 can displace IκBα from inactive RelA/p50 complexes in the cytoplasm, thereby facilitating nuclear entry in the absence of IKK activation and the subsequent induction of a subset of κB-responsive genes such as the chemokine RANTES (Yang et al., 2007). Interestingly, evidence suggests that NF-κB/STAT3 complexes are capable of binding to target sequences distinct from those utilized by either transcription factor in isolation (Yang et al., 2007; Yang & Stark, 2008). While this phenomenon has only been studied in detail for the serum amyloid A gene mobilized as part of the acute phase response, it raises the possibility that NF-κB/STAT3 complexes may be able to initiate a unique transcriptional programme, although any role in tumour progression is still unclear.

Despite strong evidence that targeting STAT3 might be a useful therapeutic strategy to suppress tumour formation, one major limitation stems from its importance in stromal as well as haematopoietic cell function. As a result, while pharmacological inhibition of STAT3 in the former may exert anti-tumourigenic effects, inhibition in the latter may actually enhance tumourigenesis, with the net result being limited efficacy. For example, homozygous deletion of the murine STAT3 gene in enterocytes has been shown to block the development of colitis-associated colon cancer (Grivennikov et al., 2009), whereas its deletion in macrophages and neutrophils results in the development of enterocolitis due to the inability of IL-10 to suppress myeloid cell function *via* STAT3 (Takeda et al., 1999). Moreover STAT3 has important pro-survival functions in healthy tissues and is central to immune cell homeostasis, making it difficult to predict the consequences of chronic global inhibition of function. Thus, alternative strategies might include specifically targeting tumourigenic cytokines that utilize STAT3, such as IL-6 or IL-23, which would spare the protective anti-inflammatory/immunosuppressive functions of STAT3-mobilising cytokines such as IL-10. Related to this point, it is been demonstrated in a mouse model of colitis-associated cancer that TGF-β-dependent suppression of T cell stimulation could enhance IL-6 expression. A concomitant reduction in membrane-bound IL-6R on epithelial cells coupled with an increase in ADAM17 in stromal tissue resulted in the generation of sIL-6R and subsequent trans-signaling (Becker et al., 2004). Tumour growth could be suppressed by the administration of sgp130Fc, which specifically blocks trans-signaling by sIL-6Rα/IL-6 but not classical IL-6 signaling through membrane-bound IL-6R. Given that a similar loss of IL-6R and upregulation of ADAM17 has been described in human colon cancer (Becker et al., 2007), selective inhibition of IL-6 trans-signaling using sgp130Fc or related therapeutics might prove a useful approach to arrest the development of colon cancer and other tumours in which a pathophysiological role for IL-6 trans-signaling could be identified.

6. Conclusions

Progress in our understanding of the molecular basis of cytokine signaling and the subsequent identification of the importance of JAK mutations and hyperactivation of STAT3 in disease states has improved our understanding of the pathogenesis of RA, PV and several cancers. A direct result of this progress has been the development and introduction of JAK inhibitors and anti-IL-6 therapies for several inflammatory and autoimmune conditions. However our understanding of JAK-STAT signaling, particularly its integration with the NF-κB and cyclic AMP signaling pathways to regulate immune and inflammatory responses, remain incomplete. Moreover, new insights continue to emerge into roles both for nuclear-localised JAKs in chromatin remodeling and cell cycle regulation (Zouein et al., 2011) and links with Rho kinase that control actomyosin contractility, cell migration and even STAT3 activation (Sanz-Moreno et al., 2011). Consequently it is likely that the first wave of therapies targeting cytokine-activated JAK-STAT pathways is the beginning of several new approaches with which to treat the wide range of diseases in which they play a critical role.

7. References

Adams, T.E., Hansen, J.A., Starr, R., Nicola, N.A., Hilton, D.J. & Billestrup, N. (1998) Growth hormone preferentially induces the rapid, transient expression of SOCS-3, a novel inhibitor of cytokine receptor signaling, *J Biol Chem.* Vol.273: 1285-1287.

Agazie, Y.M. & Hayman, M.J. (2003) Molecular mechanism for a role of SHP2 in epidermal growth factor receptor signaling,. *Mol Cell Biol.* Vol.23: 7875-7886.

Alexander, W. S. (2002) Suppressors of cytokine signaling (SOCS) in the immune system, *Nat Rev Immunol.* Vol.2: 410-416.

Alexander, W.S., Starr, R., Fenner, J.E., Scott, C.L., Handman, E., Sprigg, N.S., Corbin, J.E., Cornish, A.L., Darwiche, R., Owczarek, C.M., Kay, T.W., Nicola, N.A., Hertzog, P.J., Metcalf, D. & Hilton, D.J. (1999) SOCS1 is a critical inhibitor of interferon gamma signaling and prevents the potentially fatal neonatal actions of this cytokine, *Cell.* Vol.98: 597-608.

Alvarez, J.V. & Frank, D.A. (2004) Genome-wide analysis of STAT target genes: elucidating the mechanism of STAT-mediated oncogenesis, *Cancer Biol Ther.* Vol.3: 1045-1050.

Andersen, J.N., Mortensen, O.H., Peters, G.H., Drake, P.G., Iversen, L.F., Olsen, O.H., Jansen, P.G., Andersen, H.S., Tonks, N.K. & Moller, N.P. (2001) Structural and evolutionary relationships among protein tyrosine phosphatase domains, *Mol Cell Biol.* Vol.21: 7117-7136.

Anhuf, D., Weissenbach, M., Schmitz, J., Sobota, R., Hermanns, H.M., Radtke, S., Linnemann, S., Behrmann, I., Heinrich, P.C. & Schaper, F. (2000) Signal transduction of IL-6, leukemia-inhibitory factor, and oncostatin M: structural receptor requirements for signal attenuation, *J Immunol.* Vol.165: 2535-2543.

Atsumi, T., Ishihara, K., Kamimura, D., Ikushima, H., Ohtani, T., Hirota, S., Kobayashi, H., Park, S.J., Saeki, Y., Kitamura, Y. & Hirano, T. (2002) A point mutation of Tyr-759 in interleukin 6 family cytokine receptor subunit gp130 causes autoimmune arthritis, *J Exp Med.* Vol.196: 979-990.

Atsumi, T., Ishihara, K., Kamimura, D., Ikushima, H., Ohtani, T., Hirota, S., Kobayashi, H., Park, S.J., Saeki, Y., Kitamura, Y. & Hirano, T. (2002) A point mutation of Tyr-759 in interleukin 6 family cytokine receptor subunit gp130 causes autoimmune arthritis, *J Exp Med.* Vol.196: 979-990.

Bach, E.A., Tanner, J.W., Marsters, S., Ashkenazi, A., Aguet, M., Shaw, A.S. & Schreiber, R.D. (1996) Ligand-induced assembly and activation of the gamma interferon receptor in intact cells, *Mol Cell Biol.* Vol.16: 3214-3221.

Barford, D. & Neel, B.G. (1998) Revealing mechanisms for SH2 domain mediated regulation of the protein tyrosine phosphatase SHP-2, *Structure.* Vol.6: 249-254.

Bartoli, M., Gu, X., Tsai, N.T., Venema, R.C., Brooks, S.E., Marrero, M.B. & Caldwell, R.B. (2000) Vascular endothelial growth factor activates STAT proteins in aortic endothelial cells, *J Biol Chem.* Vol.275: 33189-33192.

Baxter, E.J., Scott, L.M., Campbell, P.J., East, C., Fourouclas, N., Swanton, S., Vassiliou, G.S., Bench, A.J., Boyd, E.M., Curtin, N., Scott, M.A., Erber, W.N., Green, A.R. & Cancer Genome Project (2005) Acquired mutation of the tyrosine kinase JAK2 in human myeloproliferative disorders, *Lancet.* Vol.365: 1054-1061.

Becker, C., Fantini, M.C., Schramm, C., Lehr, H.A., Wirtz, S., Nikolaev, A., Burg, J., Strand, S., Kiesslich, R., Huber, S., Ito, H., Nishimoto, N., Yoshizaki, K., Kishimoto, T., Galle, P.R., Blessing, M., Rose-John, S. & Neurath, M.F. (2004) TGF-beta suppresses tumor progression in colon cancer by inhibition of IL-6 trans-signaling, *Immunity.* Vol.21: 491-501.

Becker, C., Fantini, M.C., Wirtz, S., Nikolaev, A., Lehr, H.A., Galle, P.R., Rose-John, S. & Neurath, M.F. (2007) IL-6 signaling promotes tumor growth in colorectal cancer, *Cell Cycle.* Vol.4: 217-220.

Bentires-Alj, M., Kontaridis, M.I. & Neel, B.G. (2006) Stops along the RAS pathway in human genetic disease, *Nat Med.* Vol.12: 283-285.

Ben-Zvi, T., Yayon, A., Gertler, A. & Monsonego-Ornan, E. (2006) Suppressors of cytokine signaling (SOCS) 1 and SOCS3 interact with and modulate fibroblast growth factor receptor signaling, *J Cell Sci.* Vol.119: 380-387.

Bergamin, E., Wu, J. & Hubbard, S.R. (2006) Structural basis for phosphotyrosine recognition by suppressor of cytokine signaling-3, *Structure.* Vol.14: 1285-1292.

Bjorbaek, C., Buchholz, R.M., Davis, S.M., Bates, S.H., Pierroz, D.D., Gu, H., Neel, B.G., Myers, M.G., Jr. & Flier, J.S. (2001) Divergent roles of SHP-2 in ERK activation by leptin receptors, *J Biol Chem.* Vol.276: 4747-4755.

Bjorbaek, C., Elmquist, J.K., Frantz, J.D., Shoelson, S.E. & Flier, J.S. (1998) Identification of SOCS-3 as a potential mediator of central leptin resistance, *Mol Cell.* Vol.1: 619-625.

Borden, E.C., Sen, G.C., Uze, G., Silverman, R.H., Ransohoff, R.M., Foster, G.R. & Stark, G.R. (2007) Interferons at age 50: past, current and future impact on biomedicine, *Nat Rev Drug Discov.* Vol.6: 975-990.

Borland, G., Bird, R.J., Palmer, T.M. & Yarwood, S.J. (2009a) Activation of protein kinase Calpha by EPAC1 is required for the ERK- and CCAAT/enhancer-binding protein beta-dependent induction of the SOCS-3 gene by cyclic AMP in COS1 cells, *J Biol Chem.* Vol.284: 17391-17403.

Borland, G., Smith, B.O. & Yarwood, S.J. (2009b) EPAC proteins transduce diverse cellular actions of cAMP, *Br J Pharmacol.* Vol.158: 70-86.

Boyle, K., Egan, P., Rakar, S., Willson, T.A., Wicks, I.P., Metcalf, D., Hilton, D.J., Nicola, N.A., Alexander, W.S., Roberts, A.W. & Robb, L. (2007) The SOCS box of suppressor of cytokine signaling-3 contributes to the control of G-CSF responsiveness *in vivo, Blood.* Vol.110: 1466-1474.

Boyle, K. & Robb, L. (2008) The role of SOCS3 in modulating leukaemia inhibitory factor signalling during murine placental development. *J Reprod Immunol.* Vol.77: 1-6.

Brikos, C. & O'Neill, L.A. (2008) Signaling of toll-like receptors, *Handb Exp Pharmacol.* Vol.183: 21-50.

Briscoe, J., Rogers, N.C., Witthuhn, B.A., Watling, D., Harpur, A.G., Wilks, A.F., Stark, G.R., Ihle, J.N. & Kerr, I.M. (1996) Kinase-negative mutants of JAK1 can sustain interferon-gamma-inducible gene expression but not an antiviral state. *EMBO J.* Vol.15: 799-805.

Cacalano, N.A., Sanden, D. & Johnston, J.A. (2001) Tyrosine-phosphorylated SOCS-3 inhibits STAT activation but binds to p120 RasGAP and activates Ras, *Nat Cell Biol.*Vol. 3: 460-465.

Chalaris, A., Garbers, C., Rabe, B. & Rose-John, S. & Scheller, J. (2011) The soluble Interleukin 6 receptor: generation and role in inflammation and cancer, Eur J Cell Biol. Vol.90: 484-494.

Charbonneau, H., Tonks, N.K., Walsh, K.A. & Fischer, E.H. (1988) The leukocyte common antigen (CD45): a putative receptor-linked protein tyrosine phosphatase, *Proc Natl Acad Sci U S A.* Vol.85: 7182-7186.

Chen, Y., Wen, R., Yang, S., Schuman, J., Zhang, E.E., Yi, T., Feng, G.S. & Wang, D. (2003) Identification of Shp-2 as a Stat5A phosphatase, *J Biol Chem.* Vol.278: 16520-16527.

Cheng, A., Uetani, N., Simoncic, P.D., Chaubey, V.P., Lee-Loy, A., McGlade, C.J., Kennedy, B.P. & Tremblay, M.L. (2002), Attenuation of leptin action and regulation of obesity by protein tyrosine phosphatase 1B, *Dev Cell.* Vol.2: 497-503.

Croker, B.A., Krebs, D.L., Zhang, J.G., Wormald, S., Willson, T.A., Stanley, E.G., Robb, L., Greenhalgh, C.J., Forster, I., Clausen, B.E., Nicola, N.A., Metcalf, D., Hilton, D.J., Roberts, A.W. & Alexander, W.S. (2003) SOCS3 negatively regulates IL-6 signaling in vivo, *Nat Immunol.* Vol.4: 540-545.

Dalpke, A.H., Opper, S., Zimmermann, S. & Heeg, K. (2001) Suppressors of cytokine signaling (SOCS)-1 and SOCS-3 are induced by CpG-DNA and modulate cytokine responses in APCs, *J Immunol.* Vol.166: 7082-7089.

Darnell, J.E. Jr., Kerr, I.M. & Stark, G.R. (1994) Jak-STAT pathways and transcriptional activation in response to IFNs and other extracellular signaling proteins, *Science.* Vol.264: 1415-1421.

Dawson, M.A., Bannister, A.J., Göttgens, B., Foster, S.D., Bartke, T., Green, A.R. & Kouzarides, T. (2009) JAK2 phosphorylates histone H3Y41 and excludes HP1alpha from chromatin, *Nature.* Vol.461: 819-822.

Deshpande, A., Reddy, M.M., Schade, G.O., Ray, A., Chowdary, T.K., Griffin, J.D. & Sattler, M. (2011) Kinase domain mutations confer resistance to novel inhibitors targeting JAK2V617F in myeloproliferative neoplasms, *Leukemia.* doi: 10.1038/leu.2011.255.

De Souza, D., Fabri, L.J., Nash, A., Hilton, D.J., Nicola, N.A. & Baca, M. (2002) SH2 domains from suppressor of cytokine signaling-3 and protein tyrosine phosphatase SHP-2 have similar binding specificities, *Biochemistry*. Vol.41: 9229-9236.

Deb, D.K., Sassano, A., Lekmine, F., Majchrzak, B., Verma, A., Kambhampati, S., Uddin, S., Rahman, A., Fish, E.N. & Platanias, L.C. (2003) Activation of protein kinase C delta by IFN-gamma, *J Immunol.* Vol.171: 267-273.

Dewilde, S., Vercelli, A., Chiarle, R. & Poli, V. (2008) Of alphas and betas: distinct and overlapping functions of STAT3 isoforms, *Front Biosci.* Vol.13: 6501-6514.

Diamond, J.M. & Melo, J.V. (2011) Mechanisms of resistance to BCR-ABL kinase inhibitors, *Leuk Lymphoma.* Vol.52 Suppl 1: 12-22.

Ding, C., Cicuttini, F., Li, J. & Jones, G. (2009) Targeting IL-6 in the treatment of inflammatory and autoimmune diseases, *Expert Opin Investig Drugs.* Vol.18: 1457-1466.

Dovio, A., Perazzolo, L., Saba, L., Termine, A., Capobianco, M., Bertolotto, A. & Angeli, A. (2006) High-dose glucocorticoids increase serum levels of soluble IL-6 receptor alpha and its ratio to soluble gp130: an additional mechanism for early increased bone resorption, *Eur J Endocrinol.* Vol.154:745-751.

Dunn, S.L., Björnholm, M., Bates, S.H., Chen, Z., Seifert, M. & Myers, M.G. Jr. (2005) Feedback inhibition of leptin receptor/Jak2 signaling via Tyr1138 of the leptin receptor and suppressor of cytokine signaling 3. *Mol Endocrinol.* Vol.19: 925-938.

El Kasmi, K. C., Holst, J., Coffre, M., Mielke, L., de Pauw, A., Lhocine, N., Smith, A. M., Rutschman, R., Kaushal, D., Shen, Y., Suda, T., Donnelly, R.P., Myers, M.G., Jr., Alexander, W., Vignali, D.A., Watowich, S.S., Ernst, M., Hilton, D.J. & Murray, P.J. (2006) General nature of the STAT3-activated anti-inflammatory response, *J Immunol.* Vol.177: 7880-7888.

Elchebly, M., Payette, P., Michaliszyn, E., Cromlish, W., Collins, S., Loy, A.L., Normandin, D., Cheng, A., Himms-Hagen, J., Chan, C.C., Ramachandran, C., Gresser, M.J., Tremblay, M.L. & Kennedy, B.P. (1999) Increased insulin sensitivity and obesity resistance in mice lacking the protein tyrosine phosphatase-1B gene, *Science.* Vol.283: 1544-1548.

Endo, T.A., Masuhara, M., Yokouchi, M., Suzuki, R., Sakamoto, H., Mitsui, K., Matsumoto, A., Tanimura, S., Ohtsubo, M., Misawa, H., Miyazaki, T., Leonor, N., Taniguchi, T., Fujita, T., Kanakura, Y., Komiya, S. & Yoshimura, A. (1997) A new protein containing an SH2 domain that inhibits JAK kinases, *Nature.* Vol.387: 921-924.

Feldmann, M., Brennan, F. & Maini, R. (1996) Role of cytokines in rheumatoid arthritis, *Annu. Rev. Immunol.* Vol.14: 397–440.

Fourouclas, N., Li, J., Gilby, D.C., Campbell, P.J., Beer, P.A., Boyd, E.M., Goodeve, A.C., Bareford, D., Harrison, C.N., Reilly, J.T., Green, A.R. & Bench, A.J. (2008) Methylation of the suppressor of cytokine signaling 3 gene (SOCS3) in myeloproliferative disorders, *Haematologica.* Vol.93: 1635-1644.

Frobøse, H., Rønn, S.G, Heding, P.E., Mendoza, H., Cohen, P., Mandrup-Poulsen, T. & Billestrup, N. (2006) Suppressor of cytokine Signaling-3 inhibits interleukin-1 signaling by targeting the TRAF-6/TAK1 complex, *Mol Endocrinol.* Vol.20: 1587-1596.

Fu, X.Y. (1992) A transcription factor with SH2 and SH3 domains is directly activated by an interferon alpha-induced cytoplasmic protein tyrosine kinase(s), *Cell.* Vol.70: 323-335.

Fu, X.Y., Kessler, D.S., Veals, S.A., Levy, D.E. & Darnell, J.E. Jr. (1990) ISGF3, the transcriptional activator induced by interferon alpha, consists of multiple interacting polypeptide chains, *Proc Natl Acad Sci U S A.* Vol.87: 8555-8559.

Fukuda, M., Williams, K.W., Gautron, L. & Elmquist, J.K. (2011) Induction of leptin resistance by activation of cAMP-Epac signaling, *Cell Metab.* Vol.13: 331-339.

Gabay, C., Lamacchia, C. & Palmer, G. (2010) IL-1 pathways in inflammation and human diseases, *Nat Rev Rheumatol.* Vol.6: 232-241.

Gasperini, S., Crepaldi, L., Calzetti, F., Gatto, L., Berlato, C., Bazzoni, F., Yoshimura, A. & Cassatella, M.A. (2002) Interleukin-10 and cAMP-elevating agents cooperate to induce suppressor of cytokine signaling-3 via a protein kinase A-independent signal, *Eur Cytokine Netw.* Vol.13: 47-53.

Geng, H., Wittwer, T., Dittrich-Breiholz, O., Kracht, M. & Schmitz, M.L. (2009) Phosphorylation of NF-kappaB p65 at Ser468 controls its COMMD1-dependent ubiquitination and target gene-specific proteasomal elimination, *EMBO Rep.* Vol.10: 381-386.

Gerhartz, C., Heesel, B., Sasse, J., Hemmann, U., Landgraf, C., Schneider-Mergener, J., Horn, F., Heinrich, P.C. & Graeve, L. (1996) Differential activation of acute phase response factor/STAT3 and STAT1 via the cytoplasmic domain of the interleukin 6 signal transducer gp130. I. Definition of a novel phosphotyrosine motif mediating STAT1 activation, *J Biol Chem.* Vol.271: 12991-12998.

Ghoreschi, K., Laurence, A. & O'Shea, J.J. (2009) Janus kinases in immune cell signaling, *Immunol Rev.* Vol.228: 273-287.

Greenhalgh, C.J., Metcalf, D., Thaus, A.L., Corbin, J.E., Uren, R., Morgan, P.O., Fabri, L.J., Zhang, J.G., Martin, H.M., Willson, T.A., Billestrup, N., Nicola, N.A., Baca, M., Alexander, W.S. & Hilton, D.J. (2002) Biological evidence that SOCS-2 can act either as an enhancer or suppressor of growth hormone signaling, *J Biol Chem.* Vol.277: 40181-40184.

Griffiths, D.S., Li, J., Dawson, M.A., Trotter, M.W., Cheng, Y.H., Smith, A.M., Mansfield, W., Liu, P., Kouzarides, T., Nichols, J., Bannister, A.J., Green, A.R. & Göttgens, B. (2011) LIF-independent JAK signaling to chromatin in embryonic stem cells uncovered from an adult stem cell disease, *Nat Cell Biol.* Vol.13: 13-21.

Grivennikov, S., Karin, E., Terzic, J., Mucida, D., Yu, G.Y., Vallabhapurapu, S., Scheller, J., Rose-John, S., Cheroutre, H., Eckmann, L. & Karin, M. (2009) IL-6 and Stat3 are required for survival of intestinal epithelial cells and development of colitis-associated cancer, *Cancer Cell.* Vol.15: 103-113.

Grivennikov, S.I. & Karin, M. (2010) Dangerous liaisons: STAT3 and NF-kappaB collaboration and crosstalk in cancer, *Cytokine Growth Factor Rev.* Vol.21: 11-19.

Gunaje, J.J. & Bhat, G.J. (2001) Involvement of tyrosine phosphatase PTP1D in the inhibition of interleukin-6-induced Stat3 signaling by alpha-thrombin, *Biochem Biophys Res Commun.* Vol.288: 252-257.

Guschin, D., Rogers, N., Briscoe, J., Witthuhn, B., Watling, D., Horn, F., Pellegrini, S., Yasukawa, K., Heinrich, P., Stark, G.R., et al. (1995) A major role for the protein tyrosine kinase JAK1 in the JAK/STAT signal transduction pathway in response to interleukin-6, *EMBO J.* Vol.14: 1421-1429.

Haan, C., Hermanns, H.M., Heinrich, P.C. & Behrmann, I. (2000) A single amino acid substitution (Trp666Ala) in the interbox1/2 region of the interleukin-6 signal transducer gp130 abrogates binding of JAK1, and dominantly impairs signal transduction, *Biochem J.* Vol.349 :261-266.

Haan, S., Ferguson, P., Sommer, U., Hiremath, M., McVicar, D.W., Heinrich, P.C., Johnston, J.A. & Cacalano, N.A. (2003) Tyrosine phosphorylation disrupts elongin interaction and accelerates SOCS3 degradation, *J Biol Chem.* Vol.278: 31972-31979.

Haan, S., Wüller, S., Kaczor, J., Rolvering, C., Nöcker, T., Behrmann, I. & Haan, C. (2009) SOCS-mediated downregulation of mutant Jak2 (V617F, T875N and K539L) counteracts cytokine-independent signaling, *Oncogene.* Vol.28: 3069-3080.

Hanahan, D. & Weinberg, R.A. (2011) Hallmarks of cancer: the next generation, *Cell.* Vol.144: 646-674.

Hardy, M.P., Owczarek, C.M., Jermiin, L.S., Ejdeback, M. & Hertzog, P.J. (2004) Characterization of the type I interferon locus and identification of novel genes, *Genomics.* Vol.84: 331–345.

Hebenstreit, D., Wirnsberger, G., Horejs-Hoeck, J. & Duschl, A. (2006) Signaling mechanisms, interaction partners, and target genes of STAT6, *Cytokine Growth Factor Rev.* Vol.17: 173-188.

Hedvat, M., Huszar, D., Herrmann, A., Gozgit, J.M., Schroeder, A., Sheehy, A., Buettner, R., Proia, D., Kowolik, C.M., Xin, H., Armstrong, B., Bebernitz, G., Weng, S., Wang, L., Ye, M., McEachern, K., Chen, H., Morosini, D., Bell, K., Alimzhanov, M., Ioannidis, S., McCoon, P., Cao, Z.A., Yu, H., Jove, R. & Zinda, M. (2009) The JAK2 inhibitor AZD1480 potently blocks Stat3 signaling and oncogenesis in solid tumors, *Cancer Cell.* Vol.16: 487-497.

Heinrich, P.C., Behrmann, I., Haan, S., Hermanns, H.M., Müller-Newen, G. & Schaper, F. (2003) Principles of interleukin (IL)-6-type cytokine signaling and its regulation, *Biochem J.* Vol.374: 1-20.

Hemmann, U., Gerhartz, C., Heesel, B., Sasse, J., Kurapkat, G., Grötzinger, J., Wollmer, A., Zhong, Z., Darnell, J.E. Jr., Graeve, L., Heinrich, P.C. & Horn, F. (1996) Differential activation of acute phase response factor/Stat3 and Stat1 via the cytoplasmic domain of the interleukin 6 signal transducer gp130. II. Src homology SH2 domains define the specificity of stat factor activation, *J Biol Chem.* Vol.271: 12999-13007.

Hirano, T. (2010) Interleukin 6 in autoimmune and inflammatory diseases: a personal memoir, *Proc Jpn Acad Ser B Phys Biol Sci.* Vol.86: 717-730.

Hirano, T., Matsuda, T., Turner, M., Miyasaka, N., Buchan, G., Tang, B., Sato, K., Shimizu, M., Maini, R., Feldmann, M. & Kishimoto, T. (1988) Excessive production of interleukin 6/B cell stimulatory factor-2 in rheumatoid arthritis, *Eur J Immunol.* Vol.18: 1797-1801.

Hof, P., Pluskey, S., Dhe-Paganon, S., Eck, M.J. & Shoelson, S.E. (1998) Crystal structure of the tyrosine phosphatase SHP-2, *Cell.* Vol.92: 441-450.

Hookham, M.B., Elliott, J., Suessmuth, Y., Staerk, J., Ward, A.C., Vainchenker, W., Percy, M.J., McMullin, M.F., Constantinescu, S.N. & Johnston, J.A. (2007) The myeloproliferative disorder-associated JAK2 V617F mutant escapes negative regulation by suppressor of cytokine signaling 3, *Blood.* Vol.109: 4924-4929.

Huang, L.J., Constantinescu, S.N. & Lodish, H.F. (2001) The N-terminal domain of Janus kinase 2 is required for Golgi processing and cell surface expression of erythropoietin receptor, *Mol Cell.* Vol.8: 1327-1338.

Huang, Y., Qiu, J., Dong, S., Redell, M.S., Poli, V., Mancini, M.A. & Tweardy, D.J. (2007) Stat3 isoforms, alpha and beta, demonstrate distinct intracellular dynamics with prolonged nuclear retention of Stat3beta mapping to its unique C-terminal end, *J Biol Chem.* Vol.282: 34958-34967.

Ingley, E. & Klinken, S.P. (2006) Cross-regulation of JAK and Src kinases, *Growth Factors.* Vol.24: 89-95.

Irandoust, M.I., Aarts, L.H., Roovers, O., Gits, J., Erkeland, S.J. & Touw, I.P. (2007) Suppressor of cytokine signaling 3 controls lysosomal routing of G-CSF receptor, *EMBO J.* Vol. 26: 1782-1793.

James, C., Ugo, V., Le Couédic, J.P., Staerk, J., Delhommeau, F., Lacout, C., Garçon, L., Raslova, H., Berger, R., Bennaceur-Griscelli, A., Villeval, J.L., Constantinescu, S.N., Casadevall, N. & Vainchenker, W. (2005) A unique clonal JAK2 mutation leading to constitutive signaling causes polycythaemia vera, *Nature.* Vol.434: 1144-1148.

Jones, S.A., Scheller, J. & Rose-John, S. (2011) Therapeutic strategies for the clinical blockade of IL-6/gp130 signaling, *J Clin Invest.* Vol.121: 3375-3383.

Kalakonda, S., Nallar, S.C., Lindner, D.J., Hu, J., Reddy, S.P. & Kalvakolanu, D.V. (2007) Tumor-suppressive activity of the cell death activator GRIM-19 on a constitutively active signal transducer and activator of transcription 3, *Cancer Res.* Vol.67: 6212-20

Kallen, K.J. (2002) The role of transsignaling via the agonistic soluble IL-6 receptor in human diseases, *Biochim Biophys Acta.* Vol.1592: 323-343.

Kamura, T., Maenaka, K., Kotoshiba, S., Matsumoto, M., Kohda, D., Conaway, R. C., Conaway, J.W. & Nakayama, K.I. (2004) VHL-box and SOCS-box domains determine binding specificity for Cul2-Rbx1 and Cul5-Rbx2 modules of ubiquitin ligases, *Genes Dev.* Vol.18: 3055-3065.

Kamura, T., Sato, S., Haque, D., Liu, L., Kaelin, W.G., Jr., Conaway, R.C. & Conaway, J.W. (1998) The Elongin BC complex interacts with the conserved SOCS-box motif present in members of the SOCS, ras, WD-40 repeat, and ankyrin repeat families, *Genes Dev.* Vol.12: 3872-3881.

Kaplan, D.H., Greenlund, A.C., Tanner, J.W., Shaw, A.S. & Schreiber, R.D. (1996) Identification of an interferon-gamma receptor alpha chain sequence required for JAK-1 binding, *J Biol Chem.* Vol.271: 9-12.

Kaptein, A., Paillard, V. & Saunders, M. (1996) Dominant negative stat3 mutant inhibits interleukin-6-induced Jak-STAT signal transduction, *J Biol Chem.* Vol.271: 5961-5964.

Kawazoe, Y., Naka, T., Fujimoto, M., Kohzaki, H., Morita, Y., Narazaki, M., Okumura, K., Saitoh, H., Nakagawa, R., Uchiyama, Y., Akira, S. & Kishimoto, T. (2001) Signal transducer and activator of transcription (STAT)-induced STAT inhibitor 1 (SSI-1)/suppressor of cytokine signaling 1 (SOCS1) inhibits insulin signal transduction pathway through modulating insulin receptor substrate 1 (IRS-1) phosphorylation, *J Exp Med.* Vol.193:263-239.

Kralovics, R., Passamonti, F., Buser, A.S., Teo, S.S., Tiedt, R., Passweg, J.R., Tichelli, A., Cazzola, M. & Skoda, R.C. (2005) A gain-of-function mutation of JAK2 in myeloproliferative disorders, *N Engl J Med.* Vol.352: 1779-1790.

Krause, C.D., Mei, E., Xie, J., Jia, Y., Bopp, M.A., Hochstrasser, R.M. & Pestka, S. (2002) Seeing the light: preassembly and ligand-induced changes of the interferon gamma receptor complex in cells, *Mol Cell Proteomics.* Vol.1: 805-815.

Krebs, D. L. & Hilton, D. J. (2001) SOCS proteins: negative regulators of cytokine signaling, *Stem Cells.* Vol.19: 378-387.

LaFleur, D.W., Nardelli, B., Tsareva, T., Mather, D., Feng, P., Semenuk, M., Taylor, K., Buergin, M., Chinchilla, D., Roshke, V., Chen, G., Ruben, S.M., Pitha, P.M., Coleman, T.A. & Moore, P.A. (2001) Interferon-kappa, a novel type I interferon expressed in human keratinocytes, *J Biol Chem.* Vol.276: 39765-39771.

Lang, R., Pauleau, A.L., Parganas, E., Takahashi, Y., Mages, J., Ihle, J.N., Rutschman, R. & Murray, P.J. (2003) SOCS3 regulates the plasticity of gp130 signaling, *Nat Immunol.* Vol.4: 546-550.

Lee, H., Deng, J., Kujawski, M., Yang, C., Liu, Y., Herrmann, A., Kortylewski, M., Horne, D., Somlo, G., Forman, S., Jove, R. & Yu, H. (2010) STAT3-induced S1PR1 expression is crucial for persistent STAT3 activation in tumors, *Nat Med.* Vol.16: 1421-1428.

Lee, H., Herrmann, A., Deng, J.H., Kujawski, M., Niu, G., Li, Z., Forman, S., Jove, R., Pardoll, D.M. & Yu, H. (2009) Persistently activated Stat3 maintains constitutive NF-kappaB activity in tumors, *Cancer Cell.* Vol.15: 283-293.

Lehmann, U., Schmitz, J., Weissenbach, M., Sobota, R.M., Hortner, M., Friederichs, K., Behrmann, I., Tsiaris, W., Sasaki, A., Schneider-Mergener, J., Yoshimura, A., Neel, B.G., Heinrich, P.C. & Schaper, F. (2003) SHP2 and SOCS3 contribute to Tyr-759-dependent attenuation of interleukin-6 signaling through gp130, *J Biol Chem.* Vol.278: 661-671.

Levine, R.L., Wadleigh, M., Cools, J., Ebert, B.L., Wernig, G., Huntly, B.J., Boggon, T.J., Wlodarska, I., Clark, J.J., Moore, S., Adelsperger, J., Koo, S., Lee, J.C., Gabriel, S., Mercher, T., D'Andrea, A., Fröhling, S., Döhner, K., Marynen, P., Vandenberghe, P., Mesa, R.A., Tefferi, A., Griffin, J.D., Eck, M.J., Sellers, W.R., Meyerson, M., Golub, T.R., Lee, S.J. & Gilliland, D.G. (2005) Activating mutation in the tyrosine kinase JAK2 in polycythemia vera, essential thrombocythemia, and myeloid metaplasia with myelofibrosis, *Cancer Cell.* Vol.7: 387-397.

Li, C. & Friedman, J.M. (1999) Leptin receptor activation of SH2 domain containing protein tyrosine phosphatase 2 modulates Ob receptor signal transduction, *Proc Natl Acad Sci U S A.* Vol.96: 9677-9682.

Li, L., Gronning, L.M., Anderson, P.O., Li, S., Edvardsen, K., Johnston, J., Kioussis, D., Shepherd, P.R. & Wang, P. (2004) Insulin induces SOCS-6 expression and its binding to the p85 monomer of phosphoinositide 3-kinase, resulting in improvement in glucose metabolism, *J Biol Chem.* Vol.279: 34107-34114.

Li, W., Nishimura, R., Kashishian, A., Batzer, A.G., Kim, W.J., Cooper, J.A. & Schlessinger, J. (1994) A new function for a phosphotyrosine phosphatase: linking GRB2-Sos to a receptor tyrosine kinase, *Mol Cell Biol.* Vol.14: 509-517.

Lim, C.P. & Cao, X. (2006) Structure, function, and regulation of STAT proteins, *Mol Biosyst.* Vol.2: 536-550.

Liu, E., Côté, J.F. & Vuori, K. (2003) Negative regulation of FAK signaling by SOCS proteins, *EMBO J.* Vol.22: 5036-5046.

Liu, S. & Chen, Z.J. (2011) Expanding role of ubiquitination in NF-κB signaling, *Cell Res.* Vol.21: 6-21.

Loh, K., Fukushima, A., Zhang, X., Galic, S., Briggs, D., Enriori, P.J., Simonds, S., Wiede, F., Reichenbach, A., Hauser, C., Sims, N.A., Bence, K.K., Zhang, S., Zhang, Z.Y., Kahn, B.B., Neel, B.G., Andrews, Z.B., Cowley, M.A. & Tiganis, T. (2011) Elevated Hypothalamic TCPTP in Obesity Contributes to Cellular Leptin Resistance, *Cell Metab.* Vol.14: 684-699.

Lorenzen, J.A., Dadabay, C.Y. & Fischer, E.H. (1995) COOH-terminal sequence motifs target the T cell protein tyrosine phosphatase to the ER and nucleus, *J Cell Biol.* Vol.131: 631-643.

Lotfi, S., Li, Z., Sun, J., Zuo, Y., Lam, P.P., Kang, Y., Rahimi, M., Islam, D., Wang, P., Gaisano, H.Y. & Jin, T. (2006) Role of the exchange protein directly activated by cyclic adenosine 5'-monophosphate (Epac) pathway in regulating proglucagon gene expression in intestinal endocrine L cells, *Endocrinology.* Vol.147: 3727-3736.

Lupardus, P.J., Skiniotis, G., Rice, A.J., Thomas, C., Fischer, S., Walz, T. & Garcia, K.C. (2011) Structural snapshots of full-length Jak1, a transmembrane gp130/IL-6/IL-6Rα cytokine receptor complex, and the receptor-Jak1 holocomplex, *Structure.* Vol.12: 45-55.

Ma J. & Cao, X. (2006) Regulation of Stat3 nuclear import by importin alpha5 and importin alpha7 via two different functional sequence elements, *Cell Signal.* Vol.18: 1117-1126.

Madamanchi, N.R., Li, S., Patterson, C. & Runge, M.S. (2001) Thrombin regulates vascular smooth muscle cell growth and heat shock proteins via the JAK-STAT pathway, *J Biol Chem.* Vol.276: 18915-18924.

Maine, G.N. & Burstein, E. (2007) COMMD proteins and the control of the NF kappa B pathway, *Cell Cycle.* Vol.6: 672-676.

Maine, G.N., Mao, X., Komarck, C.M. & Burstein, E. (2007) COMMD1 promotes the ubiquitination of NF-kappaB subunits through a cullin-containing ubiquitin ligase, *EMBO J.* Vol.26: 436-447.

Mansell, A., Smith, R., Doyle, S.L., Gray, P., Fenner, J.E., Crack, P.J., Nicholson, S.E., Hilton, D.J., O'Neill, L.A. & Hertzog, P.J. (2006) Suppressor of cytokine signaling 1 negatively regulates Toll-like receptor signaling by mediating Mal degradation, *Nat Immunol.* Vol.7: 148-155.

Manshouri, T., Quintás-Cardama, A., Nussenzveig, R.H., Gaikwad, A., Estrov, Z., Prchal, J., Cortes, J.E., Kantarjian, H.M. & Verstovsek, S. (2008) The JAK kinase inhibitor CP-690,550 suppresses the growth of human polycythemia vera cells carrying the JAK2V617F mutation, *Cancer Sci.* Vol.99: 1265-1273.

Marin, V., Montero-Julian, F.A., Grès, S., Boulay, V., Bongrand, P., Farnarier, C. & Kaplanski, G. (2001) The IL-6-soluble IL-6Ralpha autocrine loop of endothelial activation as an intermediate between acute and chronic inflammation: an experimental model involving thrombin, *J Immunol.* Vol.167: 3435-3442.

Maritano, D., Sugrue, M.L., Tininini, S., Dewilde, S., Strobl, B., Fu, X., Murray-Tait, V., Chiarle, R. & Poli, V. (2004) The STAT3 isoforms alpha and beta have unique and specific functions, *Nat Immunol.* Vol.5: 401-409.

Marotta, L.L., Almendro, V., Marusyk, A., Shipitsin, M., Schemme, J., Walker, S.R., Bloushtain-Qimron, N., Kim, J.J., Choudhury, S.A., Maruyama, R., Wu, Z., Gönen, M., Mulvey, L.A., Bessarabova, M.O., Huh, S.J., Silver, S.J., Kim, S.Y., Park, S.Y., Lee, H.E., Anderson, K.S., Richardson, A.L., Nikolskaya, T., Nikolsky, Y., Liu, X.S., Root, D.E., Hahn, W.C., Frank, D.A. & Polyak, K. (2011) The JAK2/STAT3 signaling pathway is required for growth of $CD44^{+}CD24^{-}$ stem cell-like breast cancer cells in human tumors, *J Clin Invest.* Vol.121: 2723-2735.

Mori, H., Hanada, R., Hanada, T., Aki, D., Mashima, R., Nishinakamura, H., Torisu, T., Chien, K.R., Yasukawa, H. & Yoshimura, A. (2004) Socs3 deficiency in the brain elevates leptin sensitivity and confers resistance to diet-induced obesity, *Nat Med.* Vol.10: 739-743.

Mosenden, R. & Taskén, K. (2011) Cyclic AMP-mediated immune regulation--overview of mechanisms of action in T cells, *Cell Signal.* Vol.23: 1009-1016.

Muller-Newen, G., Kuster, A., Hemmann, U., Keul, R., Horsten, U., Martens, A., Graeve, L., Wijdenes, J. & Heinrich, P.C. (1998) Soluble IL-6 receptor potentiates the antagonistic activity of soluble gp130 on IL-6 responses, *J Immunol.* Vol.161: 6347–6355.

Murakami, M., Hibi, M., Nakagawa, N., Nakagawa, T., Yasukawa, K., Yamanishi, K., Taga, T. & Kishimoto, T. (1993) IL-6-induced homodimerization of gp130 and associated activation of a tyrosine kinase, *Science.* Vol.260: 1808-1810.

Murray, P.J. (2007) The JAK-STAT signaling pathway: input and output integration, *J Immunol.* Vol.178: 2623-2629.

Myers, M.G., Cowley, M.A. & Münzberg, H. (2008) Mechanisms of leptin action and leptin resistance, *Annu Rev Physiol.* Vol.70: 537-556.

Naka, T., Narazaki, M., Hirata, M., Matsumoto, T., Minamoto, S., Aono, A., Nishimoto, N., Kajita, T., Taga, T., Yoshizaki, K., Akira, S. & Kishimoto, T. (1997) Structure and function of a new STAT-induced STAT inhibitor, *Nature.* Vol.387: 924-929.

Narazaki, M., Witthuhn, B.A., Yoshida, K., Silvennoinen, O., Yasukawa, K., Ihle, J.N., Kishimoto, T. & Taga, T. (1994) Activation of JAK2 kinase mediated by the interleukin 6 signal transducer gp130, *Proc Natl Acad Sci U S A.* Vol.91: 2285-2289.

Narazaki, M., Yasukawa, K., Saito, T., Ohsugi, Y., Fukui, H., Koishihara, Y., Yancopoulos, G.D., Taga, T. & Kishimoto, T: (1993) Soluble forms of the interleukin-6 signal-transducing receptor component gp130 in human serum possessing a potential to inhibit signals through membrane-anchored gp130, *Blood.* Vol.82: 1120-1126.

Neel, B.G., Gu, H. & Pao, L. (2003) The 'Shp'ing news: SH2 domain-containing tyrosine phosphatases in cell signaling, *Trends Biochem Sci.* Vol.28: 284-293.

Nguyen, H., Ramana, C.V., Bayes, J. & Stark, G.R. (2001) Roles of phosphatidylinositol 3-kinase in interferon-gamma-dependent phosphorylation of STAT1 on serine 727 and activation of gene expression, *J Biol Chem.* Vol.276: 33361-33368.

Nicholson, S.E., De Souza, D., Fabri, L.J., Corbin, J., Willson, T.A., Zhang, J.G., Silva, A., Asimakis, M., Farley, A., Nash, A.D., Metcalf, D., Hilton, D.J., Nicola, N.A. & Baca, M. (2000) Suppressor of cytokine signaling-3 preferentially binds to the SHP-2-binding site on the shared cytokine receptor subunit gp130, *Proc Natl Acad Sci U S A* Vol.97: 6493-6498.

Niwa, Y., Kanda, H., Shikauchi, Y., Saiura, A., Matsubara, K., Kitagawa, T., Yamamoto, J., Kubo, T. & Yoshikawa, H. (2005) Methylation silencing of SOCS-3 promotes cell growth and migration by enhancing JAK/STAT and FAK signalings in human hepatocellular carcinoma, *Oncogene.* Vol.24: 6406-6417.

Ogura, H., Murakami, M., Okuyama, Y., Tsuruoka, M., Kitabayashi, C., Kanamoto, M., Nishihara, M., Iwakura, Y. & Hirano, T. (2008) Interleukin-17 promotes autoimmunity by triggering a positive-feedback loop via interleukin-6 induction, *Immunity.* Vol.29: 628-636.

Ohtani, T., Ishihara, K., Atsumi, T., Nishida, K., Kaneko, Y., Miyata, T., Itoh, S., Narimatsu, M., Maeda, H., Fukada, T., Itoh, M., Okano, H., Hibi, M. & Hirano, T. (2000) Dissection of signaling cascades through gp130 in vivo: reciprocal roles for STAT3- and SHP2-mediated signals in immune responses, *Immunity.* Vol.12: 95-105.

O'Neill, L.A. (2009) The interleukin-1 receptor/Toll-like receptor superfamily: 10 years of progress, *Immunol Rev.* Vol.226: 10-18.

O'Shea, J.J. & Murray, P.J. (2008) Cytokine signaling modules in inflammatory responses, *Immunity.* Vol.28: 477-487.

O'Shea, J.J., Gadina, M. & Schreiber, R.D. (2002) Cytokine signaling in 2002: new surprises in the Jak/Stat pathway, *Cell.* Vol.109: S121-131.

O'Shea, J.J., Husa, M., Li, D., Hofmann, S.R., Watford, W., Roberts, J.L., Buckley, R.H., Changelian, P. & Candotti, F. (2004) Jak3 and the pathogenesis of severe combined immunodeficiency, *Mol Immunol.* Vol.41: 727-737.

Owens, D.M. & Keyse, S.M. (2007) Differential regulation of MAP kinase signaling by dual-specificity protein phosphatases, *Oncogene* Vol.26: 3203-3213.

Paulson, M., Pisharody, S., Pan, L., Guadagno, S., Mui, A.L. & Levy, D.E. (1999) Stat protein transactivation domains recruit p300/CBP through widely divergent sequences, *J Biol Chem*. Vol.274: 25343-25349.

Pelletier, S., Duhamel, F., Coulombe, P., Popoff, M.R. & Meloche, S. (2003) Rho family GTPases are required for activation of Jak/STAT signaling by G protein-coupled receptors, *Mol Cell Biol*. Vol.23: 1316-1333.

Pierconti, F., Martini, M., Pinto, F., Cenci, T., Capodimonti, S., Calarco, A., Bassi, P.F. & Larocca, L.M. (2011) Epigenetic silencing of SOCS3 identifies a subset of prostate cancer with an aggressive behavior, *Prostate*. Vol.71: 318-325.

Piessevaux, J., Lavens, D., Montoye, T., Wauman, J., Catteeuw, D., Vandekerckhove, J., Belsham, D., Peelman, F. & Tavernier, J. (2006).Functional cross-modulation between SOCS proteins can stimulate cytokine signaling, J Biol Chem. Vol.281: 32953-32966.

Piessevaux, J., Lavens, D., Peelman, F. & Tavernier, J. (2008) The many faces of the SOCS box, *Cytokine Growth Factor Rev*. Vol.19: 371-381.

Quintás-Cardama, A., Kantarjian, H., Cortes, J. & Verstovsek, S. (2011) Janus kinase inhibitors for the treatment of myeloproliferative neoplasias and beyond, *Nat Rev Drug Discov*. Vol.10: 127-140.

Rane, S.G. & Reddy, E.P. (2002) JAKs, STATs and Src kinases in hematopoiesis, *Oncogene*. Vol.21: 3334-3358.

Robb, L., Boyle, K., Rakar, S., Hartley, L., Lochland, J., Roberts, A.W., Alexander, W.S. & Metcalf, D. (2005) Genetic reduction of embryonic leukemia-inhibitory factor production rescues placentation in SOCS3-null embryos but does not prevent inflammatory disease, *Proc Natl Acad Sci U S A*. Vol.102: 16333-16338.

Roberts, A.W., Robb, L., Rakar, S., Hartley, L., Cluse, L., Nicola, N.A., Metcalf, D., Hilton, D.J. & Alexander, W.S. (2001) Placental defects and embryonic lethality in mice lacking suppressor of cytokine signaling 3, *Proc Natl Acad Sci U S A*. Vol.98: 9324-9329.

Rossin, E.J., Lage, K., Raychaudhuri, S., Xavier, R.J., Tatar, D., Benita, Y.; International Inflammatory Bowel Disease Genetics Constortium, Cotsapas, C. & Daly, M.J. (2011) Proteins encoded in genomic regions associated with immune-mediated disease physically interact and suggest underlying biology, *PLoS Genet*. Vol.7: e1001273.

Rui, L., Yuan, M., Frantz, D., Shoelson, S. & White, M.F. (2002) SOCS-1 and SOCS-3 block insulin signaling by ubiquitin-mediated degradation of IRS1 and IRS2, *J Biol Chem*. Vol.277: 42394-42398.

Saharinen, P. & Silvennoinen, O. (2002) The pseudokinase domain is required for suppression of basal activity of Jak2 and Jak3 tyrosine kinases and for cytokine-inducible activation of signal transduction, *J Biol Chem*. Vol.277: 47954–47963.

Saharinen, P., Takaluoma, K. & Silvennoinen, O. (2000) Regulation of the Jak2 tyrosine kinase by its pseudokinase domain, *Mol Cell Biol*. Vol.20: 3387–3395.

Salmond, R.J. & Alexander, D.R. (2006) SHP2 forecast for the immune system: fog gradually clearing, *Trends Immunol*. Vol.27: 154-160.

Sands, W.A., Woolson, H.D., Milne, G.R., Rutherford, C. & Palmer, T.M. (2006).Exchange protein activated by cyclic AMP (Epac)-mediated induction of suppressor of cytokine signaling 3 (SOCS-3) in vascular endothelial cells, *Mol Cell Biol*. Vol.26: 6333-6346.

Sanz-Moreno, V., Gaggioli, C., Yeo, M., Albrengues, J., Wallberg, F., Viros, A., Hooper, S., Mitter, R., Féral, C.C., Cook, M., Larkin, J., Marais, R., Meneguzzi, G., Sahai, E. & Marshall, C.J. (2011) ROCK and JAK1 signaling cooperate to control actomyosin contractility in tumor cells and stroma, *Cancer Cell*. Vol.20: 229-245.

Sasaki, A., Inagaki-Ohara, K., Yoshida, T., Yamanaka, A., Sasaki, M., Yasukawa, H., Koromilas, A.E. & Yoshimura, A. (2003) The N-terminal truncated isoform of SOCS3 translated from an alternative initiation AUG codon under stress conditions is stable due to the lack of a major ubiquitination site, Lys-6, *J Biol Chem.* Vol.278: 2432-2436.

Sasaki, A., Yasukawa, H., Suzuki, A., Kamizono, S., Syoda, T., Kinjyo, I., Sasaki, M., Johnston, J.A. & Yoshimura, A. (1999) Cytokine-inducible SH2 protein-3 (CIS3/SOCS3) inhibits Janus tyrosine kinase by binding through the N-terminal kinase inhibitory region as well as SH2 domain, *Genes Cells.* Vol.4: 339-351.

Sawa, S., Kamimura, D., Jin, G.H., Morikawa, H., Kamon, H., Nishihara, M., Ishihara, K., Murakami, M., & Hirano, T. (2006) Autoimmune arthritis associated with mutated interleukin (IL)-6 receptor gp130 is driven by STAT3/IL-7-dependent homeostatic proliferation of CD4+ T cells, *J Exp Med.* Vol.203: 1459-1470.

Schaper, F., Gendo, C., Eck, M., Schmitz, J., Grimm, C., Anhuf, D., Kerr, I.M. & Heinrich, P.C. (1998) Activation of the protein tyrosine phosphatase SHP2 via the interleukin-6 signal transducing receptor protein gp130 requires tyrosine kinase Jak1 and limits acute-phase protein expression, *Biochem J.* Vol.335: 557-565.

Scheller, J., Chalaris, A., Schmidt-Arras, D. & Rose-John, S. (2011) The pro- and anti-inflammatory properties of the cytokine interleukin-6, *Biochim Biophys Acta.* Vol.1813:878-888.

Schuringa, J.J., Schepers, H., Vellenga, E. & Kruijer, W. (2001) Ser727-dependent transcriptional activation by association of p300 with STAT3 upon IL-6 stimulation, *FEBS Lett.* Vol.495: 71-76.

Shuai , K., Horvath, C.M., Huang, L.H., Qureshi, S.A., Cowburn, D. & Darnell, J.E. Jr. (1994) Interferon activation of the transcription factor Stat91 involves dimerization through SH2-phosphotyrosyl peptide interactions, *Cell.* Vol.76: 821-828.

Simoncic, P.D., Lee-Loy, A., Barber, D.L., Tremblay, M.L. & McGlade, C.J. (2002) The T cell protein tyrosine phosphatase is a negative regulator of janus family kinases 1 and 3, *Curr Biol.* Vol.12: 446-453.

Stahl, N., Boulton, T.G., Farruggella, T., Ip, N.Y., Davis, S., Witthuhn, B.A., Quelle, F.W., Silvennoinen, O., Barbieri, G., Pellegrini, S., et al. (1994) Association and activation of Jak-Tyk kinases by CNTF-LIF-OSM-IL-6 beta receptor components, *Science.* Vol.263: 92-95.

Stahl, N., Farruggella, T.J., Boulton, T.G., Zhong, Z., Darnell, J.E. Jr. & Yancopoulos, G.D. (1995) Choice of STATs and other substrates specified by modular tyrosine-based motifs in cytokine receptors, *Science.* Vol.267: 1349-1353.

Stark, G.R., Kerr, I.M., Williams, B.R., Silverman, R.H. & Schreiber, R.D. (1998). How cells respond to interferons, *Annu Rev Biochem.* Vol.67: 227-264.

Starr, R., Willson, T. A., Viney, E. M., Murray, L. J., Rayner, J. R., Jenkins, B. J., Gonda, T. J., Alexander, W. S., Metcalf, D., Nicola, N. A. & Hilton, D. J. (1997) A family of cytokine-inducible inhibitors of signaling, *Nature.* Vol.387: 917-921.

Stevenson, N.J., McFarlane, C., Ong, S.T., Nahlik, K., Kelvin, A., Addley, M.R., Long, A., Greaves, D.R., O'Farrelly, C. & Johnston, J.A. (2010) Suppressor of cytokine signaling (SOCS) 1 and 3 enhance cell adhesion and inhibit migration towards the chemokine eotaxin/CCL11, *FEBS Lett.* Vol.584: 4469-4474.

Symes, A., Stahl, N., Reeves, S.A., Farruggella, T., Servidei, T., Gearan, T., Yancopoulos, G. & Fink, J.S. (1997) The protein tyrosine phosphatase SHP-2 negatively regulates ciliary neurotrophic factor induction of gene expression, *Curr Biol.* Vol.7: 697-700.

Takahashi, Y., Carpino, N., Cross, J.C., Torres, M., Parganas, E. & Ihle, J N. (2003).SOCS3: an essential regulator of LIF receptor signaling in trophoblast giant cell differentiation, *EMBO J.* Vol.22: 372-384.

Takahashi-Tezuka, M., Yoshida, Y., Fukada, T., Ohtani, T., Yamanaka, Y., Nishida, K., Nakajima, K., Hibi, M. & Hirano, T. (1998) Gab1 acts as an adapter molecule linking the cytokine receptor gp130 to ERK mitogen-activated protein kinase, *Mol Cell Biol.* Vol.18: 4109-4117.

Takeda, K., Clausen, B.E., Kaisho, T., Tsujimura, T., Terada, N., Förster, I. & Akira, S. (1999) Enhanced Th1 activity and development of chronic enterocolitis in mice devoid of Stat3 in macrophages and neutrophils, *Immunity.* Vol.10: 39-49.

Tanaka, T., Soriano, M.A. & Grusby, M.J. (2005) SLIM is a nuclear ubiquitin E3 ligase that negatively regulates STAT signaling, *Immunity* Vol.22: 729-736.

Tannahill, G.M., Elliott, J., Barry, A.C., Hibbert, L., Cacalano, N.A. & Johnston, J.A. (2005) SOCS2 can enhance interleukin-2 (IL-2) and IL-3 signaling by accelerating SOCS3 degradation, *Mol Cell Biol.* Vol.25: 9115-9126.

Tanner, J,W,, Chen, W., Young, R.L., Longmore, G.D. & Shaw, A.S. (1995) The conserved box 1 motif of cytokine receptors is required for association with JAK kinases, *J Biol Chem.* Vol.270: 6523-6530.

Tartaglia, M., Mehler, E.L., Goldberg, R., Zampino, G., Brunner, H.G., Kremer, H., van der Burgt, I., Crosby, A.H., Ion, A., Jeffery, S., Kalidas, K., Patton, M.A., Kucherlapati, R.S. & Gelb, B.D. (2001) Mutations in PTPN11, encoding the protein tyrosine phosphatase SHP-2, cause Noonan syndrome, *Nat Genet.* Vol.29: 465-468.

Taylor, P.C. & Feldmann, M. (2009) Anti-TNF biologic agents: still the therapy of choice for rheumatoid arthritis, *Nat Rev Rheumatol.* Vol.5: 578-582.

Tefferi, A., Litzow, M.R. & Pardanani, A. (2011) Long-term outcome of treatment with ruxolitinib in myelofibrosis, *N Engl J Med.* Vol.365: 1455-1457.

ten Hoeve, J., de Jesus Ibarra-Sanchez, M., Fu, Y., Zhu, W., Tremblay, M., David, M. & Shuai, K. (2002) Identification of a nuclear Stat1 protein tyrosine phosphatase, *Mol Cell Biol.* Vol.22: 5662-5668.

Tiganis, T. & Bennett, A.M. (2007) Protein tyrosine phosphatase function: :the substrate perspective, *Biochem J.* Vol.402:1-15.

Umenishi, F., Narikiyo, T., Vandewalle, A. & Schrier, R.W. (2006) cAMP regulates vasopressin-induced AQP2 expression via protein kinase A-independent pathway, *Biochim Biophys Acta* Vol.1758: 1100-1105.

Ungureanu, D., Saharinen, P., Junttila, I., Hilton, D.J. & Silvennoinen, O. (2002) Regulation of Jak2 through the ubiquitin-proteasome pathway involves phosphorylation of Jak2 on Y1007 and interaction with SOCS-1, *Mol Cell Biol.* Vol.22: 3316-3326.

Ungureanu, D., Wu, J., Pekkala, T., Niranjan, Y., Young, C., Jensen, O.N., Xu, C.F., Neubert, T.A., Skoda, R.C., Hubbard, S.R. & Silvennoinen, O. (2011) The pseudokinase domain of JAK2 is a dual-specificity protein kinase that negatively regulates cytokine signaling, *Nat Struct Mol Biol.* Vol.18: 971-976.

van Boxel-Dezaire, A. H. & Stark, G. R. (2007) Cell type specific signaling in response to interferon-γ. *Curr. Top. Microbiol. Immunol.* Vol.316: 119–154.

Verstak, B., Nagpal, K., Bottomley, S.P., Golenbock, D.T., Hertzog, P.J. & Mansell, A. (2009) MyD88 adapter-like (Mal)/TIRAP interaction with TRAF6 is critical for TLR2- and TLR4-mediated NF-kappaB proinflammatory responses, *J Biol Chem.* Vol.284: 24192-24203.

Verstovsek, S., Kantarjian, H., Mesa, R.A., Pardanani, A.D., Cortes-Franco, J., Thomas, D.A., Estrov, Z., Fridman, J.S., Bradley, E.C., Erickson-Viitanen, S., Vaddi, K., Levy, R. & Tefferi, A. (2010) Safety and efficacy of INCB018424, a JAK1 and JAK2 inhibitor, in myelofibrosis, *N Engl J Med.* Vol.363: 1117-1127.

Wall, E.A., Zavzavadjian, J.R., Chang, M.S., Randhawa, B., Zhu, X., Hsueh, R.C., Liu, J., Driver, A., Bao, X.R., Sternweis, P.C., Simon, M.I. & Fraser, I.D. (2009) Suppression of LPS-induced TNF-alpha production in macrophages by cAMP is mediated by PKA-AKAP95-p105, *Sci Signal.* Vol.2: ra28.

Weber, J.D., Raben, D.M., Phillips, P.J. & Baldassare, J.J. (1997) Sustained activation of extracellular-signal-regulated kinase 1 (ERK1) is required for the continued expression of cyclin D1 in G1 phase, *Biochem J.* Vol.326: 61-68.

Wegenka, U.M., Buschmann, J., Lütticken, C., Heinrich, P.C. & Horn, F. (1993) Acute-phase response factor, a nuclear factor binding to acute-phase response elements, is rapidly activated by interleukin-6 at the posttranslational level, *Mol Cell Biol.* Vol.13: 276-288.

Williams, N.K., Bamert, R.S., Patel, O., Wang, C., Walden, P.M., Wilks, A.F., Fantino, E., Rossjohn, J. & Lucet, I.S. (2009) Dissecting specificity in the Janus kinases: the structures of JAK-specific inhibitors complexed to the JAK1 and JAK2 protein tyrosine kinase domains, *J Mol Biol.* Vol.387: 219-232.

Woolson, H.D., Thomson, V.S., Rutherford, C., Yarwood, S.J. & Palmer, T.M. (2009) Selective inhibition of cytokine-activated extracellular signal-regulated kinase by cyclic AMP via Epac1-dependent induction of suppressor of cytokine signaling-3, *Cell Signal.* Vol.21: 1706-1715.

Xia, L., Wang, L., Chung, A. S., Ivanov, S.S., Ling, M.Y., Dragoi, A.M., Platt, A., Gilmer, T.M., Fu, X.Y. & Chin, Y.E. (2002) Identification of both positive and negative domains within the epidermal growth factor receptor COOH-terminal region for signal transducer and activator of transcription (STAT) activation, *J Biol Chem.* Vol.277: 30716-30723.

Yamamoto, T., Sekine, Y., Kashima, K., Kubota, A., Sato, N., Aoki, N. & Matsuda, T (2002) The nuclear isoform of protein-tyrosine phosphatase TC-PTP regulates interleukin-6-mediated signaling pathway through STAT3 dephosphorylation, *Biochem Biophys Res Commun.* Vol.297: 811-817.

Yamauchi, K., Milarski, K.L., Saltiel, A.R. & Pessin, J.E. (1995) Protein-tyrosine-phosphatase SHPTP2 is a required positive effector for insulin downstream signaling, *Proc Natl Acad Sci U S A.* Vol.92: 664-668.

Yang, E., Wen, Z., Haspel, R.L., Zhang, J.J. & Darnell, J.E. Jr. (1999) The linker domain of Stat1 is required for gamma interferon-driven transcription, *Mol Cell Biol.* Vol.19: 5106-5112.

Yang, J. & Stark, G.R. (2008) Roles of unphosphorylated STATs in signaling, *Cell Res.* Vol.18: 443-451.

Yang, J., Liao, X., Agarwal, M.K., Barnes, L., Auron, P.E. & Stark, G.R. (2007) Unphosphorylated STAT3 accumulates in response to IL-6 and activates transcription by binding to NFkappaB, *Genes Dev.* Vol.21: 1396-1408.

Yarwood, S.J., Borland, G., Sands, W.A. & Palmer, T.M. (2008) Identification of C/EBPs as EPAC-activated transcription factors that mediate the induction of the SOCS-3 gene. *J Biol Chem.* Vol.283: 6843-6853.

Yasukawa, H., Misawa, H., Sakamoto, H., Masuhara, M., Sasaki, A., Wakioka, T., Ohtsuka, S., Imaizumi, T., Matsuda, T., Ihle, J.N. & Yoshimura, A. (1999) The JAK-binding protein JAB inhibits Janus tyrosine kinase activity through binding in the activation loop, *EMBO J.* Vol.18: 1309-1320.

Yasukawa, H., Ohishi, M., Mori, H., Murakami, M., Chinen, T., Aki, D., Hanada, T., Takeda, K., Akira, S., Hoshijima, M., Hirano, T., Chien, K.R. & Yoshimura, A. (2003) IL-6 induces an anti-inflammatory response in the absence of SOCS3 in macrophages, *Nat Immunol.* Vol.4: 551-556.

Yoon, S. & Seger, R. (2006) The extracellular signal-regulated kinase: multiple substrates regulate diverse cellular functions, *Growth Factors.* Vol.24: 21-44.

Yoshimura, A., Naka, T. & Kubo, M. (2007) SOCS proteins, cytokine signaling and immune regulation, *Nat Rev Immunol.* Vol.7: 454-465.

You-Ten, K.E., Muise, E.S., Itie, A., Michaliszyn, E., Wagner, J., Jothy, S., Lapp, W. S. & Tremblay, M.L. (1997) Impaired bone marrow microenvironment and immune function in T cell protein tyrosine phosphatase-deficient mice, *J Exp Med.* Vol.186: 683-693.

Yu, H., Pardoll, D. & Jove, R. (2009) STATs in cancer inflammation and immunity: a leading role for STAT3, *Nat Rev Cancer.* Vol.9: 798-809.

Zhang, J. G., Metcalf, D., Rakar, S., Asimakis, M., Greenhalgh, C.J., Willson, T.A., Starr, R., Nicholson, S.E., Carter, W., Alexander, W.S., Hilton, D.J. & Nicola, N.A. (2001) The SOCS box of suppressor of cytokine signaling-1 is important for inhibition of cytokine action in vivo, *Proc Natl Acad Sci U S A.* Vol.98: 13261-13265.

Zhang, J.G., Farley, A., Nicholson, S.E., Willson, T.A., Zugaro, L.M., Simpson, R.J., Moritz, R.L., Cary, D., Richardson, R., Hausmann, G., Kile, B.J., Kent, S.B., Alexander, W.S., Metcalf, D., Hilton, D.J., Nicola, N.A. & Baca, M. (1999a) The conserved SOCS box motif in suppressors of cytokine signaling binds to elongins B and C and may couple bound proteins to proteasomal degradation, *Proc Natl Acad Sci U S A.* Vol.96: 2071-2076.

Zhang, T., Kee, W.H., Seow, K.T., Fung, W. & Cao, X. (2000) The coiled-coil domain of Stat3 is essential for its SH2 domain-mediated receptor binding and subsequent activation induced by epidermal growth factor and interleukin-6, *Mol Cell Biol.* Vol.20: 7132-7139.

Zhang, X., Wrzeszczynska, M.H., Horvath, C.M. & Darnell, J.E. Jr. (1999b) Interacting regions in Stat3 and c-Jun that participate in cooperative transcriptional activation, *Mol Cell Biol.* Vol.19: 7138-7146.

Zhou, Y.J., Chen, M., Cusack, N.A., Kimmel, L.H., Magnuson, K.S., Boyd, J.G., Lin, W., Roberts, J.L., Lengi, A., Buckley, R.H., Geahlen, R.L., Candotti, F., Gadina, M., Changelian, P.S. & O'Shea, J.J. (2001) Unexpected effects of FERM domain mutations on catalytic activity of Jak3: structural implication for Janus kinases, *Mol Cell.* Vol.8: 959-969.

Zouein, F.A., Duhé, R.J. & Booz, G.W. (2011) JAKs go nuclear: Emerging role of nuclear JAK1 and JAK2 in gene expression and cell growth, *Growth Factors.* Vol.29: 245-252.

2

Interactions of the Protein Kinase A Signaling Pathway: Implications for the Treatment of Endocrine and Other Tumors

Audrey J. Robinson-White
Bethesda, Maryland, USA

1. Introduction

Cell signaling is known to be involved in the development and activation of many cancers, and has been the subject of abundant research in the field of cancer (Levitzki & Klein, 2010; Robinson-White & Stratakis, 2002). In response to extracellular and intracellular stimuli, protein kinase mediated pathways orchestrate the activation of cell signaling cascades to control cell growth, proliferation, differentiation and survival. However, there is tremendous complexity in tumor cell signaling (Iqbal et al., 2010). Signaling experiments have demonstrated that many different stimuli may activate the same pathways and many pathways interact ("cross talk") to modify and influence the outcome of a specific extracellular signal. In particular, the protein kinase A (PKA) pathway has been viewed as the central hub in the mediation of signals in endocrine tumorigenesis (Robinson-White & Stratakis, 2002). Depending on the species, tissue, cell type and ligand involved, the PKA hub interacts with high versatility with the mitogen activated protein kinase (MAPK), protein kinase B (akt/PKB), protein kinase C (PKC) and Wnt pathways to promote endocrine and other tumor cell development and progression (Almeida et al., 2010; Robinson-White & Stratakis, 2002). The purpose of this review is to highlight recent findings on the involvement of the PKA cell signaling pathway in endocrine tumorigenesis, and to define current clinical strategies and future perspectives for the treatment of endocrine and other tumors.

2. The cell signaling pathways

2.1 Description of the PKA pathway

PKA is a major factor in eukaryotic cell signaling. The PKA holoenzyme, in its inactive state, consists of two PKA isoforms (type I and type II), each in a tetramer of two homo or heterodimer regulatory subunits (RIα, RIβ, RIIα and RII β). Each R-subunit is bound to one of four catalytic subunits (Cα, Cβ, Cγ, and Prkx) (Azevedo & Stratakis, 2011). The regulatory subunits contain a dimerization/docking domain at their amino terminus and two tandem binding domains for cyclic AMP (cAMP) at their carboxyl terminus, and a linker region that contains a main docking site for the catalytic subunit (Zawadzki & Taylor, 2004). PKA may be activated (Figure 1A) indirectly by ligand binding to a heterotrimeric G-protein coupled receptor (GPCR), and/or by activation of adenyl cyclase (AC) and ATP, which stimulate the production of cAMP via the GSα subunit of the G-protein (*e.g.* GSα, Bγ); or directly by endogenous cAMP. CAMP binds to the PKA regulatory subunit leading to its subsequent activation. This, in turn, promotes dissolution of the holoenzyme and release of the catalytic subunits. The free catalytic subunits then go on to phosphorylate downstream targets that regulate effector enzymes, ion channels and transcription factors for specific genes that regulate cell growth and differentiation (Pearce et al., 2010); for example, CREB, CREM, NF-kB and other nuclear receptors (Robinson-White & Stratakis, 2002).

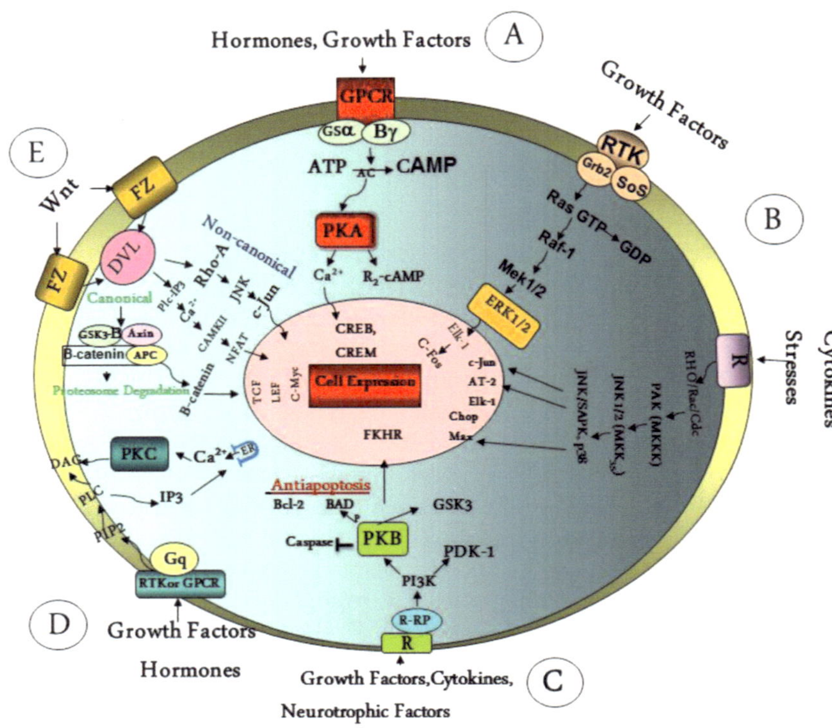

Fig. 1. General overview of the major signal transduction pathways in mammalian cells. Signal transduction in mammalian cells is controlled by five major signaling pathways (A-E), which upon activation, control gene expression and a cells' response (*e.g.* proliferation, differentiation, apoptosis, survival). Each pathway is independent, yet interacts in a complex manner with other pathways. An explanation of each pathway is given in the text.(A) Protein kinase A (PKA) pathway. (B) Mitogen-activated protein kinase (MAPK) pathway. (C) Protein kinase B (PKB) pathway. (D) Protein kinase C (PKC) pathway. (E) Wnt Pathway.

2.2 Description of other major cell signaling pathways in endocrine tissue

2.2.1 MAP-kinase signaling pathway

The MAP kinase proteins are signal transduction enzymes that mediate diverse cellular processes (*i.e.* cell proliferation, differentiation, apoptosis, survival and cytokine production)

(Robinson-White & Stratakis, 2002). The family includes more than a dozen members, arranged in separate but interacting cascades, based on sequence homology and function. These include ERK (extracellular signal-regulated kinase), JNK/SAPK (c-Jun N terminal kinase/stress activated protein kinase), and p38 MAPK. Each cascade consists of a three-core module, MKKK (MAPKK kinase), MKK (MAPK kinase) and MAPK. Upon cell activation (Figure 1B), the first core member activates the succeeding member in a serial manner (Garrington & Johnson, 1999). The final activated core member then regulates gene expression through the phosphorylation and activation of nuclear transcription factors (Cargnello & Roux, 2011) The best studied MAPK is ERK1/ERK2 (ERK1/2), 44- and 42-kDa proteins that are stimulated by a wide range of ligands (*e.g.* serum, growth factors, cytokine stresses, other mitogens). Upon ligand-induced stimulation of receptor tyrosine kinase (RTK) activity, adapter molecules (Grb2, SoS) bind to the receptors to enhance the activation of the Ras GTP-binding protein. Activated Ras binds to the first core member of the cascade, ubiquitous Raf-1 (MKKK, one of three Raf isoforms: A-Raf, B-Raf and c-Raf-1), or rap, which activates B-Raf, and transports the isoform to the plasma membrane for activation by other protein kinases (*e.g.* Src, PKC, PAK). Active Raf-1 or B-Raf dissociates from the Ras complex and phosphorylates its substrate and the second core member, MEK1/2 (MKK). Activated MEK1/2 then phosphorylates and activates the final core member, ERK1/2, which phosphorylates and activates its downstream substrates and various transcription factors (elk-1, c-fos, c/n-Myc) for expression of early response genes leading to a cell response (*e.g.* cell proliferation, differentiation, survival and apoptosis)). Other MAPK subfamilies, not as well characterized as the ERK1/2 family are the JNK/SAPK (46- and 54 kDa proteins, isoforms JNK1, JNK2 and JKN3) and p38 kinases (isoforms p38α, -β, δ, γ) . These kinases are often activated by inflammatory cytokines and cellular stresses (*e.g.* ionizing radiation, heat shock). They are associated with promotion of apoptosis and cytokine production (Robinson-White Stratakis, 2002).

2.2.2 PKB signaling pathway

PKB, a serine/threonine kinase, is named so because of its structural similarity to protein kinases A and C (Coffer & Woodgett, 1991). Its activation (Figure 1C) can occur at the plasma membrane by growth factors, cytokines and neurotrophic factors. The activated receptor phosphorylates a key regulatory protein (through monomeric Ras or Ras-related protein, R-RP), phoshatidylinositol-3 kinase (PI3K). PI3K, a heterodimer of both regulatory and catalytic subunits, has molecular masses of 85 kDa (p85) and 110 kDa (p110), respectively. Each subunit has at least five isoforms. PI3K is activated when p85 and p110 form a heterodimer that interacts with an activator protein. Binding of the activator protein causes a conformational change in the heterodimer, leading to its activation. PI3K has both lipid and protein kinase activity. It has the ability to activate several proteins such as PKB, PKC phosphoinositide-dependent kinase-1 (PDK-1), Ras and Rac cd42. It is then activated in two major steps: first, its pleckstrin homology domain binds with the main products of a PI3K, Ptdlns P and Ptdlns P2 catalyzed lipid reaction. Second, it is phosphorylated by PDK-1 kinase. Activated PKB acts to preferentially control anti-apoptotic mechanisms. It acts through several independent mechanisms: 1), it phosphorylates BAD to prevent its binding to antiapoptotic Bcl-2; 2) it inhibits caspase protease activity; and 3) it directly phosphorylates and activates cAMP-induced ribosomal protein S6 kinase (p70S6k), a key player in the PI3K/mTor/p70S6k pathway. PKB also directly phosphorylates glycogen synthetase kinase (GSK-3) to decrease its activity and prevent the induction of apoptosis. Lastly, PKB phosphorylates the winged-helix family of transcription factors, *forkhead* (FKHR) (Robinson-White & Stratakis, 2002).

2.2.3 PKC signaling pathway

PKC presents a large super family of protein kinases that comprise at least ten members or isozymes. Each isozyme has a distinct tissue distribution and functional pattern. They are grouped according to their activation requirements. Conventional PKCs (PKC-α, -B_I,-B_{II} and γ) are activated by phosphatidyl serine, Ca^{2+} and diacylglycerol (DAG). Novel PKCs (PKC-δ, -ε, -η, and -φ) do not require Ca^{2+}, while atypical PKCs (PKC-ζ and -λ) require only phosphatidyl serine. Other isozymes, the protein kinase C-related kinases (PRKs) are insensitive to Ca^{2+}, DAG and phorbol esters (Robinson-White & Stratakis, 2002). Stimulation of the PKC pathway (Figure 1D), by various agents (*e.g.* mitogenic growth factors, hormones), leads to the breakdown of plasma membrane phospholipid (PIP_2) by phospholipase C (PLC). The resultant hydrolysis generates inositol triphosphate (IP_3) and membrane located DAG. IP_3 causes the release of Ca^{2+} from the endoplasmic reticulum (ER). DAG production is transient and is frequently followed by a more sustained release following hydrolysis of phosphatidylcholine (PC) by phospholipase D. Ca^{2+} mobilization then causes PKC to bind to the plasma membrane cytosolic leaf where it is activated by DAG. PKC activation results in the activation of cell-specific transcription factors (e.g. c-Fos, c-Jun, NF-kB) for an array of cell responses (Lodish et al., 2000).

2.2.4 Wnt signaling pathway

Wnt (wingless and integration site growth factor) signal transduction is induced by lipid-modified and secreted Wnt glycoproteins. Nineteen Wnt family members and ten 7-transmembrane Frizzled (FZ) receptors exist in humans. Wnt proteins bind to receptors of the FZ family to activate both the canonical (canonical-Wnt/B-Catenin) and non-canonical pathways (Figure 1E), of which the disheveled scaffolding protein (DVL) acts as the central hub (Hu &Li, 2010). In the cytoplasm, in the absence of Wnt protein activation, B-Catenin forms a multi-protein destruction complex with adenomatous polyposis coli (APC), axin and glycogen synthase kinase -3β (GSK3-β), which leads to proteasomal degradation through β-Trep-mediated ubiquitination. Upon Wnt protein binding to activate the canonical pathway, DVL is activated. The destruction complex is disassembled, leaving free and stable B-Catenin in the cytosol. B-Catenin travels to the nucleus and forms a heterodimer with the TCF/LEF transcription factor for activation of target genes, *e.g. c-myc,* to regulate cell proliferation, stem cell maintenance or cell differentiation. Wnt proteins also stimulate B-Catenin -independent non-canonical pathways. These pathways are called the Wnt/jun N-terminal kinase (JNK) and Wnt/calcium pathways. The Wnt/JNK pathway involves the activation of small GTPases of the rho family (*e.g.* rac, cdc42) and downstream protein kinases such as JNK or rho kinase. JNK activates c-Jun, which goes into the nucleus to regulate polarized cell movement. Activation of the Wnt/calcium pathway occurs when Wnt proteins bind to FZ receptors to activate DVL and heterotrimic G-proteins that activate PLCγ and IP_3 and cause the release of Ca^{2+} from the endoplasmic reticulum. Released Ca^{2+} then activates calcium-dependent enzymes, calcium/calmodulin-dependent kinase, CaMKII, PKC or calcineurin. Calcineurin activates NFAT transcription factor and gene expression for cardiac development and hypertrophy. CaMKII activation can lead to activation of a nemo-like kinase (NLK), which prevents β-Catenin intranuclear signaling (Hu & Li, 2010; Roa & Khul, 2010).

2.3 Interaction of PKA with other signaling pathways in endocrine tissue

Signaling pathways "talk" to each other forming a complicated network of interactions that amplify and/ or dampen the activation of extracellular and intracellular signals. PKA plays the role of the central hub in endocrine signal transduction. Depending on the species, tissue, cell type and cell type involved, PKA can mediate and communicate cAMP effects to MAPK, PKC, PKB, and other pathways to modify and influence a cellular response (Robinson-White & Stratakis, 2002). We present here examples of cAMP/PKA signaling pathway interactions in endocrine tissue.

2.3.1 The hypothalamic-pituitary axis signal transduction

The hypothalamic-pituitary axis (Figure 2A) is regulated via the stimulation of gonadotropin releasing hormone (GnRH) receptors (GnRHR) and the activation of multiple G-proteins by GnRH, from hypothalamic neurosecretory cells. GnRHRs couple to heterotrimeric G-proteins and can activate PKC, PKA and MAPK simultaneously. GnRHRs activate PKC through extracellular Ca^{2+} via the G-protein Gq/11α, and activates PKA via GSα and cAMP. GnRHRs also activate GBγ-mediated PI3K and Src. While both PKC and PKA can independently activate c-Raf-1 of MAPK, PI3K can activate Ras to affect MAPK. The βγ subunit released by activation of Giα also stimulates adenyl cyclase (AC) to produce cAMP. Thus, GnRHR activation of MAPK is mediated by Gsα, Gq/11α, Giα and Gβγ (Cuny et al., 2011; Robinson-White & Stratakis, 2002). Therefore, in pituitary somatolactotroph cells, both the PKA and PKC pathways converge to activate MAPK for the regulation of gonadotropin hormone secretion and for cell proliferation. Recent studies in mouse AtT-20 corticotroph pituitary cells (not shown), suggest that PKA and the canonical Wnt signaling pathway interact to induce tumorigenesis. PKA alters the GSK-3B phosphorylation status at phosphor-GSK-3B (Ser9) thereby altering B-Catenin degradation and controlling cell proliferation (Khattak et al., 2010).

2.3.2 Thyroid Signal Transduction

In normal thyroid follicular cells, the PKA, PKC signaling pathways and protein tyrosine kinase (PTK) predominate (Figure 2B1). Thyrotrophic stimulating hormone (TSH) stimulates the thyroid via a GPCR mediated increase in cAMP levels, and activates PKA for cell differentiation. The phorbol ester, 12-O tetradecanoylphorbol 13-acetate (TPA), activates protein kinase C, and inhibits PKA to stimulate cell proliferation. Epidermal growth factor-1 (EGF-1), acting on a receptor tyrosine kinase (RTK) activates the PKC signaling pathway to induce cell proliferation. The PKA and PKC pathways interact with a high degree of antagonism and coordination. PKA inhibits thyroid cell proliferation as induced by EGF, and PKC inhibits PKA induced differentiation. Upon TSH stimulation, cAMP simultaneously stimulates PKA and influences the selection of the Ras effector (*i.e.* PI3K vs. c-Raf-1), by which PI3K is selected over c-Raf-1 as the effector. Thus, in normal thyroid, MAPK appears to be minimally (or not) involved in cell signaling. Signal transduction in the control of human thyroid carcinoma, presents a different picture (Figure 2B2). Human

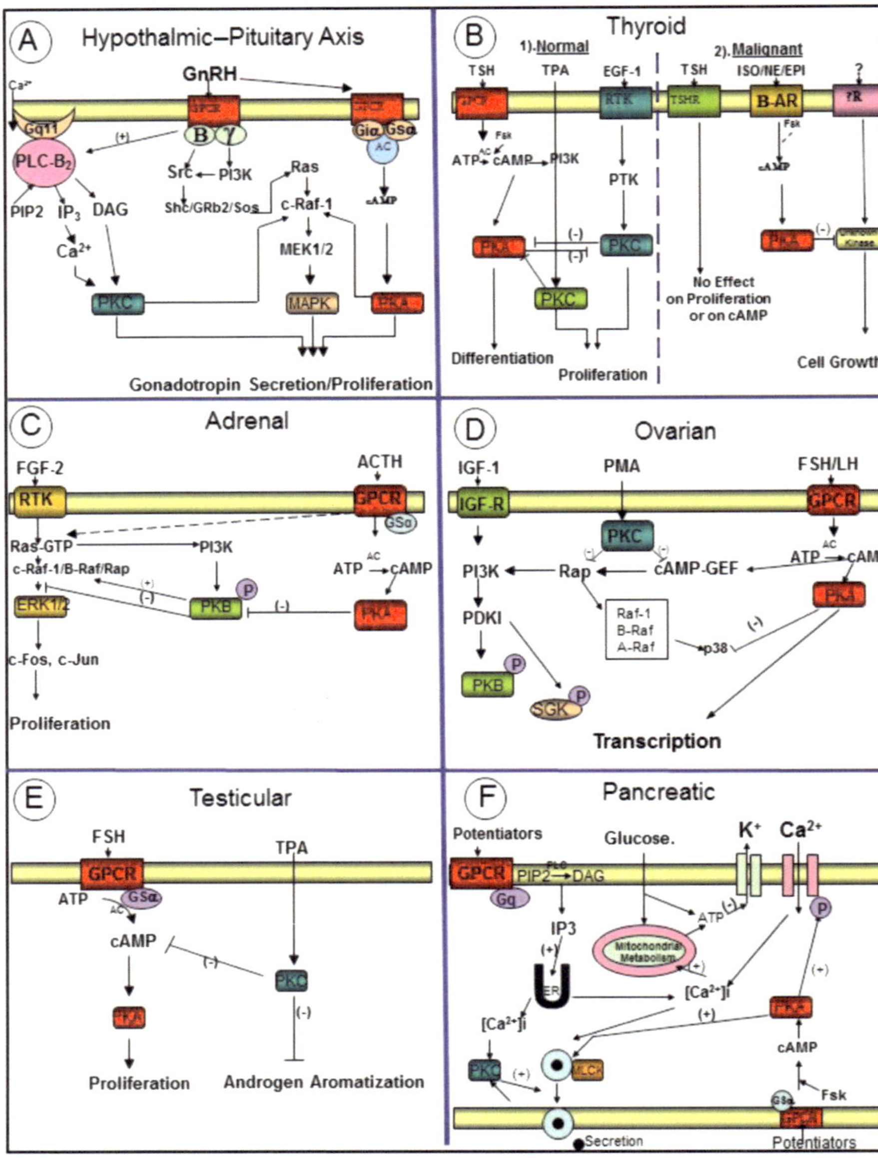

Fig. 2. Cell Signaling in Endocrine Organs. The PKA signaling pathway interacts in a complex manner with other signaling pathways in endocrine tissue (A-F), and acts as a central hub in endocrine signal transduction. An explanation of these interactions is given in the text. (A). Hypothalamic/Pituitary Axis Signal Transduction. (B) Thyroid Signal Transduction; Normal and Mutant tissue. (C) Adrenal Signal Transduction. (D) Ovarian Signal Transduction. (E) Testicular Signal Transduction (F) Pancreas Signal Transduction

thyroid papillary carcinoma cells were treated with TSH, forskolin (Fsk), 8-Br-cAMP (cAMP analogue and PKA stimulant) and the B-adrenergic receptor agonists, norepinephrine (NE), epinephrine (EPI) and isoproterenol (ISO). There was no effect on proliferation by TSH, or on the accumulation of cAMP. Fsk decreased cell growth, but had no effect on cAMP accumulation. The B-adrenergic agonists induced cAMP accumulation, and reduced cell growth via an unknown kinase. Other studies with the specific B-adrenergic receptor (B-AR) agonist, CGP-12177, showed that thyroid carcinoma cells have B-ARs that have high sensitivity to the growth inhibitory effects of cAMP (Robinson-White & Stratakis, 2002). The mechanism of this inhibition by PKA is yet to be determined.

2.3.3 Adrenocortical signal transduction

In adrenocortical signaling (Figure 2C), MAPK (ERK1/2), plays a large part in signal transduction and interacts with the PKA and PKB pathways, as seen in mouse Y1 (G0/G1 cell cycle–arrested) adrenocortical tumor cells, in normal human cortical tissue and in primary pigmented nodular adrenocortical (PPNAD) tumors. A full discussion on PPNAD will occur in later sections. In mouse Y1 cells, PKA is stimulated by GPCR activation through adrenocorticotropin (ACTH) and adenyl cyclase. Fibroblast growth factor-2 (FGF-2) acting through a receptor tyrosine kinase (RTK) activates ERK1/2 of MAPK for cell proliferation, as well as activates PI3K. PI3K causes the activation of PKB. ACTH also has a small and transitory effect on ERK1/2 early in G1, but has a large negative mitogenic effect on PKB. PKA inhibits PKB activity and its effect on ERK1/2. A similar effect on ERK1/2, to that in mouse, has been seen in normal and mutant human adrenocortical tissue (PPNAD, see Figure 3B). Therefore, PKA and PKB interact in adrenocortical tissue to either inhibit or enhance cell proliferation via ERK1/2, depending on the tissue tumor status (Robinson-White et al., 2006a; Robinson-White & Stratakis, 2002).

2.3.4 Ovarian signal transduction

Rat ovarian follicle growth and the differentiation of ovarian granulosa cells depend on sequential stimulation by FSH and LH. FSH regulates granulosa cell proliferation in small follicles. As follicles mature, FSH induces the expression of genes that encode P450 aromatase and the LH receptor. During granulosa cell development (Figure 2D), FSH and LH bind to GPCRs and stimulate the production of cAMP. CAMP then acts as a molecular switch to control several signaling pathways: In addition to the stimulation of PKA activity for the induction of transcription, it stimulates PI3K via cAMP-GEF/Ras/Rap. This action can be abolished by myristate acetate (PMA) stimulation of PKC. Insulin-like growth factor-1 (IGF-1) can also act in ovarian cells to mediate the activation of PKB, via PI3K and PDKI PKA interacts with MAPK in ovarian cells and tissues. PMA abolishes FSH, but not IGF-1 mediated PKB phosphorylation. Rap can induce the three Raf isoforms to activate the MAPK, p38, which is then inhibited by FSH/LH induced PKA activity (Robinson-White & Stratakis, 2002).

2.3.5 Testicular signal transduction

In the testis (Figure 2E), FSH stimulates PKA via the production of cAMP, to induce cell proliferation. In rat Sertoli cells, most cAMP-dependent activity is mediated by the RIIβ PKA subunit, although other subunits (RIα, RIIα, RIIβ, Cα) have been found. It has been shown that TPA induces PKC to inhibit both FSH-stimulated cAMP formation and androgen aromatization. TPA can transiently stimulate RIα and RIIβ mRNA with no effects on other PKA subunits. Rat testicular cells treated with both 8-CPTcAMP and TPA showed additive effects on stimulation of RIα mRNA and on inhibition of RIIβ mRNA levels. However, treatment with cycloheximide (a protein synthesis inhibitor) completely blocked the effect of TPA on RIIβ mRNA, but not on RIα. Therefore the inhibitory effect of PKC on RIIβ was dependent on ongoing protein synthesis. Although the precise site(s) of PKC-PKA interaction has not been determined, these data indicate that multiple and distant mechanisms are involved in the stimulation and inhibition of RIα, RIIβPKA subunit activity, respectively (Robinson-White & Stratakis, 2002).

2.3.6 Pancreatic signal transduction

Insulin secretion from the pancreas (Figure 2F) is potentiated by hormones and neurotransmitters that activate PKA and PKC through cAMP and phospholipase C (PLC), respectively, as well as via myosin light chain kinase (MLCK). High glucose concentrations generate an elevated intracellular ATP/ADP ratio that leads to closure of K^+ channels, and the subsequent opening of Ca^{2+}channels via phosphorylation by PKA, which leads to an increased intracellular Ca^{2+} load. Other potentiators of insulin secretion cause IP_3 to release Ca^{2+} from intracellular stores for the activation of PKC. Increased PKC activity induces MLCK and insulin granules to travel to the cells periphery. At the periphery, PKA plays a role in insulin secretion. Therefore release of insulin from secretory granules is orchestrated by PKA, PKC, and MLCK (Robinson-White & Stratakis, 2002).

3. PKA and apoptosis

3.1 Programmed cell death

Due to advancements in our knowledge of cell signaling pathways, programmed cell death (PCD), has recently been expanded from the inclusion of only a single cell death module (apoptosis) to include both necrosis and autophagic cell death as additional modules. Each cell death type is characterized by distinct morphological features and is regulated by different signaling pathways. Of these, necrosis is the least studied and least understood module. Necrosis was originally defined as accidental cell death that does not require gene activity. It is now recognized as a genetically controlled event (Taylor et al., 2008). It involves cell swelling, organelle dysfunction and cell lysis, caused by stress that is incompatible with cell survival. Necrotic cells trigger the immune system due to the release of danger-associated molecular patterns (DAMOs or alarmins) that stimulate recognition receptors in immune cells (Bialik et al., 2010; Degterev & Yuan, 2008). The immune system interprets the presence of necrotic cells as danger and a signal for activation of its response. In contrast to necrotic cell death, our understanding of autophagy cell death is more advanced, but not yet complete. It has been described as a pro-survival pathway that is essential for cell homeostasis and stress responses. It is mediated by ATG genes in which damaged or dysfunctional intracellular contents are engulfed by autophagosomes and degraded by lysosomal enzymes. For the purposes of this review, however, we will concentrate on the third and best studied module of PCD, apoptosis. Apoptosis is characterized by chromatin condensation and fragmentation, membrane blebbing and disintegration of the cell into smaller apoptotic bodies. It is a multistep process and involves two separate pathways, the intrinsic mitochondrial-induced pathway and the extrinsic death receptor dependent pathway. The intrinsic pathway is activated by a variety of stimuli within the cell (*e.g.* glucocorticords and staurosporin) that provoke cell stress or damage. The B cell CLL/lymphoma-2 (BCL-2) protein family is involved in this pathway. It consists of three types of proteins, anti-apoptotic proteins (*e.g.* BCL-2, BCL-XL, and MC-1), pro-apoptotic proteins (*e.g.* BAK, BAD and BAX) and BH3–only proteins (*e.g.* BID, BIM and PUMA). The pro-apoptotic members act on the mitochondria to cause the release of intramembrane space proteins, such as cytochrome C, for the formation of an apoptosome complex (*i.e.* 7 molecules of APAF-1 and 7 caspase-9 homodimers). Active caspase-9 then causes a proteolytic cascade of further effector caspase activation events which carry out cell "demolition" events, *i.e.* cell blebbing, nuclear condensation and cell fragmentation (Bialik et al., 2010). The extrinsic pathway is activated by a family of transmembrane death receptors of the tumor necrosis factor family (*e.g.* Fas/CD95, Fas/APO-1 and 2) (Walazak & Krammer, 2000). The Fas ligand binds to the receptor and Fas-associated death domain (FADD; an adaptor molecule) is recruited, followed by the recruitment of Caspase -8 and the formation of a death inducing complex (DISC). Caspase -8 then autoactivates and initiates apoptosis by the cleavage of downstream pro-caspase effector molecules (caspase-3, caspase-6 and caspase-7) as mentioned above (Bialik et al., 2010; Zimmerman et al., 2001).

3.2 Effect of PKA and its interaction with other signaling pathways on apoptosis in endocrine tissue

In endocrine tissue, PKA plays a role in regulating PCD through its effects on apoptosis. Although other modules of PCD exist in endocrine tissue, because of the high complexity of cell signaling networks and variation in pathways present in different endocrine tissue, we present examples of the interaction of PKA with other signaling pathways and the action of PKA alone only on apoptosis. PKA appears to act on apoptosis at both its upstage induction stage and downstage effector (demolition) stage. However, due to a lack of experimentation, the data does not reveal the precise molecular points of interaction of PKA within the apoptotic pathways of endocrine tissue.

3.2.1 Apoptosis in adrenal tissue

In the normal human adrenal cortex, evidence suggests that endogenous ligands (neuropeptides, NPB and NPW) of two G-protein coupling receptors, GPR7 and GPR8, stimulate glucocorticoid secretion by activating PKA and PKC signaling pathways. To study human adrenal functional regulation, human adrenocortical carcinoma-derived NCI-H29 (H295R) cells, the main model cell system for human adrenocortical tumors (Groussin et al., 2000), were employed. In H295R cells, mRNA for GPR7 and GPR8 was expressed. However, NPB and NPW did not affect secretion. Yet, both peptides enhanced cell growth by increasing proliferation and inhibiting apoptosis via stimulation of MAPK and not through PKA or PKC. The authors conclude that although GPR7 and GPR8 are expressed in H295R cells, they are a variation of those found in normal adrenal cells and may be uncoupled to PKA and PKC (Andreis et al., 2005). ACTH, the major regulator of adrenal cortex, function acts mainly through PKA (Kirschner, 2002). However, because of increasing evidence that the pro-apoptotic tumor necrosis factor-a (TNF-α) is involved in the regulation of adrenal control and function (Gonzalez-Hernandez et al. ,1996), the interaction of PKA with TNF-α on adrenocortical cell proliferation and apoptosis was studied. Activation of PKA by dibutyl cAMP [$(Bu)_2$ cAMP] inhibited proliferation and increased TNF-α-induced apoptosis. $(Bu)_2$ cAMP also increased the expression of the transcription factor c-Myc. Thus, PKA may augment TNF-α-induced apoptosis in H295R cells through increased c-Myc expression (Liu et al., 2004). In several other studies, PKA was shown to affect apoptosis in adrenal tissue. In mutant tissue, a dysregulation of the balance of the regulatory PKA subunits, RIα and RIIβ may be involved in adrenal tumorigenesis (Robinson-White et al, 2003). The impact of this imbalance on cell proliferation and on apoptosis was investigated using the specific cAMP site selective analogues, 8-Chloro cyclic-adenosine -monophosphate (8-Cl-cAMP) and 8-(4-chlorophenylthio) cAMP (8-CPT-cAMP). 8-Cl-cAMP activates both RIα and RIIβ, while 8-CPT-cAMP activates only RIIβ. Studies showed that by targeting the PKA RIIβ regulatory subunit, the balance between RIα and RIIβ is disturbed. This disturbance influences cyclin B accumulation, inactivation of CDC2 kinase, cell cycle progression at G_2 phase cell cycle arrest, and induces apoptosis (Bouizar et al, 2010). In another study, the effect of inactivation of the RIα PKA subunit on the interaction of the PKA and TGFB (stimulates SMAD proteins through type I and type II serine/threonine kinase receptors) pathways, and on apoptosis was investigated. *PRKAR1A* (specifics of which will, be given in later sections), the gene encoding the RIα subunit of PKA was silenced by siRNA. Silencing *PRKAR1A* stimulated PKA activity and increased transcriptional activity of a PKA reporter construct and expression of the endogenous PKA target, NR22. *PRKAR1A* inactivation also decreased SMAD3 mRNA and protein levels, thus altering the cells response to TGFB. Since TGFB normally stimulates apoptosis in H295R cells, this effect was counteracted by the inactivation of *PRKAR1A*. Thus, PKA and TGFB signaling pathways interact in endocrine tumorigenesis (Ragazzon et al, 2009). In another species, in mouse Y1 cells, which possess an over expressed *c-Ki-Ras* proto-oncogene and exhibits high constitutive levels of Ras-GTP, FGF2 can up-regulate Ras-GTP, activate PI3K and promote PKB phosphorylation and deactivation. This induces a mitogenic, anti-apoptotic effect in Y1 cells. ACTH activates PKA to cause a rapid dephosphorylation of PKB and a downstream de-regulation of the c-Myc protein. This data presents one mechanism of the blocking of G1-S transition in Y1 cells and induction of apoptosis (Robinson-White & Stratakis, 2002).

3.2.2 Apoptosis in thyroid tissue

Thyroid cells are highly resistant to apoptosis, which is actively suppressed by cell signaling pathways. CAMP is known to rescue cells from apoptosis as stimulated by diverse stimuli, yet the survival pathways activated by cAMP are poorly understood. The role of cAMP in thyroid cell survival was investigated in Wistar rat thyroid (WRT) follicular cells. In WRT, cAMP activates multiple signaling pathways (*e.g.* PKA, PI3K, p70S6k and Rap1), and multiple pathways modulate thyroid cell survival. Inhibition of cAMP -stimulated p70S6k, but not PI3K, abolished survival. Treatment with PKA inhibitors (H89 and RI209K) stimulated apoptosis and enhanced cell death after treatment with sodium nitroprusside (SNP). Cells expressing an activated Rap1A mutant also showed enhanced sensitivity to SNP-induced apoptosis. However, cells expressing dominant negative Rap1A were resistant to apoptosis. Although the exact mechanism has not been ascertained, these results show that PKA and Rap1 and PI3K/mTor/p70S6k pathways are involved in anti-apoptotic effects in WRT cells (Saavedra et al, 2002).

3.2.3 Apoptosis in ovarian tissue

Human ovarian surface epithelial (OSE) and Caov-3 (derived from an ovarian adenoma) cells were used to determine the involvement of PKA in the protection of ovarian cancer cells from apoptosis. Fas binds to the CD95 receptor, to stimulate the extrinsic apoptotic pathway in the ovary to subsequently activate caspase-3. Co-treatment of cells with luteinizing hormone (LH) (stimulates PKA activity) and Fas, reduced the number of apoptotic cells in a transient manner. However, LH alone had no effect on apoptosis or on proliferation. The cell permeable cAMP analogue 8-CPT-cAMP mimicked the effect of LH. This finding suggests that the PKA signaling pathway is involved in protecting OSE cells from Fas-induced apoptosis (Slot et al., 2006). Thrombopoietin (TPO), a hormone/growth factor/cytokine plays an essential role in stimulating homatopoietic stem cell function, differentiation, megakariocytosis and platelet production. It also plays a role in controlling ovarian function through various signaling pathways (JAK/STAT, MAPK and PKA). Studies with cultured porcine ovarian follicles determined that TPO regulates proliferation, apoptosis and secretion in ovarian cells and identified intracellular mediators (possibly PKA). Results showed a TPO-induced increase in PCNA (associated with proliferation), BAX (pro-apoptosis), Tyrosine kinase, cdc2/p34, PKA activity and CREB-1. The PKA blocker, KT5720 given alone, reduced BAX and TGF-2B expression and augmented PKA, CREB and oxytocin expression. TPO and K5720 together prevented/reversed the action of TPO on PKA and CREB, but not on BAX. The data suggest that PKA has a role in mediating the effect of TPO on proliferation and apoptosis in the ovary. However, the exact mechanism involved is yet to be determined (Sirotkin et al., 2004). It is well established that leptin, a product of adipose tissue, has direct effects on the control of ovarian steroidogenesis, prostaglandin secretion and IGFBP-3 and IGG-1 release. Leptin also controls the effects of LH-RH on reproduction. The role of leptin on proliferation, apoptosis and PKA was uncertain. Studies in human ovarian granulosa cells showed that leptin stimulates ovarian cell cycle proteins (PCNA and cyclin B), Bax and PKA activity. This data represents the first evidence of leptins' involvement in the control of the ovarian cell cycle, and that PKA is a potential target of leptin and mediator of ovarian cell function (*i.e.* proliferation and apoptosis) (Sirotkin et al, 2008).

3.2.4 Apoptosis in testicular tissue

The putative Leydig cell receptor, peripheral-type benzodiazepine (PBR), modulates steriodogenesis in the testis. Since benzodiazepines are extensively used in regional anesthetics, a determination was made of their peripheral effects on Leydig cell steriodogenesis and underlying transduction pathways. The effect of the benzodiazepine, Midazolam (MDZ), on primary mouse Leydig cells and MA-10 Leydig tumor cells was assessed. MDZ stimulated steriodogenesis in both cell types and induced the expression of PBR and StAR (steroidogenesis acute regulatory protein) proteins. Inhibition of PKA (by H89) and PKC (by GF109203X) decreased MDZ-induced steroid production, indicating that MDZ stimulates steroidogenesis in mouse Leydig cells and the expression of PBR and StAR proteins via PKA and PKC. Also, at high MDZ concentrations, cell round-up, membrane blebbing and later, cell death occurred in MA-10 cells. The authors conclude that MDZ can induce steriodogenesis in both cell types, and at high doses can induce apoptosis in Leydig tumor cells. Thus PKA and PKC may play a role in the induction of apoptosis in Leydig tumor cells (So et al., 2010). A growing interest exits in the regulation of mammalian fertility, by the endogenous ligand of the cannabinoid CB1 and CB2 receptors, endoicannobinoid anandamide N-arachidonoylethanolamine (AEA). Evidence has shown that endogenous cannabinoids can interfere with critical functions of mammalian and nonmammalian reproduction (*e.g.* regulation of embryo development, oviduct transport and implantation) in females, through apoptosis. However, since the impact of AEA on fertility in the male remained to be elucidated, a study using mouse primary Sertoli cells was performed to determine the effect of AEA on proliferation/apoptosis. It was shown that FSH stimulates PKA to enhance the activity of the AEA hydrolase, fatty acid amide hydrolase (FAAH), resulting in the hydrolysis of AEA. Thus, Sertoli cells are protected from the pro-apoptotic action of AEA. FSH also triggers PI3K signaling, which is required for the stimulation of cytochrome P_{450}-aromatase expression (ARO). ARO can irreversibly transform androgens into estrogens and plays a large role in male reproduction (*e.g.* spermatogenesis). FSH appears to be responsible for total number of sperm produced and not the quality of the sperm. Stimulation of FAAH activity was abrogated by inhibition of PKA by the PKA inhibitor, myristoylated amide 14-22 and the cytochrome P450 inhibitor 4-androsten-4-ol-3, 17-dione, and by the interaction of the PKC pathway with FSH-induced cAMP activity (see Figure 2E). The effects of FSH were reproduced with the PKA stimulant, $(Bu)_2$ cAMP and by estrogen. The data showed that PKA and estrogen mediate the protective effect of FSH on Sertoli cells, and may prevent apoptosis. The action of FSH via PKA and estrogen may be a means for modifying male infertility and Sertoli cell number, respectively, and points to AEA as a critical regulator of male fertility (Rossi et al., 2007).

3.2.5 Apoptosis in pancreatic tissue

Postprandial insulin secretion is regulated by hormonal factors (*i.e.* incretin hormones) released from the gut in response to nutrient ingestion. One hormone, glucose-dependent insulinotropic polypeptide (GIP) causes the release of insulin from pancreatic Beta cells in the presence of glucose via its G-protein coupled receptor. The activated GIP receptor induces stimulation of membrane bound adenyl cyclase which results in insulin secretion, influx of Ca^{2+} and cAMP production (Volz et al, 1995). Other studies showed that GIP and glucose act synergistically as anti-apoptotic factors in B-cells. Using a well differentiated B-cell line, INS-1, it was shown that mitogenic and anti-apoptotic signaling in INS-1 cells occurs in response to the synergism of GIP and glucose via multiple interacting signaling pathways (PKA, PI3K, MAPK, and PKB). The release of intracellular Ca^{2+} appeared to be at the hub of signaling. Using a panel of various pathway inhibitors, many interacting signals were seen: 1) stimulation of PKA/CREB, MAPK, and PI3K/PKB; 2) inhibition of PKA by MEK1/2 of MAPK; 3) activation of ERK1/2 of MAPK by PI3K and PKA and; 4) activation of PKB by MAPK and PKA. This study showed that GIP functions as a mitogenic and anti-apoptotic factor for B-cells by activating a tightly woven network of excitatory and inhibitory "cross-talk" of signaling pathways (Trumper et al., 2002).

4. Bilateral adrenal hyperplasia: Involvement of PKA

Bilateral adrenal hyperplasia (BAH), an adrenocorticotropin-independent form of adrenocortical disease that leads to Cushing Syndrome has been divided into two groups of disorders, micronodular and macronodular hyperplasia, based on the size of the associated nodules. The size criteria have biological significance since a continuum in the same patient is rarely found. Most patients have either micronodular or macronodular hyperplasia, and rarely both. The PKA signaling pathway and its interaction with other signaling pathways is involved in the pathogenesis of both types. Both types are, therefore, presented, however, with an emphasis on the PKA associated micronodular syndromes.

4.1 Bilateral micronodular hyperplasia

Micronodular hyperplasias consists of diseases of the adrenal cortex with multiple nodules of less than 1 cm in diameter and can be divided into 3 groups, i.e. primary pigmented nodular adrenocortical disease (c-PPNAD), usually found in the context of Carney Complex (CNC), isolated primary pigmented nodular adrenocortical disease (i-PPNAD, and isolated micronodular adrenocortical disease (i-MAD) (Stratakis, 2007; Lodish & Stratakis, 2010a).

4.1.1 Carney complex

Endocrine tumors can present as sporadic events, or as part of an inherited tumor syndrome. One such syndrome, CNC, will be highlighted in this review. CNC is a unique and relatively rare autosomal dominant multiple endocrine neoplasia syndrome first described by Dr. J. Aidan Carney and co-workers at the Mayo Clinic, is characterized by spotty skin tumors and pigmented lesions (lentigenosis), myxomas, endocrine over activity and schwannomas (Carney et al, 1986). Endocrine tumors associated with CNC include primary pigmented adrenocortical disease (PPNAD), growth hormone secreting pituitary tumors, large cell–calcifying Sertoli cell tumors, Leydig cell tumors and thyroid follicular adenomas. The clinical manifestations and molecular genetics of CNC have been well described elsewhere (Rothenbuhler & Stratakis, 2010). Genetic linkage analysis has identified two independent loci for CNC, CNC1 located on chromosome 17p22-24 and CNC2, located on chromosome 2p16 (Stratakis et al., 1996). Most cases (> 60%) of CNC are caused by inactivating mutations in the gene encoding the PKA regulatory subunit, RIα (PRKAR1A), located on chromosome 17p22-24 (Kirschner et al., 2000b) The gene responsible for CNC at locus 2p16 is as yet unknown. However, this locus is thought to contain genes responsible for at least some CNC patients, or the progression of the complex (Stratakis et al., 1996).

4.1.2 PPNAD

PPNAD is a rare bilateral adrenal disease and is the most frequent endocrine manifestation in CNC patients. From 80 to 90 % of PPNAD cases are associated with CNC. However, it may also occur independently of CNC. PPNAD received its name (Primary Pigmented Adrenocortical Disease) from the macroscopic appearance of the adrenal glands, characterized by small cortisol-producing, pigmented micronodules (black or brown, ~6mm-1 cm in diameter) set in a usually atropic adrenal cortex. The micronodules secrete cortisol independently of pituitary signaling. Adrenal atrophy is pathognomonic and reflects the autonomic function of the nodules and suppressed levels of ACTH. The combination of atrophy and nodularity gives the gland an abnormal appearance (though commonly normal in size) that is highly diagnostic. PPNAD occurs mostly in children and young adults, and peaks in the second decade. It is rare under the age of 4 years and after the age of 40. In rare cases, one or both glands may be large and contain adenomas with a calcified center. Macronodules, larger than 10 mm may also occur in older patients (Rothenbuhler et al., 2010; Stratakis, 2007). PPNAD patients present with adrenocorticotropic hormone (ACTH)-independent Cushing Syndrome (CS), which may be subclinical or cyclic, but often present with a variant CS called "Atypical" Cushing Syndrome (ACS) (Sarlis et al., 1997). Up to 90 % of CNC cases present with CS, due to PPNAD. Hypercortisolism develops progressively over years in PPNAD, or there may be a rapid outburst of excess cortisol that spontaneously regresses (Groussin et al., 2005). The 6-day Liddle test is used to establish a diagnosis in PPNAD, where patients show a "paradoxical" increase in 24-hour urinary free cortisol (UFC) and/or 17-hydroxysteroids. The increase is progressive and peaks on day 2 of high-dose dexamethasone administration in patients with CS. Patients with ACS tend to have normal or near normal 24-hour UFC production with an occasional interruption of days or weeks of hypercortisolism (Lodish & Stratakis, 2010a; Rothenbuhler & Stratakis, 2010; Stratakis, 2007). Most Patients with PPNAD have germline mutations in PRKAR1A and in some cases mutations in genes encoding the cAMP-binding phosphodiesterases (PDEs), PDE11A and PDE8B. It should be noted that PRKAR1A and the PDE genes PDE11A and PDE8B control the action and levels of cAMP in the cell with a final impact on cell proliferation (Horvath et al., 2006, 2008a).

4.1.3 Isolated PPNAD (i-PPNAD)

PPNAD, typically observed in CNC, can also present as a sporadic and isolated disease (i-PPNAD). It occurs in children and young and middle aged adults. It is characterized by micro-adrenomatous hyperplasia with mostly internodular atrophy and nodular pigment (lipofusin). It is autosomal dominant but is not associated with CNC nor does it present any other CNC manifestations. It also has no obvious family history, which would be suggestive of CNC. In a minority of patients, however, other familial cases of PPNAD are known, suggesting a genetic cause (Bourdeau et al., 2003). It is caused by inactivating mutations in PRKAR1A of chromosome 17p22-24, and possibly mutations of gene(s) found in the 2p12-16 locus. Inactivating mutations of phosphodiesterase 11A4 (PDE11A) are also involved. In most cases (~ 65%), these are de novo mutations, which explains the lack of family history (Rothenbuhler et al., 2010; Stratakis, 2007).

4.1.4 Isolated Micronodular Adrenal Disease (i-MAD)

I-MAD occurs mostly in children and young adults. It occurs with hyperplasia of the surrounding zona fasciculate with absent or limited nodular pigment. It is autosomal dominant and isolated and does not occur in the context of CNC. I-MAD is caused by mutations in PDE11A and PDE8B and possibly genes of the 2p12-p16 locus (Stratakis, 2007). It was initially thought that i-MAD patients were CNC patients, but it is now clear that it is not the same as PPNAD (Gunther et al., 2004).

4.2 Bilateral macronodular hyperplasia

Bilateral macronodular hyperplasia, another form of BAH, also causes Cushing Syndrome, but has no other clinical findings. Macronodular hyperplasia is characterized by multiple adrenocortical nodules of greater than 1Cm each, and can be divided into three groups. The first two groups include bilateral macro-adrenomatous hyperplasia (BMAH) and BMAH of childhood (c-BMAH). BMAH occurs in middle age and c-BMAH occurs in infants, and very young children. In both groups, adenomas (~2-3) with intranodular atrophy and an occasional microadenoma can be seen in histological preparations. BMAH is caused by mutations in menin (the ubiquitous nuclear protein coded for by the MEN1 gene), APC, (Adematous Polyposis Coli tumor suppressor gene), GNAS (codes for GSα Gs protein subunit), and FH (codes for the tricarboxylic acid cycle enzyme fumarate hydralase). C-BMAD is caused by mutation in GNAS (Stratakis, 2007). The third group, ACTH-independent macronodular adrenocortical hyperplasia (AIMAH) is also known as massive macronodular disease (MMAD; AIMAH/MMAD) occurs always in middle age. It is a cause of CS due to large bilateral cortisol secreting macronodules. The nodules cause a large increase in the size (up to 100 times the normal size) and weight of the adrenal glands (Stratakis et al., 2001). In this group, adenomatous hyperplasia with internodule hyperplasia of the zona fasciculate has been observed. MMAD is autosomal dominant and isolated. It is associated with ectopic GPCRs, WISP-2 and Wnt–signaling, and genes of the chromosome 17q22-24 locus (Stratakis et al., 2007).

5. Pituitary tumors in CNC

Pituitary adenoma is a benign neoplasia with excess proliferation of each subtype of pituitary cell. The tumors can give rise to severe illness due to hormonal excess or to visual disturbance due to mass effect. This tumor can present as an isolated event or as part of an inherited syndrome (Tichomirowa et al., 2009; Zhang & Nose, 2011). Currently, at least 5% of all patients with pituitary adenoma have been found to have a family history, mainly due to multiple endocrine neoplasia type 1 (MEM1) and CNC (Tichomirowa et al., 2009). These patients have growth Hormone (GH) as well as prolactin (PRL) secreting adenomas, however, PRL-secreting adenomas predominate in MEM1. In CNC, GH-producing tumors are seen in ~10% of patients. Acromegaly with elevated Insulin-like growth factor (IGF1) and GH levels, and subtle hyperprolactinemia, can be seen in up to 75 % of patients. However, clinical acromegaly is rare (Rothenbuhler et al., 2010; Zhang & Nose, 2011). Lesions in the pituitary gland range from pituitary cell hyperplasia to multiple microadenomas to invasive macroadenomas. CNC-related acromegaly is distinguished microscopically by multifocal hyperplasia of somatomammotropic cells (Zhang & Nose 2011, Pack et al., 2000; Stergiopoulos et al., 2004). Studies with a tissue-specific knockout (KO) mouse, pointed to the Prkar1a gene as the causative agent. These studies showed that complete loss of Prkar1a allows the formation of pituitary tumors and abnormalities of the GH-axis, with close analogy to tumors in human patients with CNC (Zhirong et al., 2007).

6. PKA mutations

6.1 PRKAR1A

The gene that codes for the RIα subunit of PKA is a key component of the cAMP/PKA signaling pathway, located in the q22-24 region of chromosome 17. The finding of tumor specific loss of heterozygosity (LOH) within the q22-24 locus (Krischner et. al, 2000a) and the complete or partial loss of RIα in CNC tumors suggested that PRKAR1A is a tumor suppressor gene (Kirschner et al., 2000a). PRKAR1A is mutated in almost half of CNC patients. Mutant PRKAR1A codes for the only PKA subunit that has been shown to lead to human disease. Its genomic region is approximately 21 kb-long, and the open reading frame contains 11 exons that code for a protein that has 384 amino acids. Over one hundred disease causing pathogenic sequence variants have been identified (Rothenbuhler & Stratakis, 2010). These have been extensively recorded elsewhere (Horvath et al., 2010). Most PRKAR1A mutations are small deletions and insertions, rearrangements, or base substitutions (Rothenbuhler & Stratakis, 2010). Large deletions can also occur, however, rarely (Horvath et al., 2008b). However, the most frequently found PRKAR1A mutation is a 2 bp deletion at position 578 in exon 4B of PRKAR1A (c-578delTG). Over 70 % of CNC patients have a classical phenotype with a PRKAR1A mutation that leads to a premature stop codon and subsequently non-sense mediated mRNA decay (NMD), and thus PRKAR1A haploinsufficiency (Bertherat et al., 2009; Kirschner et al., 2000a; Stratakis et al., 2001). Less frequently, mutations do escape NMD, causing the expression of abnormal RIα proteins (Groussin et al., 2006; Meoli et al., 2008). PRKAR1A haploinsufficiency leads to excess cAMP/PKA signaling in affected tissues, and increased cell proliferation and survival (Robinson-White et al., 2006b), through an increase in total cAMP-stimulated PKA activity. Two mechanisms have been proposed to explain the increase in signaling: 1) RIα-haploinsufficiency leads to a higher intracellular C to R subunit ratio, and therefore, increased availability of the free catalytic subunits that phosphorylate downstream targets. 2). RIα- haploinsufficiency leads to an upregulation of other PKA subunits including PRKAR1B and PRKAR2A or PRKAR2B, depending on the tissue type. It is also possible that the other regulatory subunits do not act as effectively as RIα in controlling cAMP signaling (Rothenbuhler, 2010).

6.2 M1V PRKAR1A

Another mutation has recently been found in the PRKAR1A gene, M1V PRKAR1A, caused by a mutation in the initiation codon of PRKAR1A. This mutation results in a phenotype characterized by PPNAD alone, and sometimes mild ACS. There is significant variation in age of onset and clinical severity, but no other classical manifestations of CNC. Mutant mRNA is expressed equally with wild type levels. However, the mutant protein is not expressed in cells. In a cell free system, the RIα regulatory subunit was shorter than the wild type protein. This observation has highly significant implications for RIα's role in adrenal function and tumorigenesis (Pereira et al., 2010).

7. Mechanism of action of mutant PRKAR1A in endocrine tumorigenesis

The Mechanism of action of PRKAR1A inactivation was investigated using a model cell system, endocrine tissue and mouse models, to determine the manner in which mutant PRKAR1A affects cell proliferation, cell survival and apoptosis, leading to tumorigenesis.

7.1 A Cellular Model System

In varied cell types, including B and T-lymphocytes (Robinson-White et al., 2003), RIα type PKA interacts with the ERK1/2 cascade of MAPK at c-Raf-1, causing a cell-type specific inhibition of MAPK and of cell proliferation. Since RIα is present in B-lymphocytes of CNC patients (Kirschner et al., 2000b), and these cells carry the c.578delTG inactivating PRKAR1A mutation, B-lymphocytes from CNC patients were used as a model cell system (Figure 3A) to determine if a MAPK/PKA interaction could be responsible for CNC tumorigenesis. In PRKAR1A-mutant cells (Mutant), PKA activity, both at baseline and after stimulation with cAMP was increased. Quantitative mRNA analysis indicated that type I PKA subunits (RIα and RIβ) were the main subunits expressed in both normal and mutant cells. However, RIα was decreased in mutant cells. The cell and pathway specific stimulant, lysophosphatidic acid (LPA), stimulated ERK1/2 levels in both cell types with greater stimulation in mutant cells. Forskolin (Fsk; activates adenyl cyclase for the production of cAMP) and Isoproternol (ISO) stimulated PKA activity, inhibited LPA-induced ERK1/2 activity and cell proliferation in normal cells, but stimulated these parameters in mutant cells. These data were replicated in a pituitary tumor cell line carrying the c.578delTG mutation and in COS-7 cells transfected with an in vitro construct bearing the RIα 184-236 mutation that leads to increased PKA-mediated physphorylation (Robinson-White et al., 2003). The differences in the effect of PKA stimulants on normal and mutant B-lymphocytes, was postulated to occur due to the sensitivity of the Raf isoforms to PKA. In cells having both B and c-Raf-1, PKA inhibits c-Raf-1 and stimulates B-Raf to inhibit and stimulate cell proliferation, respectively, through ERK1/2 (Erhardt et al., 1995). The action of PKA on B-Raf, in some cells, is through activation of the G-protein, Rap1 (Vossler et al., 1997). In normal cells, a balance exists between the stimulation and inhibition of ERK1/2 activity by PKA, with inhibition being more dominant. This dominance is due to the presence of a greater number (70-80%) of the type I PKA isoform than the type II isoform (20-25 %) in lymphocytes (Schmitd & Stork, 2000), and a higher affinity of cAMP for the RIα subunit (Amieux et al., 1997). In RIα-haploinsufficient (PRKAR1A-mutant) lymphocytes, as in CNC-affected cells, the balance is thought to shift towards activation of another PKA subunit (perhaps the RIβ subunit) (Robinson-White et al., 2003). Thus, the underlying mechanism was suggested to be through biochemical compensation or substitution of one subunit (e.g. RIα) for another (e.g. RIβ), as seen in other cell systems (Amieux et al., 1997), which may lead to cell proliferation instead of inhibition, as seen in normal cells (Robinson-White et al., 2003). The data suggested that PKA acts to inhibit ERK1/2 activity (through inhibition of c-Raf-1) and cell proliferation in normal cells, but stimulate ERK1/2 activity (through B-Raf) and cell proliferation in mutant cells. The authors hypothesized that the reversal of PKA-mediated inhibition of MAPK by mutant-PRKAR1A may contribute to CNC tumorigenesis (Figure 3A). These studies were extended to confirm the previous data and to determine the depth of the interaction of the PKA and MAPK signaling pathways in CNC cells. An analysis of the effect of PKA stimulation on components of the ERK1/2 cascade (B and C-Raf-1, and MEK1); on the ERK1/2 activating transcription factor c-Myc, on cell cycle progression and proliferation and on apoptosis was performed. Data in mutant cells was compared to that in normal cells from matched controls. Both Fsk and ISO inhibited cell proliferation in normal cells, but stimulated proliferation in mutant cells. Analysis of the phosphorylation patterns of B and c-Raf-1 in mutant cells upon stimulation by both Fsk and ISO showed that phosphorylation of B-Raf is increased, whereas phosphorylation of c-Raf-1 is inhibited. In normal cells, both B and c-Raf-1 phosphorylation is inhibited. This implied a switch from inhibition to stimulation of ERK1/2 in mutant cells. Likewise, in mutant cells, phosphorylation of MEK1/2 was also increased (Robinson-White et al., 2006b), possibly due to high levels of phosphorylated B-Raf in mutant cells. Fsk also increased LPA-induced stimulation of c-Myc in mutant cells, but c-Myc phosphorylation was inhibited in normal cells. The increased phosphorylation of B-Raf, MEK1/2, ERK1/2 and c-Myc in mutant cells suggested that PRKAR1A inactivation leads to c-Myc activation. C-Myc can respond to mitogenic signals

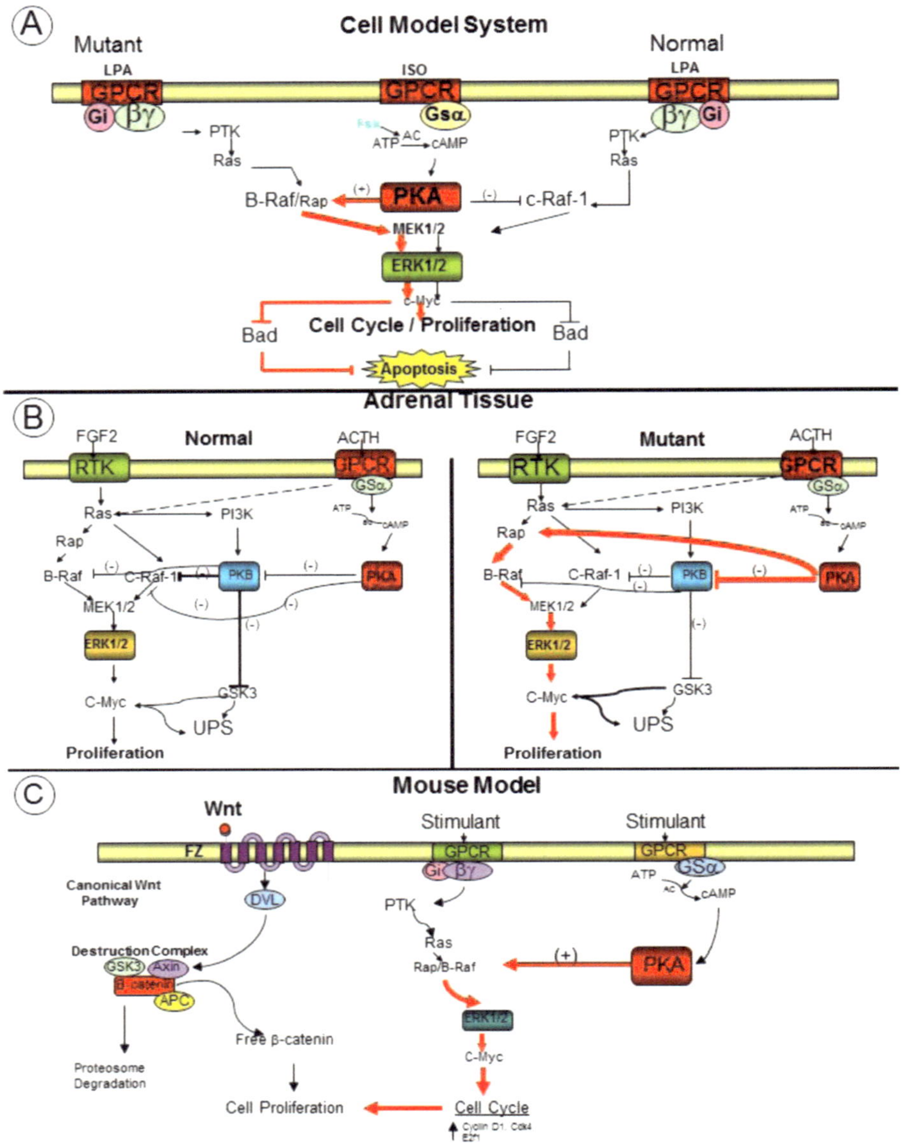

Fig. 3. Mechanism of action of Mutant-*PRKAR1A* in Endocrine Tumorigenesis. **(A). Signaling in normal and *PRKAR1A*-mutant B Lymphocytes.** Lysophosphatidic acid (LPA), and isoproterenol (ISO) stimulate G-protein coupled receptors (GPCRs) in B- lymphocytes. LPA activates cytosolic protein tyrosine kinase (PTK) via βγ G-proteins, to stimulate Ras, followed by the activation of B-Raf and c-Raf-1. ISO, via GSα and adenyl cyclase (AC) activates cAMP/PKA, or AC is activated directly by forskolin (Fsk). In normal lymphocytes, PKA inhibits the ERK1/2 signaling pathway at c-Raf-1, but stimulates Rap/B-raf in *PRKAR1A*- mutant cells, to inhibit and stimulate the cell cycle and proliferation, respectively. ERK1/2, in both cell types, phosphorylates and inhibits BAD of the intrinsic apoptotic pathway. ERK1/2 due to the enhanced ERK1/2 activity in mutant cells, inhibition

of apoptosis is greater in these cells. Line thickness indicates degree of action. **(B) Hypothetical Signaling Mechanism in PPNAD Adrenocortical Tissue.** 1) In both normal and *PRKAR1A*-mutant adrenocortical tissue, fibroblast growth factor2 (FGF2) activates receptor tyrosine kinase (RTK) to stimulate Ras, followed by the sequential activation of B-Raf and c-Raf-1, MEK1/2 and ERK1/2 of the MAPK signaling pathway. Activated ERK1/2 then activates c-Myc, leading to cell proliferation. ERK1/2 is also transiently activated by ACTH, at a point upstream of MEK1/2. 2) Ras stimulates PI3K, which phosphorylates/ activates the PKB pathway. PKB, to a small extent, then inhibits B-Raf and c-Raf-1 in mutant tissue, but inhibits these isoforms to a larger extent (mostly c-Raf-1) in normal tissue. 3) ACTH, via a GSα type G-protein and adenyl cyclase (AC), stimulates cAMP production and PKA activity. In normal tissue, PKA inhibits PKB, c-Raf-1 and B-Raf. A different effect occurs in mutant tissue. PKA greatly inhibits PKB, but has a large stimulatory effect on Rap, which overrides the inhibitory effect that PKB has on the ERK1/2 cascade. PKA, thus, greatly enhances the activity of the ERK1/2 cascade and cell proliferation in mutant tissue. 4) The phosphorylation/deactivation of GSK3 by PKB is reduced in mutant tissue due to a less active PKB, allowing more c-Myc to be targeted for degradation by the ubiquitin proteosome system (UPS). Line thickness indicates degree of activity. **(C) Signaling in a *Prkar1a* mouse model** 1) Wnt proteins bind to a 7-transmembrane Fizzled (FZ) receptor to activate the canonical Wnt pathway. Upon binding of Wnt, the disheveled scaffolding protein (DVL) is activated, causing the disassembly of the destruction complex (GSK3, Axin, B-catenin and APC). Free β-catenin travels to the nucleus to activate target genes for cell proliferation. 2) PKA stimulates the ERK1/2 cascade of MAP kinase. PKA, in *PRKAR1A*-haploinsufficient cells, stimulates ERK1/2 (at rap/B-raf) to increase c-Fos and c-Myc activity and to promote activation of cell cycle cyclin D1, cdk4 and E2f (Go/G1 of the cell cycle) and facilitates cell cycle progression. β-catenin and ERK1/2 may act synergistically to promote cell proliferation.

and move cells from Go to G1 and/or from G1 to S phases of the cell cycle. Cell cycle analysis by Flow Cytometry indicated an increased rate of cell cycle transition by synchronized mutant cells from Go/G1 to S phase, even in the absence of PKA stimulants. The data suggested that this fast transition rate leads to increased cell proliferation in mutant cells. Next, the possibility that apoptosis could account for the decrease in proliferation seen in normal cells was investigated. Staurosporin and CH11 were used as control stimulants for the intrinsic and extrinsic apoptotic pathways, respectively. Both normal and mutant cells were relatively resistant to staurosporin-induced apoptosis with significantly less apoptosis in mutant cells. On stimulation by ISO, apoptosis in mutant cells was again significantly less. The mediator of this decreased apoptosis was found to be the pro-apoptotic protein, BAD. Since ERK1/2 is known to phosphorylate/inactivate BAD, the data suggested that the higher levels of ERK1/2 in mutant cells induced by PRKAR1A deficiency resulted in a small but significant inhibition of the intrinsic apoptotic pathway, contributing to cell survival. The extrinsic apoptotic pathway was induced by the Fas/CD95 receptor agonist CH11. Like staurosporin and ISO, apoptosis was less pronounced in mutant cells. This data was confirmed by increased levels of cleaved caspase-3 in normal versus mutant cells, indicating the increased ability of normal cells to activate the effector caspases leading to apoptosis. In general, this data indicated that the large decrease in cell proliferation in normal cells when stimulated by Fsk and ISO was not due to apoptosis or necrosis, but suggested that PRKAR1A inactivation leads to significant changes in both apoptotic pathways that control enhancement of mutant cell survival. This study suggested that the balance between cell proliferation and death that occurs in tissue homeostasis (Fulda & Debatin, 2004) may be disturbed in PRKAR1A-defficent cells, and presents a model (Figure 2A) by which partial inactivation of PRKAR1A may increase cell cycle progression, proliferation and survival (Robinson-White et al., 2006b).

7.2 Adrenal tissue

An investigation was made to determine how PKA and its subunits act in the presence of PRKAR1A mutations in adrenocortical tissue. As a result of this investigation, a hypothetical cell proliferation pathway was designed (Figure 3B). This pathway was devised mainly from studies on the activity of PKA and its subunits, ERK1/2, other immunoassays and immunohistochemistry in adrenocortical samples from normal subjects and from patients with germline PRKAR1A mutations (Robinson-White et al., 2006a). Patient samples included multiple cortical-producing adenomas, single cortical-producing adenomas, micronodular hyperplasia, and i-PPNAD. However, the majority of samples were from patients with PPNAD or CNC. Data obtained from other studies on Y1 adrenocortical tumor cells was also included in the formulation of the hypothetical pathway. Since, in many cell types (Jun et al., 1999), the PKB signaling pathway inhibits both the B and c-Raf-1 kinase activity of MAPK, data from PKB was included. Data in adrenocortical tissue showed increased total PKA activity that was associated with mutant PRKAR1A. Quantitative mRNA analysis and immunoblotting showed a 2-fold and 1.8-fold decrease in RIα mRNA and protein expression, respectively; and an increase in levels of other PKA subunits. ERK1/2 immunoassays showed a 2-fold and 6-fold decrease in baseline ERK1/2 (unphosphorylated/non-activated ERK1/2) and corresponding increases in phosphorylated/activated ERK1/2 (p-ERK1/2) in mutant-PRKAR1A samples. Other components of the ERK1/2 cascade, i.e. basal B-raf kinase activity, phosphorylated-MEK1/2 (p-MEK1) were increased in mutant tissue, as well as levels of the phosphorylated transcription factor c-Myc (p-c-Myc). There was no difference in levels of phosphorylated c-Raf-1 (p-c-raf-1) in both tissue types. Likewise, levels of phosphorylated PKB (p-PKB) were not increased. These data were all supported by immunohistochemistry studies (Robinson-White et al., 2006a). Data showed that Y1 cells possess an amplified and constitutively expressed c-Ki-Ras that elicits a mitogenic response to fibroblast growth factor2 (FGF-2) through a tyrosine kinase receptor (RTK). ERK1/2 is then rapidly activated for the induction of transcription factors (c-Fos and c-Jun), the phosphorylation of c-Myc, expression of cyclins D and E, and phosphorylation of the retinoblastoma protein; all leading to the activation of the cell cycle, DNA synthesis and cell proliferation. Phosphorylated PKB (p-PKB) can enhance the mitogenic response to FGF2 by phosphorylating and inactivating glycogen synthetase kinase (GSK) and preventing the targeting of c-Myc for degradation by the ubiquitin proteasome system (UPS) (Lepique et al., 2004). PKB has also been shown to inhibit B-Raf and c-Raf-1 kinase activity in many mammalian cell types (Guan et al., 2000; Jun et al., 1999). However, basal B-Raf is normally 4-fold greater than basal c-Raf-1 (Mercer & Pritchard, 2003). Y1 cells can also exert a strong anti-mitogenic effect, mediated by PKA acting to dephosphorylate p-PKB. This releases PKBs' inhibitory effect on GSK3 (Lepique et al., 2004), leading to the degradation of c-Myc by UPS. Thus, FGF-2 S-phase entry can be blocked by a PKA-dependent process. S-phase entry may also be blocked by PKA in normal adrenocortical tissue through an inhibition by PKA of c-Raf-1 kinase, and a subsequent decrease in phosphorylated MEK1/2, ERK1/2 and c-Myc, which may lead to a decrease in cell proliferation (Robinson-White et al., 2006a). Therefore, the collection of evidence presented suggested that cell proliferation in normal and mutant-PRKAR1A adrenocortical tissue results from the convergence of at least three signaling pathways, PKA, PKB and MAPK, and are consistent with data found in other human tissue with RIα deficiency (Robinson-White et al., 2003; Robinson-White et al., 2006b; Vossler et al. 1997). In general, the following cell proliferation pathway was presented (Figure 3B 1 and 2): 1). In Normal and PRKAR1A-mutant tissues, FGF-2/RTK induced stimulation of ERK1/2 leads to cell proliferation through activation of the transcription factor c-Myc. In normal tissue (Figure 3B 1), PKA (as activated by ACTH at a GPCR) inhibits PKB to a lesser degree than in mutant-PRKAR1A tissue, allowing activated PKB to inhibit ERK1/2 at B-Raf and c-Raf-1 kinase levels, as well as to directly inhibit c-Raf-1. In mutant-PRKAR1A containing tissue (Figure 3B 2), PKA has a greater inhibitor effect on PKB than in normal tissue and a large stimulatory effect on Rap, which then directly activates B-Raf kinase, for the stimulation of ERK1/2. The net effect of these interactions is the stimulation of cell proliferation in mutant-PRKAR1A tissues.

7.3 Mouse models

For a better understanding of the mechanism by which mutant PRKAR1A causes disease in CNC, several investigative approaches have been made using mouse model systems. Studies with RIα knockout mice confirmed the role of RIα as the key compensatory regulatory subunit of PKA activity in tissue where the three other regulatory subunits are expressed. RIα knockout caused deficits in the morphogenesis of the embryonic germ layer, and early embryonic lethality due to failed cardiac morphogenesis. Embryonic germ layer failure could be rescued by crossing RIα with Cα knockout mice, suggesting that inappropriately regulated PKA C-subunit activity is responsible for the phenotype (Amieux & McKnight, 2002). In other studies, transgenic mice (tTA/X2AS), carrying an antisense transgene for the mouse Prkarla exon 2(X2AS) developed thyroid follicular hyperplasia, adenomas, adrenocortical hyperplasia and other PPNAD-like features. Allelic loss of the mouse chromosome 11 Prkarla locus, increase in type II PKA activity and greater RIIB protein levels were associated with these lesions. The authors conclude that the tTA/X2AS mouse, having a down-regulated Prkarla gene, replicates findings in CNC patients. This data supports the role of RIα as a tumor suppressor gene (Griffin et al., 2004). In another study, conventional and conditional null alleles for Prkarla were developed in mice. Prkaria$^{+/-}$mice developed a spectrum of tumors that overlapped those observed in CNC patients, e.g., those arising in cAMP-responsive tissues (bone, Schwann and thyroid follicular cells). This suggested that a complete loss of Prkarla plays a role in tumorigenesis. Likewise, tissue specific ablation of Prkarla from facial neural crest cells caused the formation of schwannomas, although with aberrant differentiation than those seen in CNC. Since data with the Prkarla$^{+/-}$ mouse suggested the importance of increased cAMP/PKA signaling for tumor formation, in vitro studies were performed in Prkarla $^{-/-}$ mouse embryonic fibroblasts and in transformed adrenal PRKAR1A-haploinsufficent cells. These studies showed that dysregulation of cyclins and E2F1 (cell cycle factors and mediators of proliferation via defective RIα, respectively), were key changes in the process of immortalization of the two cell lines. These observations confirmed the identity of PRKAR1A as a tumor suppressor gene (Kirschner et al, 2005; Nadulla & Kirschner, 2005;

Nesterova et al., 2008). Prkar1a$^{+/-}$ mice, however, did not develop skin and other CNC tumors (e.g., heart myxomas and pituitary adenomas); therefore, an investigation was made to determine if the Prkar1a defect is a generic, but a weak tumorigenic signal that depends on tissue-specific or other factors. Since p53 (product of the Trp53 gene) and Rb1 (retinoblastoma; product of the Rb1 gene) are essential for cell cycle control, mouse models were bred to determine if the tumorigenic properties of Prkar1a-haploinsufficiency emulate more accurately the CNC phenotype in the background Trp53 or Rb1 (Almeida et al., 2010). Prkar1a mice were bred within the Trp53$^{+/-}$ or Rb1$^{+/-}$ background, or treated with a skin carcinogenic protocol, including 7, 12-dimethylbenz (a) anthracene (DMBA; causes irreversible activation of the HRas oncogene) and 12-O-tetradecanoylphorbol-13-acertate (TPA; promotes the formation of papillomas). The data showed that Prkaria$^{+/-}$ Trp53$^{+/-}$ and Prkar1a$^{+/-}$ Rb1$^{+/-}$ mice developed more sarcomas and endocrine tumors, respectively than Trp53$^{+/-}$ and Rb1$^{+/-}$ mice. The authors suggest that Prkaria-haploinsufficiency has a synergistic and not an additive effect when combined with Tr53$^{+/-}$ and Rb1$^{+/-}$ defects in mice. A genome wide transcriptome profiling of tumors, produced along with qRT-PCR arrays, immunohistochemistry and cell cycle analysis of siRNA down-regulation of cell cycle genes, indicated that Wnt-signaling is the main signaling pathway activated by abnormal cAMP/PKA signaling. They conclude that in the mouse, Prkar1a-haploinsufficiency is a relatively weak tumorigenic signal that can act synergistically with other tumor suppressor gene defects or chemicals to induce tumors, mostly through Wnt-signaling and cell cycle dysregulation (Figure 3C) (Almeida et al., 2010). From these studies in mice, Wnt signaling appears to be the major signaling pathway for PRKAR1A-induced tumorigenesis. However, since Wnt–induced cell proliferation has been shown to be mediated via the ERK pathway in mouse NIH3T3 cells (Kim & Choi, 2007), the possibility exist that inter-pathway crosstalk (above the level of the cell cycle) between PKA/MAPK/Wnt may also occur in mice and in PRKAR1A-mutant cells and tissues. Differences in pathway interactions may also occur in humans vs. mice, as well as between different cell and tissue types, bringing about a much more complex picture.

8. Targeting PKA activity in endocrine and other cell types

8.1 Effect of 8-Cl-cAMP and 8-Cl- Adenosine (8-Cl-ADO) on cell proliferation and apoptosis

Recent investigations have provided a plethora of information on the pathophysiology and genetics of CNC tumors. However, effective treatment strategies for many tumors of CNC are not available. PKA affects cell proliferation in many CNC cell types and is thus a potential target for treatment of CNC tumors. PKA activity is stimulated by cAMP and its cAMP analogues. One such analogue, the 8-Cl- derivative of cAMP (8-Cl-CAMP), the most potent of the cAMP analogues initially tested (Katsaros et al., 1987), and its dephosphorylated metabolite, 8-Cl-ADO are known inhibitors of cancer cell proliferation. However, their mechanisms of action were debated. Questions asked were: 1) does 8-CL-ADO have inhibitory effects on cell growth in human thyroid, HeLa and other tumor cell types; 2) does 8-Cl-ADO act on adenosine receptors to stimulate adenyl cyclase and PKA activity; 3) are the inhibitory effects of 8-Cl-cAMP due to its by-product 8-Cl-ADO; and 4) does 8-Cl-ADO exert its inhibitory effects by changing the RI to RII subunit ratio as has been suggested for 8-Cl-cAMP (Katsaros et al., 1987; Robinson-White et al., 2008). In [^{3}H] thymidine uptake and Cell Titer 96 AQ (MST) cell proliferation assays using human thyroid tumor cells (WRO), HeLa cells and other cancer cell types (colon carcinoma/ARO and melanoma/ NPA) both 8-CL-cAMP and 8-Cl-ADO inhibit cell growth and proliferation (IC_{50} values ranged from 0.5 μM to 1.7 μM for 8-CL-ADO and 0.55 μM to 4.4 μM with 8-Cl cAMP, depending on the cell type). The inhibition by 8-Cl-ADO was decreased with time (beginning on the 5th day of incubation). To determine if this decrease was due to intracellular or extracellular 8-CL-ADO metabolism, preincubation experiments were performed. When 8-Cl-ADO was preincubated without cells for 7 days (37°C), and cells were then incubated using the same previously preincubated solution for 5 days; or the culture was replenished with fresh drug on day 4, the decrease in inhibition was eliminated, with no difference seen in inhibition with the preincubated drug and the replenished drug. This suggested that 8-Cl-ADO is not metabolized in the media at 37°C (Robinson-White et al., 2008), but, as others have reported (Halgren et al., 1998; Langeveld et al., 1992) may be degraded within the cell with time, by intracellular adenosine deaminase to an inactive product, 8-Cl-inosine. Studies with adenosine receptor agonists (adenosine and NECA) showed no effect on cell proliferation and the adenosine receptor antagonists, XAC and MRS 1523, did not affect 8-Cl-ADO-induced inhibition, excluding the possibility that 8-Cl-ADO's effect is by an effect on adenosine receptors. The question as to whether 8-Cl-cAMP acts as a pro-drug and is metabolized to 8-Cl-ADO by extracellular phosphodiesterase and 5' nucleotidase, was addressed when cells were incubated with 8-Cl-ADO, 8-Cl-cAMP and 8-Cl-cAMP plus the phosphodiesterase inhibitor 3-isobutytl-1-methylxanthaine (IBMX) in media containing serum or with 8-CL-cAMP in serum-free media. The inhibition by 8-Cl-cAMP was greatly reduced by IBMX (to 62%) and further reduced in serum free media (to 75%). This data suggested that 8-Cl-cAMP is metabolized to 8-Cl-ADO, and the inhibitory effect of 8-Cl-cAMP is due to its by-product, 8-Cl-ADO. The involvement of 8-Cl-ADO in PKA cell signaling was examined. 8-Cl-ADO altered PKA activity in cell types in a differential manner, with a small increase in activity in thyroid cells. No reduction occurred in the ratio of RI to RII subunits, in all, suggesting that PKA and the R subunits have no direct effect on the inhibition by 8-Cl-cAMP and 8-Cl-ADO. To examine the mechanism of the inhibition by the two drugs, the effect of 8-Cl-ADO on apoptosis and on the cell cycle was measured by Flow Cytometry using the Annexin V/7AAD assay and BrdU incorporation, respectively. Apoptosis was induced in all cell lines, in the rank order in cells matching the IC_{50} values in MTS assays. Both drugs also caused rapid and late apoptosis/necrosis. Cell cycle analysis showed that 8-Cl-ADO caused an arrest of the cell cycle with an accumulation of cells in G2/M (NPA) and G1/S (WRO and ARO) phases. The data suggested that 8-Cl-ADO (or 8-Cl-cAMP) may be a potent chemotherapeutic agent for the treatment of thyroid and other cancers and its' effect will not depend on PKA signaling (Robinson-White et al., 2008).

8.2 PKA-dependent and Independent effects of 8-Cl-ADO on proliferation and apoptosis

Because PRKAR1A mutations are present in B-lymphocytes of CNC patients (Kirschner et al, 2000b), B-lymphocytes from CNC patients and their normal matched controls were used in a multi-parametric approach (i.e. growth and proliferation assays, PKA and PKA subunit assays, cAMP and cAMP-binding assays and apoptosis assays) to study the effects of 8-Cl-ADO on growth, cell proliferation and apoptosis; and on the relationship between these parameters and PKA signaling. Since adenosine (a structural analogue of 8-CL-ADO) is transported across the lymphocyte plasma membrane by specialized membrane nucleoside carrier proteins (equilibrate nucleoside transporters; ENTs) by facilitated diffusion for subsequent metabolism and to induce apoptosis, the possibility that 8-Cl-ADO is also transported across the plasma membrane to induce apoptosis was explored. 8-Cl-ADO inhibited cell proliferation in a time and concentration-dependent manner, with greater inhibition in mutant cells. IC_{50} values were higher (1μM and 2μM in normal and mutant cells, respectively) than those previously reported for other cell types. The manner in which 8-Cl-ADO affects PKA signaling was explored. Cyclic AMP levels were increased, while [^{3}H]-cAMP binding was decreased in both cell types. However, binding experiments suggested a time-required down-regulation of cAMP binding. PKA subunit levels were altered with an overall reduction in the RI/RII subunit ratio, whereas levels of the catalytic subunit Cα were 2-fold that of RII. Basal and cAMP-stimulated PKA activity was increased by 8-Cl-ADO (Robinson-White et al., 2009). These data suggested that 8-Cl-ADO can change the site selectivity of cAMP toward R-subunits, as shown for 8-Cl-cAMP (Ally et al., 1988; Rohlff et al., 1993). Experiments with the ENT protein inhibitor dipyridamole (DIP) and 8-Cl-ADO indicated a competition between 8-Cl-ADO and DIP for the ENT protein and suggested that 8-Cl-ADO is transported into lymphocytes by facilitated diffusion. Also, an analysis of components of the intracellular adenosine metabolic pathway, as affected by 8-Cl-ADO, showed that adenosyl-homocysteine (ADOHcy, the main product of adenosine metabolism) alone inhibited cell proliferation, while the inhibition by 8-Cl-ADO in both cell types was increased. This inhibition was greater in mutant cells. Likewise, 8-Cl-ADO also increases levels of p53, another component of adenosine metabolism. Since adenosine metabolism results in apoptosis, whether or not 8-CL-ADO induces apoptosis in B-lymphocytes and if this apoptosis is reversed by DIP was explored. Apoptosis was induced by 8-Cl-ADO, was reversed by DIP and was greater in mutant cells. As in proliferation studies, the effect indicated a competition by DIP and 8-Cl-ADO for the ENT protein. Together, the data suggested that 8-Cl-ADO is transported in and metabolized to ADOHcy to induce apoptosis in B-lymphocytes. Data also indicated that other cancer cell types (e.g. MCF-7 breast carcinoma and SW-13 adrenal cortex carcinoma cells, HeLa, NPA, WRO, and ARO) transport 8-Cl-ADO via ENTs and may indicate a general mechanism of action of 8-CL-ADO on cell proliferation. 8-Cl-ADO also inhibits ISO-induced GPCR activation of cell proliferation, LPA-stimulation of RTK-induced proliferation, as well as pERK1/2 levels in lymphocytes, but had only a small and nonspecific inhibitory effect on A1, A2a, and A2b adenosine receptors, and no effect on A3 receptors, suggesting that 8-Cl-ADO can also inhibit proliferation induced by different GPCRs and proliferation induced by other receptor types (e.g. RTK). The authors present a hypothesis (Figure 4) for the mechanism of action of 8-Cl-ADO on cell proliferation. The inhibition may be due to: 1) stimulation of an ENT protein to transport 8-Cl-ADO intracellularly for metabolism to adenosyl-homocysteine (ADOHcy) to induce apoptosis; 2), intracellular 8-Cl-ADO may stimulate PKA activity. However, the ability of transported 8-Cl-ADO to inhibit proliferation would depend on the cell's ability to use the PKA signaling pathway to induce proliferation (Robinson-White et al, 2008); 3) PKA activity may be increased by the selective binding of 8-Cl-ADO to R-subunits; 4) 8-Cl-ADO may inhibit GPCR-and RTK induced proliferation; and 5) ENT proteins may transport 8-Cl-ADO in other cell types to inhibit proliferation. Additionally, the effect of 8-Cl-ADO on PKA may be overlooked because of the large and direct effects of transported 8-Cl-ADO on proliferation and on the induction of apoptosis. The conclusions from this work may have applications in solid tumors of CNC and even in non-PRKAR1A-mutant systems (Robinson-White et al., 2009).

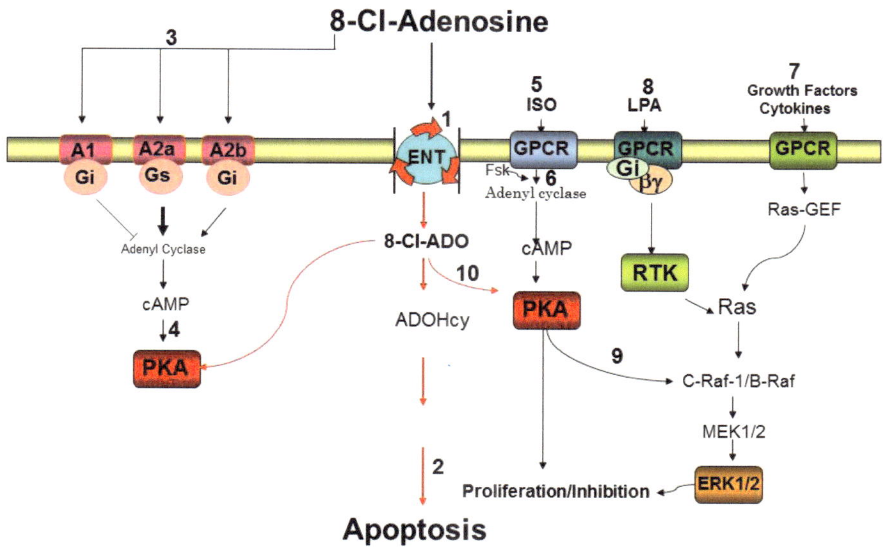

Fig. 4. Mechanism of the Inhibition by 8-Cl-Adenosine (8-Cl-ADO) on proliferation in B Lymphocytes 1), 8-C-ADO is transported into the cell by an *es*-type equilibrate nucleoside transporter protein (ENT) and metabolized to adenosyl-homocysteine (ADOHcy) to (2) induce apoptosis. 3), Extracellular 8-CL-ADO alters the action of adenosine receptors (A1, A2a and A2b) that (4) activate or inhibit PKA. 5), PKA is activated extracellularly at G-protein coupled receptors (GPCR) by isoproterenol (ISO) or intracellularly at (6) adenyl cyclase by forskolin (Fsk). 7), Growth factors or cytokines stimulate GPCRs to activate the ERK1/2 cascade of the MAPK signaling pathway. Lysophosphatidic acid (LPA) stimulates GPCRs that interact with receptor tyrosine kinase (RTK) that activates ERK1/2. 9), PKA alters c-Raf-1 or B-Raf activity to inhibit or stimulate, respectively, cell proliferation. 10), 8-CL-ADO alters the activity of PKA.

9. Future perspectives

9.1 Targeted therapy and drug design for endocrine tumors

Targeted therapy refers to the use of anti-cancer drugs that are designed to interfere with a specific molecular target in a cell signaling pathway that plays a crucial role in tumorigenesis and in its progression. Appropriate targets within the pathways must be identified based on a detailed understanding of the molecular mechanisms underlying the cancer progression. Targeted therapy has improved the therapeutic landscape among solid tumors and offers new treatment strategies beyond surgery, conventional chemotherapy and radiation. Use of small molecular compounds that regulate oncogenic transduction pathways and monoclonal antibodies have emerged as promising treatment options in solid tumors.

9.1.1 Tyrosine kinase inhibitors

Of the greater than 500 different protein kinases in humans, only 91 are tyrosine protein kinases and are serine/threonine/kinases. A subset of serine/threonine kinases (e.g., such as PKB, ERK1/2, PKC), are involved in signal transduction as key regulators of cell proliferation, differentiation and apoptosis (Levitzki & Klein, 2010). Of the small molecular inhibitors that affect signaling pathways, tyrosine kinase inhibitors (TKIs) are designed to affect tyrosine kinase-dependent oncogenic pathways. TKIs selectively inhibit tyrosine kinase activity and provide a relatively high therapeutic window and low toxicity, in comparison with conventional chemotherapy. Some TKIs currently in use for the treatment of endocrine cancers include erlotinib, axitinib and vatalanib for pancreatic cancer, axitinib and vandetanib, imatinib and motesanib for thyroid cancer, while others are in ongoing clinical trials. There is conservation in the structure of the ATP binding site in all protein kinases. This allows TKIs to act as small molecules with structural similarity to ATP, and to bind to and disrupt tyrosine kinases catalytic activity. However, the homology of structure may also afford TKIs the ability to act against a broad range of protein kinases, to affect multiple pathways and bring about important side effects. Recently endocrine-related side effects have been seen, some of which include alterations in thyroid function, bone metabolism, growth, gonadal function, fetal development, glucose metabolism and adrenal function (Lodish & Stratakis, 2010b). Although TKIs offer promising treatment options, further investigation is needed to determine the molecular mechanisms underlying endocrine dysfunction, as well to bypass TKI related side effects to obtain wider use of TKIs in the treatment of cancer.

9.1.2 Monoclonal antibodies

A number of monoclonal antibodies (mAbs) have been developed for cancer therapy in recent years and many more are in clinical trials. Monoclonal antibodies interfere with cancer progression by three basic mechanisms: inhibiting signal transduction, antibody-dependent cellular cytotoxicity (ADCC) and complement-dependent cytotoxicity (CDC). The hallmark of an effective therapeutic mAb is its ability to target antigens that are crucial for tumor maintenance, inhibit signaling by that antigen and induce cytotoxicity. Antibodies inhibit signal transduction by binding to, e.g. cell surface receptors, or by preventing ligand binding to specific receptors, to prevent the activation of signaling pathways. ADCC occurs when mAbs bind to tumor cells and to Fc receptors on immune cells (e.g., neutrophils, mononuclear phagocytes, but mainly to natural killer cells). This activates the cells to engulf and destroy the bound tumor cell. The same mAbs can also activate the complement system to kill cells directly, by the formation of a "membrane attack complex" (MAC). MAC produces a pore within the tumor cell membrane, which leads to tumor cell destruction (Levitzki & Klein, 2010). Over expression or activation of epidermal growth factor receptors (EGFRs) has been reported in many cancers, including ovarian cancer. Cetuximab (Erbitux®), panitumumab (Vestibix®) and trastuzumab (Herceptin®) are three EGFR targeting mAbs that are licensed for clinical use. Cetuximab blocks EGFR ligand binding and induces receptor internalization and degradation. It inhibits tumor growth and proliferation via Go/G1 cell cycle arrest, induction of apoptosis and inhibition of metastasis and enhances radiosensitivity (Stoffel, 2010). Trastuzumab is so far, the most successful anti-EGFR antibody, having been used for HER2 positive breast cancer. It binds to the juxtamembrane portion of the HER2 receptors' extracellular domain, to prevent the activation of its intracellular tyrosine kinase domain. The endothelial growth factor (VEGF) can increase tumor vasculature to enhance tumor growth, and is also implicated in various other aspects of tumorigenesis (e.g. angiogenesis, mitogenesis and permeability enhancing effects). Bevacizumab (Avastin®) is the most advanced VEGF targeted agent for treatment of solid tumors (Levitzki & Klein, 2010, Stoffel, 2010), and the first VEGF inhibitor approved by the FDA (2004) as a first-line treatment of metastatic colorectal cancer. It binds to and neutralizes all VEGFA type isoforms, preventing binding to VEGF tyrosine kinase receptors. Combination therapy with mAbs and other anti-tumor agents in clinical trials showed, in most cases, an increase in response rate over monotherapy. However, median survival showed no significant improvement. Clinical use of mAbs has also shown that antibodies can have deleterious side effects, mainly because the same antigens present on cancer cells are also present on normal cells (Stoffel, 2010). Investigations are ongoing for the improvement of (Vossler et al., 1997) mAb therapy for efficacy and the relief of side effects.

9.2 Viral therapy

Advances in cancer biology and virology has allowed the compilation of a growing arsenal of naturally occurring and genetically engineered oncolytic viruses (OV) to treat human cancer. These viruses give a promise of killing cancer cells with low toxicity to normal tissue. They act by infecting tumor cells and selectively replicating inside cells, achieved by modification of the viral genome for targeting of specific tumor cell molecules or signal transduction pathways. Infected cells then become "factories" for the replication, subsequent release of new viruses, and a cycle of infection-replication-release ensues. This cycle is repeated indefinitely to effectively kill the tumor mass. Adenoviruses (Ad), Herpes simplex virus (HSV), and vaccinia (VV; the prototype poxvirus) are the most studied viruses and have shown promising results in clinical trials. Ad infects cells via receptor mediated endocytosis using two separate receptors. The virus first binds to the primary receptor, the host cell receptor (Coxsakie-adenovirus receptor), and is then internalized by binding of its viral capsid with host avB integrins. It enters the cell via endosomes where it is dissembled and lysed. Ads have also been engineered to selectively infect and lyse tumor cells for the purpose of targeting specific cancer cell mutations. Like Ad, HSV are genetically engineered for tumor-selective replication (i.e., it only targets tumor cells), accomplished by deletion of viral genes to restrict replication to tumor cells. VV relies on its own encoded proteins as well as host proteins and functions for its life cycle. This allows for rapid and efficient translation, transcription and replication in poxvirus "factories', in the host cell cytoplasm. VV can enter and replicate in many cell types from various species, although it has a propensity toward human cells. It enters the cell via an endosome or through the plasma membrane. Since VV is a large (250 x 350 mm) virus it may require a leaky vasculature for extravasations from the vasculature into tumor tissues. Tumors are known for their leaky vasculature and production of vascular endothelial cell growth factor (VEGF). With this in mind, a number of strategies are being investigated to enhance replication in tumors cells, and concurrently reduce replication in normal cells. Another new and exciting approach to viral therapy is the use of carrier cells to deliver tumor-specific OV to tumors. These are autogenous cells, e.g. mesenchymal progenitor stem cells and immune cells that have a great potential to by-pass pre-exiting immunity and to reduce "off target toxicity" as well. In all, viral therapy has greatly progressed in recent years, and much success has indicated possible widespread future use. However, there are still unmet requirements to overcome, 1) overcoming pre-existing host immunity; 2) since only small fractions of carrier cells reach the tumor target, strategies for enhanced trafficking should be investigated and, 3) there is a need for increased OV distribution throughout the tumor tissue for more effective therapy (Guo et al., 2008).

9.3 Modulation of apoptosis

Apoptotic pathways have recently garnered considerable attention in the development of cancer treatment strategies, and studies have focused on key regulatory points in these pathways. In preclinical and clinical investigations, pro-apoptotic and anti-apoptotic Bcl-2 proteins, that control the intrinsic pathway, have been used as drug targets (Elkholi et al., 2011). In particular, studies on the pro-apoptotic BH3-only proteins have shown alterations in protein expression in several cancers, (e.g., colon, lung and other cancers). For example, use of Oblimersen, designed to target Bcl-2 mRNA (Klasa et al., 2002), is one of the first attempts to regulate Bcl-2 proteins. However, treatment has not been successful, possibly due to Oblimersen's targeting of only BCL-2 protein, with no effect on other anti-apoptotic proteins, and in some cases an enhanced effect on other proteins that increase BCL-2. To remedy this problem, several small molecules (BH3 mimetics) have been developed to functionally mimic the BH3 domain of Bcl-2. These proteins bind within the hydrophobic groove of Bcl-2 and inhibit its activity. Obatoclax®, the first pan anti-apoptotic Bcl-2 protein inhibitor, not only inhibits Bcl-2, Bcl-xL and MC-1, but presumably other anti-apoptotic proteins. However, Obatoclax® may also have Bcl-2–independent targets, decreasing its specificity. The development of synthetic protein mimetics and the manipulation of these peptides (e.g. "stapling to maintain α-helicity") are valuable tools for designing cancer therapeutics and studying the exact nature of protein-protein interactions, as well as for the development of models to target individual proteins (Elkholi et al, 2011). The MAP kinase, ERK1/2, has been shown to phosphorylate and deactivate the pro-apoptotic protein BAD for inhibition of apoptosis in PRKAR1A-mutant B-lymphocytes (Robinson-White et al., 2006b). Studies on the pro-apoptotic ability of cAMP/PKA showed that PKA induces G1 phase cell cycle arrest and apoptosis in wild type S49 cells by up-regulation of pro-apoptotic BIM, but not in deathless (D) S49 lymphoma cells which lack cAMP promoted apoptosis. Up-regulation of BIM may, thus, be an important element of cAMP/PKA mediated apoptosis (Zambon et al., 2011). Therefore, the manipulation of various proteins, small molecular inhibitors and signaling pathways can be used in therapeutic strategies to regulate apoptosis. However, the specific protein, signaling pathway, stimulant and cell and tissue type must be considered. Further investigation into the mechanisms involved in the interactions of these proteins and signaling pathways with apoptosis in cancer cells will certainly bring to light other therapeutic targets to benefit cancer patients.

9.4 Challenges of anticancer efforts

Tremendous advances in certain areas of cancer diagnosis and treatment have occurred within the last 20 years. However, we still face many challenges in the treatment of most cancer types. More in-depth research and improved treatment strategies are needed to overcome these obstacles.

9.4.1 Lack of biomarkers

There is a need for reliable biomarkers for endocrine and other cancers. Diagnosis and treatment of certain cancers are still impeded by the lack of specific and sensitive biomarkers that correlate with disease burden and respond to surgical and systemic treatment. For example, pancreatic adenocarcinoma, the fourth highest cause of cancer-related deaths in the United States, has the most aggressive presentation, resulting in a very short median survival time, and one of the lowest survival rates for solid tumors. At the present time there is still no reliable method for early detection, and less than 10% of cases are diagnosed at an early stage. Because surgical resection remains the only option for a potential cure and most patients present with inoperable disease, early and accurate diagnosis is essential. Carbohydrate antigen 19-9, i.e. sialylated Lewis antigen, is the only widely used biomarker for pancreatic cancer (sensitivity 80.8%, specificity 89%). It, however, is expressed in other cancers and in other benign diseases, and is thus an ineffective screening tool in the general population (Castellanos et al., 2011). Recently, a novel urine steroid metabolomics approach was designed to detect adrenocortical carcinoma (ACC) and to differentiate ACC from adrenocortical adenoma (ACA). The method is based on gas chromatography/mass spectrometry steroid profiling, followed by machine learning analysis. Recent data showed that 9 steroids differentiate ACC from ACA (specificity= sensitivity = 90%). Thus, urine steroid metabolism is both sensitive and specific and has great promise for differentiating benign from malignant disease (Arlt et al., 2011). Epithelial ovarian cancer (EOC) is the most common gynecological cancer and the ninth most common cancer overall. Diagnostic and treatment options lag due to poor characterization of disease progression, disease heterogeneity and lack of early detection biomarkers. Early detection and treatment of EOC would significantly benefit from specific and sensitive routine screening tests on biofluids. Over 200 potential biomarkers for EOC have been identified in the past twenty years. Of the many biomarker candidates, only a small number are validated with clinical samples, due to the lack of biomaterials that are linked with accurate clinical data. Research is now geared toward identifying more stratified biomarkers to accurately follow disease progression as well as efficacy of treatment (Elschenbroich et al., 2011).

9.4.2 Resistance to drug therapy

Resistance to drug therapy is a major cause of treatment failure. For example, relapse after successful treatment of patients with chronic myeloid leukemia with the TKI, imatinib, occurred due to the emergence of cells having a second-site mutation in the BCR/ABL oncoprotein, which prevented imatinib binding. Likewise, second-site mutations were identified in epidermal growth factor receptors (EGFRs) in non-small cell lung carcinoma. These cells acquired resistance to the TKI, erlotinib. Many second-site mutations exist prior to therapy; therefore there is a need for mutational analysis of the cancerous tissue and combination drug therapy in the form of a "drug cocktail" as the first line of therapy to prevent the emergence of resistant clones. Studies have also shown that drug efficacy can be compromised by mutations in other un-related genes. Mutations in genes such as PTEN and p53 can control the outcome of targeted therapy. The mechanism of this reaction is unknown. Another cause of relapse after successful treatment is the ability of cancer cells to adapt to treatment. Cancer cells can re-wire signal transduction networks to become less dependent on the molecular target (McCormick, 2011). The mechanism that drives this is part of a process called "oncogene addition" (Weinstein & Joe, 2008). Cancer cells are addicted to and depend on a particular oncogene for survival. For example, certain cancer cells that depend on, e.g. mutated B-Raf are addicted to MEK. These cells are exquisitely sensitive to MEK1/2 inhibitors. But other tumor cells or normal cells are not. The confounding aspect of this observation is that "oncogene addicted" cancer cells, after treatment can lose their addition and fail to respond to the same therapy (McCormick, 2011). Some cancer cells can switch to alternate signaling pathways. Successful treatment of estrogen positive breast cancer with aromatase inhibitors, was eventually decreased due to adaptive changes in these cells; leading to a decrease in the expression of estrogen receptor α and aromatase, and a switch to use of the MAPK signaling pathway (Brodie et al., 2011). Other outcomes of targeted therapy have also left investigators perplexed, and go beyond the present level of understanding of signaling pathways. In particular, patients who have K-Ras mutations and were treated with EGFR inhibitors plus chemotherapy, showed increased tumor progression in spite of the combined therapy (McCormick, 2011). In all, these examples of events occurring in cancer cells, allow us to realize the extent of the adaptive ability of cancer cells to resist therapeutic strategies. Therefore a comprehensive knowledge of all mutations and signaling pathways in a tumor should be sought to predict a drug response.

9.4.3 Complexity of signaling

Most anticancer strategies focus on attacking specific targets in a given biological pathway to achieve target specificity and to minimize toxicity. However, due to resistance to therapy, this approach has not produced the desired effect, i.e. eradication of all tumor cells. The major problem has been our incomplete understanding of the multiplicity and complexity of genetic and epigenetic changes, and the redundancies and "cross-talk" found in key signaling pathways occurring in the majority of cancer cells, as well as the ability of cancer cells to use alternative signaling pathways to resist therapy (Azmi et al., 2010). We are now aware that much more knowledge should be gathered on tumor genetics and the complexities of tumor signaling pathways and pathway interactions versus that in normal cells for the development of more refined therapeutic strategies. Instead of a single-pathway targeted approach or the use a single therapeutic moiety, innovative approaches (e.g. combination therapy with chemotherapy, mAb and TKI moieties) are needed to overcome these obstacles to treatment efficacy.

9.4.4 Lack of individualized therapy

The one size fits all approach to cancer therapeutics is based on standards of care, obtained by epidemiological studies of large cohorts. It is a reactive approach that focuses on treatment of signs and symptoms. It does not take into account the vast underlying biological heterogeneity of tumors, host genetics, and inter-individual variations in drug response. The need for a more tailored approach to treatment is apparent. Likewise, biomarkers do not predict which patients will respond positively, which are non-responders and which patients will experience adverse reactions for the same medications and dose. At present, dosage must be optimized by a trial and error method. Recent advances in medical genetics and molecular profiling, e.g. genetic and proteomic profiling, genetic testing and improvement in treatment response assessment, has allowed the advancement of the field of personalized medicine. Personalized medicine, considers each persons unique clinical, genetic, genomic and environmental information (Shastry, 2006). The implementation of personalized medicine would allow the selection of the right drugs and manner of treatment for a given patient, and to define lifelong health maintenance strategies.

10. Conclusions

In the last 25 years, an understanding of endocrine tumorigenesis has greatly advanced, due to the increase in our knowledge of underlying molecular signaling pathways and cellular and molecular genetics. In endocrine and other cells, various signaling pathways communicate ("talk") to each other forming a highly complex network of interactions that amplify or dampen extracellular and intracellular signals. The PKA signaling pathway appears to be at the hub of signaling in endocrine tissue. It interacts with versatility with the MAPK, PKC, PKB, Wnt and other signaling pathways, as well as with apoptotic pathways, in a manner that depends on the species, tissue, cell type and ligand involved. From pathophysiological and other studies, endocrine tumors have been well characterized and classified. Of these tumors, we have focused here, in particular, on CNC and its main clinical manifestation, PPNAD. Mutations in the gene encoding the RIα subunit of PKA, *PRKAR1A*, have been shown to be central to the development of tumors of CNC. Through studies in *PRKAR1A* mutant cells and tissues and in mouse models, we are beginning to understand *PRKAR1A*s role in the intricate molecular mechanisms involved in CNC tumor formation Despite the fact that effective treatment strategies for many tumors of CNC are currently lacking, studies with the cAMP analogue, 8-Cl-adenosine, have shown that this drug may be a useful therapeutic agent for the treatment of CNC tumors. Although the future looks bright for the use of other highly innovative approaches in the treatment of endocrine and other tumors, it is apparent that there are still many challenges to overcome before these strategies use be widely implemented. More comprehensive research into the complexity of extracellular/intracellular signaling, inherent biological heterogeneity of tumors, host genetics and inter-individual variations in host response should be undertaken to overcome these challenges.

11. Acknowledgments

The author wishes to thank Dr. Wolfgang W. Leitner, NIAID, National Institutes of Health, Bethesda, Maryland, USA, for his helpful suggestions during the preparation of this manuscript.

12. References

Ally, S.; Tortora, G.; Clair, T.; Grieco, D.; Merlo, G.; Katsaros, D.; Ogreid, D.; Doskeland, S.O.; Jahnsen, T. & Cho-Chung, Y.S. (1988). Selective modulation of protein kinase isozymes by the site-selective analogue 8-chloroadenosine 3′, 5′-cyclic monophosphate provides a biological means for control of human colon cancer cell growth. *Proceedings of the National Academy of Science USA*, Vol. 85, No. 17, (September 1988), pp. 6319-6322, ISSN 0027-8424

Almeida, M.Q.; Muchow, M.; Boikos, S.; Bauer, A.J.; Griffin, K.J.; Tsang, K.M.; Cheadle, C.; Watkins, T.; Wen, F.; Starost M.F, Bossis, I.; Nesterova, M. & Stratakis, C.A. (2010) Mouse *Prkar1a* haploinsufficiency leads to an increase in the Trp53+/- or Rb1+/- backgrounds and chemically induced skin papillomas by dysregulation of the cell cycle and Wnt signaling. *Human Molecular Genetics,* Vol. 19, No. 8, (April 2010), pp. 1387-1398, ISSN 0964-6906

Amieux, P.S. & McKnight, G.S. (2002). The essential role of RI alpha in the maintenance of regulated PKA activity. *Annals of the New York Acadamy of Science,* Vol. 968, (June 2002), pp. 75-95, ISSN 007-8923

Amieux, P.S.; Cummings, D.E. ; Motamed, K.; Brandon, E.P.; Wailes, L.A.; Le, K.; Idzerda, R.L. & McKnight, G.S. (1997). Compensatory regulation of RIα protein levels in protein kinase A mutant mice. *Journal of Biological Chemistry,* Vol. 272, No. 7, (February 1997), pp. 3993-3998, ISSN 002-9258

Andreis, P.G.; Rucinski, M.; Neri, G.; Conconi, M.T.; Petrelli, L.; Parnigotto, P.P.; Malendowicz, L.K. & Nussdorfer, G.G. (2005). Neuropeptides B and W enhance the growth of human adrenocortical carcinoma-derived NCI-H295 cells by exerting MAPK p42/p44-mediated proliferogenic and antiapoptotic effects. *International Journal of Molecular Medicine,* Vol. 16, No. 6, (December 2005), pp. 1201-1208, ISSN 1107-3756

Arlt, W.; Biehl, M.; Taylor, A.E.; Hahner, S.; Libe, R.; Hughes, B.A.; Schneider, P.; Smith, D.J.; Stiekema, H.; Krone, N.; Porfiri, E.; Opocher, G.; Bertherat, J.; Mantero, F.; Allolio, B.; Terzolo, M.; Nightingale, P. Sackleton, C.H.; Bertagna, X.; Fassnacht, M. & Stewart, P.M,. (2011). Urine steroid metabolomics as a biomarker tool for detecting malignancy in adrenal tumors. *Journal of Clinical Endocrinology and Metabolism, n.d. ISSN 0021-972x*

Azevedo, M. & Stratakis, C.A. (2011). The transcriptome that mediates increased cAMP signaling in *PRKAR1A* defects and other settings. *Endocrine Practice,* Vol. 31, No. 4, (August 2011), pp. 1-19, ISSN 1530-891X

Azmi, A.S.; Wang, Z.; Philip, P.A.; Mohammad, R.M, & Sarkar, F.H. (2010). Proof of concept: network and systems biology approaches aid in the discovery of potent anticancer drug combinations. *Molecular Cancer Therapeutics,* Vol. 9, No. 12, (December 2010), pp 3137-3144, ISSN 1535-7163

Bertherat, J.; Horvath, A.; Groussin, L.; Grabers, S.; Boikos, S.; Cazabat, L.; Liba, R.; Rene-Corail, F.; Stergiopoulos, S.; Bourdeau, I.; Bei, T.; Clauser, E.; Calender, A.; Kirschner, L.S.; Bertagna, X.; Carney, J.A. & Stratakis, C.A. (2009). Mutations in regulatory subunit type 1A of cyclic adenosine-5'-monophosphate-dependent protein kinase *(PRKAR1A)*: phenotype analysis in 353 patients and 80 different genotypes. *Journal of Clinical Endocrinology and Metabolism,* Vol. 94, No. 6, (June 2009), pp. 2085-2091, ISSN 0021-972X

Bialik, S.; Zalckvar, E.; Ber, Y.; Rubinstein, A.D. & Kimchi, A. (2010). Systems biology analysis of programmed cell death. *Trends in Biochemical Science,* Vol. 35, No. 10, (October 2010), pp. 556-564, ISSN 0968-0004

Bouizar, Z.; Ragazzon, B.; Viou, L.; Hortane, M.; Bertherat, J. & .Rizk-Rabin, M. (2010). 8Cl-cAMP modifies the balance between PKAR1 and PKAR2 and modulates the cell cycle, growth and apoptosis in human adrenocortical H295R cells. *Journal of Molecular Endocrinology,* Vol. 44, 6, (June 2010), pp. 331-347, ISSN 095-5041

Bourdeau, I.; LaCroix, A.; Schurch, W.; Caron, P.; Antakly, T. & Stratakis, C.A. (2003). Primary pigmented nodular adrenocortical disease: paradoxical responses of cortisol secretion to dexamethasone occur in vitro are associated with increased expression of the glucocoid receptor. *Journal of Clinical Endocrinology and Metabolism*, Vol. 88, No. 8, (August 2003), pp. 3931-3937, ISSN 0021-972X

Brodie, A. & Sabnis, G. (2011). Adaptive changes result in activation of alternate signaling pathways and acquisition of resistance to aromatase inhibitors. *Clinical Cancer Research*, Vol. 17, No. 13, (July 2011), pp. 4208-4213, ISSN 1078- 0432

Cargnello, M. & Roux, P.P. (2011). Activation and function of MAPKs and their substrates, the MAPK-activated protein kinases. *Microbiology and Molecular Biology Review*, Vol. 75, No.1, (March 2011), pp. 50-83, ISSN 1092-2172

Carney, J.A.; Hruska, L.S.; Beauchamp, G.D. & Gordon, H. (1986). Dominant inheritance of the complex of myxomas, spotty pigmentation and endocrine overactivity. *Mayo Clinic Proceedings*, Vol. 61, No. 3, (March 1985), pp. 165-172, ISSN 0025-6196

Castellanos, E.; Berlin, J. & Cardin, D.B. (2011). Current treatment options for pancreatic carcinoma. *Current Oncology Reports*, Vol. 13, No. 3, (June 2007), pp. 195-205, ISSN 1523-3790

Coffer, P.J. & Woodgett, J.R. (1991). Molecular cloning and characterization of a novel putative protein-serine kinase related to the cAMP-dependent and protein kinase C families. *European Journal of Biochemistry*, Vol. 201, No. 2 (October 1991), pp. 475-481, ISSN 0014-2956

Cuny, T.; Gerard, C.; Saveanu, A.; Barlier, A. & Enjalbert, A. (2011). Physiopathology of somatolactotroph cells: from transduction mechanisms to cotargeting therapy. *Annals of the New York Academy of Science*, Vol. 1220 (March 2011), pp. 60-70, ISSN 007-8923

Degterev, A. & Yuan, J. (2008). Expansion and evolution of cell death programmes. *Molecular and cellular Biology*, Vol. 9, No. 5, (May 2008), pp. 378-390, ISSN 072-7306

Elkholi, R.; Floros, K.V. & Chipuk, J.E. (2011). The role of BH3-only proteins in tumor cell development, signaling, and treatment. *Genes and Cancer*, Vol. 2, No. 5, (May 2011), pp. 523-537, ISSN 1447-6019

Elschenbroich, S; Ignatchenko, V Clarke, B.; Kalloger, S.E.; Boutros, P.C.; Gramolini, A.O.; Shaw, P.; Jurisca, I. & Kislinger, T. (2011). In-depth proteomics of ovarian cancer ascites: combining shotgun and selected reaction monitoring mass spectrometry. *Journal of Proteome Research*, Vol. 10, No. 5, (May 2011), pp. 2286-2299, ISSN 1535-3893

Erhardt, P.; Troppmair, J.; Rapp, U.R. & Cooper, G.M. (1995). Differential regulation of Raf-1 and B-Raf and Ras-dependent activation of mitogen activated protein kinase by cyclic AMP in PC12 cells. *Molecular Cell Biology*, Vol. 15, No. 10, (October 1995), pp. 5524-5530, ISSN 0270-7306

Fulda, S. & Debatin, K.M. (2004). Apoptosis signaling in tumor therapy. *Annals of the New York Academy of Science*, Vol. 1028, (December 2004), pp. 150-156, ISSN 0007-8932

Garrington, T.P. & Johnson, G.L. (1999). Organization and regulation of mitogen-activated protein kinase signaling pathways. *Current Opinions in Cell Biology*, Vol. 11, (April 1999), pp. 211-218, ISSN 0955-0674

Griffin, K.J.; Kirschner, L.S.; Matyakhina, L.; Stergiopoulos, S.G.; Robinson-White, A.; Lenherr, S.M.; Weinberg, F.D.; Clafin, E.S.; Bastista, D.; Bourdeau, I.; Voutetakis,

A.; Sandrini, F.; Meoli, E.M.; Bauer, A.J.; Cho-Chung, Y.S.; Bornstain, S.R. & Stratakis, C.A. (2004). A transgenic mouse bearing an antisense construct of regulatory subunit type IA of protein kinase A develops endocrine and other tumors: comparison with Carney complex and other *PRKAR1A*-induced lesions. *Journal of Medical Genetics,* Vol.41, No. 12, (December 2004), pp. 923-931, ISSN 0303-7207

Gonzalez- Hernandez, J.A.; Ehrhart-Bornstein, M.; Spath-Schwalbe, E.; Scherbaum, W.A. & Bornstein, S.R. (1996). Human adrenal cells express tumor necrosis factor-a messenger ribonucleic acid: evidence for paracrine control of adrenal function. *Journal of Clinical Endocrinology and Metabolism,* Vol. 81, No.2, (February 1996), pp. 807-813, ISSN 0021- 972X

Groussin, L.; Cazabat, L.; Rene-Corail, F.; Jullian, E. & Bertherat, J. (2005). Adrenal pathophysiology: lessons from the Carney complex. H*ormone Research,* Vol. 64, No. 3, (September 2005), pp. 132-139, ISSN 0301-0163

Groussin, L.; Horvath, A.; Jullian, E.; Boikos, S.; Rene-Corail, F.; Lefebvre, H.; Cephise-Velayoudom, F.L.; Vantyghem, M.C.; Chanson, P.; Conte-Devoix, B.; Lucas, M.; Gentil, A.; Malchoff, C.D.; Tissier, F.; Carney, J.A.; Bertagna, X., Stratakis, C.A. & Bertherat J. (2006). A *PRKAR1A* mutation associated with primary pigmented nodular adrenocortical disease in 12 kindreds. *Journal of Clinical Endocrinology and Metabolism,* Vol. 91, No. 5, (May 2006), pp. 1943-1949, ISSN 0021- 972X

Groussin, L.; Massias, J.F.; Bertagna, X & Bertherat, J. (2000). Loss of expression of the ubiquitous transcription factor cAMP response element-binding protein (CREB) and compensatory over expression of the activator CREMtau in the human adrenocortical cancer cell line H295R. *Journal of Clinical Endocrinology and Metabolism,* Vol. 85, No.1, (January 2000), pp. 345-354, ISSN 0021-972X

Guan, K-L.; Figueroa, C.; Brtva, T.R.; Zhu, T; Taylor, J. & Barber, T.D. & Vojtex A.P. (2000). Negative regulation of the serine/threonine kinase B-raf by akt. *Journal of Biological Chemistry,* Vol. 275, No. 35, (September 2000), pp. 27354-27359, ISSN 0021- 9258

Guo, Z.S.; Thorne, S.H. & Bartlett, D.L. (2008). Oncolytic virotherapy: molecular targets in tumor-selective replication and carrier cell-mediated delivery of oncolytic viruses. *Biochemica Biophysica Acta* , Vol. 1752, No. 2, (April 2008), pp. 217-231, ISSN 006-3002

Gunther, D.F.; Bourdeau, I.; Matyakhina, L.; Cassarino, D.; Kleiner, D.E.; Griffin, K.; Courkoutsakis, N.; Abu-Asab, M.; Tsokos, M.; Keil, M.; Carney, J.A. & Stratakis, C.A. (2004). Cyclical Cushing Syndrome presenting in infancy: an early form of primary pigmented nodular adrenocortical disease, or a new entity? *Journal of Clinical Endocrinology and Metabolism,* Vol. 89, No. 7, (July 2004), pp. 3173-3182, ISSN 0021-972X

Halgren, R.G.; Traynor, A.E.; Pillay, S.; Zell, J.L.; Heller, K.E.; Krett, N.L. & Rosen, S.T. (1998). 8-Cl-cAMP cytotoxicity in both steroid sensitive and insensitive multiple myeloma cell lines is mediated by 8 Cl-adenosine. *Blood,* Vol. 92 , No. 8, (October 1998), pp. 2893-2880, ISSN 0006-4971

Horvath, A.; Boikos, S.; Giatzakis.; C Robinson-White, A.; Groussin, L.; Griffin, K.J.; Stein, E.; Levin, E.; Delimpasi, G.; Hsiao, H-P.; Keil, M.; Heyerdahl, S.; Matyakhina, L.; Liba, R.; Fratticci, A.; Kirschner, L.S.; Cramer, K.; Gaillard, R.C.; Bertagna, X.; Carney, J.A.; Bertherat, J.; Bossis, I. & Stratakis, C.A. (2006). A genome wide scan identifies mutations in the gene encoding phosphodiesterase 11A4 (PDE11A) in individuals with adrenocortical hyperplasia. *Nature Genetics*, Vol. 38, No. 7, (July 2006), pp. 794-800, ISSN 1061-4036

Horvath, A.; Mericq, V. & Stratakis, C.A. (2008a). Mutation in *PDE8B*, a cyclic AMP-specific phosphodiesterase in adrenal hyperplasia. *New England Journal of Medicine,* Vol. 358, No. 7, (February 2008), pp. 750-752, ISSN 0028-4793

Horvath, A.; Boikos, S.; Giatzakis, C.; Levine, E.; Weinberg, F.; Meoli, E.; Robinson-White, A.; Siegel, J.; Soni, P.; Groussin, L.; Matyakhina, L.; Veerma, S.; Remmers, E,; Nesterova, M.; Carnet, J.A.; Bertherat, J. & Stratakis, C.A. (2008b). Large deletions of the *PRKAR1A* gene in Carney complex. *Clinical Cancer Research,* Vol. 14, No. 2, (January 2008), pp. 388- 395, ISSN 1078–0432

Horvath, A.; Bertherat, J.; Groussin, L.; Guillard-Bataille, M.; Tsang, K.; Cazabat, L.; Liba, R.; Remmers, E.; Rene-Corail, F.; Faucz, F.R.; Clauser, E.; Calender, A.; Bertagna, X.; Carney, J.A. & Stratakis, C.A. (2010). Mutations and polymorphisms in the gene encoding regulatory subunit type 1-alpha of protein kinase A (*PRKAR1A*): an update. *Human Mutation, Vol.* 31, No. 4, (April 2010), pp. 369-379, ISSN 1059-7794

Hu, T. & Li, C. (2010). Convergence between Wnt-B-catenin and EGFR signaling in cancer. *Molecular Cancer*, Vol. 9 (September 2010), pp.236-242, ISSN 1476-4598

Iqbal, J.; Sun, L. & Zaidi, M. (2010). Complexity in signal transduction. *Annals of the New York Academy of Science*, Vol . 1192 (April 2010), pp. 238-244, ISSN 007-8923

Jun,T.; Gjoerup, O. & Roberts, T.M. (1999). Tangled webs: evidence of cross-talk between c-Raf-1 and akt. *Science STKE,* Vol.13, No. 13, (December 1999), pp. 1-4, ISSN 0831-7445

Katsaros, D.; Tortora, G.; Tagliaferra, P.; Clair, T.; Ally, S.; Neckers, L.; Robins, R.K. & Cho-Chung, Y.S. (1987). Site selective cAMP analogues provide a new approach in the control of cancer cell growth. *FEBs Letters,* Vol. 223, No. 1, (October 1987), pp. 97-103, ISSN 0014–5793

Khattak, M.N, Buchfelder, M., Kleindienst, A. Schofl, C., & Kremenevskaja, N. (2010). CRH and SRIF have opposite effects on the Wnt/B-catenin signaling pathway through PKA/GSK-3B in corticotroph pituitary cells. *Cancer Investigation*, Vol. 28, No. 8 (October 2010), pp. 797-805, ISSN 0735-7907

Kim, S-E & Choi, K-Y. (2007). EGF receptor is involved in Wnt3a-mediated proliferation and motility of NIH3T3 cells via ERK pathway activation. *Cell Signaling,* Vol. 19, No. 7, (July 2007), pp. 1554-1564, ISSN 1876-3901

Kirschner, L.S. (2002). Signaling pathways in adrenocortical cancer. *Annals of the New York Academy of Science*, Vol. 968 (June 2002), pp. 222-239, ISSN 007-8923

Kirschner, L.S.; Carney, J.A.; Pack, S.D.; Taymans, S.E.; Giatzakis, C.; Cho, Y.S.; Cho-Chung, Y.S. & Stratakis, C.A. (2000a). Mutations of the gene encoding the protein kinase A type I-alpha regulatory subunit in patients with the Carney complex. *Nature Genetics,* Vol. 26, No. 1, (September 2000), pp. 89-92, ISSN 1061-4036

Kirschner, L.S.; Sandrini, F.; Monbo, J.; Jing, L.P.; Carney, J.A. & Stratakis, C.A. (2000b). Genetic heterogeneity and spectrum of mutations of the *PRKAR1A* gene in patients with the Carney complex. *Human Molecular Genetics,* Vol. 9, No. 20, (December 2000), pp. 3037-3046, ISSN 0964-6906

Kirschner, L.S.; Kusewitt, D.F.; Matyakhina, L.; Towns, W.H. II; Carney, J.A.; Westphal, H. & Stratakis, C.A. (2005). A mouse model for the Carney complex-tumor syndrome develops neoplasia in cyclic AMP-responsive tissues. *Cancer Research,* Vol. 65, No. 11 (June 2005), pp. 4056-4514, ISSN 008-5472

Klasa, R.J.; Gillum, A.M.; Klem, R.E. & Frankel, S.R. (2002). Oblimersen Bcl-2 antisense: facilitating apoptosis in anticancer treatment. *Antisense Nucleic Acid Drug Development,* Vol. 12, No. 3, (June 2002), pp. 193-213, ISSN 1087-2906

Langeveld, C.H.; Jongenelen, C.A.; Heimans, J.J. & Stoff, J.C. (1992). Growth inhibition of human glioma cells induced by 8- chloroadenosine, an active metabolite of 8-chloro cyclic adenosine 3':5'-monophosphate. *Cancer Research,* Vol. 52, No. 14, (July 1992), pp. 3994-3999, ISSN 0008-5472

Lepique, A.P.; Moraes, M.S.; Rocha, K.M.; Eichler, C.B.; Haji, G.N.M.; Schwindt, T.T. & Armelin, H.A. (2004). c-Myc protein is stabilized by fibroblast growth factor 2 and destabilized by ACTH to control cell cycle in mouse Y1 adrenocrotical cells. *Journal of Molecular Endocrinology,* Vol. 33, No. 3, (December 2004), pp. 623-638, ISSN 0952-5041

Levitzki, A. & Klein, S. (2010). Signal transduction therapy of cancer. *Molecular Aspects of Medicine,* Vol. 31, No. 4, (August 2010), pp. 287-329, ISSN 0098-2997

Liu, J.; Li, X-D.; Ora, A.; Heikkila, P.; Vaheri, A. & Voutilainen, R. (2004). cAMP-dependent protein kinase activation inhibits proliferation and enhances apoptotic effect of tumor necrosis factor-alpha in NCI-H295R adrenocortical cells. *Journal of Molecular Endocrinology,* Vol. 33, No. 2 (October 2004), pp. 511-522, ISSN 095—5041

Lodish, H.; Berk, A.; Zipursky, S.L.; Matsudaira, P.; Baltimore, D.; Darnell, J. (2000). Cell-to-cell signaling: hormones and receptors. In: *Molecular Cell Biology,* H. Lodish (ed), pp. 848-909, W.H. Freeman, ISBN -10:0-7167-3136-3, New York.

Lodish, M.B. & Stratakis, C.A. (2010a). Rare and unusual endocrine cancer syndromes with mutated genes. *Seminars in Oncology,* Vol. 37, No. 6, (December 2010), pp. 680-690, ISSN 0093-7754

Lodish, M.B. & Stratakis, C.A. (2010b). Endocrine side effects of broad-acting kinase inhibitors. *Endocrine Related Cancer, Vol..* 17, No. 3, (September 2010), pp. R233-R244, ISSN 1351-0088

McCormick, F. (2011). Cancer therapy based on oncogene addition. *Journal of Surgical Oncology, Vol.* 103, No. 6, (May 2011), pp. 464-467, ISSN 0022-4790

Meoli, E.; Bossis, I.; Cazabet, I.; Mavrakis, M.; Horvath, A.; Stergiopoulos, S.; Shiferaw, M.L.; Fumey, G.; Perlemoine, K.; Muchow, M.; Robinson-White, A.; Weinberg, F.; Nesterova, M.; Patronas, Y,; Groussin, L.; Bertherat, J. & Stratakis, C.A. (2008). Protein kinase A effects of an expressed PRKAR1A mutation associated with aggressive tumors. *Cancer Research,* Vol. 68, No. 9, (May 2008), pp. 3133-3141, ISSN 0008-5472

Mercer, K.E. & Pritchard, C.A. (2003). Raf Proteins and cancer: B-Raf identified as a mutational target. *Biochemica Biophys Acta,* Vol. 1653, No. 1 (June 2003), pp. 25-40, ISSN 006-3002

Nadulla, K.S. & Kirschner, L.S. (2005). Disruption of protein kinase A regulation causes immortalization and dysregulation of d-type cyclins. *Cancer Research,* Vol. 65, No. 22, (November 2005), pp. 10307-10315, ISSN 0008-5472

Nesterova, M.; Bossis, I.; Wen, F.; Horvath, A.; Matyakhina, l. & Stratakis, C.A. (2008). An immortalized human cell line bearing a *PRKAR1A*-inactivating mutation: effects of over expression of the wild-type allele and other protein kinase A subunits. *Journal of Clinical Endocrinology and Metabolism,* Vol. 93, No. 2, (February 2008), pp. 565-571, ISSN 0021-972X

Pack, S.D.; Kirschner, L.S.; Pak, E.; Zhuang, Z.; Carney, J.A. & Stratakis, C.A. (2000). Genetic and histologic studies of somatomammotropic pituitary tumors in patients with the "complex of spotty skin pigmentation, myxomas, endocrine overactivity and schwannomas "(Carney complex). *Journal of Clinical Endocrinology and Metabolism,* Vol. 85, No. 10, (October 2000), pp.3860-3865, ISSN 0021-972X

Pearce, LR.; Komander, D. & Alessi, D.R. (2010). The nuts and bolts of AGC proteins kinases. *Nature Reviews of Molecular and Cellular Biology,* Vol. 11, No. 1, (January 2010), pp.9-22, ISSN 1471-0072

Pereira, A.M.; Hes, F.J.; Horvath, A.; Woortman, S.; Greene, E.; Bimpaki, E.I.; Alatsatianos, A.; Boikos, S.; Smit, J.W.; Romijn, J.A.; Nesterova, M. & Stratakis, C.A. (2010). Association of the MIV *PRKAR1A* mutation with primary pigmented nodular adrenocortical disease in two large families. *Journal of Clinical Endocrinology and Metabolism,* Vol. 95, No. 1, (January 2010), pp. 338-342, ISSN 0021-972X

Ragazzon, B.; Cazabat, L.; Rizk-Rabin, M.; Assie, G.; Groussin, L.; Fierrard, H.; Perlemoine, K.; Martinez, A. & Bertherat, J. (2009). Inactivation of the Carney complex gene 1 (Protein Kinase A regulatory subunit 1A) inhibits SMAD3 expression and TGFβ-stimulated apoptosis in adrenocortical cells. *Cancer Research,* Vol. 69, No.18, (September 2009), pp. 7278-7284, ISSN 0008-5472

Rao, T.P. & Kuhl, M. (2010). An updated overview on Wnt signaling pathways. A prelude for more. *Circulation Research,* Vol. 106, No. 12, (June 2010), pp. 1798-1806, ISSN 0065-8499

Robinson-White, A. & Stratakis, C.A. (2002). Protein kinase A signaling "cross talk" with other pathways in endocrine cells. *Annals of the New York Academy of Science,* Vol. 968, (June 2002), pp. 256-270, ISSN 007-8923

Robinson-White, A.; Meoli, E.; Stergiopoulos, .S.; Horvath, A.; Boikos, S.; Bossis, I. & Stratakis, CA. (2006a). *PRKAR1A* mutations and protein kinase A interactions with other signaling pathways in the adrenal cortex. *Journal of Clinical Endocrinology and Metabolism,* Vol. 91, No. 6, (June 2006), pp. 2380-2388, ISSN 0021-972X

Robinson-White, A.J.; Leitner, W.W.; Aleem, E.; Kaldis, P.; Bossis, I. & Stratakis, C.A. (2006b). *PRKAR1A* inactivation leads to increased proliferation and decreased apoptosis in human B-lymphocytes. *Cancer Research,* Vol. 66, No. 21, (January 2006), pp. 10603-10612, ISSN 0008-5472

Robinson-White, A.; Hundley, T.R.; Shiferaw, M.; Bertherat, J.; Sandrini, F. & Stratakis, C.A. (2003). Protein kinase A activity in *PRKAR1A*-mutant cells, and regulation of mitogen-activated protein kinases ERK1/2. *Human Molecular Genetics,* Vol. 12, No. 13, (July 2003), pp. 1475-1484, ISSN 0964-6906

Robinson-White, A.J.; Hsiao, H-P.; Leitner, W.W.; Greene, E.; Bauer, A.; Krett, N.L.; Nesterova, M. & Stratakis, C.A. (2008). Protein Kinase A independent inhibition and induction of apoptosis in human thyroid cells by 8-Cl-Adenosine. *Journal of Clinical Endocrinology and Metabolism,* Vol. 93, No. 3, (March 2008), pp. 1020-1029, ISSN 0021-972X

Robinson-White, A.J.; Bossis, I.; Hsiao, H-P.; Nesterova, M.; Leitner, W.W. & Stratakis, C.A. (2009). 8-Cl-Adenosine inhibits proliferation and causes apoptosis in B-lymphocytes via protein kinase A-dependent and independent effects:

implications for treatment of Carney complex-associated tumors. *Journal of Clinical Endocrinology and Metabolism,* Vol. 94, No. 10, (October 2009), pp. 4061-4069, ISSN 0021-972X

Rohlff, C.; Clair, T. & Cho-Chung, Y.S. (1993). 8-Cl-cAMP induces truncation and down-regulation of the RIα subunit and up-regulation of the RIIB-subunit of cAMP-dependent protein kinase leading to type II holoenzyme-dependent growth inhibition and differentiation of HL-60 leukemia cells. *Journal of Biological Chemistry,* Vol. 268, No. 8, (March 1993), pp. 5774-5782, ISSN 0021-9258

Rossi, G.; Gasperi, V.; Paro, R.; Barsacchi, D.; Cecconi, S. & Maccarrone, M. (2007). Follicle-stimulating hormone activates fatty acid amide hydrolase by protein kinase A and aromatase-dependent pathways in mouse primary Sertoli cells. *Endocrinology,* Vol. 148, No. 3, (March 2007), pp. 1431-1439, ISSN 0013-7227

Rothenbuhler, A. & Stratakis, C.A. (2010). Clinical and molecular genetics of Carney complex. *Best Practice and Research Clinical Endocrinology and Metabolism,* Vol. 24, No. 3, (June 2010), pp. 389-397, ISSN 1521-690X

Saavedra, A.P.; Tsygankova, O.M.; Prendergast, G.V.; Dworet, J.H.; Cheng, G. & Meinkoth, J.L. (2002). Role of cAMP, PKA and Rap1A in thyroid follicular cell survival. *Oncogene,* Vol. 21, No. 5, (January 2001), pp. 778-788, ISSN 0950-9232

Sarlis, N.J.; Chrousos, G.P.; Doppman, J.L.; Carney, J.A. & Stratakis, C.A. (1997). Primary pigmented nodular adrenocortical disease: reevaluation of a patient with Carney complex 27 years after unilateral adrenalectomy. *Journal of Clinical Endocrinology and Metabolism,* Vol. 82, No. 4, (April 1007), pp. 1274-1278, ISSN 0021-972X

Schmitd, J.M. & Stork, P.J.S. (2000). B2-adrenergic receptor activates extracellular signal – regulated kinases (ERKs) via the small G protein Rap1 and the serine threonine B-raf. *Journal of Biological Chemistry,* Vol. 275, No. 33, (August 2000), pp. 25342-25350, ISSN 0021-9258

Shastry, B.S (2006). "Pharmacogenetics and the concept of individualized medicine." *Pharmacogenomics Journal,* Vol. 6, No. 1, (January 2006), pp. 16-21, ISSN 1462-2416

Sirotkin, A.V.; Sanislo, P.; Schaeffer, H-J.; Florkovicova, I.; Kotwica, J.; Bulla, J.& Hetenyi, L. (2004) Thrombopoietin regulates proliferation, apoptosis, secretory activity and intracellular messengers in porcine ovarian follicular cells: involvement of protein kinase A. *Journal of Endocrinology,* Vol.183, No. 3, (December 2004), pp. 595-604, ISSN 0022- 0795

Sirotkin, A.V.; Mlyncek, M.; Makarevich, A.V.& Florkovicova, I. (2008). Leptin affects proliferation, apoptosis and protein kinase A-related peptides in human ovarian granulosa cells. *Physiological Research,* Vol. 57, No. 3, (December 2007), pp. 437-442, ISSN 0862-8408

Slot, K.A, deBoer-Brouwer, M.; Houweling, M.; Vaandrager, A.B.; Dorrington, J.H.; Teerds, K.J. (2006). Luteinizing hormone inhibits Fas-induced apoptosis in ovarian surface epithelial cell lines. *Journal of Endocrinology, Vol.* 188, No. 2, (February 2006), pp. 227-239, ISSN 0022-0795

So, E.C.; Chang, Y.T.; Hsing, C.H.; Poon, P.W.; Leu, S-F.& Huang, B-M. (2010). The effect of medazolam on mouse Leydig cell steroidogenesis and apoptosis. Toxicology Letters, Vol. 192, No. 2 (February 2010), pp. 169-178, ISSN 0378-4274

Stergiopoulos, S.G.; Abu-Asab, M.S.; Tsokos, M. & Stratakis, C.A. (2004). Pituitary pathology in Carney complex patients. *Pituitary*, Vol. 7, No. 2, (April 2004), pp. 73-82, ISSN 1386-341X

Stratakis, C.A. (2007). Adrenocortical tumors, primary pigmented adrenocortical disease (PPNAD)/Carney complex, and other bilateral hyperplasias: The NIH studies. *Hormone and Metabolism Research*, Vol. 39, No. 6, (June 2007), pp. 467-473, ISSN 0018-5043

Stratakis, C.A.; Carney, J.A.; Lin, J.P.; Papanicolaou, D.A.; Karl, M.; Kastner, D.L.; Pras, E. & Chrousos, G.P. (1996). Carney complex, a familial multiple neoplasia and lentigenois syndrome. Analysis of 11 kindreds and linkage to the short arm of chromosome 2. *Journal of Clinical Investigation*, Vol. 97, No. 3, (February 1996), pp. 699-705, ISSN 0921-9738

Stratakis, C.A.; Kirschner, L.S. & Carney, J.A. (2001). Clinical and molecular features of the Carney complex; diagnostic criteria and recommendations for patient evaluation, *Journal of Clinical Endocrinology* and *Metabolism*, Vol. 86, No. 9, (September 2001), pp. 4041-4046, ISSN 0021-972X

Stoffel, A. (2010). Targeted therapies for solid tumors: Current status and future perspectives. *BioDrugs*. Vol. 24, No. 5, (October 2010), pp. 303-316, ISSN1173-8804

Taylor, R.C.; Cullen, S.P. & Martin, S.J. (2008). Apoptosis: controlled demolition at the cellular level. *Molecular and Cellular Biology*, Vol. 9, No. 3, (March 2008), pp. 231-241, ISSN 072-7306

Tichomirowa, M.A.; Daly, A.F. & Beckers, A. (2009), Familial pituitary adenomas. *Journal of Internal Medicine, Vol.* 266, No. 1, (April 2009), pp. 5-18, ISSN 0954-6820

Trumper, A.; Trumper, K. & Horsch, D. (2002). Mechanisms of mitogenic and anti-apoptotic signaling by glucose- dependent insulinotropic polypeptide in β(INS-1) cells. *Journal of Endocrinology*, Vol. 174, No. 2, (August 2002), pp. 233-246, ISSN 0022-0795

Volz, A.; Goke, R.; Lankat-Buttqereit, B.; Fehmann, H.C.; Bode, H.I. & Goke, B. (1995). Molecular cloning, functional expression, and signal transduction of the GIP-receptor cloned from a human Insulinoma. *FEBS Letters*, Vol. 373, No. 1, (August 1995), pp. 23-29, ISSN 0014-5793

Vossler, M.R.; Yao, H.; York, R.D.; Pan, M.G.; Rim, C.S. & Stork, P.J.s. (1997). CAMP activates MAP kinase and Elk-1 through a B-raf- and Rap1-dependent pathway. *CELL, Vol.* 89, No. 1, (April 1997), pp. 73-82, ISSN 0092-8674

Walczak, H. & Krammer, P.H. (2000). The CD95 (APO-1/Fas) and the TRAIL (APO-2l) apoptosis systems. *Experimental Cell Research*, Vol. 256, No. 1, (April 2000), pp. 58-66, ISSN 0014-4827

Weinstein, IB. & Joe, A. (2008). Oncogene addiction. *Cancer Research*, Vol. 68, No. 9, (May 2008), pp. 3077-3080, ISSN 0008- 5472

Zambon, A.C.; Wilderman, A.; Ho, A. & Insel, P.A. (2011). Increased expression of the pro-apoptotic protein BIM, a mechanism for cAMP/Protein kinase A (PKA)-induced apoptosis of immature T cells. *Journal of Biological Chemistry*, Vol. 286, No. 38, (September 2011), pp. 33260-33267, ISSN 0021-9258

Zawadzki, K.M. & Taylor, S.S. (2004). cAMP-dependent protein kinase regulatory subunit type IIB: active site mutations define an isoform-specific network for allosteric signaling by cAMP. *Journal of Biological Chemistry*, Vol. 279, No.8, (February 2004), pp. 7029-7036, ISSN 0021-9258

Zhang, Y. & Nose, V. (2011). Endocrine tumors as part of inherited tumor syndromes. *Advances in Anatomic Pathology, Vol.* 18, No. 3, (May 2011); pp. 9206-217, ISSN 1072-4109

Yin, Z.; Williams-Simons, L.; Parlow, A.F.; Asa, S. & Kirschner, L.S. (2007). Pituitary-specific knockout of the Carney complex gene *Prkar1a* leads to pituitary tumorigenesis. *Molecular Endocrinology,* Vol.22, No. 2, (February 2004), pp. 380-387, ISSN 0888-8809

Zimmerman, K.C., Bonzon, C. & Green, D.R. (2001). The machinery of programmed cell death. *Pharmacology Therapy,* Vol.92, No. 1, (October 2001), pp. 57-70, ISSN 0163-7258

3

The Role of Tpl2 Protein Kinase in Carcinogenesis and Inflammation

Katie DeCicco-Skinner[1], Monika Deshpande[2] and Jonathan Wiest[2]
[1]*Department of Biology, American University, Washington, DC,*
[2]*Laboratory of Cancer Biology and Genetics,*
National Cancer Institute,
National Institutes of Health, Bethesda, MD
USA

1. Introduction

The mitogen activated protein kinase (MAPK) cascade is a family of serine-threonine protein kinases controlling a vast array of cellular responses. Among the oldest signaling pathways known, the MAPK cascade is evolutionarily conserved, being found in species ranging from yeast to humans (Widmann *et al.*, 1999). All eukaryotic cells possess multiple MAPK signaling pathways. In mammals, four major subfamilies (termed classical MAPKs) have been identified including extracellular signal-regulated kinase 1/2 (ERK1/2), c-jun N-terminal protein kinases (JNK/SAPK), p38 isoforms (p38 α, β, γ, or δ), and ERK5 (Dhanasekaran and Johnson, 2007). These multiple parallel MAPK cascades all operate through a three tiered system characterized by consecutive phosphorylation events. Each of the pathways is equipped to respond to a remarkably diverse array of stimuli, including cytokines, growth factors, irradiation, hormones, injury, and stress (Cargnello and Roux, 2011). The binding of stimuli to receptors triggers the initial phosphorylation events. These events transform inactive MAP3K into active MAP3K. Subsequently, active MAP3K phosphorylates, and thereby activates, downstream MAP2K (MEK) proteins. The active MAP2K proteins then phosphorylate MAPK (ERK) molecules. These MAPK proteins then regulate the transcription of a diverse group of genes involved in inflammation, cell survival, cell division, development, metabolism, differentiation, motility, and repair (Cargnello and Roux, 2011).

Additionally, over the last two decades atypical pathways including ERK 3/4, ERK 7/8 and Nemo-like kinase (NLK) have been discovered (Cargnello and Roux, 2011). Similar to classical MAPK family members, these proteins phosphorylate target substrates on serine or threonine residues. However, they are not organized into the classical three-tiered model and contain a less well conserved motif on the activation loop of the kinase domain.

Due to the diverse effects that MAPK signaling pathways have on cellular processes, the classical MAPK pathways have been intensely studied to assess their role in human disease. Detailed reviews for each pathway were previously published (Weston and Davis, 2007; Wagner and Nebreda, 2009; Cargnello and Roux, 2011; Schindler *et al*, 2007; Nishimoto and Nishida, 2006). Therefore, this chapter will focus on one protein kinase, *Tpl2,* within the

MAPK family and how altered expression (either underexpression or overexpression) of this kinase is linked to inflammation, cancer and other disease states.

2. The MAP3K8/Tpl2 gene

MAP3K8 (known as *Cot* or Tpl2 in rodents) is a MAP3K protein first isolated from thyroid tumors as a gene capable of inducing morphological transformation of NIH3T3 and SHOK cells (Miyoshi *et al.*, 1991). Structurally, the open reading frame of the *Tpl2* gene encodes a 467 amino acid polypeptide (Aoki *et al.*, 1991). Provirus integration occurring in the last intron of the *Tpl2* gene in rodents generates a mutant protein 36 amino acids shorter than the full length version and also is an activating mutation (Patriotis *et al.*, 1993; Erny *et al.*, 1996; Ceci *et al.*, 1997). This truncated *Tpl2* gene has enhanced kinase activity (Gandara *et al.*, 2003). Additionally, analysis of the N-terminal region of this gene reveals three alternate translation start sites. Consequently, the polypeptides translated from the *Tpl2* gene will contain 467, 464, or 438 amino acids (Aoki *et al.*, 1991). The 467- and the 464-amino acid proteins are detected as a single 58-kDa band by Western Blot analysis and the 438 amino acid protein is detected as a 52-kDa band.

Functionally, some differences exist between these two forms of the protein. The 58-kDa protein has a shorter half-life (10 minutes) than the 52-kDa protein (30 minutes). Thus, the amino-terminal domain contains sequences important for Tpl2 stability. Additionally, cell lines transformed with constructs only expressing the 58-kDa Tpl2 protein have higher tumor incidence and reduced latency when compared to cell lines transformed with the 52-kDa protein indicating that the N-terminal region of the protein is important in regulating function of this gene (Aoki *et al.*, 1991).

Tpl2 has the ability to cross-talk with numerous signaling pathways. In unstimulated cells Tpl2 forms a ternary complex with Nuclear factor kappa B p105 (NF-κB p105) and A20-binding inhibitor of NF-κB2 (ABIN-2) (Lang *et al.*, 2004). This complex increases Tpl2 stability and renders Tpl2 inactive towards its main downstream substrate MEK, the ERK kinase. Upon activation by inflammatory stimuli, Tpl2 becomes phosphorylated by IκB kinase (IKK), releasing both Tpl2 and NF-κB to participate in downstream signaling. In addition to participating in ERK signaling, Tpl2 can cross-talk with multiple MAPK signaling pathways including p38, JNK, ERK 1/2, and ERK 5 (Salmeron *et al.*, 1996, Patriotis *et al.*, 1993, Chiariello *et al.*, 2000) (Figure 1). Additionally, wildtype and mutant Tpl2 activates the transcription factor NFAT in response to stimulation through the T cell receptor (TCR) and causes heightened activation of NF-κB (Tsatsanis *et al.*, 1998, Belich *et al.*, 1999).

3. Tpl2 regulation involves NF-κB

Our understanding of Tpl2 regulation was greatly enhanced by the identification of the interaction between p105 and Tpl2 (Belich *et al.*, 1999). p105 is a NF-κB precursor protein which also acts as a NF-κB-inhibitory protein by sequestering NF-κB complexes in the cytoplasm of cells (Rice *et al.*, 1992). Stoichiometric binding of Tpl2 to the C-terminus of p105 occurs within two distinct regions of Tpl2. The C-terminal region of Tpl2 (residues 398-467) binds to a region adjacent to the ankyrin repeats on p105 (residues 497 to 534), while the kinase domain of Tpl2 binds to the death domain on p105 (Beinke *et al.*, 2003). In steady state conditions, the entire pool of Tpl2 is associated with p105, irrespective of the cell type, whereas only about one third of the p105 pool is occupied by binding to Tpl2 (Belich, et al., 1999).

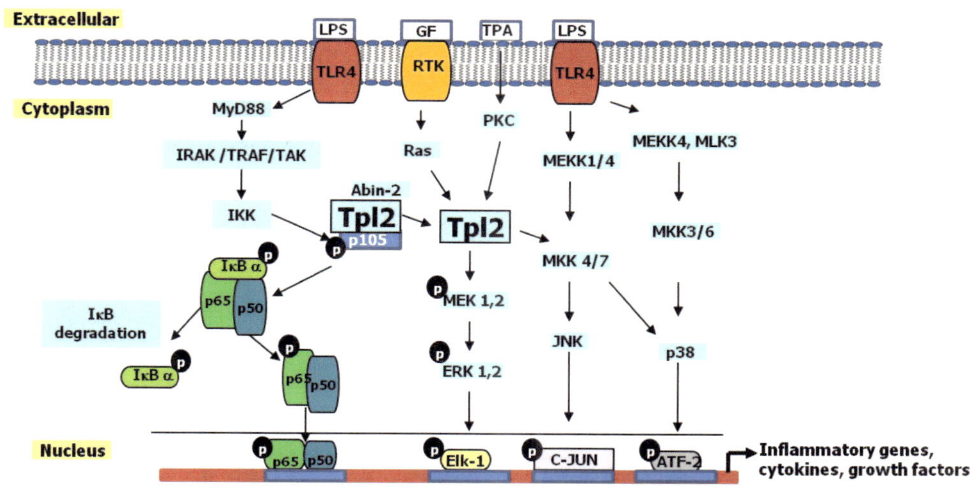

Fig. 1. Tpl2 signaling pathways. Tpl2 interacts with numerous signaling pathways including ERK, JNK, p38, and NF-κB. Tpl2 is normally held in complex with Abin-2 and p105. Upon activation with a stimulus, IKK phosphorylates p105, freeing Tpl2 to participate in ERK, JNK and p38 signaling and p50 to heterodimerize with other NF-κB family members.

Functionally, this interaction between Tpl2 and p105 is significant. p105 binding to Tpl2 inhibits Tpl2 kinase activity and thus prevents Tpl2 from being able to activate downstream signaling cascades. Despite this ability to inhibit Tpl2 kinase activity, p105 improves Tpl2 solubility and stability (Beinke, *et al.*, 2003; Jia, *et al.*, 2005). Tpl2, when not bound to p105, is unstable, appears to have difficulty folding and can form insoluble aggregates (Jia, *et al.*, 2005). Despite the fact that Tpl2 binds p105, p105 is not necessary for *Tpl2* gene expression. Bone-marrow derived macrophages (BMDMs) and splenocytes that lack p105 have normal *Tpl2* gene expression levels (Waterfield, *et al.*, 2003). However, Tpl2 protein levels in these cells are significantly reduced. This reduction in Tpl2 protein expression is correlated with a subsequent reduction in LPS induced ERK signaling (Waterfield *et al.*, 2003). This indicates that binding to p105 not only confers Tpl2 stability, but it also renders it unable to phosphorylate MEK, possibly by blocking access to MEK (Waterfield *et al.*, 2003; Bienke *et al.*, 2003). The inactivity of Tpl2 in the complex may function as a mechanism to prevent unregulated Tpl2 activity in unstimulated cells.

As discussed later in this chapter, truncation of Tpl2 in its C-terminus greatly enhances the catalytic activity and oncogenic potential of Tpl2 in T cell lymphomas (Ceci, et al., 1997). Since the interaction of Tpl2 with p105 requires the C terminus of Tpl2 in vitro, truncation of Tpl2 could theoretically change the association between Tpl2 and p105. In 2003, Beinke addressed this issue and found that truncation of Tpl2 did not hinder the ability of this protein to bind to p105 (Beinke, *et al.*, 2003). However, it did block the ability of p105 to negatively regulate Tpl2 kinase activity. The heightened Tpl2 kinase activity found in Tpl2 truncated cells can result in persistent phosphorylation of targets such as MEK, ultimately leading to increased gene expression of numerous genes involved in cell cycle progression (Matsumura, *et al.*, 2009). Additionally, increased Tpl2 kinase activity can be oncogenic, as aberrant MEK/ERK activity has been identified in several cancer types (Matsumura, *et al.*, 2008).

In 2004, a third protein, ABIN-2, was reported to associate with the Tpl2/p105 complex (Lang, *et al.*, 2004). Tpl2 and ABIN-2 bind to similar regions on the C-terminal region of p105. Both Tpl2 and ABIN-2 bind to the death domain and bind to residues 497-538. However, ABIN-2 also needs to bind to the PEST region of p105 for optimal association. It is reported that in order to maintain steady state protein levels of Tpl2, ABIN-2 is necessary (Lang, *et al.*, 2004). However, it is not involved in maintaining protein levels of p105. Whereas, the functional significance of the role of ABIN-2 in the Tpl2/p105 complex has not been fully elucidated, it has been shown that the solubility of ABIN-2 is increased by binding to p105. Therefore, similar to what was described above for Tpl2, it appears that ABIN-2 binding to p105 is required to ensure correct protein folding.

Tpl2 is activated in response to various proinflammatory stimuli such as lipopolysaccharide (LPS), tumor necrosis factor (TNF) and CD40 ligand. LPS stimulation results in IKK-induced p105 proteolysis, releases Tpl2 from p105, renders Tpl2 enzymatically active, and increases downstream MEK kinase activity (Beinke *et al.*, 2004; Waterfield *et al.*, 2003). This newly liberated Tpl2, which is now susceptible to rapid degradation by the proteosome, will phosphorylate substrates in the ERK and JNK pathways. Additionally, p105 is subsequently degraded into p50 by the proteosome (Beinke, *et al.*, 2004). The p50 subunit can now dimerize with other NF-κB family members and translocate to the nucleus where the active NF-κB complex can regulate over 400 genes. The overall result is an upregulation of diverse genes involved in growth, differentiation, and inflammation.

4. Tpl2 knockout mice

Despite the significant function Tpl2 has in eliciting cellular responses, Tpl2$^{-/-}$ animals appear phenotypically normal (Dumitru *et al.*, 2000). Immune organs including bone marrow, thymus, spleen, and lymph nodes all develop normally and production of T and B cell subsets are unaltered. Additionally, when subjected to challenge by T cell-dependent and independent antigens such as KLH and LPS-TNP, Tpl2$^{-/-}$ mice produce normal antibody responses (Dumitru *et al.*, 2000).

Over the last decade, numerous laboratories have used a variety of disease models to assess the response of Tpl2$^{-/-}$ mice to bacterial, viral, or carcinogenic challenge. However, the results have shown differential responses to challenge, with the absence of Tpl2 sometimes promoting a disease state and sometimes protective against the disease. These differing responses are most likely due to variation in stimuli used and is cell-type specific. As described below, most laboratories have used lipopolysaccharide (LPS), which binds through TLR4, as the major stimulus in Tpl2$^{-/-}$ mice. However, targeting different receptors through the use of other stimulators such as 12-O-tetradecanoylphorbol-13-acetate (TPA), Poly IC, or CpG receptors can yield different responses. Thus, because Tpl2 can signal through numerous pathways, analysis of the effects of underexpression or overexpression of this protein kinase is quite convoluted.

4.1 Evidence supporting that Tpl2$^{-/-}$ mice exhibit lower levels of inflammatory markers

Several laboratories have reported a reduction in pro-inflammatory cytokines in Tpl2$^{-/-}$ mice. Dumitru reported that TNF-α is markedly lower in Tpl2$^{-/-}$ macrophages stimulated with LPS, providing Tpl2$^{-/-}$ mice with resistance to LPS/D-Galactosamine stimulated endotoxic shock (Dumitru *et al.*, 2000). Others have reported similar findings, deficits in TNF- α, Cox-2, PGE_2, and CXCL1 in LPS-stimulated macrophages (Eliopoulos *et al.*, 2002; López-Peláez *et al.*, 2011). In LPS-stimulated Tpl2$^{-/-}$ macrophages, this impaired secretion of TNF- α and PGE_2 is triggered by a defect in ERK activation. Additionally, in bone marrow-derived dendritic cells (BMDCs) and macrophages, Tpl2$^{-/-}$ mice have significantly lower expression of IL-1β in response to LPS, poly IC and LPS/MDP (Mielke *et al.*, 2009). Consequently, these mice displayed significantly higher mortality and bacterial burden when infected with *L. monocytogenes*. Recently, it was reported that obese Tpl2$^{-/-}$ mice have a significant reduction in TNF- α and monocyte chemoattractant protein-1 (MCP-1) in adipose tissue compared to WT controls (Perfield, *et al.*, 2011). This reduction in inflammatory markers in Tpl2$^{-/-}$ mice was associated with an improvement in whole-body insulin resistance. Because inhibition of *Tpl2* is associated with a reduction in inflammatory markers in LPS-stimulated cells, several pharmaceutical companies have developed small molecule inhibitors of Tpl2 to assess whether inhibition of this protein can be beneficial for chronic inflammatory disorders such as rheumatoid arthritis, inflammatory bowel disease, and pancreatitis. Details on these studies are outlined later in this chapter.

4.2 Tpl2$^{-/-}$ mice can display enhanced pro-inflammatory profiles

Contradictory to the details above, our laboratory and others have reported heightened inflammatory responses in Tpl2$^{-/-}$ mice. López-Peláez found that Tpl2 deficiency in LPS-stimulated macrophages was associated with an increase in Nitric Oxide Synthase 2 expression (López-Peláez *et al.*, 2011). Similarly, Zacharioudaki found that expression of IRAK-M, whose function is to compete with IL-1R-associated kinase (IRAK) family of kinases, is decreased in Tpl2$^{-/-}$ macrophages (Zacharioudaki *et al.*, 2009). This deficit in IRAK-M in Tpl2$^{-/-}$ macrophages was associated with elevations in TNF and IL-6 and a hyperresponsiveness to LPS.

Our laboratory examined the inflammatory response in TPA-treated Tpl2$^{-/-}$ or wild type mouse skin (DeCicco-Skinner, *et al.*, 2011). We found significantly higher numbers of neutrophils in Tpl2$^{-/-}$ mice, achieving levels 35 times greater than control animals within 12 hours (DeCicco-Skinner *et al.*, 2011). Additionally, the skin of Tpl2$^{-/-}$ mice exhibited significant increases in the number of mast cells, $CD3^+$ cells and the pro-inflammatory cytokines IL-12 and IL-1 β.

The heightened inflammation found in the skin of Tpl2$^{-/-}$ mice may be initiated by aberrant inflammatory signaling pathways. Numerous inflammatory pathways are associated with Tpl2 activity including ERK, JNK, p38 and NF-κB. We used selective inhibitors specific to each of these pathways both *in vitro* and *in vivo* to assess differential responses in keratinocytes from Tpl2$^{-/-}$ or wildtype mice and found that a dysregulation in NF-κB activation and signaling is the primary defect in Tpl2$^{-/-}$ keratinocytes (DeCicco-Skinner, *et al.*, 2011). We confirmed that keratinocytes from Tpl2$^{-/-}$ mice have basal and TPA-treated NF-κB activity levels significantly greater than wildtype mice using an NF-κB reporter assay. Moreover, whereas treatment with an NF-κB inhibitor could block edema and neutrophil infiltration in wildtype mice, the inhibitor was unable to block the exaggerated NF-κB response in the Tpl2$^{-/-}$ mice. Additionally, using confocal microscopy we found high levels of active p65 NF-κB present in the nucleus of Tpl2$^{-/-}$ keratinocytes prior to TPA treatment that dramatically increased within 15-30 minutes of TPA treatment. The increase in active NF-κB may be explained by lower levels of inhibitory proteins that normally sequester NF-κB in the cytoplasm of cells. In this regard we found that wildtype mice had significantly higher protein levels of the NF-κB inhibitors IκB alpha and IκB beta compared to Tpl2$^{-/-}$ animals. Others have reported similar findings as our laboratory (Mielke *et al.*, 2009; López-Peláez *et al.*, 2011). Mielke reported that Tpl2$^{-/-}$ bone marrow-derived dendritic cells (BMDCs) displayed increased p65 NF-κB both basally and after stimulation with LPS when compared to wildtype controls (Mielke, *et al.*, 2009). Similarly, they found that increased IκB degradation may be partially responsible for the heightened NF-κB levels. Moreover, López-Peláez demonstrated that while degradation of the NF-κB inhibitor IκB alpha was similar between Tpl2$^{-/-}$ and wildtype BMDMs, Tpl2$^{-/-}$ cells exhibited a slower recovery in IκB alpha protein levels, remaining low at 4 hours post LPS treatment (López-Peláez *et al.*, 2011).

The data associating the absence of *Tpl2* with a heightened NF-κB activation, opposes what others have found. Previous reports have shown LPS stimulation leads to equal induction of NF-κB in Tpl2$^{-/-}$ and control macrophages (Dumitru *et al.*, 2000). Others have shown that Tpl2$^{-/-}$ mouse embryo fibroblasts and Jurkat T cells with kinase-deficient Tpl2 have a reduction in NF-κB activation but only when certain stimuli are administered (Lin *et al.*, 1999; Das *et al.*, 2005). Thus the relationship between Tpl2 and NF-κB still remains an enigma and more work is needed to clarify the nuances of this interaction.

It is possible that whether Tpl2 serves more of a pro-inflammatory or anti-inflammatory role may depend on cell type and stimuli. As previously mentioned, Tpl2 exhibits broad specificity because it interacts with numerous signaling pathways. Most of the previous reports that found Tpl2 to be proinflammatory used stimuli functioning through TLRs. It should be noted that the two stage skin carcinogenesis model used in our model employs TPA as a tumor promoter signaling through protein kinase C (PKC). PKC is a multigene family of serine/threonine kinases (Griner and Kazanietz, 2007). Six isoforms of PKC are expressed in mouse keratinocytes, namely, α,δ, ε, μ, η, ζ (Reddig *et al.*, 1999). Each isoform contributes differently to normal keratinocyte growth and differentiation, and, if inappropriately regulated, certain isoforms are associated with disease pathogenesis. It is currently unknown if one or more PKC isoforms are dysregulated in Tpl2$^{-/-}$ mice.

4.3 Tpl2$^{-/-}$ mice have alterations in T cell polarization

Tpl2 is reportedly activated in rodents by provirus integration in retrovirus-induced T cell lymphomas (Ceci *et al.*, 1997). Therefore, due to this connection with T cells, several laboratories have investigated if Tpl2$^{-/-}$ mice have alterations in T cell profiles in response to antigenic challenge. Watford, *et al.* reported that *Tpl2* was induced by IL-12 in human and mouse T cells in a Stat4-dependent manner (Watford *et al.*, 2008). Tpl2$^{-/-}$ mice have impaired host defense against *Toxoplasma gondii,* reduced parasite clearance, and decreased IFN-γ production. Additionally, ovalbumin-immunized Tpl2$^{-/-}$ mice challenged with ovalbumin developed more severe bronchoalveolar eosinophilic inflammation than wildtype controls (Watford *et al.*, 2010). Tpl2$^{-/-}$ mice were polarized toward a Th2 response, exhibiting heightened IL-4 and IL-10 signaling. In contrast, Sugimoto, *et al.* reported a significant increase in IL-12 and a decrease in CpG-DNA treated Tpl2$^{-/-}$ macrophages and dendritic cells, polarizing the mice toward a Th1 profile (Sugimoto *et al.*, 2004). Additionally, it has been reported that IL-23, a cytokine necessary for expansion and stabilization of Th17 cells, is reduced in Tpl2-/- macrophages indicating Tpl2 may be involved in balancing Th1/Th17 differentiation (Kakimoto, *et al.*, 2010).

5. Tpl2 and carcinogenesis

The exact role of Tpl2 in carcinogenesis is also convoluted. Early reports demonstrated *Tpl2* truncation of the C-terminus, resulting from provirus insertion, to be associated with T-cell lymphoma (Ceci *et al.*, 1997). Additionally, overexpressing the truncated form of this gene was associated with rodent T-cell lymphomas and an accelerated cell cycle in T lymphocytes.

Subsequently, several reports found elevated MAP3K8 activity in a number of human cancers including breast, endometrial, thymomas, lymphomas, lung, Hodgkin's disease, and nasopharyngeal carcinoma (Salmeron *et al.*, 1996; Sourvinos *et al.*, 1999; Patriotis *et al.*, 1993,; Tsatsanis *et al.*, 2000; Eliopoulos *et al.*, 2002; Clark *et al.*, 2004). In lymphomas the enhanced MAP3K8 activity could modulate ERK1/2 effector proteins such as p70 S6K and Jun B, resulting in an increased progression through the cell cycle (López-Peláez *et al.*, 2011; Fernández *et al.*, 2011). Recently it was reported that overexpression of Tpl2 induced androgen-dependent prostate cancer growth, and that MAP3K8 was upregulated in prostate cancers in mouse models and clinical specimens (Jeong *et al.*, 2011). Additionally, heightened *MAP3K8* expression levels are correlated with acquired resistance to drug therapy in melanoma (Johannessen *et al.*, 2010). The physiological role of Tpl2 in the etiology of human cancers remains a mystery. Overexpression of *Tpl2/MAP3K8 in vivo* is weakly oncogenic and *Tpl2/MAP3K8* mutations in human cancers are rarely found (Ceci *et al.*, 1997; Clark et al. 2004, Vougioukalaki *et al.*, 2011).

In contrast to the oncogenic role of *Tpl2,* recent evidence suggests that under certain conditions Tpl2 may serve a tumor suppressor role. Using Tpl2-/- mice crossed with the T cell receptor transgene, Tsatsanis found that 81% of mice developed T cell lymphomas within 6 months whereas lymphomas were absent from wildtype mice (Tsatsanis *et al.*, 2008). This heightened carcinogenesis was due to chronic stimulation of splenic T cells in Tpl2-/- mice, for anti-CD3/anti-CD28 stimulated Tpl2-/- splenocytes have increased proliferation and reduced expression of a negative regulator of T cell activation, CTLA4 .

Moreover, our laboratory recently reported that *Tpl2* can act as a tumor suppressor gene in chemically-induced skin cancer (DeCicco-Skinner *et al.*, 2011). In our model, Tpl2-/- or wildtype mice were subjected to a two-stage dimethylbenzanthracene/ 12-O-tetradecanoylphorbol-13-acetate (TPA) mouse skin carcinogenesis model. We found that Tpl2-/- mice developed a significantly higher incidence of tumors (80%) than wildtype mice (17%), as well as reduced tumor latency and a significantly higher number of total tumors (113 vs 6). Additionally, keratinocytes from Tpl2-/- mice were found to have an accelerated cell cycle as demonstrated *in vivo* through heightened Ki67 staining.

A skin grafting protocol has been used by our laboratory to further determine the *in vivo* phenotype of keratinocytes and fibroblasts from Tpl2-/- and wildtype mice. In agreement with our two-stage skin carcinogenesis model, we found that nude mice grafted with fibroblasts and v-ras^{Ha} -transduced keratinocytes from Tpl2$^{-/-}$ mice developed tumors nine times larger than mice grafted with fibroblasts and v-ras^{Ha} -transduced keratinocytes from wildtype mice. Additionally, mice grafted with mixed genotypes (WT fibroblasts/KO keratinocytes or KO fibroblasts/WT keratinocytes) developed tumors that were intermediate in size suggesting a defect in Tpl2 signaling in both fibroblasts and keratinocytes was responsible for the increased tumor formation.

Despite the fact that *Tpl2* can serve as an oncogene, and sometimes a tumor suppressor gene in mice and rats, the physiological response of *Tpl2* in the etiology of human cancers remains a mystery. Our laboratory has previously identified a *Tpl2* truncation mutant in a primary human lung adenocarcinoma, but analyses of other lung cancers suggest this isn't a common occurrence (Clark *et al.*, 2004). Additionally, others have searched the Sanger Institute COSMIC database for *Tpl2* gene mutations and identified only one human tumor sample (out of 477) that contained a point mutation in the *Tpl2* gene (Gantke *et al.*, 2011; Parsons *et al,* 2008). The significance of this point mutation in the development or progression of the patient's glioblastoma multiforme is unknown.

6. Tpl2 inhibitors

As described above, *Tpl2* can interact with several inflammatory pathways. It activates the MEK/ERK pathway and plays a pivotal role in the production and signaling of TNF. Tpl2 also regulates TNF-R signaling and the production of type 1 interferon and other pro-inflammatory cytokines such as IL-1β. In macrophages stimulated by LPS, Tpl2 activates the MAPK pathway. Overexpression of Tpl2 activates the MAPK signaling pathways, NFAT, and induces IL-2 production which further activates NF-κB via the NF-κB inducing kinase (NIK) (Patriotis *et al.*, 1994; Salmeron *et al.*, 1996; Tsatsanis *et al.*, 1998a and 1998b; Lin *et al,* 1999). Analysis of Tpl2$^{-/-}$ mice revealed a role for Tpl2 in TLR, IL-1, CD40, TCR, and also reduction of LPS-induced TNF-α production (Dumitru *et al.*, 2000; Tsatsanis *et al.*, 2008; Eliopoulos *et al.*, 2003). TNF-α, a pro-inflammatory cytokine, is involved in the initiation and progression of inflammatory diseases such as rheumatoid arthritis (RA), psoriasis, inflammatory bowel disease, and multiple sclerosis (Feldmann *et al.*, 2005). TNF-α also stimulates the production of other proinflammatory cytokines such as interleukin-1 (IL-1) and interleukin- 6 (IL-6), and induces the activity of enzymes such as the matrix metalloproteases (MMPs), all of which contribute to RA. Inhibition of TNF-α with a number of biological agents that bind and neutralize TNF-α has offered a significant therapeutic advantage in treating RA (Jarvis & Faulds, 1999,; Elliot *et al.*, 2008; Bang & Keating, 2004). The therapeutics included soluble TNF receptor and monoclonal antibodies for the neutralization of TNF-α, such as etanercept (soluble TNFRII-Fc; Enbrel), infliximab (Remicade), and adalimumab (anti-TNFα antibodies; Humira) and have improved the quality of life of RA patients.

The success of these drugs has validated the use of anti-TNF-α therapies for treating arthritic and inflammatory diseases. It has also necessitated the search for additional small molecule inhibitors with similar or related mechanisms of action. Major pharmaceutical companies have focused on the inhibition of TNF-α production through either inhibiting TNF-α converting enzyme or the disruption of signal transduction pathways that include several members of the MAPK family. However, the therapeutic agents described above require administration via injection or infusion. Therefore, identification of an orally available small molecule therapy would provide an additional benefit to patients. Since Tpl2 inhibition can decrease TNF-α levels, it has become an attractive target for treating RA and an orally available small molecule inhibiting TNF-α synthesis and/or signaling could have widespread therapeutic potential. Since Tpl2 ablation protects mice from TNF-α -induced inflammatory bowel disease and pancreatic and pulmonary inflammation, Tpl2 inhibition is an attractive strategy for treating these diseases.

Research studying the effect of Tpl2 inhibition on disease outcome is still in its infancy because only a handful of small molecule inhibitors of Tpl2 have been identified. A few factors contribute to the paucity of Tpl2 inhibitors. First, no crystal structure is available for Tpl2. Second, Tpl2 has low homology to other kinases. Although this is an advantage in selecting unique targets, it also presents significant challenges for structure-based drug design, especially when coupled with the lack of crystal structure. On the other hand, Tpl2 is not inhibited by staurosporine, a nonspecific kinase inhibitor, and it is the only known human kinase that has a proline instead of a conserved glycine in the glycine-rich ATP binding loop (Luciano *et al.,* 2004; Patriotis, *et al.,* 2003). These factors may lead to the discovery of highly selective inhibitors. Nonetheless, several strategies have been used to synthesize and/or isolate Tpl2 inhibitors.

One natural product inhibitor of Tpl2 has been identified. Luteolin [2-(3,4-dihydroxyphenyl)-5,7-dihydroxy-4-chromenone] is a flavonoid found abundantly in green peppers, perilla, celery, and chamomile tea. Luteolin has been shown to inhibit TNF-α-induced COX-2 expression in mouse skin epidermal cells by inhibiting Tpl2 activity (Kim *et al.,* 2011).

Several synthetic inhibitors of Tpl2 have also been characterized with 1,7-Naphthyridine-3-carbonitriles 1 and quinoline- 3-carbonitrile 2 as the earliest classes of compounds reported as reversible and ATP-competitive Tpl2 kinase inhibitors. These Tpl2 inhibitors demonstrate a good correlation between cell free Tpl2 inhibition with p-MEK and TNF inhibition in cell-based assays. Thieno [2,3-c]pyridines and pyrimidines are the other two classes of Tpl2 inhibitors studied recently. 1,7-napthyridine- 3-carbonitriles and quinoline-3-carbonitriles, indazoles, and thienopyridines are the most common Tpl2 inhibitors. These classes of compounds will be discussed below.

Gavrin et al., (2005) first described a series of 6-substituted-4-anilino-[1,7]-naphthyridine-3-carbonitriles as Tpl2 inhibitors evaluated in cell-free, cellular, and blood environments. One of these compounds, 1,7-naphtyridine- 3-carbonitrile, was used successfully to inhibit RANKL-induced osteogenesis through the induction of transcription factors c-Fos and NFATc1 in RANKL treated cells, unraveling the pivotal role played by Tpl2 in osteoclastogenesis (Hirata *et al.,* 2010). The same inhibitor subsequently demonstrated that LPS-induced TNF-α production via the Tpl2-MEK signaling is regulated by modulating the levels of tip-associated protein (Hirata *et al.,* 2010a). It appears that Tpl2 might be involved in not only the post-transcriptional but also the translational stages of TNF-α production. Some of the carbonitriles were further tested in primary human monocytes and blood. Inhibition of Tpl2 led to a decrease in production of TNF-α and other pro-inflammatory factors. In synoviocytes, the inhibitors blocked ERK activation, COX-2 expression, and the production of IL-6, IL-8, PGE_2, MMP-1 and MMP-3, further validating the role of these inhibitors as therapeutics (Hall *et al.,* 2007).

A series of quinoline-3-carbonitriles were synthesized and tested in cell-based phosphorylation assays in LPS-treated human monocytes. Compounds were identified that selectively inhibited pMEK formation and LPS/D-Gal-induced TNF-α release (Hu *et al.*, 2006). Another series of compound synthesis involved studying 8-chloro-4-(3-chloro-4-fluorophenylamino)-6-((1-(1-ethylpiperidin-4-yl)-1*H*-1,2,3-triazol-4-yl)methylamino)quinoline-3-carbonitriles that inhibited TNF-α production in an LPS-stimulated rat inflammation model. This was an important step in Tpl2 inhibitor identification since this compound was orally efficacious and demonstrated potential for oral treatment for RA (Wu *et al.*, 2009).

Structure based design efforts yielded a series of novel Indazole-based compounds. These compounds showed promising results as Tpl2 inhibitors as tested in LANCE based and cell-based pERK assays. Some of the top compounds inhibited pERK activity in monocytes and had an IC_{50} of 3.2μM. They exhibited good kinase selectivity against certain kinases but not against others. Efforts to improve their kinase selectivity profile are under way (Hu *et al.*, 2011).

Another novel series of 7-amino substituted thieno[2,3-*c*]pyridines was identified as Tpl2 inhibitors with an IC_{50} of 1μM. These compounds were studied for their Tpl2 inhibition and selectivity against kinases. They successfully and selectively inhibited the Tpl2 pathway following LPS stimulation in macrophages and BMDM cells in a dose-dependent manner and showed good selectivity over MEK and ERK (Cusac *et al.*, 2009). More recently, a high through-put screening of a compound library yielded thieno[3,2-d]pyrimidines as promising Tpl2 kinase inhibitors. They had an IC_{50} as low as 0.18 μM, exhibited good binding to the Tpl2 protein as shown by biacore-based competitive binding studies, and showed good selectivity against kinases (Ni *et al.*, 2011).

7. Conclusion

There has been an explosion of interest in Tpl2 inhibitors to treat RA and other inflammatory and autoimmune diseases. However, similar to the description above for the role of Tpl2 in carcinogenesis, it may be an oversimplification of the situation to assume that inhibition of Tpl2 would reduce inflammation. We have found thirty times higher neutrophil numbers in TPA-treated skin from Tpl2 knockout mice when compared to wildtype TPA-treated mouse skin. Additionally, we have found higher edema, inflammatory enzymes and cytokines in Tpl2 knockout mice (DeCicco-Skinner *et al.*, 2011). Moreover, in contrast to the description above in RA-FLS cells, we have identified significantly elevated COX-2, PGE_2 and PGE_2 receptor levels in Tpl2 knockout keratinocytes with the elevated COX-2 contributing to increased skin tumorigenesis (unpublished findings).

In summary, more research needs to be conducted to clarify the role that *Tpl2* plays in immune responses, carcinogenesis, and inflammatory conditions. We caution that *Tpl2* gene regulation needs to be better understood before inhibitors are used systemically. *Tpl2* appears to have divergent roles in different cells and tissues, and therefore drugs used to selectively inhibit *Tpl2* in the hopes of controlling inflammation may actually exacerbate inflammation and carcinogenesis in certain tissues such as skin.

8. Acknowledgements

This research was supported in part by the Intramural Research Program of the National Cancer Institute, National Institutes of Health and by Grant Number UA5CA152907 from the National Cancer Institute. The content is solely the responsibility of the authors and does not necessarily represent the official views of the National Cancer Institute or the National Institutes of Health.

9. References

Aoki, M.; Akiyama, T.; Miyoshi, J. & Toyoshima, K. (1991). Identification and characterization of protein products of the cot oncogene with serine kinase activity. *Oncogene*. Vol.6, 1515-1519.

Babu, GR.; Jin, W.; Norman,L.; Waterfield, M.; Chang, M.; Wu, X.; Zhang, M. & Sun, SC. (2006). Phosphorylation of NF-kappaB1/p105 by oncoprotein kinase Tpl2: implications for a novel mechanism of Tpl2 regulation. Biochim. Biophys. Acta. Vol. 1763, 174–181.

Bang, LM. & Keating, GM. (2004). Adalimumab: a review of its use in rheumatoid arthritis. *BioDrugs*. Vol. 18, 121-139.

Beinke, S.; Deka, J.; Lang, V.; Belich, MP.; Walker, PA.; Howell, S.; Smerdon, SJ.; Gamblin, SJ. & Ley, SC. (2003). NF-kappaB1 p105 negatively regulates TPL-2 MEK kinase activity. *Mol Cell Biol*. Vol. 23, 4739-4752.

Beinke, S.; Robinson, MJ.; Salmeron, A.; Hugunin, M.; Allen, H. & Ley, SC. (2004) Lipopolysaccharide activation of the TPL-2/MEK/extracellular signal-regulated kinase mitogen-activated protein kinase cascade is regulated by IκB kinase-induced proteolysis of NF-κB1 p105. *Mol Cell Biol*. Vol. 24, 9658–9667.

Belich, MP.; Salmerón, A.; Johnston, LH. & Ley, SC. (1999). TPL-2 kinase regulates the proteolysis of the NF-kappaB-inhibitory protein NF-kappaB1 p105. *Nature*. Vol.397, 363-368.

Cargnello, M. & Roux, PP. (2011). Activation and function of the MAPKs and their substrates, the MAPK-activated protein kinases. *Microbiol Mol Biol Rev. Vol.* 75, 50-83.

Ceci, JD.; Patriotis, CP.; Tsatsanis, C.; Makris, AM.; Kovatch, R.; Swing, DA.; Jenkins, NA.; Tsichlis, PN. & Copeland, NG. (1997). Tpl-2 is an oncogenic kinase that is activated by carboxy-terminal truncation. *Genes Dev*. Vol. 11, 688-700.

Chiariello, M.; Marinissen, MJ. & Gutkind, JS. (2000). Multiple mitogen-activated protein kinase signaling pathways connect the Cot oncoprotein to the c-jun promoter and to cellular transformation. *Mol Cell Biol*. Vol. 20, 1747-1758.

Clark, AM.; Reynolds, SH.; Anderson, M. & Wiest JS. (2004). Mutational activation of the MAP3K8 protooncogene in lung cancer. *Genes Chromosomes Cancer*. Vol. 41,99-108.

Das, S.; Cho, J.; Lambertz, I.; Kelliher, MA.; Eliopoulos, AG.; Du, K.& Tsichlis, PN. (2005). Tpl2/cot signals activate ERK, JNK, and NF-kappaB in a cell-type and stimulus-specific manner. *J. Biol Chem*. Vol. 280, 23748-23757.

DeCicco-Skinner, KL.; Trovato, EL.; Simmons, JK.; Lepage, PK. & Wiest JS. (2011). Loss of tumor progression locus 2 (tpl2) enhances tumorigenesis and inflammation in two-stage skin carcinogenesis. *Oncogene*. Vol. 30, 389-397.

Dhanasekaran, DN. & Johnson, GL. (2007). MAPKs: function, regulation, role in cancer and therapeutic targeting. *Oncogene*. Vol. 26, 3097-3099.

Dumitru, CD.; Ceci, JD.; Tsatsanis, C.; Kontoyiannis, D.; Stamatakis, K.; Lin, JH.; Patriotis, Jenkins, NA.; Copeland, NG.; Kollias, G. & Tsichlis, PN. (2000). TNF-alpha induction by LPS is regulated posttranscriptionally via a Tpl2/ERK-dependent pathway. *Cell*. Vol. 103, 1071-1083.

Eliopoulos, AG.; Dumitru,CD.; Wang, CC.; Cho, J. & Tsichlis, PN. (2002). Induction of COX-2 by LPS in macrophages is regulated by Tpl2-dependent CREB activation signals. EMBO J. Vol. 21, 4831–4840.

Eliopoulos, AG.; Wang, CC.; Dumitru, CD. & Tsichlis PN. (2003). Tpl2 transduces CD40 and TNF signals that activate ERK and regulates IgE induction by CD40. *EMBO J*. Vol. 22, 3855-3864.

Elliot, MJ.; Maini, RN.; Feldmann, M.; Long-Fox, A.; Charles, P.; Katasikis, P.; Brennan, FM.; Bijl, H.; Ghrayeb, J. & Woody, JN. (2008). Treatment of rheumatoid arthritis with chimeric monoclonal antibodies to tumor necrosis factor alpha. *Arthritis Rheum*. Vol. 58(2 Suppl), S92-S101.

Erny, KM.; Peli, J.; Lambert, JF.; Muller, V. & Diggelmann H. (1996). Involvement of the Tpl-2/cot oncogene in MMTV tumorigenesis. *Oncogene*. Vol. 13, 2015-2020.

Feldmann, M.; Brennan, FM.; Foxwell, BM.; Taylor, PC.; Williams, RO. & Maini RN. (2005). Anti-TNF therapy: where have we got to in 2005? *J Autoimmun*. Vol. 25 Suppl, 26-28.

Fernández,M.; Manso, R.; Bernáldez, F.; López, P.; Martín-Duce, A. & Alemany, S. (2011). Involvement of Cot activity in the proliferation of ALCL lymphoma cells. *Biochem Biophys Res Commun*. Vol. 411, 655-660.

Gandara, ML.; Lopez, P.; Hernando, R.; Castano, JG. & Alemany S. (2003). The COOH-terminal domain of wild-type Cot regulates its stability and kinase specific activity. *Mol Cell Biol*. Vol. 23, 7377-7390.

Gantke, T.; Sriskantharajah, S. & Ley SC. (2011). Regulation and function of TPL-2, an IκB kinase-regulated MAP kinase kinase kinase. *Cell Res*. Vol. 21, 131-145.

Gavrin, LK.; Green, N.; Hu, Y.; Janz, K.; Kaila, N.; Li, HQ.; Tam, SY.; Thomason, JR.; Gopalsamy, A.; Ciszewski, G.; Cuozzo, JW.; Hall, JP.; Hsu, S.; Telliez, JB. & Lin LL. (2005). Inhibition of Tpl2 kinase and TNF-alpha production with 1,7-naphthyridine-3-carbonitriles: synthesis and structure-activity relationships. *Bioorg Med Chem Lett*. Vol. 15, 5288-5292.

Griner, EM. & Kazanietz, MG. (2007). Protein kinase C and other diacylglycerol effectors in cancer. *Nat Rev Cancer*. Vol.7, 281-294.

Hall, JP.; Kurdi, Y.; Hsu, S.; Cuozzo, J.; Liu, J.; Telliez, JB.; Seidl, KJ.; Winkler, A.; Hu, Y.; Green, N.; Askew, GR.; Tam, S.; Clark, JD. & Lin LL. (2007). Pharmacologic inhibition of tpl2 blocks inflammatory responses in primary human monocytes, synoviocytes, and blood. *J. Biol. Chem*. Vol. 282, 33295–33304.

Hirata, K.; Taki, H.; Shinoda, K.; Hounoki, H.; Miyahara, T.; Tobe, K.; Ogawa, H.; Mori, H. & Sugiyama, E. (2010). Inhibition of tumor progression locus 2 protein kinase suppresses receptor activator of nuclear factor-kappaB ligand-induced osteoclastogenesis through down-regulation of the c-Fos and nuclear factor of activated T cells c1 genes. *Biol Pharm Bull*. Vol. 33, 133-137.

Hirata, K.; Miyashiro, M.; Ogawa, H.; Taki, H.; Tobe, K. & Sugita T. (2010). Inhibition of tumor progression locus 2 protein kinase decreases lipopolysaccharide-induced tumor necrosis factor alpha production due to the inhibition of the tip-associated protein induction in RAW264.7 cells. *Biol Pharm Bull.* Vol. 33, 1233-1237.

Hu, Y.; Green, N.; Gavrin, LK.; Janz, K.; Kaila, N.; Li, HQ.; Thomason, JR.; Cuozzo, JW.; Hall, JP.; Hsu, S.; Nickerson-Nutter, C.; Telliez, JB.; Lin, LL. & Tam S. (2006). Inhibition of Tpl2 kinase and TNFalpha production with quinoline-3-carbonitriles for the treatment of rheumatoid arthritis. *Bioorg Med Chem Lett.*Vol. 16, 6067-6072.

Hu, Y.; Cole, D.; Denny, RA.; Anderson, DR.; Ipek, M.; Ni, Y.; Wang, X.; Thaisrivongs, S.; Chamberlain, T.; Hall, JP.; Liu, J.; Luong, M.; Lin, LL.; Telliez, JB. & Gopalsamy A. (2011). Discovery of indazoles as inhibitors of Tpl2 kinase. *Bioorg Med Chem Lett.* Vol. 21, 4758-4761.

Jarvis B & Faulds D. (1999). Etanercept: a review of its use in rheumatoid arthritis. *Drugs.* Vol. 57, 945-966.

Jeong, JH.; Bhatia, A.; Toth, Z.; Oh, S.; Inn, KS.; Liao, CP.; Roy-Burman, P.; Melamed, J.; Coetzee, GA. & Jung JU. (2011). TPL2/COT/MAP3K8 (TPL2) activation promotes androgen depletion-independent (ADI) prostate cancer growth. *PLoS One.* Vol. 6, e16205.

Jia, Y.; Quinn, CM.; Bump, NJ.; Clark, KM.; Clabbers, A.; Hardman, J.; Gagnon, A.; Kamens, J.; Tomlinson, MJ.; Wishart, N. & Allen, H. (2005). Purification and kinetic characterization of recombinant human mitogen-activated protein kinase kinase kinase COT and the complexes with its cellular partner NF-κB1 p105. *Arch Biochem Biophys.* Vol. 441, 64–74.

Johannessen, CM.; Boehm, JS.; Kim, SY.; Thomas, SR.; Wardwell, L.; Johnson, LA.; Emery, CM; Stransky, N.; Cogdill, AP.; Barretina, J.; Caponigro, G.; Hieronymus, H.; Murray, RR.; Salehi-Ashtiani, K.; Hill, DE.; Vidal, M.; Zhao, JJ.; Yang, X.; Alkan, O.; Kim, S.; Harris, JL.; Wilson, CJ.; Myer, VE.; Finan, PM.; Root, DE.; Roberts, TM.; Golub, T.; Flaherty, KT.; Dummer, R.; Weber, BL.; Sellers, WR.; Schlegel, R.; Wargo, JA.; Hahn, WC. & Garraway LA. (2010). COT drives resistance to RAF inhibition through MAP kinase pathway reactivation. *Nature.* Vol. 468, 968-972.

Kakimoto, K.; Musikacharoen, T.; Chiba, N.; Bandow, K.; Ohnishi, T. & Matsuguchi T. (2010). Cot/Tpl2 regulates IL-23 p19 expression in LPS-stimulated macrophages through ERK activation. *J Physiol Biochem.* Vol. 66, 47-53.

Kim, JE.; Son, JE.; Jang, YJ.; Lee, DE.; Kang, NJ.; Jung, SK.; Heo, YS.; Lee, KW. & Lee, HJ. (2011). Luteolin, a novel natural inhibitor of tumor progression locus 2 serine/threonine kinase, inhibits tumor necrosis factor-alpha-induced cyclooxygenase-2 expression in JB6 mouse epidermis cells. *J Pharmacol Exp Ther.* Vol. 338, 1013-1022.

Lang, V.; Symons. A.; Watton ,SJ.; Janzen, J.; Soneji, Y.; Beinke, S.; Howell, S. & Ley SC. (2004). ABIN-2 forms a ternary complex with TPL-2 and NF-kappa B1 p105 and is essential for TPL-2 protein stability. *Mol Cell Biol.* Vol. 24, 5235-5248.

Lin, X.; Cunningham, ET Jr.; Mu, Y.; Geleziunas, R. & Greene WC. (1999). The proto-oncogene Cot kinase participates in CD3/CD28 induction of NF-kappaB acting through the NF-kappaB-inducing kinase and IkappaB kinases. *Immunity.* Vol. 10, 271-280.

López-Peláez, M.; Soria-Castro, I.; Boscá, L.; Fernández, M. & Alemany, S. (2011). Cot/tpl2 activity is required for TLR-induced activation of the Akt p70 S6k pathway in macrophages: Implications for NO synthase 2 expression. *Eur J Immunol.* Vol. 41, 1733-1741.

Luciano, BS.; Hsu, S.; Channavajhala, PL.; Lin, LL. & Cuozzo, JW. (2004). Phosphorylation of threonine 290 in the activation loop of Tpl2/Cot is necessary but not sufficient for kinase activity. *J Biol Chem.* Vol. 279, 52117-52123.

Matsumura, I.; Mizuki, M.; Kanakura, Y. (2008). Roles for deregulated receptor tyrosine kinases and their downstream signaling molecules in hematologic malignancies. *Cancer Sci.* Vol. 99, 479-485.

Mielke, LA.; Elkins, KL.; Wei, L.; Starr, R.; Tsichlis, PN.; O'Shea, JJ. & Watford, WT. (2009). Tumor progression locus 2 (Map3k8) is critical for host defense against Listeria monocytogenes and IL-1 beta production. *J Immunol.* Vol. 183, 7984-7993.

Miyoshi, J.; Higashi, T.; Mukai, H.; Ohuchi, T. & Kakunaga T. (1991). Structure and transforming potential of the human cot oncogene encoding a putative protein kinase. *Mol Cell Biol.* Vol. 11, 4088-4096.

Ni, Y.; Gopalsamy, A.; Cole, D.; Hu, Y.; Denny, R.; Ipek, M.; Liu, J.; Lee, J.; Hall, JP.; Luong, M.; Telliez, JB. & Lin LL. (2011). Identification and SAR of a new series of thieno[3,2-d]pyrimidines as Tpl2 kinase inhibitors. *Bioorg Med Chem Lett.* Vol. 21, 5952-5956.

Nishimoto, S.; Nishida, E. (2006). MAPK signalling: ERK5 versus ERK 1/2. *EMBO Rep.* Vol. 7, 782-6.

Parsons, DW.; Jones, S.; Zhang, X.; Lin, JC.; Leary, RJ.; Angenendt, P.; Mankoo, P.; Carter, H.; Siu, IM.; Gallia, GL.; Olivi, A.; McLendon, R.; Rasheed, BA.; Keir, S.; Nikolskaya, T.; Nikolsky, Y.; Busam, DA.; Tekleab, H.; Diaz, LA Jr.; Hartigan, J.; Smith, DR.; Strausberg, RL.; Marie, SK.; Shinjo, SM.; Yan, H.; Riggins, GJ.; Bigner, DD.; Karchin, R.; Papadopoulos, N.; Parmigiani. G., Vogelstein, B.; Velculescu, VE. & Kinzler, KW. (2008). An integrated genomic analysis of human glioblastoma multiforme. *Science.* Vol. 321,1807-1812.

Patriotis, C.; Makris, A.; Bear, SE. & Tsichlis PN. (1993). Tumor progression locus 2 (Tpl-2) encodes a protein kinase involved in the progression of rodent T-cell lymphomas and in T-cell activation. *Proc Natl Acad Sci U S A* Vol. 90, 2251-2255.

Patriotis, C.; Makris, A.; Chernoff, J. & Tsichlis PN. (1994). Tpl-2 acts in concert with Ras and Raf-1 to activate mitogen-activated protein kinase. *Proc Natl Acad Sci U S A.* Vol. 91, 9755-9759.

Perfield, JW 2nd.; Lee, Y.; Shulman, GI.; Samuel, VT.; Jurczak, MJ.; Chang, E.; Xie, C.; Tsichlis, PN.; Obin, MS. & Greenberg, AS. (2011). Tumor progression locus 2 (TPL2) regulates obesity-associated inflammation and insulin resistance. *Diabetes.* Vol. 60, 1168-1176.

Reddig, PJ.; Dreckschmidt, NE.; Ahrens, H.; Simsiman, R.; Tseng, CP.; Zou, J.; Oberley, TD. & Verma, AK. (1999). Transgenic mice overexpressing protein kinase Cdelta in the epidermis are resistant to skin tumor promotion by 12-O-tetradecanoylphorbol-13-acetate. *Cancer Res.* Vol. 59, 5710-5718.

Rice, NR. ; MacKichan, ML. & Israel, A. (1992). The precursor of NF-κκB p50 has IκκB-like functions. *Cell.* Vol.71, 243-253.

Salmeron, A.; Ahmad, TB.; Carlile, GW.; Pappin, D.; Narsimhan, RP. & Ley SC. (1996). Activation of MEK-1 and SEK-1 by Tpl-2 proto-oncoprotein, a novel MAP kinase kinase kinase. *EMBO J.* Vol. 15, 817-826.

Schindler, JF.; Monahan, JB.; Smith, WG. (2007). P38 pathway kinases as anti-inflammatory drug targets. *J. Dent Res.* Vol. 86, 800-11. Sourvinos, G.; Tsatsanis, C. & Spandidos, DA. (1999). Overexpression of the Tpl-2/Cot oncogene in human breast cancer. *Oncogene*, Vol. 18, 4968-4973.

Sugimoto, K.; Ohata, M.; Miyoshi, J.; Ishizaki, H.; Tsuboi N.;, Masuda ,A.; Yoshikai, Y.; Takamoto, M.; Sugane, K.; Matsuo, S.; Shimada, Y. & Matsuguchi T. (2004). A serine/threonine kinase, Cot/Tpl2, modulates bacterial DNA-induced IL-12 production and Th cell differentiation. *J Clin Invest.*,Vol. 114, 857-866.

Tsatsanis, C.; Patriotis, C.; Bear, SE. & Tsichlis, PN. (1998a). The TPL-2 proto-oncoprotein activates the nuclear factor of activated T cells and induces interleukin 2 expression in T cell lines. *Proc Natl Acad Sci USA,* Vol. 95, 3827–3832.

Tsatsanis, C.; Patriotis, C. & Tsichlis PN. (1998b). Tpl-2 induces IL-2 expression in T-cell lines by triggering multiple signaling pathways that activate NFAT and NF-kappaB, *Oncogene* Vol. 17, 2609-2618.

Tsatsanis, C. & Spandidos, DA. (2000). The role of oncogenic kinases in human cancer. *Int J. Mol Med,* Vol. 5, 583-90.

Tsatsanis, C.; Vaporidi, K.; Zacharioudaki, V.; Androulidaki, A.; Sykulev, Y.; Margioris, AN. & Tsichlis PN. (2008). Tpl2 and ERK transduce antiproliferative T cell receptor signals and inhibit transformation of chronically stimulated T cells. *Proc Natl Acad Sci USA.,* Vol. 105, 2987-2992.

Vougioukalaki, M.; Kanellis, DC.; Gkouskou, K. & Eliopoulos, AG. (2011). Tpl2 kinase signal transduction in inflammation and cancer. *Cancer Lett.* Vol. 304, 80-89.

Wagner, EF; Nebreda, AR. (2009). Signal integration by JNK and p38 MAPK pathways in cancer development. *Nat Rev Cancer.* Vol. 9, 537-49.

Waterfield, MR.; Zhang, M.; Norman, LP. & Sun SC. (2003). NF-kappaB1/p105 regulates lipopolysaccharide-stimulated MAP kinase signaling by governing the stability and function of the Tpl2 kinase. *Mol Cell.* Vol. 11, 685-694.

Watford, WT.; Hissong, BD.; Durant, LR.; Yamane, H.; Muul, LM.; Kanno, Y.; Tato, CM.; Ramos, HL.; Berger, AE.; Mielke, L.; Pesu, M.; Solomon, B.; Frucht, DM.; Paul, WE.; Sher, A.; Jankovic, D.; Tsichlis, PN. & O'Shea JJ. (2008). Tpl2 kinase regulates T cell interferon-gamma production and host resistance to Toxoplasma gondii. *J. Exp Med* Vol. 205, 2803-2812.

Watford, WT.; Wang, CC.; Tsatsanis, C.; Mielke, LA,; Eliopoulos, AG.; Daskalakis, C.; Charles, N.; Odom, S.; Rivera, J. O'Shea, J. & Tsichlis, PN. (2010). Ablation of tumor progression locus 2 promotes a type 2 Th cell response in Ovalbumin-immunized mice. *J Immunol.* Vol. 184, 105-113.

Weston, CR.; Davis, RJ. (2007). The JNK signal transduction pathway. *Curr Opin Cell Biol.* Vol. 19, 142-9.

Widmann, C.; Gibson, S.; Jarpe, MB.; & Johnson GL. (1999). Mitogen-activated protein kinase: conservation of a three-kinase module from yeast to human. *Physiol Rev.* Vol. 79, 143-180.

Wu, J.; Green, N.; Hotchandani, R.; Hu, Y.; Condon, J.; Huang, A.; Kaila, N.; Li, HQ.; Guler, S.; Li, W.; Tam, SY.; Wang, Q.; Pelker, J.; Marusic, S.; Hsu, S.; Perry Hall, J.;

Telliez, JB.; Cui, J. & Lin LL. (2009). Selective inhibitors of tumor progression loci-2 (Tpl2) kinase with potent inhibition of TNF-alpha production in human whole blood. *Bioorg Med Chem Lett.* Vol. 19, 3485-3488.

Zacharioudaki, V.; Androulidaki, A.; Arranz, A.; Vrentzos, G.; Margioris, AN. & Tsatsanis, C. (2009). Adiponectin promotes endotoxin tolerance in macrophages by inducing IRAK-M expression. *J Immunol.* 182, 6444- 6451.

4

MEK1/2 Inhibitors to Treat Dilated Cardiomyopathy Caused by *LMNA* Mutations

Antoine Muchir[1,2]
[1]*Department of Medicine,*
[2]*Department of Pathology and Cell Biology,*
College of Physicians and Surgeons, Columbia University, New York, NY
USA

1. Introduction

Inherited dilated cardiomyopathies are a major cause of heart disease in human, often with an onset in adolescence or early adult life. Despite technological advances that foster early diagnosis and alleviation of some symptoms, inherited dilated cardiomyopathies remains a critical unsolved problem for public health. The recent years provided some clues to explain the pathogenesis of inherited dilated cardiomyopathies, which might open new and encouraging perspectives for clinical trials.

2. *LMNA* dilated cardiomyopathy

Cardiomyopathy is an anatomic and pathologic condition associated with muscle dysfunction of the heart. Dilated cardiomyopathy, the most common form, is characterized by an increase in both myocardial mass and volume, which compromises cardiac contractility and ultimately results in reduced left ventricular function (Luk et al. 2008). Dilated cardiomyopathy is the third leading cause of heart failure in the United States behind coronary artery disease and hypertension. Genetically inherited forms of dilated cardiomyopathy have been identified in 30% of patients presenting with this disease (Michels et al., 1992). Many other acquired conditions may result in an identical clinical presentation and pathological function, which include alcohol-induced cardiomyopathy, hypertension, chronic anemia, ischemic cardiomyopathy, valvular diseases and viral myocarditis (Maron et al., 2006).

Inherited dilated cardiomyopathies are caused by mutations in genes that encode components of a wide variety of cellular components and pathways, including the nuclear envelope, contractile apparatus and the force transduction apparatus (Morita et al., 2005). The generation of contractile force by the sarcomere and its transmission to the extracellular matrix are the fundamental functions of cardiac cells. Inadequate performance in either components of this structural cellular network leads to cardiac remodeling and ultimately to dilated cardiomyopathy. Defects in generating force are typically due to a loss of integrity of the sarcomere unit. Mutations in the loci coding for β-cardiac myosin heavy chain, actin (Olson et al., 1998) and cardiac troponin T (Li et al., 2001) have been identified to disrupt force generation (Kamisago et al, 2000). Defects in the force transmission occur when there is impairment in the propagation of force from the sarcomere to the sarcolemma (Schonberger et al. 2001). Mutations in α-tropomyosin (Olson et al. 2001), dystrophin (Muntoni et al., 1993), desmin (Li et al 1999) and δ-sarcoglycan (Tsubata et al. 2000) have been identified to disrupt force transmission.

Among the causing genes, *LMNA* mutations encoding proteins of the inner nuclear membrane, have also been found associated to dilated cardiomyopathy (Fatkin et al. 1999). This finding raises the possibility that the nuclear envelope may play an important function as mechanosensor in cardiomyocyte (Nikolova et al. 2004, Lammerding et al. 2004). *LMNA* mutations appear to be responsible for approximately 8% of cases of inherited cardiomyopathy (Taylor et al., 2003), which strongly suggest that *LMNA* may be the most prevalent dilated cardiomyopathy gene. *LMNA* dilated cardiomyopathy is characterized by cardiac dilatation and impaired systolic function. In addition, affected patients exhibit early conduction defects before the left ventricular dysfunction and dilatation stages. The onset of symptoms in *LMNA* dilated cardiomyopathy is variable, ranging from the first to sixth decade of life and occurring most frequently in the third decade (mean age = 38 years) (Ben Yaou et al., 2006). There are high rates of life-threatening arrhythmias (abnormal electrical conduction), gradually worsening and leading to sudden death (Sanna et al. 2003). *LMNA* dilated cardiomyopathy has a more aggressive course than other inherited dilated cardiomyopathies. While sudden death from arrhythmias may be prevented by implantation of a pacemaker and/or implantable defibrillator, the progressive heart failure eventually becomes resistant to treatment (Golzio et al. 2007, van Berlo et al. 2005, Meune et al., 2006). No drugs are curative and heart transplantation is frequently necessary.

3. A-type nuclear lamins

LMNA, located on human chromosome 1q21.2-21.3, encodes A-type lamins. Lamin A and lamin C are the major A-type lamins expressed in somatic cells. They arise via alternative splicing of pre-mRNA encoded by exon 10 (Lin & Worman 1993). Lamin A is synthesized as a precursor, prelamin A, which has a unique C-terminal amino acid tail that triggers a series of enzymatic reactions to yield lamin A. Two other genes in the mammalian genome, *LMNB1* and *LMNB2*, respectively encode lamins B1 and B2. Lamins A and C are widely expressed in most differentiated somatic cells but lacking from early embryos and some undifferentiated cells whereas lamins B1 and B2 are expressed in all or most somatic cells. However, there are little data and no systemic studies on the differences in the relative amounts of lamins A, C, B1 and B2 expression. Lamins are intermediate filaments proteins that polymerize to form the nuclear lamina, a fibrous meshwork underlining the inner nuclear membrane of most eukaryotic cells (Fisher et al. 1986, McKeon et al. 1986, Aebi et al. 1986). The nuclear lamina is attached to the inner nuclear membrane via interactions with integral proteins and to the chromatin. More recently, it has been demonstrated that lamin A/C also have interactions with the cytoskeleton, through a multi-protein complex called "LINC" (LInker of Nucleoskeleton and Cytoskeleton) (Stewart et al. 2007). One function of the lamina is to provide structural support to the nucleus. Nuclear lamins have also been implicated in processes such as chromatin organization, gene regulation, DNA replication and RNA splicing (Dechat et al. 2008). However, the specific mechanistic roles of lamins in these processes, particularly in a cell or tissue type-specific context, remain obscure.

4. Pathogenesis of *LMNA* dilated cardiomyopathy

The pathogenesis of *LMNA* dilated cardiomyopathy remains a puzzle in medical genetics. Mouse models have been extremely helpful in deciphering critical mechanisms, which could partially explain the pathogenesis of the disease as well as for proposing potential innovative pharmacological therapies. Using a murine model of *LMNA* dilated cardiomyopathy, we recently brought some insights into the molecular pathogenesis of this disease, which have paved the way to potential therapies. To approach the issue of understanding the pathogenesis of *LMNA* dilated cardiomyopathy, we studied the transcriptome from hearts of *Lmna* H222P mice (a mouse model of *LMNA* dilated cardiomyopathy), using the Affymetrix® array technology. Male *Lmna* H222P mice develop cardiac chamber dilation, decreased left ventricle ejection fraction and hypokinesis detectable by echocardiography at 8 to 10 weeks of age (Arimura et al. 2005). To avoid interference caused by fibrotic cells and nonspecific tissue damage in hearts from older *Lmna* H222P mice, we initially analyzed samples from mice at 10 weeks of age where there were no detectable cardiac histological abnormalities. We analyzed gene ontology terms applied to genes, to identify functional classes of genes differentially expressed in hearts of *Lmna* H222P mice compared with those expressed in controls. Analysis using functional class scoring improves sensitivity by statistically evaluating genes in biologically meaningful groups. Genes encoding proteins in mitogen-activated protein kinase (MAPK) signaling pathway demonstrated significantly altered expression in hearts of *Lmna* H222P mice (Muchir et al. 2007) (Figure 1). Because enhanced activity of the Extracellular signal-regulated kinase1/2 (ERK1/2), a branch of MAPK signaling pathways, has been formerly shown to be causing cardiomyopathy, we focused subsequent experiments on analyzing this signaling in tissues form *Lmna* H222P mice and in cultured cells. We then demonstrated an aberrant activation of ERK1/2 signaling in hearts from *Lmna* H222P mice, as early as 4 weeks of age (Muchir et al. 2007). Our work proved that the activation of ERK1/2 signaling pathway preceded the cardiac dysfunction of *Lmna* H222P mice and that it is a consequence of alterations in A-type lamins and not secondary to non-specific affects.

5. MEK-ERK signaling pathway

MAPK signaling pathways are major information highways from extracellular mitogens, growth factors and cytokines at the cell surface to the nucleus to control gene expression (Davis, 1993). These signaling pathways control complex cellular programs, such as embryogenesis, differentiation, proliferation and cell death, in addition to short-term changes required for mechanical stress response and acute hormonal responses. The output of these pathways is transduced via MAPK family members that phosphorylate and regulate a wide array of substrates including transcription factors, cytoskeletal elements and other protein kinases (Seger & Krebs, 1995). Stimulation of many receptor classes can activate ERK1/2 including receptors with intrinsic tyrosine kinase activity, cytokine receptors and G-protein-coupled receptors (GPCR), including those coupling via G-proteins of the $G_{q/11}$, $G_{i/o}$ and G_s family. In the heart, ERK1/2 stimulation has been shown by fibroblast growth factor, insulin-like growth factor-1, estrogen, neuregulin-1, atrial natriuretic peptide, α1- and β-adrenoceptor agonists. Moreover, cardiac ERK1/2 can be activated independently than GPCR receptors, via mechanical stress, osmotic shock.

The MAPKs are activated by protein kinase cascades comprising at least three enzymes acting in series. ERK are activated directly by ERK kinases (MEK), which are dual specificity protein kinases that generally recognize only certain MAPKs as substrates. MEK are activated by MEK kinases, a structurally diverse group of kinases with less predictable specificities. MEK1/2, which activate ERK1/2, have very narrow substrate specificity. It is assumed, from lack of evidence to the contrary, that ERK1/2 are the only substrates of MEK1/2. Activated ERK1/2 kinases phosphorylate and activate a variety of substrates. All these substrates can be categorized into several groups including: transcription factors (Atf2, Elk1, c-Fos…), protein kinases and phosphatases (FAK1, MLCK, PAK1,…), cytoskeletal and scaffold proteins (dystrophin, Tau, Synaptin…), receptors and signaling molecules (EGFR, PLCg,…) and apoptosis-related proteins (Bad, Calpain, caspase 9, …). Some of the substrates can be found in the cytosol (paxilin, calnexin…), in agreement with the role of ERK1/2 in the regulation of both cytosolic and nuclear processes

6. Pharmacological therapy

Because we found abnormal activation of ERK1/2 signaling pathway in hearts of *Lmna* H222P mice, we hypothesized that pharmacological inhibition of this signaling pathway would prevent the cardiac deterioration. We treated *Lmna* H222P mice with PD098059, a tool compound that inhibits MEK1/2. We administered PD098059 or placebo (dimethylsulfoxide; DMSO) (daily, intraperitoneal injection) to *Lmna* H222P mice. We first treated male *Lmna* H222P mice starting at 8 weeks of age, prior to the onset of clinically detectable cardiac abnormalities, and analyzed them at 16 weeks (Muchir et al. 2009). Pathological dilatation of the cardiac left ventricle is often associated with fibrosis, and reactivation of a fetal gene expression program characterized by increased levels of atrial natriuretic peptide, brain natriuretic peptide, and β-myosin light and heavy chains. Accordingly, in hearts from untreated *Lmna* H222P mice and those treated with vehicle (DMSO), expression of mRNAs encoding natriuretic peptide precursors as well as mRNAs encoding myosin light chains were significantly increased. We showed that after treatment with PD098059, the cardiac expression of these mRNAs was significantly lowered compared to vehicle-treated *Lmna* H222P mice (Muchir et al. 2009). Similarly, we also demonstrated that *Lmna* H222P mice treated with PD98059 had a lower degree of cardiac fibrosis than the *Lmna* H222P mice treated with the vehicle. After 8 weeks of treatment with DMSO or PD98059 *Lmna* H222P mice were anesthetized and the cardiac dimensions and function measured by cardiac ultrasound. M-mode transthoracic echocardiography showed increased left ventricle end-diastolic diameter and left ventricle end-systolic diameter in *Lmna* H222P mice treated with DMSO compared with control mice. *Lmna* H222P mice treated with PD98059 had significantly smaller left ventricle end-systolic diameters compared to the DMSO-treated mice (Table 1). Cardiac fractional shortening and ejection fraction were reduced in *Lmna* H222P mice compared to control mice but increased in the *Lmna* H222P mice treated with PD98059.

As treatment of cardiomyopathy in human subjects may more likely be administered after the onset of symptoms or detectable cardiac abnormalities, we next treated mice with PD98059 starting at 16 weeks of age, when male *Lmna* H222P mice have left ventricular dilatation and an ejection fraction approximately 70 percent that of wild type mice, and analyzed the mice at 20 weeks (Wu et al. 2011). Treatment with PD98059 prevented left ventricular end-systolic dilatation, increased ejection fraction (Table 1), blocked increased cardiac expression of RNAs encoding natriuretic peptide precursors and reversed the induction of elements of the "fetal gene program" compared to placebo-treated mice. As significant cardiac fibrosis occurs in end-stage dilated cardiomyopathy, particularly *LMNA* dilated cardiomyopathy, we also examined cardiac fibrosis after treatment. *Lmna* H222P mice treated with PD98059 had a lower degree of cardiac fibrosis than the *Lmna* H222P mice treated with placebo. Overall, this work showed that inhibiting ERK1/2 signaling had positive effects on cardiac biochemistry and physiology in a mouse model of *LMNA* dilated cardiomyopathy, (Figure 1).

	8-16 weeks				*16-20 weeks*			
Genotype (Treatment)	n	LVEDD (mm)	LVESD (mm)	EF (%)	n	LVEDD (mm)	LVESD (mm)	EF (%)
Lmna$^{+/+}$	13	3.3 ± 0.1	2.0 ± 0.1	76.8 ± 2.0	12	3.5 ± 0.1	2.1 ± 0.1	73.2 ± 1.2
Lmna$^{H222P/H222P}$ (DMSO)	15	3.6 ± 0.1 *	2.7 ± 0.1***	56.9 ± 2.9 ***	12	4.4 ± 0.1 *	3.5 ± 0.1 ***	42.6 ± 3.6 ***
Lmna$^{H222P/H222P}$ (PD98059)	7	3.1 ± 0.2	1.8 ± 0.2 ##	73.5 ± 4.7 ##	19	3.6 ± 0.1	2.4 ± 0.1 ###	65.5 ± 2.6 ##

LVEDD, left ventricular end-diastolic diameter; LVESD, left ventricular end-systolic diameter; EF, ejection fraction.
Values are means ± standard errors.
Comparison between DMSO-treated *Lmna*$^{H222P/H222P}$ and *Lmna*$^{+/+}$ mice was performed using Student unpaired t-test, *P<0.05, ***P<0.0005.
Comparison between PD98059-treated *Lmna*$^{H222P/H222P}$ and DMSO-treated *Lmna*$^{H222P/H222P}$ mice was performed using Student unpaired t-test, ##P<0.005, ###P<0.0005.

Table 1. Echocardiographic data for *Lmna*$^{+/+}$ mice and *Lmna*$^{H222P/H222P}$ mice treated with vehicle (DMSO) or MEK1/2 inhibitor (PD98059) between 8-16 weeks of age and 16-20 weeks of age.

7. MEK1/2 inhibitors

In the field of target identification there has been a great deal of enthusiasm for identifying novel drug targets based on knowledge of key signal transduction components and their link to human disease. As signaling disorders represent a major cause for the pathological states and as most of the recently validated target molecules of drug research are signal transduction kinases, signal transduction therapy has become one of the most important areas of drug research (Keri et al. 2006, Levitzki 1996). Approximately 25% of the druggable genome consists of kinases involved in signal transduction. However, only a handful of kinases inhibitors are being used in clinical practice (Margutti & Laufer 2007). This remains then a wide perspective for drug discovery. The common feature conserved throughout the entire protein kinase family is the catalytic domain. The chemical activity of a kinase involves removing a phosphate group from ATP and covalently attaching it to a free hydroxyl group. Most kinases act on both serine and threonine, others act on tyrosine, and a number act on all three (dual-specificity kinases), like MEK1/2. The fact that kinases share a highly homologous catalytic domain, and the common co-substrate ATP, initially led to the assumption that protein kinases constitute a non-druggable family of protein kinases. A

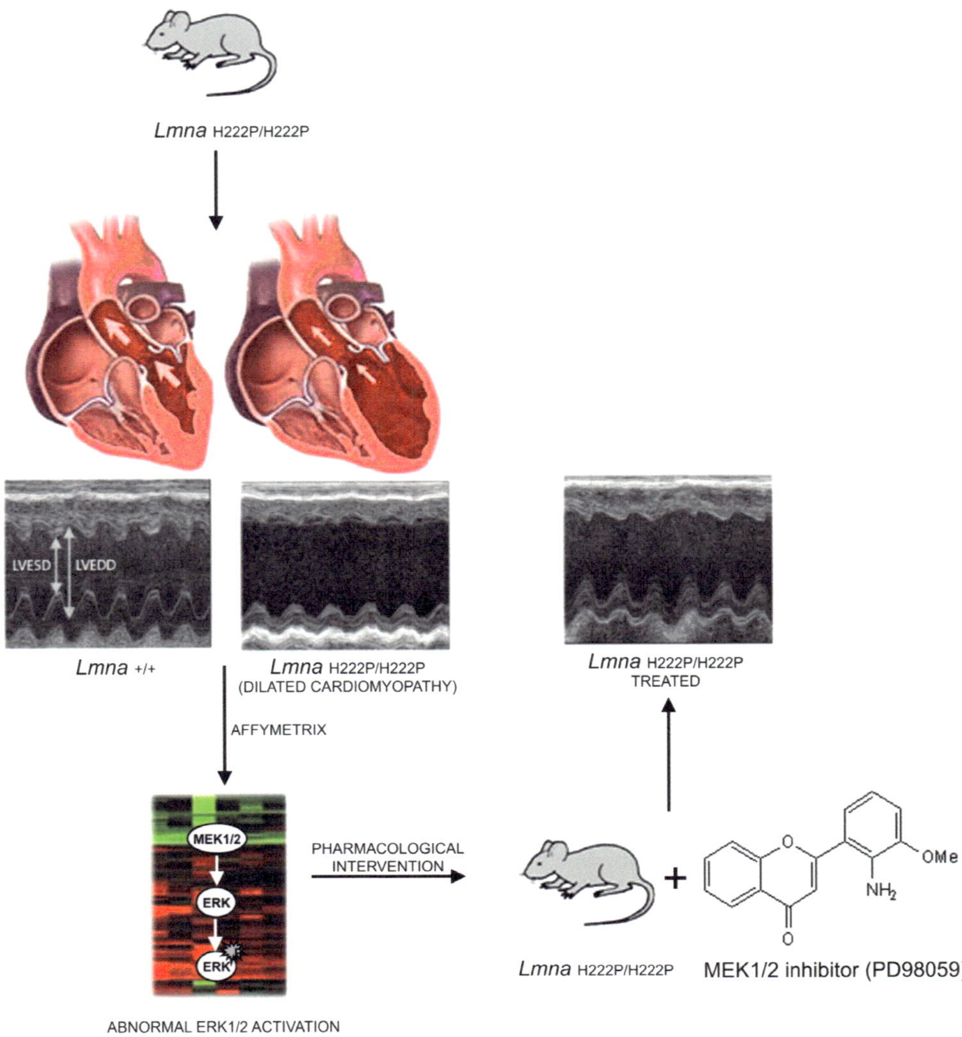

Fig. 1. Study from *Lmna*$^{H222P/H222P}$ mice suggests that activation of ERK1/2 underlies the development of *LMNA* dilated cardiomyopathy. *Lmna*$^{H222P/H222P}$ mice develop cardiac chamber dilation and decreased left ventricle ejection fraction. Affymetrix approach showed that ERK1/2 signaling pathways is abnormally activated in the heart of *Lmna*$^{H222P/H222P}$ mice compared to *Lmna*$^{+/+}$ mice, before any detectable sign of cardiac deterioration. Pharmacological intervention using PD98059, an inhibitor of MEK1/2, the kinase that activate ERK1/2, improves the cardiac function of *Lmna*$^{H222P/H222P}$ mice.

type of inhibitor, allosteric (non-ATP competitive), could potentially solve the selectivity issues related to protein kinase inhibition. PD098059 (Dudley et al. 1995) and U0126 (Favata et al. 1998) were among the first specific MEK1/2 allosteric inhibitors described. Although they have been extremely useful in the in vitro study of MAPK signaling, they have not been pursued in clinical development because of poor pharmacologic characteristics.

Several allosteric MEK1/2 inhibitors are in clinical and pre-clinical development. Some of the MEK1/2 inhibitors that have been used in human subjects include:

CI-1040 (PD184352), a benzhydroxamate from Pfizer, was the first small-molecule MEK1/2 inhibitor that proceeded to clinical testing. It was developed based on compounds and structures identified during the screening that led to the identification of PD098059, but had improved potency and selectivity (Barrett et al. 2008). Cl-1040 is an oral MEK1/2 inhibitor with promising pre-clinical activity that led to its clinical development (Sebolt-Leopold et al. 1999). It underwent phase I testing in 77 patients with advanced solids tumors (LoRusso et al. 2005). CI-1040 was well tolerated with no grade IV toxicities and only a limited number of grade III toxicities. The majorities of toxicities (98%) were grade I/II and included diarrhea (43%), fatigue (30%), rash (18%) and nausea (16%). Antitumor activity was seen in 1 pancreatic cancer patient who had a partial response lasting 12 months. Nineteen (25%) subjects had stable disease for 3 months, and this observation was commonly associated with symptomatic benefit. On the basis of these results, a multicenter, parallel arm phase II study of CI-1040 was performed in patients with advanced breast, colon, pancreatic, and non-small cell lung cancer. CI-1040 was relatively well tolerated, with 19% experiencing grade III toxicities and no patients having grade IV toxicities. The toxicities included diarrhea, nausea, fatigue, rash, edema, abdominal pain, anorexia, and facial edema. However, no patients had a partial or complete response and the trial was closed. It appears that CI-1040 will not be further developed in these tumor types.

PD0325901 is a second-generation oral MEK1/2 inhibitor subsequently developed by Pfizer. Relatively minor changes distinguish the chemical structure of PD0325901 from that of CI-1040. The cyclopropylmethoxy group of CI-1040 was replaced with a R-dihydroxy-propoxy group and the 2-chloro substituent of CI-1040 was replaced with a 2-fluoro group on the second aromatic ring. Nevertheless, these minor structural changes imparted significant increases in potency with PD0325901 (Brown et al. 2007). Pre-clinical findings of significantly improved pharmacologic and pharmaceutical properties of PD0325901 were determined to hold promise for the use of the compound as a therapeutic agent. The first-in-human trial of PD0325901 employed an open-label, dose-escalating design in 41 patients with advanced colon, melanoma, and non-small cell lung cancer. Adverse events were observed, including rash (49%), diarrhea (49%), fatigue (34%), visual disturbance (34%), nausea (29%), edema (29%), pruritus (14%), anemia (11%) and dyspepsia (11%). The compound underwent a phase II testing in patients with advanced colon, melanoma and non-small cell lung cancer. More concerns have been focused on neurologic adverse event (confusion and hallucination) in three patients observed in the first trial, leading to putting a halt on the human trials using PD0325901. It is unclear if the drug caused these adverse events.

ARRY-142886/AZD6244/selumetinib (Array Biopharma/AstraZeneca) is a potent, highly specific MEK1/2 inhibitor. ARRY-142886/AZD6244 has undergone phase I testing in a trial of 57 patients with solid tumors. Hypoxia, rash, diarrhea, nausea, fatigue and blurred vision have been documented as the most common treatment-related toxicities in this study (grade I to III). Thirty-nine patients completed the study and 19 of them had stabilization of their disease after the treatment, 9 remained stable for five or more months. These promising results triggered a phase II study, which is currently under investigation.

ARRY-162, ARRY-300 and other Array MEK inhibitors (Array Biopharma/Novartis) are potent, highly specific MEK1/2 inhibitors. ARRY-162 is currently in phase I development for cancer. Previously, it failed to meet efficacy endpoints in phase II studies in rheumatoid arthritis. ARRY-162 is an orally active, potent, selective, non-ATP-competitive inhibitor of MEK 1/2.

RDEA119 (Ardea Biosciences/Bayer) is another highly selective MEK1/2 inhibitor. Pre-clinical and clinical results suggest that RDEA119 has favorable properties, including oral dosing, excellent selectivity and limited retention in the brain, which, in turn, may result in a reduced risk of central nervous system side effects. In preclinical studies, RDEA119 has demonstrated synergistic activity when used in combination with multiple anti-cancer agents in a wide range of tumor cell lines. Ardea Biosciences initiated a phase I clinical of RDEA119 trial in 60 patients with advanced cancer.

8. Discussion

Less than a decade ago the kinases constituting mammalian MAPK pathways were identified through intense efforts to understand the molecular events underlying cellular responses to extracellular signals. During this decade the kinases constituting ERK1/2 signaling pathways have come to be appreciated as key cellular signal transducers and thus attractive targets for drug development. Successful drug development has required the demonstration that the difficulties presented by a large gene family with a highly conserved catalytic core could successfully be targeted with specific and potent small-molecule inhibitors. These efforts are now beginning to bear fruit with the initiation of clinical trials in multiple human diseases. It is currently unclear whether it will be efficacious in *LMNA* dilated cardiomyopathy. The relevance of pre-clinical to basic research to human clinical protocols is still relatively unclear. Nevertheless, the outcome of clinical trials of compounds inhibiting ERK1/2 signaling pathways is of significant interest to both the basic and the clinical scientific communities focusing on *LMNA* dilated cardiomyopathy. Their positive outcome would be a triumph of translating basic scientific understanding of cellular function into successful human therapies.

9. References

Aebi, U., Cohn, J., Buhle, L. & Gerace, L. (1986). The nuclear lamina is a meshwork of intermediate-type filaments. *Nature* vol.323, No.6088, pp.560-564, ISSN 0028-0836.

Arimura, T., Helbling-Leclerc, A., Massart, C., Varnous, S., Niel, F., Lacène, E., Fromes, Y., Toussaint, M., Mura, A.M., Keller, D.I., Amthor, H., Isnard, R., Malissen, M., Schwartz, K. & Bonne, G. (2005). Mouse model carrying H222P-Lmna mutation develops muscular dystrophy and dilated cardiomyopathy similar to human striated muscle laminopathies. *Hum Mol Genet* Vol.14, No.1, pp.155-169, ISSN 0964-6906.

Barrett, S.D., Bridges, A.J., Dudley, D.T., Saltiel, A.R., Fergus, J.H., Flamme, C.M., Delaney, A.M., Kaufman, M., LePage, S., Leopold, W.R., Przybranowski, S.A., Sebolt-Leopold, J., van Becelaere, K., Doherty, A.M., Kennedy, R.M., Marston, D., Howard Jr, W.A., Smith, Y., Warmus, J.S. & Tecle, H. (2008) The discovery of the benzhydroxamate MEK inhibitors CI-1040 and PD 0325901. *Bioorg Med Chem Lett.* Vol.18, No.24, pp.6501-6504, ISSN 0960-894X.

Ben Yaou, R., Gueneau, L., Demay, L., Stora, S., Chikaoui, K., Richard, P. & Bonne, G. (2006) Heart involvement in lamin A/C related diseases. *Arch Mal Coeur Vaiss* vol.99, No.9, pp.848-855, ISSN 0003-9683.

Brown, A.P., Carlson, T.C.G., Loi, C.M. & Graziano, M.J. (2007) Pharmacodynamic and toxicokinetics evaluation of the novel MEK inhibitor, PD0325901, in the rat following oral and intravenous administration. *Cancer Chemother Pharmacol* vol.59, No.5, pp.671-679, ISSN 0344-5704.

Davis, R.J. (1993) The mitogen-activated protein kinase signal transduction pathway. *J Biol Chem* vol.268, No.20, pp.14553-14556, ISSN 0021-9258.

Dechat, T., Pfleghaar, K., Sengupta, K., Shimi, T., Shumaker, D. K., Solimando, L. & Goldman, R. D. (2008). Nuclear lamins: major factors in the structural organization and function of the nucleus and chromatin. *Genes Dev* vol.22, No.7, pp.832-853, ISSN 0890-9369.

Dudley, D.T., Pang, L., Decker, S.J., Bridges, A.J. & Saltiel, A.R. (1995) A synthetic inhibitor of the mitogen-activated protein kinase cascade. *Proc. Natl. Acad. Sci. USA* vol.92, No.17, pp.7686-7689, ISSN 0027-8424.

Favata, M.F., Horiuchi, K.Y., Manos, E.J., Daulerio, A.J., Stradley, D.A., Feeser, W.S., Van Dyk, D.E., Pitts, W.J., Earl, R.A., Hobbs, F., Copeland, R.A., Magola, R.L., Scherle, P.A. & Trzasko, J.M. (1998) Identification of a novel inhibitor of mitogen-activated protein kinase kinase. *J Biol Chem* vol.273, No.29, pp.18623-18632, ISSN 0021-9258.

Fisher, D. Z., Chaudhary, N. & Blobel, G. (1986) cDNA sequencing of nuclear lamins A and C reveals primary and secondary structural homology to intermediate filament proteins. *Proc Natl Acad Sci USA* vol.83, No.17, pp.6450-6454, ISSN 0027-8424.

Golzio, P.G., Chiribiri, A. & Gaita, F. (2007) "Unexpected sudden death avoided by implantable cardioverter-defibrillator in Emery-Dreifuss patient. *Europace* vol.9, No.12, pp.1158-1160, ISSN 1099-5129.

Kamisago, M., Sharma, S.D., DePalma, S.R., Solomon, S., Sharma, P., McDonough, B., Smoot, L., Mullen, M.P., Woolf, P.K., Wigle, E.D., Seidman, J.G. & Seidman, C.E. (2000) Mutations in sarcomere protein genes as a cause of dilated cardiomyopathy. *N Engl J Med* vol.343, No.23, pp.1688-1696, ISSN 0028-4793.

Keri, G., Orfi, L., Eros, D., Hegymegi-Barakonyi, B., Szantai-Kis, C., Horvath, Z., Waczek, F., Marosfalvi, J., Szabadkai, I., Pato, J., Greff, Z., Hafenbradl, D., Daub, H., Muller, G., Klebl, B. & Ullrich, A. (2006) Signal transduction therapy with rationally designed kinase inhibitors. *Curr Signal Transduct Ther* vol.1, No.1, pp.67-95, ISSN 1574-3624.

Lammerding, J., Schulze, P.C., Takahashi, T., Kozlov, S., Sullivan, T., Kamm, R.D., Stewart, C.L. & Lee, R.T. (2004) Lamin A/C deficiency causes defective nuclear mechanics and machanotransduction. *J Clin Invest* vol.113, No.3, pp.370-378, ISSN 0021-9738.

Levitzki A (1996) Targeting signal transduction for disease therapy. *Curr Opin Cell Biol.* vol.8, No.2, pp.239-44, ISSN 0955-0674.

Li, D., Tapscoft, T., Gonzalez, O., et al. (1999) Desmin mutations responsible for idiopathic dilated cardiomyopathy. *Circulation* vol.100, No.5, pp.461-464, ISSN 0009-7322.

Li, D., Czernuszewicz, G.Z., Gonzales, O., Tapscoft, T., Karibe, A., Durand, J.B., Brugada, R., Hill, R., Gregoritch, J.M., Anderson, J.L., Quinones, M., Bachinski, L.L. & Roberts, R. (2001) Novel cardiac troponin T mutation as a cause of familial dilated cardiomyopathy. *Circulation* vol.104, No.18, pp.2188-2193, ISSN 0009-7322.

Lin, F., & Worman, H. J. (1993). Structural organization of the human gene encoding nuclear lamin A and nuclear lamin C. *J Biol Chem* vol.268, No.22, pp.16321-16326, ISSN 0021-9258.

LoRusso, P., Adjei, A., Varterasian, M., Gadgeel, S., Reid, J., Mitchell, D.Y., Hanson, L., DeLuca, P., Bruzek, L., Piens, J., Asbury, P., Van Becelaere, K., Herrera, R., Sebolt-Leopold, J. & Meyer, M.B. (2005) Phase I and pharmacodynamic study of the oral MEK inhibitor CI-1040 in patients with advanced malignancies. *J Clin Oncol* vol.23, No.23, pp.5281-5293, ISSN 0732-183X.

Luk, A., Ahn, E., Soor, G.S. & Butany, J. (2009) Dilated cardiomyopathy: a review. *J Clin Pathol* vol.62, No.3, pp.219-225, ISSN 0002-9173.

Margutti, S. & Lauger, S.A. (2007) Are MAP kinases drug targets? Yes, but difficult ones. *Chem Med Chem* vol.2, No.8, pp.1116-1140, ISSN 1860-7187.

Maron, B.J., Towbin, J.A., Thiene, G., Antzelevitch, C., Corrado, D., Arnett, D., Moss, A.J., Seidman, C.E. & Young, J.B. (2006) Contemporary definitions and classification of the cardiomyopathies: an American Heart Association Scientific Statement from the Council on Clinical Cardiology, Heart Failure and Transplantation Committee; Quality of Care and Outcomes Research and Functional Genomics and Translational Biology Interdisciplinary Working Groups; and Council on Epidemiology and Prevention *Circulation* vol.113, No.14, pp.1807-1816, ISSN 0009-7322.

McKeon, F. D., Kirschner, M. W. & Caput, D. (1986). Homologies in both primary and secondary structure between nuclear envelope and intermediate filament proteins. *Nature* vol.319, No.6053, pp.463-468, ISSN 0028-0836.

Meune, C., Van Berlo, J. H., Anselme, F., Bonne, G., Pinto, Y. M. & Duboc, D. (2006). Primary prevention of sudden death in patients with lamin A/C gene mutations. *N Engl J Med* vol.354, No.2, pp.209-210, ISSN 0028-4793.

Michels, V.V., Moll, P., Miller, F.A, Tajik, A.J., Chu, J.S., Driscoll, D.J., Burnett, J.C., Rodeheffer, R.J., Chesebro, J.H. & Tazelaar, H.D. (1992) The frequency of familial dilated cardiomyopathy in a series of patients with idiopathic dilated cardiomyopathy. *N Engl J Med* vol.326, No.2, pp.77-82, ISSN 0028-4793.

Morita, H., Seidman, J. & Seidman, C.E. (2005) Genetic causes of human heart failure. *J Clin Invest* vol.115, No.3, pp.518-526, ISSN 0021-9738.

Muchir, A., Pavlidis, P., Decostre, V., Herron, A. J., Arimura, T., Bonne, G. & Worman, H. J. (2007). Activation of MAPK pathways links *LMNA* mutations to cardiomyopathy in Emery-Dreifuss muscular dystrophy. *J Clin Invest* vol.117, No.5, pp.1282-1293, ISSN 0021-9738.

Muchir, A., Shan, J., Bonne, G., Lehnart, S.E. & Worman, H.J. (2009) Inhibition of extracellular signal-regulated kinase signaling to prevent cardiomyopathy caused by mutation in the gene encoding A-type lamins. *Hum Mol Genet* vol.18, No.2, pp.241-247, ISSN 0964-6906.

Muntoni, F., Cau, M., Ganau, A., Congliu, R., Arvedi, G., Mateddu, A., Marrosu, M.G., Cianchetti, C., Realdi, G., Cao, A. & Melis, M.A. (1993) Brief Report: deletion of the dystrophin muscle-promoter region associated with X-linked dilated cardiomyopathy. *N Engl J Med* vol.329, No.13, pp.921-925, ISSN 0028-4793.

Nikolova, V., Leimena, C., McMahon, A.C., Tam, J.C., Chandar, S., Jogia, D., Kesteven, S.H., Michalicek, J., Otway, R., Verheyen, F., Rainer, S., Stewart, C.L., Martin, D.,

Feneley, M.P. & Fatkin, D. (2004) Defects in nuclear structure and function promote dilated cardiomyopathy in lamin A/C-deficient mice. *J Clin Invest* vol.113, No.3, pp.357-369, ISSN 0021-9738.

Olson, T.M., Michels, V.V., Thibodeau, S.N., Tai, Y.S. & Keating, M.T. (1998) Actin mutations in dilated cardiomyopathy, a heritable form of heart failure. *Science* vol.280, No.5364, pp.750-752, ISSN 0036-8075.

Olson, T.M., Kishimoto, N.Y., Whitby, F.G., & Michels, V.V. (2001) Mutations that alter the surface charge of alpha-tropomyosin are associated with dilated cardiomyopathy. *J Mol Cell Cardiol* vol.33, No.4, pp.723-732, ISSN 0022-2828.

Tan, W., DePrimo, S., Krishnamurthi, S.S. et al. (2007) Pharmacokinetic (PK) and pharmacodynamic (PD) results of a phase I study of PD-0325901, as second generation oral MEK inhibitor, in patients with advanced cancer. *Mol Cancer Ther* vol.6, pp.3648 (abstract), ISSN 1535-7163.

Taylor, M.R.G., Fain, P.R., Sinagra, G., Robinson, M.L., Robertson, A.D., Carniel, E., Di Lenarda, A., Bohlmeyer, T.J., Ferguson, D.A., Brodsky, G.L., Boucek, M.M., Lascor, J., Moss, A.C., Li, W.L.P., Stetler, G.L., Muntoni, F., Bristow, M.R. & Mestroni, L. (2003) Natural history of dilated cardiomyopathy due to lamin A/C gene mutations. *J Am Coll Cardiol* vol.41, No.5, pp.771-780, ISSN 0735-1097.

Tsubata, S., Bowles, K.R., Vatta, M., Zintz, C., Titus, J., Muhonen, L., Bowles, N.E. & Towbin, J.A. (2000) Mutations in the human delta-sarcoglycan gene in familial and sporadic dilated cardiomyopathy. *J Clin Invest* vol.106, No.5, pp.655-662, ISSN 0021-9738.

Sanna, T., Dello Russo, A., Toniolo, D., Vytopil, M., Pelargonio, G., De Martino, G., Ricci, E., Silvestri, G., Giglio, V., Messano, L., Zachara, E. & Bellochi, F. (2003) Cardiac features of Emery-Dreifuss muscular dystrophy caused by lamin A/C gene mutations. *Eur Heart J* vol.24, No.24, pp.2227-2236, ISSN 0195-668X.

Schonberger, J. & Seidman, C.E. (2001) Many roads lead to a broken heart: The genetics of dilated cardiomyopathy. *Am J Hum Genet* vol.69, No.2, pp.249-260, ISSN 0002-9297.

Sebolt-Leopold, J.S., Dudley, D.T., Herrera, R., Van Becelaere, K., Wiland, A., Gowan, R.C., Tecle, H., Barrett, S.D., Bridges, A., Przybranowski, S., Leopold, W.R., & Saltiel, A.R. (1999) Blockade of the MAP kinase pathway suppresses growth of colon tumors in vivo. *Nat Med* vol.5, No.7, pp.810-816, ISSN 1078-8956.

Seger, R. & Krebs, E.G. (1995) The MAPK signaling cascade. *FASEB J* vol.9, No. 9, pp.726-735, ISSN 0892-6638.

Stewart, C.L., Roux, K.J. & Burke, B. (2007) Blurring the boundary: the nuclear envelope extends its reach. *Science* vol.318, No.5855, pp.1408-1412, ISSN 0036-8075.

Van Berlo, J.H., de Voogt, W.G., van der Kooi, A.J., van Tintelen, J.P., Bonne, G., Ben Yaou, R., Duboc, D., Rossenbacker, T., Heidbuchel, H., de Visser, M., Crijns, H.J.G.M. & Pinto, Y.M. (2005) Meta-analysis of clinical characteristics of 299 carriers of *LMNA* gene mutations: de lamin A/C mutations portend a high risk of sudden death? *J Mol Med* vol.83, No.1, pp.79-83, ISSN 1432-1440.

Wu, W., Shan, J., Bonne, G., Worman, H.J. & Muchir, A. (2010) Pharmacological inhibition of c-Jun N-terminal kinase signaling prevents cardiomyopathy caused by mutation in LMNA gene. *Biochim Biophysi Acta* vol.1802, No.7-8, pp.632-638, ISSN 0006-3002.

Wu, W., Muchir, A., Shan, J., Bonne, G., Worman, H.J. (2011) Mitogen-activated protein kianse inhibitors improve heart function and prevent fibrosis in cardiomyopathy caused by mutation in lamin A/C gene. *Circulation* vol.123, No.1, pp.53-61, ISSN 0009-7322.

Yoon, S. & Seger, R. (2006) The extracellular signal-regulated kinase: Multiple substrates regulate diverse cellular functions. *Growth Factors* vol.24, No.1, pp.21-44, ISSN 0897-7194.

5

Signaling Pathways Coupled to Activation of the Kinin B1 Receptor

Pamela Ehrenfeld, Carlos D. Figueroa,
Kanti D. Bhoola and Carola E. Matus
Laboratory of Cellular Pathology, Institute of Anatomy, Histology and Pathology, Universidad Austral de Chile, Valdivia
Chile

1. Introduction

1.1 The field of kinins and their receptors

Kinins are a group of bioactive peptides, which are formed by an endogenous enzymatic cascade consisting of precursor substrates called kininogens and the proteolytic kallikrein enzymes (kininogenases). So far, two kinin-releasing enzymes (kininogenases) have been characterized, plasma (hKB1/*LKLB1*) and tissue (hK1/*KLK1*) kallikreins (Fig. 1). Both enzymes are serine proteases that are found in glandular cells, neutrophils, and biological fluids. Kininogens are multifunctional proteins involved in cascade reactions during inflammation (Bhoola et al., 1992), and more recently in carcinogenesis (Bhoola et al., 1992; 2001; Leeb-Lundberg et al., 2005). Kinins namely bradykinin (BK) and kallidin (Lys-BK) are among the most potent pro-inflammatory vasoactive peptides generated during tissue injury and noxious stimulation. BK and Lys-BK undergo metabolic degradation at a variable rate by amino-, carboxy- and endopeptidases found in tissues and biological fluids. The most physiologically relevant enzymes are carboxypeptidase N (from plasma) and carboxypeptidase M (from cell membranes), which remove the carboxy-terminal Arg present at end of the kinin molecule, resulting in the formation of the active metabolites des-Arg^{9}-BK (DBK) and des-Arg^{10}-kallidin, known also as Lys-des-Arg^{9}-BK (LDBK) (Fig. 1). Furthermore, neutral endopeptidase 24.11 (CD10, enkephalinase) and angiotensin I converting enzyme, act as dipeptidyl carboxypeptidases by removing the dipeptide Phe^{8}-Arg^{9} from the carboxy terminus end of BK or Lys-BK (Couture et al., 2004). It is important to mention that endopeptidase 24.15 and angiotensin I converting enzyme cleave the dipeptide Ser^{6}-Pro^{7} from bradykinin 1-7 (BK1-7) to produce bradykinin 1-5 (BK1-5), one of the final metabolites of BK and DBK that possesses the longer half-life of this peptide family. The half-life of kinins depends on the rate and site of destruction, most rapid in the circulation, but less so in the extracellular fluid space and by cells. Actually, in plasma the half-life of BK and LDBK is short (15 to 20 sec) whereas BK1-5, considered in the past an inactive fragment, is considered to have a half-life of 86 to 101 min (Shima et al., 1992; Murphey et al., 2000; 2006; Morinelli et al., 2002).

Kinins are positively charged peptides that influence tissues and cells by stimulating two pharmacologically distinct G protein-coupled receptors (GPCRs). Kinin receptors are

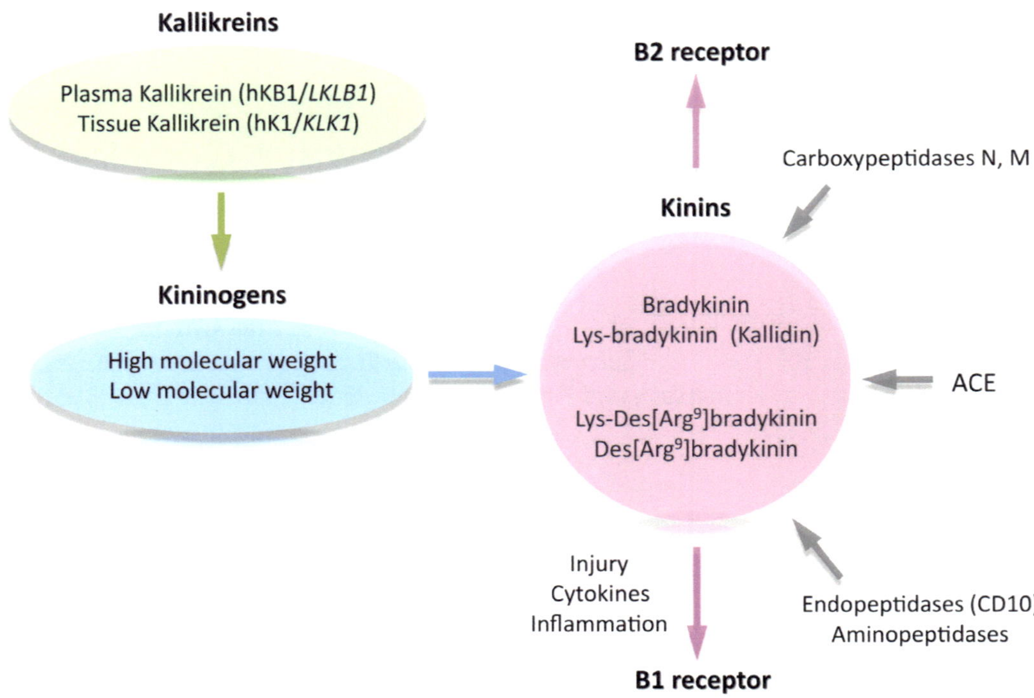

Fig. 1. Major components of the kinin system. The kininogenases plasma and tissue kallikreins hydrolize either high or low molecular weight kininogens to release the kinin domain. Carboxypeptidases remove the Arg9 from bradykinin or Lys-bradykinin to generate B1 receptor agonists.

situated on the plasma membranes of many cell types and are coupled principally to Gαq and Gαi, (Austin et al., 1997; Regoli et al., 2001). These receptors are designated as kinin B1 (B1R) or B2 (B2R) receptors (Fig.1). B2R is a preformed receptor, widely distributed and activated by the parent molecules BK and Lys-BK. B1R is activated by DBK and LDBK, two kinins that lack the Arg^9 at the carboxy-terminus of the parental BK or LBK molecule (Bhoola et al., 1992; 2001; Leeb-Lundberg et al., 2005).

B1R is generally expressed at low levels in normal tissues, but is strongly up-regulated during inflammation, tissue injury, cancer, exposure to bacterial endotoxins such as lypopolysaccharide, growth factors (epidermal growth factor and endothelial growth factor) cytokines (principally interleukin-lβ, interleukin-8 and tumor necrosis factor-α) or even its own agonist, LDBK (Marceau et al., 1998; Yang et al., 2001). For this reason, there is an increase in the number of B1 binding sites in inflamed or carcinogenic tissue. Moreover, an up-regulation of the kinin B1R has been found in several tumors, immune-modulated disorders such as rheumatoid arthritis, transplant rejection, glomerulonephritis and in human fibrotic lung tissue.

Additional differences between both types of kinin receptors reside in the fact that the B2R is internalized and extensively recycled to the cell surface by its agonist, whereas the B1R is up-regulated in inflammatory disorders and cancer, and its expression is controlled by signaling pathways such as stress mitogen-activated protein and nuclear factor kappa B

(NF-κB) (Marceau et al., 2002). In fact, sequence analyses show the presence of a transcriptional regulatory site for NF-κB in the promoter region of the B1R (Bachvarov et al., 1996; Ni et al., 1998) a finding that may explain why expression of the B1R gene is tissue and cell specific (Yang et al., 2001).

The different cell types in which expression and activation of kinin receptors has been determined include endothelial, epithelial and neural cells, smooth muscle cells, neutrophils, lymphocytes, monocytes, keratinocytes, chondrocytes and fibroblasts. Depending upon the cell type involved, kinin receptors activate different intracellular signaling pathways, which regulate processes such as, cell proliferation, differentiation, migration, vascular permeability, contraction of smooth muscle, excitation of nerve endings and release of a variety of biologically active secondary mediators. In this chapter, we will focus primarily on the most important signaling pathways triggered after B1R stimulation.

2. Structure of kinin B1 receptor

In general, the information on the structure of kinin receptors has been obtained by chemical cross-linking experiments, mutagenesis approaches, classical pharmacological assays, and studies with domain-specific anti-receptor antibodies (Prado et al., 2002; Blaukat, 2004). All these approaches have shown that B1R belongs to the GPCR family and that it is integrated into the plasma membrane. The amino-terminus end is exposed to the extracellular space and contains (three) consensus sites of the Asn-Xaa-Ser/Thr type for N-glycosilation and the most of them are clustered on their N-terminal domains. Is noteworthy that the kinin-binding site is located at the amino-terminus end of the third extracellular loop. According to Kang et al.,(2005) the presence of multiple bands observed for the recombinant expressed human B1R is strongly reminiscent of patterns of the partially glycosylated B2R. However, a B1R in which the N-terminal domain had been truncated to remove putative N-glycosylation sites migrated as a homogeneous specie (Kang et al., 2005). Furthermore, the treatment of the hemagglutinin-tagged human B1R, expressed in HEK293 cells, with N-glycosidase F resulted in the conversion of the receptor from a heterogeneous specie migrating at 35 to 45 kDa to a relatively homogeneous one migrating at 37 kDa. We have observed a similar electrophoretic pattern of the B1R protein in human neutrophils (Ehrenfeld et al., 2006), breast cancer cells (Molina et al., 2009), endothelial cells and fibroblasts (our group, unpublished data). It is postulated that glycosylation probably increases the hydrophilicity of the extracellular portions of B1R affecting or regulating ligand affinity, efficient G protein coupling, maturation (folding, stabilization), intracellular trafficking and receptor oligomerization or receptor degradation (Menke et al., 1994; MacNeil et al., 1995; Leeb-Lundberg et al., 2005). Even more, it has been demonstrated that kinin-mediated mitogenic signaling and prostate cell growth is blocked by B1 and B2 receptor antagonists indicating that these effects depend on both kinin receptors. These results provide evidence for the existence of B1R/B2R heterodimers in PC3 prostate cancer cells and demonstrate that antagonism of one receptor interferes with the ability of the other, possibly at the level of receptor-Gαq protein coupling (Barki-Harrington et al., 2003). It is known that B1R can dimerize, oligomerize or heterodimerize with the B2R or other molecule. Actually, it is known that B1R heterodimerizes with membrane carboxypeptidase M, facilitating receptor signaling via carboxypeptidase M-mediated conversion of bradykinin or kallidin to a des-[Arg^9]kinin. This critical interaction potentiates B1R signaling and uncovers a new mode of GPCR activation by a cell surface peptidase (Zhang et al., 2011). Therefore, it is clear that knowledge of the molecular structure has been important in determining B1R function.

3. Classical signaling pathways activated by kinin B1 receptor

In general, B1R activates most of the signaling pathways activated by B2R (reviewed by Marceau et al., 1998 and Leeb-Lundberg et al., 2005). Although both kinin receptors are coupled to similar signal transduction pathways, the differences in patterns of signaling are due to different degrees of short-term regulation that include both receptor desensitization and internalization (Leeb-Lundberg et al., 2005). The B2R is stable in the absence of its agonist, but it is rapidly desensitized after ligand stimulation by a mechanism that includes recruitment of β-arrestin 2, internalization in a clathrin-dependent manner and recycling upon agonist treatment (Enquist et al., 2007). The B1R is not phosphorylated either under basal conditions or in response to agonist (Blaukat et al., 1999) and desensitizes slightly upon further stimulation with its agonist (Mathis et al., 1996). Actually, the B1R is constitutively internalized in the absence of agonist via a clathrin-dependent pathway, do not recruit β-arrestin 2, bind G protein-coupled receptor sorting protein and go to lysosomes for degradation. Binding of agonist to the B1R inhibits its constitutive internalization thereby reducing the rate of spontaneous clearance of receptors from the cell membrane and at the same time delaying B1R degradation (Enquist et al., 2007). This behavior is according with B1R up-regulation by cytokines and other inflammatory stimuli that result in a high number of receptors on the cell surface, available for ligand binding (Fig. 2).

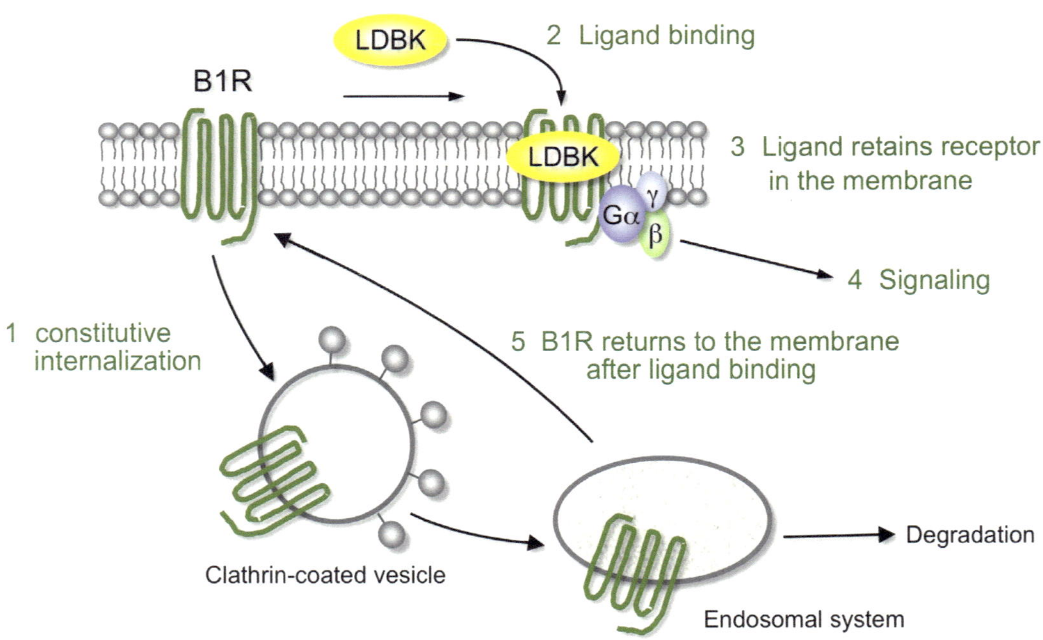

Fig. 2. Regulation of B1R levels on the cell membrane. The B1R is constitutively internalized, but the number of B1R molecules on the cell membrane increases after binding of its ligand.

It is known that B1R can be coupled to different pathways according to the cell type involved. Signaling of the B1R results in the activation of protein kinase C (PKC) and tyrosine kinase cascades, coordinated with activation of the mitogen-activated protein kinase (MAPK) pathway and NF-κB. B1R activation also stimulates phosphatidilinositol hydrolysis in smooth muscle cells leading to mobilization of intracellular Ca^{2+}, phospholipase C or phospholipase A, and appears to induce biosynthesis and release of prostaglandins (Bhoola et al., 1992; 2001; Leeb-Lundberg et al., 2005). In vascular smooth muscle cells, B1R stimulation produces a significant dependence on extracellular Ca^{2+} and a transient increase in phosphatidilinositol hydrolysis that is more sustained than that generated by the B2R (Tropea et al., 1993). Nevertheless, in the rat renal cortical collecting duct cells, keratinocytes, breast cancer cells and neutrophils B1R does not induce intracellular Ca^{2+} mobilization (Ehrenfeld et al., 2006; Matus et al., 2008; Molina et al., 2009)

Another characteristic of B1R is its capacity to regulate the cell cycle. In some cases, B1R is proliferative as seen in fibroblasts, vascular smooth muscle cells, estrogen-sensitive breast cancer cells, and this response involves MAPK activation (Marceau and Tremblay, 1986; Christopher et al., 2001; Molina et al., 2009). In vascular smooth muscle cells, B1R is induced only in response to injury, regulates proliferation by pathways that include activation of cholera toxin-sensitive Gαq, PKC, Src kinase and MAPK. In estrogen-sensitive breast cancer cells, Molina et al.,(2009) showed that nanomolar concentrations of the B1R agonist produced an increase in BrdU incorporation as measured in a proliferation assay. The use of inhibitors of MEK, the kinase which phosphorylates ERK1/2 MAPK, such as PD98059 or UO126, decrease the phosphorylation of ERK1/2 and completely abolished the incorporation of BrdU. At the same time, we demonstrated that this effect was dependent on epidermal growth factor receptor (EGFR) transactivation because the use of AG1478 blunted the activation of this MAPK. In addition, we have observed the proliferative effect of B1R stimulation on MDA-MB-231 estrogen-insensitive breast cancer cells (unpublished data). On the contrary, an antiproliferative effect has been observed in vascular smooth muscle cells, probably due to prolonged activation of MAPK and increased p27Kip1 activity (Dixon et al., 2002).

Studies on the involvement of B1R in cell migration have shown that in primary cultures of arterial smooth muscle cells, activation of B1R inhibited cell migration, an effect that involves activation of PI3K, but not nitric oxide or prostanoid release (Morissette et al., 2006). In contrast, activation of B1R induces the migration of human PC3 prostate cancer cells via activation of focal adhesion kinase (FAK), an important kinase involved in cytoskeletal reorganization and cell migration (Taub et al., 2003). On the other hand, studies performed in human neutrophils show that B1R stimulates leukocyte chemotaxis, though the pathways implicated in this process are still unknown (Ahluwalia & Perretti, 1996; Paegelow *et al.*, 2002; Ehrenfeld *et al.*, 2006). Moreover, it is known that release of the kinin moiety from kininogens sited on the neutrophil surface by enzymatic action of the classical tissue kallikrein (hK1) results in opening of junctions between the endothelial cells, thereby causing plasma extravasation (Stuardo et al., 2004). B1R modulates the release of matrix metalloproteases from both human neutrophils (Ehrenfeld et al., 2009) and breast cancer cells (Ehrenfeld et al., 2011), an event that may contribute to the extracellular matrix remodeling in processes such as inflammation, wound healing and cancer. Part of the effects produced by B1R in neutrophils and cancer cells include activation of ERK1/2 and p38 MAPK and transactivation of EGFR in breast cancer cells, an event that usually results in the activation of MAPK.

Phosphorylation of a 125-kDa protein, following stimulation of B1R expressed in human keratinocytes was identified as FAK (Yurko et al., 2001; Matus et al., 2008). The activation of this kinase after stimulation of B1R has also been observed in immortalized HaCaT keratinocytes (unpublished data) and in human PC3 prostate cancer cells (Taub et al., 2003). Moreover, we have shown that treatment of keratinocytes with herbimycin before stimulation with B1R agonists reduces FAK phosphorylation (Matus et al., 2008). Thus, it is likely that B1R activation may contribute to cell motility in several cell types. Another signaling pathway used by B1R includes activation of PKC. Overexposure of human keratinocytes to phorbol 12-myristate 13-acetate (PMA) or the preincubation with GF109203X, a potent and selective inhibitor of PKCs, demonstrated that phosphorylation of the EGFR was greatly reduced, corroborating the involvement of PKC (Matus et al., 2008).

4. Mechanisms of EGFR transactivation by kinin B1 receptor

It is known that various stress factors such as ultraviolet light and ionizing radiation can activate receptor tyrosine kinases (RTKs), like EGFR, in the absence of ligand. Also, activation of GPCRs, like the B1R, can activate RTKs by a mechanism that has been called "transactivation" by Alex Ullrich's group (Daub et al., 1996). They found that several GPCR agonists were able to activate the EGFR that in turn acts as a signal transducer for GPCRs. The EGFR belongs to a family of type I RTKs that comprise four members: EGFR (ErbB1/HER1), ErbB2 (HER2/Neu), ErbB3 (HER3) and ErbB4 (HER4) (reviewed by Liebmann, 2011). Activation of these receptors occurs by dimerization after ligand binding or by high receptor density on the cell membrane. When the ligand binds to its receptor, it induces the formation of homo and heterodimers activating the intrinsic tyrosine kinase domain that results in phosphorylation of specific tyrosine residues forming part of the cytoplasmic receptor tail. These residues serve as docking sites for a variety of signaling molecules (Endoh et al., 2009). Overexpression of EGFR is an indication of poor prognostic in multiple tumor types. EGFR autophosphorylation in sites such as Tyr845 are directly phosphorylated by Src kinase family (Wetzker & Böhmer, 2003) and GPCR agonists induce activation of Src, which directly phosphorylates EGFR by ligand-independent pathways (Biscardi et al., 1999; Tice et al., 1999). Furthermore, phosphorylation of EGFR at Tyr845 in the kinase domain is implicated in stabilizing the activation loop, maintaining the active state enzyme, and providing a binding surface for substrate proteins among which are the MAPK pathways (Cooper & Howell, 1993; Hubbard et al., 1994). EGFR and cytoplasmic tyrosine kinase c-Src cooperate in several cellular functions such as proliferation and apoptosis. Boerner et al.,(2005) showed that c-Src-dependent phosphorylation of Tyr845 of EGFR is required for the DNA synthesis induced by activation of some G protein-coupled receptors in murine fibroblasts and breast cancer cells. The role of Tyr845 in DNA synthesis and cell proliferation was demonstrated by the microinjection of phosphoTyr845-containing peptide in these cells, which was able to ablate EGF-stimulate S-phase entry in both cell systems. This finding suggests that this residue is an important regulator of DNA synthesis induced by mitogens like EGF. Sato et al., (2003) using A431 carcinoma cells demonstrated that the expression of adaptor protein p52shc, or stimulation with EGF or H_2O_2 leads to phosphorylation of EGFR on Tyr845. Phosphorylation in this residue was inhibited by PP2, but not by AG1478, and is associated with Src activation and phosphorylation of activators of transcription type 3 and 5 (STAT 3/5). This effect was inhibited by introduction of an antibody against phosphorylated Tyr845 or by transfection of a dominant-negative for c-Src into the cells. Moreover, the co-incubation of puried c-Scr and EGFR leads to phosphorylation of Tyr845 *in vitro*. Altogether, these results demonstrate that c-Src can directly phosphorylate EGFR on Tyr845. Similarly, different cell lines and animal studies have shown that MAPK and signal transducers like STAT-3 are important mediators of EGFR signaling after phosphorylation of different tyrosine residues in liver cells and hepatocellular carcinoma. Interestingly, it has been shown that homogenates of hepatocellular carcinomas present phosphorylation at Tyr845, but no EGFR phosphorylation at Tyr998, Tyr1045 or Tyr1068 (Kannangai et al., 2006). On the other hand, the SH2 domain of PLCγ binds at phospho-Tyr992, resulting in activation of PLCγ-mediated downstream signaling (Emlet et al., 1997). Phosphorylation of this residue was demonstrated in 38 of the 39 cases of esophageal carcinomas by inmunohistochemistry with an anti-phosphoTyr992-EGFR antibody (Miyawaki et al., 2008). Immunoreactivity to phosphorylated Tyr992 was mainly associated with areas of severe dysplasia and microinvasive foci adjacent to invasive carcinoma, suggesting a role of phosphorylated Tyr992-EGFR in invasion. Moreover, phosphorylation of Tyr992, in responses of ionizing radiation in chinese hamster ovary or A431 cells, is related to activation of PLCγ and increases when the phosphatase SHP2 is blocked (Sturla et al., 2005).

Phosphorylation of EGFR at Tyr1045 creates a major docking site for c-Cbl, an adaptor protein that leads to receptor ubiquitination and degradation following EGFR activation (Levkowitz, et al.,1999; Ettenberg, et al.,1999). The GRB2 adaptor protein binds activated EGFR at phospho-Tyr1068 (Rojas et al., 1996). Interestingly, Yamauchi et al.,(1998) demonstrated that binding of growth hormone to its receptor, which belongs to the cytokine receptor superfamily, activates Janus kinase tyrosine kinase, STAT proteins and MAPK that regulate expression of c-fos. This activation depends on phosphorylation of Tyr1068 by Janus kinase tyrosine kinase 2, providing docking sites for Grb2 and Shc protein adaptor and activating MAPK and gene expression. Phosphorylated Tyr1148 and Tyr1173 provide a docking site for the Shc scaffold protein and induce MAPK activation (Zwick, et al.,1999). Phosphorylation of these tyrosine residues is a key step for signaling activation and inhibition of intrinsic tyrosine kinase activity by highly potent and selective inhibitors such as AG1478 (tyrphostin), that block receptor phosphorylation and the subsequent signaling pathways triggered by EGFR (Fig. 3).

In several cell types, the EGFR transactivation induced by GPCRs is mediated by the release of EGFR ligands such as heparin-binding EGF-like growth factor (HB-EGF), transforming growth alpha (TGF-α) and/or amphiregulin (Fig. 3). These ligands are generated by activation of the ADAM family (a disintegrin and metalloprotease) of zinc-dependent metalloproteases. Members of the ADAM family like ADAM10, ADAM12 and ADAM17 mediate GPCR-induced EGFR transactivation in different cell types (reviewed by Rozengurt et al., 2007). The mechanism by which members of the ADAM family are activated has been suggested to depend on the GPCR and the cell type involved. Activation can be through reactive oxygen species, PKC and Src, PI3K or ERKs (review by Ohtsu et al., 2006; Lemjabbar-Alaoui er al, 2011; Maretzky et al., 2011; Sun et al., 2010). In addition, different reports indicate that the use of Src inhibitors such as PP2 and GM6001, a broad spectrum metalloprotease inhibitor, partially block the MAPK or Akt pathways (Chen et al., 2011; Cramer et al., 2001; Mugabe et al., 2010; Stirnweiss et al., 2006) suggesting that GPCRs do not necessarily transactivate the EGFR to activate signaling pathways. Through

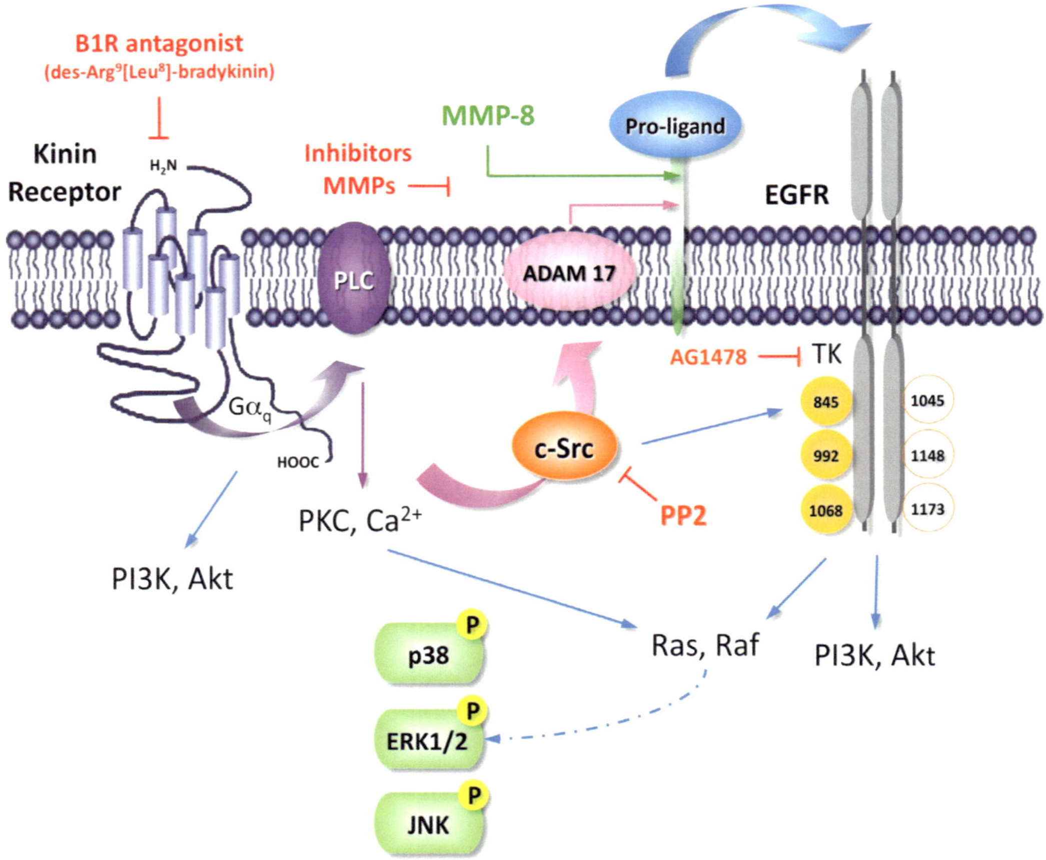

Fig. 3. Schematic representation of EGFR transactivation by a GPCR like the B1R. Stimulation of the B1R by its natural agonist may trigger specific signaling pathways and also produce activation of matrix metalloproteases (MMP-8, ADAM17) to release membrane-bound proligands that after being released will stimulate the EGFR. Stimulation of B1R also results in c-Src phosphorylation that may in turn produce phosphorylation of specific tyrosine residues (Tyr845, Tyr992, Tyr1068) of EGFR. Phosphorylation of other tyrosine residues such as Tyr1045, Tyr1148 and Tyr1173 has not been yet described. Transactivation of EGFR is used by the B1R to activate specific signaling pathways like p38, ERK1/2 and JNK MAPK. PLC, Phospholipase C.

transactivation, GPCRs can regulate cell growth, cell differentiation, survival and migration and play important roles in pathophysiological processes such as embryo development, wound healing and cancer progression.

Few studies have considered the transactivation of EGFR by either kinin B1 or B2 receptors. Studies performed in different cell types show that B2R activation triggers EGFR transactivation. The phosphorylation of ERK1/2 MAPK, produced by stimulation of COS-7 cells with a B2R agonist involves the independent activation of PKC and EGFR transactivation (Adomeit et al., 1999). Similar results in keratinocytes and renal IMCD-3 cells show that MAPK phosphorylation was partially and completely dependent on EGFR transactivation, respectively (Mukhin et al., 2006 and Vidal et al., 2005). It has been shown that bradykinin can produce activation of p60Src and Src-dependent phosphorylation on Tyr845 of EGFR as well as recruitment of PLCγ in primary cultures of rat adrenal chromaffin cells and PC12 cells (Hur et al., 2004). Moreover, Yang et al., (2005) demonstrated, using Western blot and [^{3}H]thymidine incorporation, that bradykinin induces proliferation of vascular smooth muscle cells mediated, at least in part, through activation of kinases of the Src family, EGFR transactivation, and PI3K-AKT pathways. Studies directed to elucidate the transactivation mechanism induced by B2R have shown that depending on the cell type, involve ADAM17 or metalloprotease 8 to cleave EGFR ligands (Dey et al., 2010; Methner et al., 2009). Reports regarding these events and the B1R are scarce. Matus et al., (2008) showed that stimulation of the kinin B1R in human keratinocytes produced the phosphorylation of tyrosine residues in a protein with a molecular mass of 170 kDa, that was later identified as EGFR. In these cells, the B1R induces the phosphorylation of Tyr845, Tyr992 and Tyr1068 residues on the EGFR molecule, an effect that was blocked by AG1478, a specific inhibitor of the EGFR tyrosine kinase activity. Similar results were obtained in the HaCaT keratinocyte cell line when EGFR transactivation by the B1R was analyzed (Matus et al., unpublished data).

Studies performed on estrogen-sensitive breast cancer cells strongly suggest that the proliferative effect induced by the B1R depends on the activity of EGFR and subsequent ERK1/2 MAPK phosphorylation (Molina et al., 2009). We have also reported that release of metalloproteases in MCF-7 and MDA-MB-231 breast cancer cells was blocked by AG1478, an observation that was confirmed by transfection of breast cancer cells with the dominant negative EGFR mutant HERCD533 (Ehrenfeld et al., 2011). The transactivation mechanism involved in EGFR activation by B1R agonists has not been elucidated yet, but work performed by our group in human HaCaT keratinocytes, MCF-7 breast cancer cell has visualized that B1R stimulation produces Src phosphorylation, that is blocked by the specific Src inhibitor, PP2 (Figs. 4 and 5). Interestingly, Src phosphorylates specifically the Tyr845 residue present in the active site of EGFR, therefore when we inhibited Src activity, phosphorylation of EGFR in Tyr845 was also inhibited (our group, unpublished data (Fig. 3).

5. Potential usefulness of currently available kinin antagonists and kinase inhibitors in pathological processes

The importance of kinin receptors, especially the B1R, is illustrated by many publications that demonstrate their involvement in different pathological processes such as cancer, and especially breast cancer as mentioned early in the text. From a functional point a view, both B1 and B2 receptors are central players in the aetiology of pain, inflammation and cancer. Thus, the use of antagonists or inhibitors directed to specific intracellular pathways may be a useful approach to understand the mechanisms of particular pathological processes and then to promote them as useful pharmacological agents. Frequently, B2R is associated with the acute phase of inflammation and noniception, whereas the B1R after its up-regulation by inflammatory mediators is more relevant during chronic or persistent inflammation. For this reason, the use of antagonists of both peptidic and non-peptidic nature or blockade of kinase pathways triggered by activation of kinin receptors may become important clinical tools for treatment of persistent inflammation, cancer and pain, especially when no other therapy is available or provides beneficial effects (Campos et al., 2006).

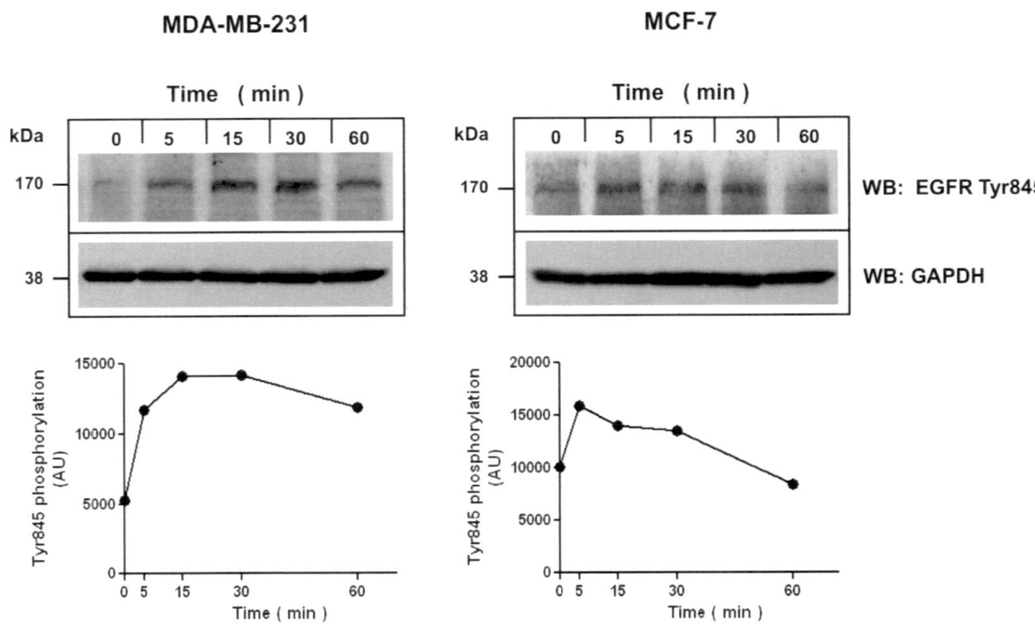

Fig. 4. The kinin B1R stimulation triggers transactivation of EGFR in estrogen-insensitive (MDA-MB-231) and estrogen-sensitive (MCF-7) breast cancer cells. Cells were cultured in DMEM-F12 or DMEM without phenol red, synchronized for 24 h and then stimulated with 10 nM Lys-des[Arg^9]bradykinin for different periods of time. When stimulation was completed, the cells were homogenized in the presence of protease inhibitors, proteins separated by polyacrylamide gel electrophoresis and then transferred to immobilon membranes. Representative Western blots performed with a monoclonal antibody that detects phosphorylation on Tyr845 residue of the EGFR are shown. AU, arbitrary intensity units.

IL-4 is a cytokine with an important role in asthma and chronic obstructive pulmonary disease that promotes eosinophilic inflammation and mucus hypersecretion. In this regard, Bryborn et al.,(2004) demonstrated that long-term exposure to IL-4 increases contractile response of bronchial smooth muscle induced by B1 or B2 receptor agonists. This effect appears to be mediated via an up-regulation of B1R and MAPK pathways since the pre-treatment with SP600125, a c-Jun N-terminal kinase inhibitor or SB203580, a p38 MAPK inhibitor or PD98059, an inhibitor of ERK1/2 MAPK phosphorylation, inhibited that response. Similarly, IL-1β induces the up-regulation of B1R and B2R in mouse tracheal smooth muscle and increases the contractile responses induced by the respective kinin agonists (Zhang et al., 2007a). The effect was inhibited by SP600125 and TAT-TI-JIP, two JNK inhibitors, but not by inhibitors of ERK1/2 MAPK pathways (Zhang et al., 2007b). The importance of B1R in inflammatory reactions was also investigated by the group of Lin et al.,(2010) who demonstrated the involvement of B1R in cigarette smoke-induced airway inflammation by a mechanism that includes the participation of IL-1β. In fact, rat lung slices treated with total particulate matter of cigarette smoke for 24 h showed an enhanced expression of B1R and IL-1β. At the same time, Xu et al.,(2010) reported that nicotine induces airway hyperresponsiveness via transcriptional up-regulation of B1R and B2R and activation of JNK, but not ERK1/2 or p38 MAPK. The use of SP600125 a specific JNK inhibitor and

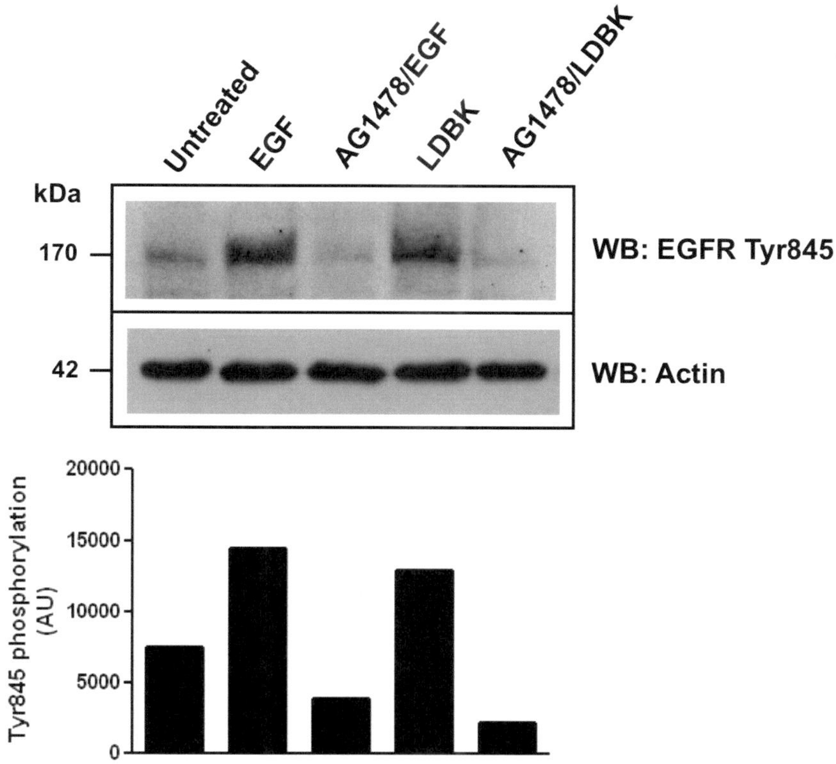

Fig. 5. Stimulation of the kinin B1R triggers transactivation of EGFR in human HaCaT keratinocytes. The epidermal cells were pre-incubated with 1 μM of AG1478 for 30 min or directly stimulated with 100 nM Lys-des[Arg^9]bradykinin for 15 min or 2.5 ng/ml EGF for 10 min. Proteins were separated by polyacrylamide gel electrophoresis and Western blots were performed using a monoclonal antibody that detects phosphorylation on Tyr845 residue of the EGFR molecule. A representative experiment is shown. AU, arbitrary intensity units.

YM976 a specific phosphodiesterase inhibitor abolished the nicotine-induced effects on kinin receptor-mediated contractions and reverted the enhanced receptor mRNA expression. Other reports demonstrate that the increased expression of B1R relays on oxidative stress, IL-1β and NF-κB, but not on cyclooxygenase or TNF-α. These results were obtained using the human alveolar epithelial A549 cell line treated with particulate matter. The increase of B1R mRNA was prevented by co-treatments with N-acetyl-l-cysteine (a potent antioxidant), diphenyleneiodonium (NADPH oxidase inhibitor), IL-1Ra (IL-1 receptor antagonist) and SN-50 (specific inhibitor of NF-κB activation), but not by pentoxifylline (TNF-α release inhibitor), indomethacin and niflumic acid (COX-1 and 2 inhibitors)(Talbot et al., 2011).

In the skin, the mRNAs for hK1 tissue kallikrein and B1R have been reported in both normal human skin and in biopsies of patients with some skin diseases (Schremmer-Danninger et al.,1995; 1998; 2004; Matus et al.,2008; Pietrovski et al., 2009). Furthermore, the expression of B1R mRNA and protein has been demonstrated in keratinocytes of normal human epidermis, in primary cultures and in the immortalized HaCaT cell line (Matus et al.,2008). Additionally, the expression of B1R in skin may be enhanced when this organ is exposed to many noxious agents present in our environment. In cutaneous models of inflammation developed in mice, peptidic and non-peptidic kinin receptor antagonists have been effective in reducing cutaneous neurogenic inflammation (Pietrovski et al., 2009). Recently, Pietrovski et al.,(2011) demonstrated that both B1R and B2R participate in the hyperproliferative and inflammatory responses of mouse skin suggesting that the use of kinin receptor antagonists may be useful in the treatment of some skin disorders.

Another disease in which the role of B1R has been considered is diabetes. Dias & Couture (2011) used the B1R antagonist, SSR240612 in a rat model of diabetes and insulin resistance, based on the hypothesis that B1R activates signaling pathways that lead to increased oxidative stress and promote insulin resistance (Dias et al., 2010). The use of such an experimental design reversed the effects of glucose on the expression of IL-1β, TNF-α, and macrophage CD68 that are linked to the deposition of adipose tissue. Given the association between chronic inflammation and insulin resistance, inhibition of B1R activity may contribute to increase sensitivity to insulin and to prevent obesity. In general, diabetes induces sensory polyneuropathy associated with pain (allodynia and termal hyperalgesia), conditions that are difficult to treat. For this reason, Dias et al.,(2007) tested the non-peptide and orally active B1R antagonist SSR240612 in a rat model of insulin resistance and tactile cold allodynia. Their results are the first to associate the B1R with allodynia, pain that was alleviated by treatment with SSR240612, probably by direct inhibition of B1R affecting the spinal cord and/or afferent sensory nerve excitation. Hawkinson et al.,(2007) probed other B1R antagonist named benzamide B1R antagonist 7-chloro-2-[3-(9-pyridin—4-yl-3,9-diaza-spiro[5.5]undecanecarbonyl)phenyl]-2,3-dihydro-isoindol-1-one (ELN441958). This antagonist reduces in a dose-dependent manner the carrageenan-induced thermal hyperalgesia in a rhesus monkey tail-withdrawal model.

In renal disease, B1R has been associated with tissue inflammation and renal fibrosis. Focal and segmental glomeruloesclerosis is one of the most important causes of end-stage renal failure and up-regulation of B1R in podocyte injury has been demonstrated by Pereira et al.,(2011). Despite this report, there are not additional studies that examine the use of specific antagonists or inhibitors for specific signaling pathways activated by the B1R in this disease. Respect to the participation of the kinin B1R in cancer, only a few reports indicates the importance of interrupting its intracellular signaling; the major efforts have been orientated to the B2R. Thus, many studies have tried to elucidate the importance of B2R in cancer and the consequence of using specific antagonists to reduce cell proliferation, enhance apoptosis or decrease the secretion of metalloproteases (Chan et al., 2002; Stewart et al.,2002; 2003; 2005; Jutras et al., 2010; Wang et al., 2010). Although one of these studies demonstrate that both PGE_2 and BK stimulate invasion of head and neck squamous carcinoma cells via EGFR, the treatment of these cells with the B2R antagonist CU201 resulted in growth inhibition (Thomas et al., 2006). The combination of CU201 with the EGFR small-molecule inhibitor erlotinib resulted in additive inhibitory effects on cell growth in the same type of cells *in vitro*.

6. Conclusion

Despite the high number of reports demonstrating the involvement of B1R in the regulation of various diseases, studies on the use of specific kinin antagonists and/or kinase inhibitors as potential therapeutic drugs has been limited. Therefore, further validation of kinin receptors as important therapeutic targets is crucial. Even though EGFR tyrosine kinase inhibitors such as erlotinib and gefitinib interfere with ATP binding and have been used in clinical trials, inhibiting tyrosine kinase activity and subsequently inhibiting signal transduction pathways triggered by EGFR such as MAPK, PI3K/Akt, STAT and PLCγ/PKC. Activation of these pathways also influences cell proliferation, survival and the metastatic potential of tumor cells. Others EGFR tyrosine kinase inhibitors (PD15035, AG1478 and PKI 166) have been tested in animal model as mice with Type 2 Diabetic, mice with high-fat diet-fed or rats with diabetic nephropathy. Results of experiments indicate that these inhibitors restore ischemia-induced neovascularization and blood flow recovery in type 2 diabetic mice, improves glucose tolerance, insulin sensitivity and reduces subclinical inflammation in high-fat diet-fed mice and attenuates the development diabetic nephropathy in rats. (Advani et al., 2011; Choi et al., 2011; Prada et al., 2009).

Despite the success of inhibitors reported in clinical trials, some patients still develop resistance to anti-EGFR therapy. In addition, there are two mutations of EGFR that have been described. Mutations of the exon coding for tyrosine kinase domain (18 to 21) and truncating mutations (exons 2 to 7) that involves signaling pathways. The first group of mutations is involved in the failure of tyrosine kinase inhibitors whereas the second group of mutation leads to resistance of drugs (Colabufo, 2011). However, an alternative approach for overcoming resistance may be to simultaneously target additional pathways such as VEGF/VEGFR, and/or inhibit parallel signaling pathways that complement those activated by EGFR (e.g., mesenchymal-epithelial transition factor and mTOR) (Giaccone & Wang, 2011) and we suggest, by the evidence expose in this chapter, may include the B1R, or B2R. Furthermore, understanding of the intracellular signaling pathways initiated by kinin receptors linked to EGFR and their involvement in the causation and progression of most human neoplasias (e.g., non-small cell lung cancer, colorectal cancer, glioblastoma, breast cancer and squamous cell carcinomas) suggests that these signaling pathways could be used as targets for development of novel drugs that inhibit EGFR activation. Currently there are insufficient studies on the inhibition of EGFR transactivation triggered by B1R to elucidate whether the interaction between EGFR and B1R could provide novel molecular agents for prevention or treatment of human cancer.

7. Acknowledgments

The authors wish to thank the financial support of grants 11090292 (PE) and 1110464 (CDF) from FONDECYT (Chile) and grant 24090029 from CONICYT (CEM).

8. References

Adomeit, A.; Graness, A.; Gross, S.; Seedorf, K.; Wetzker, R. & Liebmann, C. (1999). Bradykinin B(2) receptor-mediated mitogen-activated protein kinase activation in COS-7 cells requires dual signaling via both protein kinase C pathway and epidermal growth factor receptor transactivation. *Molecular & Cellular Biology*, Vol.19, N°8, (August 1999), pp. 5289-1297, ISSN 0270-7306

Advani, A.; Wiggins, KJ.; Cox, AJ.; Zhang, Y.; Gilbert, RE. & Kelly, DJ. (2011). Inhibition of the epidermal growth factor receptor preserves podocytes and attenuates albuminuria in experimental diabetic nephropathy. *Nephrology (Carlton)*, Vol.16, N°6, (August 2011), pp. 573-581, ISSN 1320-5358

Ahluwalia, A. & Perretti, M. (1996). Involvement of bradykinin B1 receptors in the polymorphonuclear leukocyte accumulation induced by IL-1 beta in vivo in the mouse. *Journal of Immunology*, Vol.156, N°1, (January 1996), pp. 269-274, ISSN 0022-1767

Austin, CE.; Faussner, A.; Robinson, HE.; Chakravarty, S.; Kyle, DJ.; Bathon, JM. & Proud D. (1997). Stable expression of the human kinin B1 receptor in Chinese hamster ovary cells. Characterization of ligand binding and effector pathways. *Journal of Biological Chemistry* Vol.272, N° 17, (April 1997), pp. 11420-11425, ISSN 0021-9258

Bachvarov, DR.; Hess, JF.; Menke, JG.; Larrivée, JF. & Marceau, F. (1996). Structure and genomic organization of the human B1 receptor gene for kinins (BDKRB1). *Genomics*, Vol.33, N°3, (May 1996), pp. 374-381, ISSN 0888-7543

Barki-Harrington L., Bookout AL., Wang G., Lamb ME., Leeb-Lundberg LM. & Daaka Y. (2003). Requirement for direct cross-talk between B1 and B2 kinin receptors for the proliferation of androgen-insensitive prostate cancer PC3 cells. *Biochemistry*, Vol.2, Pt2, (April 2003), pp. 581-587, ISSN 0006-2960

Bhoola KD.; Figueroa CD. & Worthy K. (1992). Bioregulation of kinins: kallikreins, kininogens, and kininases. *Pharmacological Reviews*, Vol.44, N°1 (March 1992), pp. 1-80, ISSN 0031-6997

Bhoola KD.; Ramsaroop, R.; Plendl, J.; Cassim, B.; Dlamini, Z. & Naicker, S. (2001). Kallikreins and kinin receptor expression in inflammation and cancer. *Biological Chemistry* Vol.382, N°1, (January 2001), pp. 77-89, ISSN 1431-6730

Biscardi, JS.; Maa, MC.; Tice, DA.; Cox, ME.; Leu, TH. & Parsons, SJ. (1999). c-Src-mediated phosphorylation of the epidermal growth factor receptor on Tyr845 and Tyr1101 is associated with modulation of receptor function. *Journal of Biological Chemistry*, Vol.274, N° (March 1999), pp. 8335-8343, ISSN 0021-9258

Blaukat A. (2004). Identification of G-protein-coupled receptor phosphorylation sites by 2D phosphopeptide mapping. *Methods in Molecular Bio*logy, Vol. 259, pp. 283-297, ISSN 1064-3745

Boerner, JL.; Biscardi, JS.; Silva, CM. & Parsons, SJ. (2005). Transactivating agonist of the EGF receptor require Tyr 845 phosphorylation for induction of DNA synthesis. *Molecular Carcinogenics*, Vol.44, N°4, (December 2005), pp. 262-273, ISSN 0899-1987

Bryborn, M.; Adner, M. & Cardell, LO. (2004). Interleukin-4 increases murine airway response to kinins, via up-regulation of bradykinin B1-receptors and altered signalling along mitogen-activated protein kinase pathways. *Clinical & Experimental Allergy*, Vol.34, N°8, (August 2004), pp. 1291-1298, ISSN 0954-7894

Campos, MM.; Leal, PC.; Yunes, RA. & Calixto, JB. (2006). Non-peptide antagonists for kinin B1 receptors: new insights into their therapeutic potential for the management of inflammation and pain. *Trends in Pharmacological Sciences*, Vol.27, N°12, (December 2006), pp. 646-651, ISSN 0165-6147

Chan, DC.; Gera, L.; Stewart, JM.; Helfrich, B.; Zhao, TL.; Feng, WY.; Chan, KK.; Covey, JM. & Bunn, PA Jr. (2002). Bradykinin antagonist dimer, CU201, inhibits the growth of human lung cancer cell lines in vitro and in vivo and produces synergistic growth inhibition in combination with other antitumor agents. *Clinical Cancer Research*, Vol.8, N°5, (May 2002), pp. 1280-1287, ISSN 1078-0432

Chen, H.; Ma, N.; Xia, J.; Liu, J. & Xu, Z. (2011). β2-Adrenergic Receptor-induced Transactivation of EGFR and PDGFR via Src kinase promotes rat cardiomyocytes survival. *Cell Biology International*, (September 2011), [Epub ahead of print], doi:10.1042/CBI20110162, ISSN 1065-6995

Choi, SK.; Galán, M.; Partyka, M.; Trebak, M.; Belmadani, S. & Matrougui, K. (2011). Chronic inhibition of epidermal growth factor receptor tyrosine kinase and extracellular signal-regulated kinases 1 and 2 (ERK1/2) augments vascular response to limb ischemia in type 2 diabetic mice. *American Journal of Pathology*, (November 2011), [Epub ahead of print], ISSN 0002-9440

Christopher, J.; Velarde, V. & Jaffa, AA. (2001). Induction of B(1)-kinin receptors in vascular smooth muscle cells: cellular mechanisms of map kinase activation. *Hypertension*, Vol.38, N°3 Pt2, (September), pp. 602-605, ISSN 0194-911X

Colabufo, NA.; Contino, M.; Niso, M.; Berardi, F.; Leopoldo, M. & Perrone, R. (2011). EGFR tyrosine kinase inhibitors and multidrug resistance: perspectives. *Frontiers in Bioscience*, Vol.16, (January 2011), pp. 1811-1823, ISSN 1093-9946

Cooper, JA. & Howell, B. (1993). The when and how of Src regulation. *Cell*, Vol.73, N°6, (June 1993), pp. 1051-1054, ISSN 0092-8674

Couture, R. & Girolami, JP. (2004) Putative roles of kinin receptors in the therapeutic effects of angiotensin 1-converting enzyme inhibitors in diabetes mellitus. *European Journal of Pharmacology*, Vol.500, N°1-3, (October 2004), pp. 467-485, ISSN 0014-2999

Cramer, H.; Schmenger, K.; Heinrich, K.; Horstmeyer, A.; Boning, H.; Breit, A.; Piiper, A.; Lundstrom, K.; Muller-Esterl, W.; Schroeder, C. (2001). Coupling of endothelin receptors to the ERK/MAP kinase pathway. Roles of palmitoylation and Gαq. *Eur J Biochem*, 268:5449-5459.

Daub, H.; Weiss, FU.; Wallasch, C.; Ullrich, A. (1996). Role of transactivation of the EGF receptor in signalling by G-protein-coupled receptors. *Nature*. 8; 379 (6565):557-560.

Dey, M.; Baldys, A.; Sumter, DB.; Göoz, P.; Luttrell, LM.; Raymond, JR.; Göoz, M. (2010). Bradykinin decreases podocyte permeability through ADAM17-dependent epidermal growth factor receptor activation and zonula occludens-1 rearrangement. *J Pharmacol Exp Ther*. 334:775-783.

Dias, JP.; Couture, R. (2011). Blockade of kinin B(1) receptor reverses plasma fatty acids composition changes and body and tissue fat gain in a rat model of insulin resistance. *Diabetes Obes Metab*. 24. doi: 10.1111/j.1463-1326.2011.01521.x. [Epub ahead of print]

Dias, JP.; Ismael, MA.; Pilon, M.; de Champlain, J.; Ferrari, B.; Carayon, P.; Couture, R. (2007). The kinin B1 receptor antagonist SSR240612 reverses tactile and cold allodynia in an experimental rat model of insulin resistance. *Br J Pharmacol*. 152:280-287.

Dias, JP.; Talbot, S.; Sénécal, J.; Carayon, P.; Couture, R. (2010). Kinin B1 receptor enhances the oxidative stress in a rat model of insulin resistance: outcome in hypertension, allodynia and metabolic complications. *PLoS One*. ;5:e12622.

Dixon, BS.; Evanoff, D.; Fang, WB.; Dennis, MJ. (2002) Bradykinin B1 receptor blocks PDGF-induced mitogenesis by prolonging ERK activation and increasing p27Kip1. *Am J Physiol Cell Physiol*. 283:C193-203.

Ehrenfeld, P.; Conejeros, I.; Pavicic, MF.; Matus, CE.; Gonzalez, CB.; Quest, AF.; Bhoola, KD.; Poblete, MT.; Burgos, RA.; Figueroa, CD. (2011). Activation of kinin B1 receptor increases the release of metalloproteases-2 and -9 from both estrogen-sensitive and -insensitive breast cancer cells. *Cancer Lett*. 301:106-118.

Ehrenfeld, P.; Matus, CE.; Pavicic, F.; Toledo C.; Nualart, F.; Gonzalez, CB.; Burgos, RA.; Bhoola, KD.; Figueroa, CD. (2009). Kinin B_1 receptor activation turns on exocytosis of matrix metalloprotease-9 and myeloperoxidase in human neutrophils: involvement of mitogen-activated protein kinase family. *J Leukoc Biol*. 86: 1179-1189.

Ehrenfeld, P.; Millan, C.; Matus, CE.; Figueroa, J.; Burgos, R.; Nualart, F.; Bhoola, KD.; Figueroa, CD. (2006). Activation of kinin B_1 receptors induces chemotaxis of human neutrophils. *J Leukoc Biol*. 80:117-124.

Endoh, H.; Ishibashi, Y.; Yamaki, Ei.; Yoshida, T.; Yajima, T.; Kimura, H.; Kosaka, T.; Onozato, R.; Tanaka, S.; Mitsudomi, T.; Kuwano, H. (2009). Immunohistochemical analysis of phosphorylated epidermal growth factor receptor might provide a surrogate marker of EGFR mutation. *Lung Cancer*. 63: 241-246.

Enquist, J.; Skröder, C.; Whistler, JL.; Leeb-Lundberg, LM. (2007). Kinins promote B2 receptor endocytosis and delay constitutive B1 receptor endocytosis. *Molecular Pharmacology*. Vol 71, N°2, (February 2007), pp.494-507, ISSN 1521-0111.

Emlet, DR.; Moscatello, DK.; Ludlow, LB.; Wong, AJ. (1997). Subsets of epidermal growth factor receptors during activation and endocytosis. *J Biol Chem*. 272:4079-4086.

Ettenberg, SA.; Keane, MM.; Nau, MM.; Frankel, M.; Wang, LM.; Pierce, JH.; Lipkowitz, S. (1999). cbl-b inhibits epidermal growth factor receptor signaling. *Oncogene*. 18:1855-1866.

Giaccone, G.; Wang, Y. (2011). Strategies for overcoming resistance to EGFR family tyrosine kinase inhibitors. *Cancer Treat Rev*. 37:456-464.

Hawkinson, JE.; Szoke, BG.; Garofalo, AW.; Hom, DS.; Zhang, H.; Dreyer, M., Fukuda, JY., Chen L.; Samant, B.; Simmonds, S.; Zeitz, KP.; Wadsworth, A:, Liao, A.; Chavez, RA.; Zmolek, W.; Ruslim, L.; Bova, MP.; Holcomb, R.; Butelman, ER.; Ko, MC.; Malmberg AB. (2007). Pharmacological, pharmacokinetic, and primate analgesic efficacy profile of the novel bradykinin B1 Receptor antagonist ELN441958. *J Pharmacol Exp Ther*. 322:619-630.

Hubbard, SR.; Wei, L.; Ellis, L.; Hendrickson, WA. (1994). Crystal structure of the tyrosine kinase domain of the human insulin receptor. *Nature*. ;372:746-754.

Hur, EM.; Park, YS.; Lee, BD.; Jang, IH.; Kim, TD.; Suh, PG.; Ryu, SH.; Kim, KT. (2004).Sensitization of epidermal growth factor-induced signaling by bradykinin is mediated by c-Src. Implications for a role of lipid microdomains. *J Biol Chem*. 279: 5852-5860.

Jutras, S.; Bachvarova, M.; Keita, M.; Bascands, JL.; Mes-Masson, AM.; Stewart, JM.; Buchvarov, D. (2010). Strong cytotoxic effect of the bradykinin antagonist BKM-570 in ovarian cancer cells-analysis of the molecular mechanism of its antiproliferative action. *FEBS J*. 277:5146-5160. doi: 10.1111/j.1742-4658.2010.07928.

Kannangai, R.; Sahin, F.; Torbenson, MS. (2006). EGFR is phosphorylated at Tyr 845 in hepatocellular carcinoma. *Mod Pathol.* 19: 1456-1461.

Kang, DS.; Gustafsson, C.; Mörgelin, M.; Leeb-Lundberg, LM. (2005). B1 bradykinin receptor homo-oligomers in receptor cell surface expression and signaling: effects of receptor fragments. *Mol Pharmacol.* 67:309-318.

Leeb-Lundberg, LM.; Marceau, F.; Müller-Esterl, W.; Pettibone, DJ.; Zuraw, BL. (2005). International union of pharmacology. XLV. Classification of the kinin receptor family: from molecular mechanisms to pathophysiological consequences. *Pharmacol Rev.* 57: 27-77.

Lemjabbar-Alaoui, H.; Sidhu, SS.; Mengistab, A.; Gallup, M.; Basbaum, C. (2011). TACE/ADAM-17 phosphorylation by PKC-epsilon mediates premalignant changes in tobacco smoke-exposed lung cells. *PLoS One.* 15;6(3):e17489.

Levkowitz, G.; Waterman, H.; Ettenberg, SA.; Katz, M.; Tsygankov, AY.; Alroy, I.; Lavi, S.; Iwai, K.; Reiss, Y.; Ciechanover, A.; Lipkowitz, S.; Yarden, Y. (1999). Ubiquitin ligase activity and tyrosine phosphorylation underlie suppression of growth factor signaling by c-Cbl/Sli-1. *Mol Cell.* 4:1029-40.

Liebmann, C. (2011). EGF receptor activation by GPCRs: an universal pathway reveals different versions. *Mol Cell Endocrinol.* 331:222-231.

Lin, JC.; Talbot, S.; Lahjouji, K.; Roy, JP.; Sénécal, J.; Couture, R.; Morin, A. (2010). Mechanism of cigarette smoke-induced kinin B(1) receptor expression in rat airways. *Peptides.* 31:1940-1945.

MacNeil, T.; Bierilo, KK.; Menke, JG.; Hess, JF. (1995) Cloning and pharmacological characterization of a rabbit bradykinin B1 receptor. *Biochim Biophys Acta.* 1264:223-228.

Marceau, F.; Bachvarov, DR. (1998). Kinin receptors. *Clin Rev Allergy Immunol.* 16:385-401.

Marceau, F.; Sabourin, T.; Houle, S.; Fortin, JP.; Petitclerc, E.; Molinaro, G.; Adam A. (2002) Kinin receptors: functional aspects. *Int Immunopharmacol.* 2:1729-1739.

Marceau, F.; Tremblay B. (1986) Mitogenic effect of bradykinin and of des-Arg9-bradykinin on cultured fibroblasts. *Life Sci.* 39:2351-2358.

Maretzky, T.; Zhou, W.; Huang, XY.; Blobel, CP. (2011). A transforming Src mutant increases the bioavailability of EGFR ligands via stimulation of the cell-surface metalloproteinase ADAM17. *Oncogene.* 3;30(5):611-618.

Mathis, SA.; Criscimagna, NL.; Leeb-Lundberg, LM. (1996). B1 and B2 kinin receptors mediate distinct patterns of intracellular Ca2+ signaling in single cultured vascular smooth muscle cells. *Molecular Pharmacology.* Vol 50, N°1, (July 1996), pp. 128-39, ISSN 1521-0111.

Matus, CE.; Ehrenfeld, P.; Pavicic, F.; Sarmiento, JM.; Astroza, A.; Sanchez, T;. Salem, C.; Concha, M.; Vidal, MA.; Gonzalez, CB.; Figueroa, CD. (2008). Activation of kinin B receptor triggers differentiation of cultured human keratinocytes. *Br J Dermatol.* 159:792-803.

Menke, JG.; Borkowski, JA.; Bierilo, KK.; MacNeil, T.; Derrick, AW.; Schneck, KA.; Ransom, RW.; Strader, CD.; Linemeyer, DL.; Hess, JF. (1994) Expression cloning of a human B1 bradykinin receptor. *J Biol Chem.* 269:21583-21586.

Methner, C.; Donat, U.; Felix, SB.; Krieg, T. (2009). Cardioprotection of bradykinin at reperfusion involves transactivation of the epidermal growth factor receptor via matrix metalloproteinase-8. *Acta Physiol (Oxf).* 197:265-271.

Miyawaki, M.; Hijiya, N.; Tsukamoto, Y.; Nakada, C.; Kawahara, K.; Moriyama, M. (2008). Enhanced phosphorylation of the epidermal growth factor receptor at the site of tyrosine 992 in esophageal carcinomas. *APMIS*. 116: 1097-1106.

Molina, L.; Matus, CE.; Astroza, A.; Pavicic, F.; Tapia, E.; Toledo, C.; Perez, JA.; Nualart, F.; Gonzalez, CB.; Burgos, RA.; Figueroa, CD.; Ehrenfeld, P.; Poblete, MT. (2009). Stimulation of the bradykinin B(1) receptor induces the proliferation of estrogen-sensitive breast cancer cells and activates the ERK1/2 signaling pathway. *Breast Cancer Res Treat*. 118:499-510.

Morinelli, TA.; Meier, GP.; Webb, JG.; Jaffa, AA.; Privitera, PJ.; Margolius, HS. (2002). Utilization of a radioimmunoassay to detect the generation of Arg-Pro-Pro-Gly-Phe, a stable endproduct of bradykinin metabolism (from cultured rat aortic smooth muscle cells exposed to bradykinin). *Int Immunopharmacol*. 2:1995-2003.

Morissette G, Sabourin T, Adam A, Marceau F. (2006). Inhibition of human and rabbit arterial smooth muscle cell migration mediated by the kinin B1 receptor: role of receptor density and released mediators. *Can J Physiol Pharmacol*, Vol. 84, N°11, (November 2006), pp1107-1119, ISSN

Mugabe, BE.; Yaghini, FA.; Song, CY.; Buharalioglu, CK.; Waters, CM.; Malik, KU. (2010). Angiotensin II-induced migration of vascular smooth muscle cells is mediated by p38 mitogen-activated protein kinase-activated c-Src through spleen tyrosine kinase and epidermal growth factor receptor transactivation. *J Pharmacol Exp Ther*. 332:116-124.

Mukhin, YV.; Gooz, M.; Raymond, JR.; Garnovskaya, MN. (2006). Collagenase-2 and -3 mediate epidermal growth factor receptor transactivation by bradykinin B2 receptor in kidney cells. *J Pharmacol Exp Ther*. 318:1033-1043.

Murphey, LJ.; Hachey, DL.; Oates, JA.; Morrow, JD.; Brown NJ. (2000) Metabolism of bradykinin In vivo in humans: identification of BK1-5 as a stable plasma peptide metabolite. *J Pharmacol Exp Ther*. 294:263-269.

Murphey, LJ.; Malave, HA.; Petro, J.; Biaggioni, I.; Byrne, DW.; Vaughan, DE.; Luther, JM.; Pretorius, M.; Brown, NJ. (2006) Bradykinin and its metabolite bradykinin 1-5 inhibit thrombin-induced platelet aggregation in humans. *J Pharmacol Exp Ther*. 318:1287-1292.

Ni, A.; Chai, KX.; Chao, L.; Chao, J. (1998). Molecular cloning and expression of rat bradykinin B1 receptor. *Biochim Biophys Acta*. 1442:177-185.

Ohtsu, H.; Dempsey, PJ.; Eguchi, S. (2006). ADAMs as mediators of EGF receptor transactivation by G protein-coupled receptors. *Am J Physiol Cell Physiol*. 291(1):C1-10.

Paegelow, I.; Trzeczak, S.; Böckmann, S.; Vietinghoff, G. (2002) Migratory responses of polymorphonuclear leukocytes to kinin peptides. *Pharmacology*. 66:153-161.

Pereira, RL.; Buscariollo, BN.; Corrêa-Costa, M.; Semedo, P.; Oliveira, CD.; Reis, VO.; Maquigussa, E.; Araújo, RC.; Braga, TT.; Soares, MF.; Moura, IC.; Malheiros, DM.; Filho, AP.; Keller, AC.; Câmara, NO. (2011). Bradykinin receptor 1 activation exacerbates experimental focal and segmental glomerulosclerosis. *Kidney Int*. 79:1217-1227. doi: 10.1038/ki.2011.14. Epub 2011 Mar 16.

Pietrovski, EF.; Otuki, MF.; Regoli, D.; Bader, M.; Pesquero, JB.; Cabrini, DA.; Zampronio, AR. (2009). The non-peptide kinin receptor antagonists FR 173657 and SSR

240612: preclinical evidence for the treatment of skin inflammation. *Regul Pept.* 152:67-72.

Pietrovski, EF.; Paludo, KS.; Mendes, DA.; Guimarães, Fde S.; Veiga, SS.; Buchi, Dde F.; Fonseca, RG.; Zampronio, AR.; Bader, M.; Pesquero, JB.; Ferreira, J.; Otuki, MF.; Cabrini, DA. (2011). B1 and B2 kinin receptor participation in hyperproliferative and inflammatory skin processes in mice. *J Dermatol Sci.* 64:23-30.

Prada, PO.; Ropelle, ER.; Mourão, RH.; de Souza, CT.; Pauli, JR.; Cintra, DE.; Schenka, A.; Rocco, SA.; Rittner, R.; Franchini, KG.; Vassallo, J.; Velloso, LA.; Carvalheira, JB.; Saad, MJ. (2009). EGFR tyrosine kinase inhibitor (PD153035) improves glucose tolerance and insulin action in high-fat diet-fed mice. *Diabetes.* 58:2910-2919.

Prado, GN.; Taylor, L.; Zhou, X.; Ricupero, D.; Mierke, DF.; Polgar, P. (2002) Mechanisms regulating the expression, self-maintenance, and signaling-function of the bradykinin B2 and B1 receptors. *J Cell Physiol.* 193:275-286.

Regoli, D.; Rizzi, A.; Perron, SI.; Gobeil, F Jr. (2001). Classification of kinin receptors. *Biol Chem.* 382:31-35.

Rojas, M.; Yao, S.; Lin, YZ. (1996). Controlling epidermal growth factor (EGF)-stimulated Ras activation in intact cells by a cell-permeable peptide mimicking phosphorylated EGF receptor. *J Biol Chem.* 271:27456-27461.

Rozengurt, E. (2007). Mitogenic signaling pathways induced by G protein-coupled receptors. *J Cell Physiol.* 213:589-602.

Schremmer-Danninger, E.; Heinz-Erian, P.; Töpfer-Petersen, E.; Roscher, AA. (1995). Autoradiographic localization and characterization of bradykinin receptors in human skin. *Eur J Pharmacol.* 283:207-216.

Sato, K.; Nagao, T.; Iwasaki, T.; Nishihira, Y. & Fukami, Y. (2003). Src-dependent phosphorylation of the EGF receptor Tyr-845 mediates Stat-p21waf1 pathway in A431 cells. *Genes to cells : devoted to molecular & cellular mechanisms,* Vol 8, N°12, (December 2008), pp. 995-1003, ISSN 1356-9597

Schremmer-Danninger, E.; Heinz-Erian, P.; Töpfer-Petersen E. & Roscher AA. (1995). Autoradiographic localization and characterization of bradykinin receptors in human skin. *European journal of pharmacology,* Vol 283, N° 1-3 (September 1995), pp. 207-216, ISSN 0014-2999

Schremmer-Danninger, E.; Toepfer-Petersen, E.; Fritz H. & Roscher AA. (1998). Bradykinin-induced tyrosine phosphorylation of proteins in cultured human keratinocytes. *Biological research,* Vol 31, N°3, (1998), pp. 189-198, ISSN 0716-9760

Schremmer-Danninger, E.; Naidoo, S.; Neuhof, C.; Valeske, K.; Snyman, C.; Sander, C.; Bhoola, KD. & Neuhof H. (2004). Visualisation of tissue kallikrein, kininogen and kinin receptors in human skin following trauma and in dermal diseases. *Biological chemistry,* Vol 385, N° 11 , (Nov 2004), pp.1069-1076, ISSN 1431-6730

Shima, C.; Majima, M. & Katori, M. (1992) A stable metabolite, Arg-Pro-Pro-Gly-Phe, of bradykinin in the degradation pathway in human plasma. *Japanese journal of pharmacology,* Vol 60, N°2, (October 1992), pp.111-119, ISSN 0021-5198

Stewart, JM. (2003). Bradykinin antagonists as anti-cancer agents. *Current pharmaceutical design.* Vol 9, N°25, (2003), pp.2036-2042, ISSN 1381-6128

Stewart, JM.; Gera, L.; Chan, DC.; Bunn, PA Jr.; York, EJ.; Simkeviciene, V. & Helfrich B. (2002). Bradykinin-related compounds as new drugs for cancer and inflammation.

Canadian journal of physiology and pharmacology, Vol 80, N°25, (April 2002), pp. 275-280, ISSN 0008-4212

Stewart, JM.; Gera, L.; Chan, DC.; York, EJ.; Simkeviciene, V.; Bunn, PA Jr. & Taraseviciene-Stewart L. (2005). Combination cancer chemotherapy with one compound: pluripotent bradykinin antagonists. *Peptides*, Vol 26, N°8, (August 2005), pp.1288-1291, ISSN 0196-9781

Stirnweiss, J.; Valkova, C.; Ziesché, E.; Drube, S. & Liebmann, C. (2006). Muscarinic M2 receptors mediate transactivation of EGF receptor through Fyn kinase and without matrix metalloproteases. *Cell Signal*ing, Vol 18, N° 8, (August), pp. 1338-1349, ISSN 0898-6568

Stuardo, M.; Gonzalez, CB.; Nualart, F.; Boric, M.; Corthorn, J.; Bhoola, KD. & Figueroa, CD. (2004) Stimulated human neutrophils form biologically active kinin peptides from high and low molecular weight kininogens. *Journal of leukocyte biology*, Vol 75, N°4, (April 2004), pp. 631-640, ISSN 0741-5400

Sturla, LM.; Amorino, G.; Alexander, MS.; Mikkelsen, RB.; Valerie, K. & Schmidt-Ullrichr, RK. (2005). Requirement of Tyr-992 and Tyr-1173 in phosphorylation of the epidermal growth factor receptor by ionizing radiation and modulation by SHP2. *The Journal of Biological Chemistry*, Vol 280, N° 15 , (April 2005), pp. 14597-14604, ISSN 0021-9258

Sun, C.; Wu, MH.; Guo, M.; Day, ML.; Lee, ES. & Yuan, SY. (2010). ADAM15 regulates endothelial permeability and neutrophil migration via Src/ERK1/2 signalling. *Cardiovascular Research, Vol* 87, N°2, (2010), pp. 348-355, ISSN 0008-6363

Talbot, S.; Lin, JC.; Lahjouji, K.; Roy, JP.; Sénécal, J.; Morin, A. & Couture, R. (2011). Cigarette smoke-induced kinin B1 receptor promotes NADPH oxidase activity in cultured human alveolar epithelial cells. *Peptides*, Vol 32, N° 7, (May 2011), pp.1447-1456, ISSN 0196-9781

Taub, JS.; Guo, R.; Leeb-Lundberg, LM.;, Madden JF. & Daaka, Y. (2003). Bradykinin receptor subtype 1 expression and function in prostate cancer. *Cancer Research*, Vol 63, N°9, (May 2003), pp. 2037-2041, ISSN 0008-5472

Thomas, SM.; Bhola, NE.; Zhang, Q.; Contrucci, SC.; Wentzel, AL. & Freilino, ML. (2006). Cross-talk between G protein-coupled receptor and epidermal growth factor receptor signaling pathways contributes to growth and invasion of head and neck squamous cell carcinoma. *Cancer Research*, Vol 66, N°24, (2006), pp.11831-11839, ISSN 0008-5472

Tice, DA.; Biscardi, JS.; Nickles, AL. & Parsons, SJ. (1999). Mechanism of biological synergy between cellular Src and epidermal growth factor receptor. *Proceedings of the National Academy of Sciences of the United States of America*, Vol 96, N°4, (February 1999), pp.1415–1420, ISSN 0027-8424

Tropea, MM.; Gummelt, D.; Herzig, MS. & Leeb-Lundberg, LM. (1993). B1 and B2 kinin receptors on cultured rabbit superior mesenteric artery smooth muscle cells: receptor-specific stimulation of inositol phosphate formation and arachidonic acid release by des-Arg9-bradykinin and bradykinin. *The Journal of pharmacology and experimental therapeutics*, Vol 264, N°2, (February 1993), pp.930-937, ISSN 0022-3565

Vidal, MA.; Astroza, A.; Matus, CE.; Ehrenfeld, P.; Pavicic, F.; Sanchez, T.; Salem, C.; Figueroa, J.; Concha, M.; Gonzalez, CB. & Figueroa, CD. (2005). Kinin B2 receptor-coupled signal transduction in human cultured keratinocytes. *The Journal of investigative dermatology,* Vol 124, N° 1, (January 2005), pp.178-186, ISSN 0022-202

Wang, J.; Krishnamoorthi, V.; Wang E.; Yang C.; Baptista, D.; Wu, X.; Liu, M.; Gardner, M.; Elkins P.; Hines, J. & Liu, P. (2010). LC/MS characterization of impurities and degradation products of a potent antitumor peptidic dimer, CU201. *Journal of pharmaceutical and biomedical analysis,* Vol 51, N°4, (March 2010), pp. 824-833, ISSN 0731-7085

Wetzker, R. & Böhmer, FD. (2003). Transactivation joins multiple tracks to the ERK/MAPK cascade. *Nature reviews. Molecular cell biology,* Vol 4, N°8, (August 2010), pp. 651-657, ISSN 1471-0072

Xu, CB.; Lei, Y.; Chen, Q.; Pehrson, C.; Larsson, L. & Edvinsson, L. (2010). Cigarette smoke extracts promote vascular smooth muscle cell proliferation and enhances contractile responses in the vasculature and airway. *Basic and clinical pharmacology and toxicology,* Vol 107, N°6, (December 2010), pp.940-948, ISSN 1742-7835

Yang, CM.; Lin, MI.; Hsieh, HL.; Sun, CC.; Ma, YH. & Hsiao, LD. (2005). Bradykinin-induced p42/p44 MAPK phosphorylation and cell proliferation via Src, EGF receptors, and PI3-K/Akt in vascular smooth muscle cells. *Journal of cellular physiology,* Vol 203, N°3, (June 2005), pp. 538-546, ISSN 0021-9541

Yang, X.; Taylor, L.; Yu, J.; Fenton, MJ. &Polgar, P. (2001). Mediator caused induction of a human bradykinin B1 receptor minigene: participation of c-Jun in the process. *Journal of cellular biochemistry,* Vol 82, N°1, (April 2001), pp.163-170, ISSN 0730-2312

Yamauchi, T.; Ueki, K.; Tobe, K.; Tamemoto, H.; Sekine, N.; Wada, M.; Honjo, M.; Takahashi, M.; Hirai, H.; Tsushima, T.; Akanuma, Y.; Fujita, T.; Komuro, I.; Yazaki, Y. & Kadowaki, T. (1998). Growth hormone-induced tyrosine phosphorylation of EGF receptor as an essential element leading to MAP kinase activation and gene expression. *Endocrine journal,* Vol, 45, N°, (April 1998), pp. S27-31, ISSN 0918-8959

Yurko, MA.; O'Toole, EA. & Woodley, DT. (2001). Phosphorylation of focal adhesion kinase (pp125(FAK)) is increased in human keratinocytes induced to migrate by extracellular matrices. *Journal of cellular physiology,* Vol.188, N°1, (July 2001), pp.24-32, ISSN 0021-9541

Zhang, Y.; Adner, M. & Cardell, LO. (2007a). IL-1beta-induced transcriptional up-regulation of bradykinin B1 and B2 receptors in murine airways. *American journal of respiratory cell and molecular biology.* Vol.36, N°6, (June 2007), pp.697-705, ISSN 1044-1549

Zhang, Y.; Brovkovych, V.; Brovkovych, S.; Tan, F.; Lee, BS.; Sharma, T. & Skidgel, RA. (2007b). Dynamic receptor-dependent activation of inducible nitric-oxide synthase by ERK-mediated phosphorylation of Ser745. *Journal Biological Chemistry,* Vol. 282, N°44, (November 2007), pp.32453-32461, ISSN 0021-9258

Zhang, X.; Tan, F.; Brovkovych, V.; Zhang, Y. & Skidgel, RA. (2011). Cross-talk between carboxyeptidase M and the kinin B1 receptor mediates a new mode of G protein-coupled receptor signaling. *Journal Biological Chemistry,* Vol. 286, N° 27, (May 2011), pp.18547-18561, ISSN 0021-9258

Zwick, E.; Hackel, PO.; Prenzel, N.; Ullrich, A. (1999). The EGF receptor as central transducer of heterologous signalling systems. *Trends Pharmacol*ogy Science, Vol.20, N° 10 , (October 1999), pp.408-412, ISSN:0165-6147

6

Multiple Kinase Involvement in the Regulation of Vascular Growth

Shaquria P. Adderley, Chintamani N. Joshi,
Danielle N. Martin, Shayna Mooney and David A. Tulis
*East Carolina University, Brody School of Medicine,
Department of Physiology , Greenville NC,
USA*

1. Introduction

The initial discovery of protein phosphorylation as a regulatory mechanism for the control of glycogen metabolism has led to intense interest of protein phosphorylation in regulating protein function (Cohen *et al.*, 2001). Kinases play a variety of roles in many physiological processes within cells and represent one of the largest families in the human genome with over 500 members comprising protein serine/threonine, tyrosine, and dual-specificity kinases (Manning *et al.*, 2002). Phosphorylation of proteins is one of the most significant signal transduction mechanisms which regulate intracellular processes such as transport, growth, metabolism, apoptosis, cystoskeletal arrangement and hormone responses (Bononi *et al.*, 2011; Heidenreich *et al.*, 1991; Manning *et al.*, 2002; Pawson *et al.*, 2000). As such, abnormal phosphorylation of proteins can be either a cause or a consequence of disease. Kinases are regulated by activator and inhibitor proteins, ligand binding, and phosphorylation by other proteins or via autophosphorylation (Hanks *et al.*, 1991; Hug *et al.*, 1993; Scott, 1991; Taylor *et al.*, 1990; Taylor *et al.*, 1992). Since kinases play key functions in many cellular processes, they represent an attractive target for therapeutic interventions in many disease states such as cancer, inflammation, diabetes and arthritis (Cohen *et al.*, 2010; Fry *et al.*, 1994; Karin, 2005; Mayers *et al.*, 2005). In particular, the serine/threonine family of kinases comprises approximately 125 of the 500 family of kinases and includes the cyclic adenosine monophosphate (cAMP)-dependent protein kinase (PKA), the cyclic guanosine monophosphate (cGMP)-dependent protein kinase (PKG), and protein kinase C (PKC). These kinases are implicated in the regulation of cell growth and are the focus of this current study.

Nitric oxide (NO) is the primary endogenous activator of soluble guanylate cyclase (sGC) and cGMP generation in the vasculature with PKG historically acknowledged as its primary downstream kinase (Ignarro, 1991; Ignarro *et al.*, 1985). However, this area of research is controversial since the effects of NO are tightly regulated and specific to certain sub-cellular microenvironments. The NO-sGC-cGMP signaling pathway is of major importance in the cardiovascular system where it modulates functions such as vascular smooth muscle cell (VSMC) relaxation and growth (Ignarro, 1991; Ignarro *et al.*, 1986; Ignarro *et al.*, 1985; Lucas *et al.*, 2000). Vascular remodeling is associated with injury and the pathogenesis of disorders such as atherosclerosis and it is well established that the proliferation and migration of VSMCs play a major role in these processes (Gibbons *et al.*, 1994; Herity *et al.*, 1999; Shi *et al.*, 1996). The molecular mechanisms of cGMP signaling are not well understood, since studies suggest that NO can both promote and inhibit pathological vascular remodeling (Kawashima *et al.*, 2001; Ozaki *et al.*, 2002; Rudic *et al.*, 1998). Furthermore, recent studies provide sound evidence for NO-independent heme oxygenase (HO) and carbon monoxide (CO) as physiologically relevant signals capable of activating sGC and promoting cGMP signal transduction (Christodoulides *et al.*, 1995; Siow *et al.*, 1999; Tulis *et al.*, 2001a; Tulis *et al.*, 2001b; Tulis *et al.*, 2005). BAY 41-2272 (BAY) is a prototype of a novel class of sGC modulators that potentiate both NO- and CO-mediated signaling and that has potential for use in clinical trials for the treatment of cardiovascular disease, yet mechanisms of action for BAY and similar agents still remain to be determined (Boerrigter *et al.*, 2007; Mendelev *et al.*, 2009; Roger *et al.*, 2010). One study suggests that BAY may have phosphodiesterase (PDE) 5 inhibitory activity at high concentrations (Mullershausen *et al.*, 2004); however this is still controversial. BAY interacts with sGC via a different mechanism than that of NO; therefore, the amount of cGMP generated and subsequent targets that are activated may be distinct leading to different responses than that of traditional NO-based signaling. This is important and a perfect example of a drug-induced versus a cell-specific response.

Individual drugs interact with cells differently depending on the cellular microenvironments within the cell. Recent reports demonstrate the ability of cell- and agonist-specific responses following ligand stimulation due to intracellular localization of proteins including kinases which operate largely to ensure signal specificity (Di Benedetto *et al.*, 2008; Lissandron *et al.*, 2006; Xiang, 2011). An increase in cyclic nucleotide signaling can affect multiple signaling pathways because of activation of such kinases in particular intracellular domains. Hence, compartmentalization of proteins is very critical in cellular/biomolecular signaling in order to ensure a specific response to a particular stimulus in cells.

VASP, a cytoskeletal protein that belongs to the Ena/VASP family, was originally characterized as a substrate for cyclic nucleotide-dependent PKA and PKG (Krause *et al.*, 2002; Reinhard *et al.*, 2001). At least four distinct phosphorylation sites on VASP (Ser_{157}, Ser_{239}, Thr_{278}, Ser_{322}) have been identified (Butt *et al.*, 1994; Chitaley *et al.*, 2004; Thomson *et al.*, 2011). Despite the conventional thought that phosphorylation of VASP at Ser_{239} and Ser_{157} occurs by PKG and PKA, respectively (Butt *et al.*, 1994), more recent studies suggest that crosstalk exists and that Ser_{157}, Ser_{239} and Thr_{278} can be phosphorylated by both PKA and PKG. However, phosporylation of VASP by PKC and AMP kinase is more selective (Benz *et al.*, 2009; Blume *et al.*, 2007; Zhuang *et al.*, 2004). The phosphorylation pattern of VASP is complex and dynamic and occurs in site-specific order with different priorities (Abel *et al.*, 1995; Butt *et al.*, 1994; Zhuang *et al.*, 2004). Functionally, VASP plays an important role in regulating cytoskeletal dynamics and processes such as cell adhesion and migration (Kwiatkowski *et al.*, 2003; Mitchison *et al.*, 1996; Vasioukhin *et al.*, 2000), and phosphorylation of VASP has been shown to be important physiologically as it is involved in these processes. It was reported that VASP phosphorylation at Ser_{157} is required for the growth stimulatory effect and phosphorylation at Ser_{239} is required for the growth inhibitory effect of NO on SMCs (Chen *et al.*, 2004). Although VASP phosphorylation continues to be used as a biochemical marker for selective kinase activation, its precise cellular and molecular functions remain to be fully determined.

Previously, we demonstrated in both rat commercial (A7R5) VSMCs and rat primary VSMCs that BAY increases cyclic nucleotides and inhibits vascular growth through processes involving VASP phosphorylation (Joshi *et al.*, 2011; Mendelev *et al.*, 2009). In this report we demonstrate for the first time that BAY increases the activity of protein kinases and that this system has the ability to regulate growth of VSMCs. Novel results include demonstration that kinase activity from PKA, PKG and PKC has the capacity to increase phosphorylation of cellular proteins including VASP. Notably, the implication of multiple kinases in the regulation and activation of VASP provides sound evidence for potential crosstalk among otherwise distinct kinases within the cell. Functionally, strong evidence is presented linking cyclic nucleotide-mediated kinase activation and VASP phosphorylation with inhibition of vascular smooth muscle growth through modulation of proliferation and migration.

2. Methods

2.1 Rat primary VSMC culture

Following established procedures (Liu *et al.*, 2009; Tulis *et al.*, 2002), thoracic aorta VSMCs were obtained from male Sprague-Dawley rats (100-125 grams) and cultured in Dulbecco's modified Eagle's medium supplemented with 10% fetal bovine serum (FBS) and Primocin (100mg/L) at 37°C in 95% air/5% CO_2. Cells were split and propagated through passage 6 (unless otherwise specified). For select experiments higher passage cells were used between passage 7 and 10. All studies abided by the guidelines of the Institutional Animal Care & Use Committee and conformed to the Guide for the Care and Use of Laboratory Animals (US National Institutes of Health, Publication No. 85-23, revised 1996).

2.2 In-cell western

Phosphorylation of VASP was examined by use of an established cell-based immunocytochemical assay for monitoring kinase signaling pathways (Chen *et al.*, 2005). Rat VSMCs were seeded (~20,000 cells/well) in 96 well plates and once adhered, pretreated with either kinase and PDE inhibitors or vehicle for 30 minutes after which they were stimulated with either BAY for 60 minutes or cyclic nucleotide analogs (8Br-cAMP; 10μM or 8Br-cGMP; 10μM, Tocris) for 30 minutes. The inhibitors used for In-Cell Westerns were: adenosine 3′,5′-cyclic monophosphate-dependent protein kinase inhibitor (5-24), (PKI; 10μM), a selective PKA inhibitor (Enzo); DT-2 trifluoroacetate salt, (DT2; 10μM), a selective PKG inhibitor (Sigma-Aldrich); calphostin C (CALC; 100nM), a selective inhibitor of PKC (Enzo); cilostazol (CILO; 10 μM), a selective PDE3 inhibitor (Sigma-Aldrich); rolipram (ROL; 10 μM), a selective PDE4 inhibitor (Tocris); zaprinast (ZAP; 10 μM; MP Biochemicals) and vardenifil (VAR; 50 nM; Toronto Research Chemicals), selective PDE5 inhibitors. The vehicle for CILO, ROL, CALC and ZAP was DMSO ($\leq$ 0.5%). After incubation, media was removed and cells were fixed with 4% formalin in PBS for 20 minutes. Cells were washed and permeabilized with PBS containing 0.1% Triton X-100, followed by 1X PBS + 0.1% Tween-20. Cells were blocked with IR blocking solution (Odyssey) for 3 hours and then incubated with primary antibodies directed against VASP at $pSer_{239}$ (1:500; Cell Signaling), VASP at $pSer_{157}$ (1:500; Cell Signaling) or α-tubulin (1:500; Sigma) overnight at 4°C. Cells were washed with PBS + 0.1% Tween-20 followed by incubation with two secondary antibodies: IRDye 800CW

(1:500; Odyssey) and Alexa Fluor 680 for 1 hour. Cells were washed with PBS containing 0.1% Tween-20 followed by plain PBS. The plate was allowed to dry overnight and scanned using the appropriate channels for detection with the Odyssey Imager.

2.3 Western blot

Rat VSM cells were used between passages 4 and 6 for all experiments. Cells were lysed in buffer (50mM Tris, pH 6.8; 1% SDS; 0.1% Triton; protease inhibitor cocktail (Roche); phosphate inhibitor cocktail (Santa Cruz)) and sonicated on ice followed by centrifugation (14000g, 20 min). Protein concentration of whole cell lysates was determined by the BCA assay (Pierce). Proteins were separated on SDS-PAGE gels and transferred to Immobilon-P membranes (Millipore) using Mini-Protean Electrophoresis and Mini Trans-Blot Cells (Bio-Rad). Blots were blocked with 5% non-fat milk in TTBS (25mM Tris/Tris-HCl; 0.0027M KCl; 0.13M NaCl; 0.1% Tween 20) or 5% BSA in TTBS for 1 hour and incubated with primary antibodies directed against protein kinase G (PKG) and protein kinase A (PKA) (1:1000; Abcam), phospho-PKA_{Ser338} (1:1000; Biosource), VASP, phospho-$VASP_{Ser157}$, and phospho-$VASP_{Ser239}$ (1:1000; Cell Signaling) and α-tubulin (1:1000; Sigma) overnight at 4 °C. Blots were washed with TTBS, incubated with peroxidase-linked secondary antibodies (1:5000) for 1 hour, visualized using Pierce ECL Western blotting kit, documented using an Alpha Imager 2200, and analyzed with ImageJ 1.40g software (NIH).

2.4 Hemocytometry assay

Rat VSMCs were seeded in 24 well plates (40,000 cells/well) and after adherence were quiesced overnight in 0.2% FBS. Cells were then pretreated with PKI, DT2 or vehicle for 30 mins, prior to stimulation with BAY (10μM) for 48 or 72 hrs. At the end of the time point, cells in each well were trypsinized and the cell count estimated using a hemocytometer at 10X magnification.

2.5 MTT assay

Rat VSMCs were seeded in 96 well plates and pretreated with PKI, DT2 or vehicle for 30 mins, prior to stimulation with BAY (10μM) for 24hrs. After treatment media was removed and MTT was added to each well and incubated for 4 hours at 37°C. The MTT was removed, MTT solvent added and absorbance read after 15 mins.

2.6 BrdU assay

Rat VSMCs were seeded in 96 well plates and pretreated with PKI, DT2, CALC or vehicle for 30 mins, prior to stimulation with BAY (10μM) for 48hrs. After treatment media was removed and BrdU labeling solution was added to each well and incubated for 3 hours at 37°C. Next the labeling solution was removed, cells were fixed and the anti-BrdU antibody added for 90 mins. Cells were washed, the substrate was then added, and absorbance was read after 30 mins.

2.7 Kinase assay

Rat VSMCs were seeded in 6 well plates and grown to confluence. The cells were then pretreated with PKI, DT2, CALC or vehicle for 30 mins, prior to stimulation with BAY

(10μM) for 1 hour. The cells were lysed, protein was isolated and protein concentration was determined via a BSA assay. Kinase activity was determined using an activity kit (Enzo Life Sciences). Briefly, samples were added to the wells, the reaction was initiated with ATP, and cells were incubated at 30°C for 90 mins. A phosphospecific substrate antibody was added and incubated at RTP for 60 mins. Next the plate was washed and the HRP conjugate was added and incubated for 30 mins. Finally the plate was washed and substrate added for 30 mins and absorbance was read.

2.8 Scratch wound assay

Rat VSMCs were seeded in 24 well plates. Cells were injured with a 1 mm scrape and the images captured. Immediately cells were then treated with PKI, DT2, CALC or vehicle for 30 mins, prior to stimulation with BAY (10μM) for 6 or 16 hrs. Afterward the images were captured and distances migrated were measured.

2.9 Data analysis

Statistical significance between experiments was determined using an analysis of variance (ANOVA). In the event that the *F* ratio indicated a significant $P<0.05$ change occurred, a post hoc test was performed to identify individual differences. Results were reported as mean ± the standard error of the mean (SEM).

3. Results

3.1 The effect of kinase inhibitors on BAY-induced VASP phosphorylation

We recently demonstrated that BAY, a potent stimulator of sGC, increased phosphorylation of VASP in a site-specific fashion in rat A7R5 VSMCs (Mendelev *et al.*, 2009) and in rat primary VSMCs (Joshi *et al.*, 2011). In the current study we sought to identify kinases responsible for this selective phosphorylation of VASP induced by BAY. In primary VSMCs, BAY (10 μM) significantly increased VASP phosphorylation at both Ser_{157} and Ser_{239} (Figures 1A, 1B, respectively). Addition of PKI (10 μM) or CALC (100 nM) slightly (non-significantly) increased BAY-induced phosphorylation of VASP at Ser_{157} compared to effects of BAY alone (Fig. 1A), whereas CALC slightly (non-significantly) reduced VASP phosphorylation at Ser_{239} compared to BAY alone (Fig. 1B).

Due to the relative ineffectiveness of select kinase inhibition on site-specific VASP phosphorylation, the possibility of crosstalk between these kinases was examined. Intriguingly, the combination of PKI and DT2 increased the phosphorylation induced by BAY at Ser_{157} (Fig 2A), while the combination of PKI and CALC markedly increased the phosphorylation induced by BAY at Ser_{239} (Fig. 2B).

3.2 Effect of PDE inhibition and kinase activity on VASP phosphorylation

In order to determine if kinase inhibition affects the ability of PDEs to regulate VASP, VSMCs were pretreated with inhibitors of kinases and PDEs prior to stimulation with BAY. In the presence of BAY and DT2, the PDE 3 inhibitor CILO and the PDE 5 inhibitors VAR or ZAP, significantly increased phosphorylation at Ser_{157} compared to BAY alone (Fig. 3A). PKI increased Ser_{157} phosphorylation induced by BAY (Fig. 3B); however, PDE inhibition

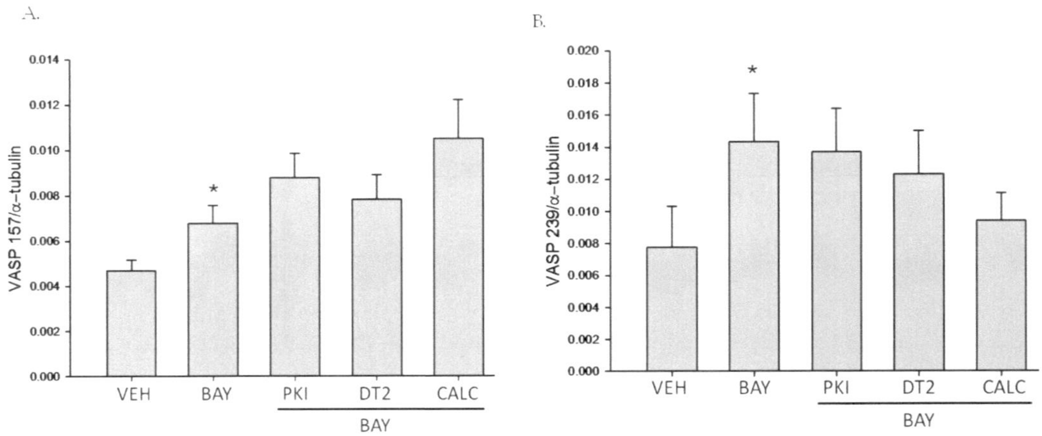

Fig. 1. (A) Effect of kinase inhibitors on BAY 41-2272 (BAY)-induced VASPSer$_{157}$ phosphorylation. VSMCs were pretreated with PKI, DT2, CALC or vehicle for 30 min prior to stimulation with BAY (10uM) for 60 minutes (n=15). (B) Effect of kinase inhibitors on BAY-induced VASPSer$_{239}$ phosphorylation. VSMCs were pretreated with PKI, DT2, CALC or vehicle for 30 min prior to stimulation with BAY (10uM) for 60 minutes (n=15). Values are the means ± SE. * = different from vehicle ($P<0.05$).

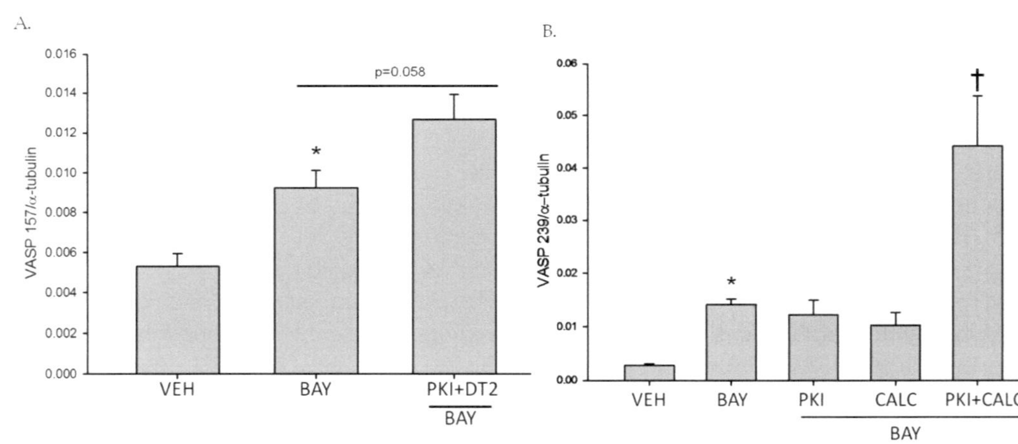

Fig. 2. (A) Effect of PKI and DT2 on BAY-induced VASPSer$_{157}$ phosphorylation. VSMCs were pretreated with PKI and DT2 or vehicle for 30 min prior to stimulation with BAY (10uM) for 60 minutes (n=5). (B) Effect of PKI and CALC on BAY-induced VASPSer$_{239}$ phosphorylation. VSMCs were pretreated with PKI or CALC individually, PKI+CALC or vehicle for 30 min prior to stimulation with BAY (10uM) for 60 minutes (n=5). Values are the means ± SE. * = different from vehicle ($P<0.05$); † = different from vehicle and BAY ($P<0.05$).

had no further effect on Ser$_{157}$ phosphorylation. BAY-induced Ser$_{239}$ phosphorylation was potentiated only in the presence of PKI and the PDE 5 inhibitor ZAP (Fig 3C). Neither an inhibitor of PKA nor PKG and PDEs had any effect on Ser$_{239}$ in any other instance (data not shown).

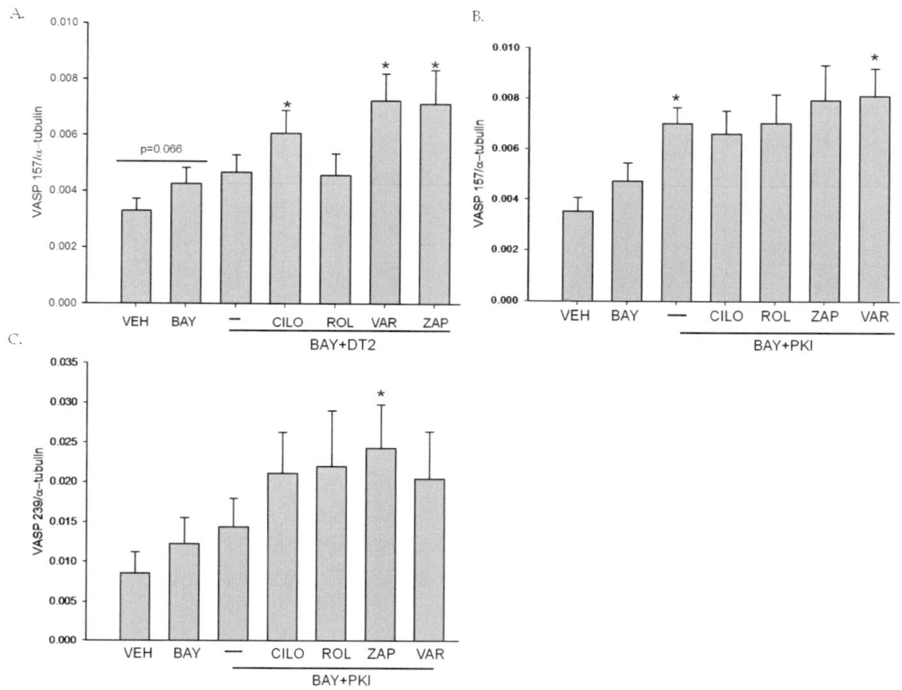

Fig. 3. (A) Effect of DT2 and PDE inhibitors on BAY-induced VASPSer$_{157}$ phosphorylation. VSMCs were pretreated with DT2 alone or CILO, ROL, VAR, or ZAP in the presence of DT2 or vehicle for 30 min prior to stimulation with BAY (10uM) for 60 minutes (n=13). (B) Effect of PKI and PDE inhibitors on BAY-induced VASPSer$_{157}$ phosphorylation. VSMCs were pretreated with PKI alone, CILO, ROL, VAR, or ZAP in the presence of PKI, or vehicle for 30 min prior to stimulation with BAY (10uM) for 60 minutes (n=8). (C) Effect of PKI and PDE inhibitors on BAY 41-2272 (BAY)-induced VASPSer$_{239}$ phosphorylation. VSMCs were pretreated with PKI alone, CILO, ROL, VAR, or ZAP in the presence of PKI, or vehicle for 30 min prior to stimulation with BAY (10uM) for 60 minutes (n=14). Values are the means ± SE. * = different from BAY (P<0.05).

3.3 Effect of kinase inhibitors on cyclic nucleotide analog-induced VASP phosphorylation

Since BAY increases cAMP and cGMP (Joshi *et al.*, 2011; Mendelev *et al.*, 2009; Stasch *et al.*, 2001; Stasch *et al.*, 2009) as well as the activity of their respective kinases (Joshi *et al.*, 2011), and considering that both kinases can phosphorylate VASP at either kinase-preferred site (Butt *et al.*, 1994), cyclic nucleotide analogs were used to determine the kinase(s) specifically involved in VASP phosphorylation. VSMCs were incubated with selective kinase inhibitors prior to addition of a cAMP or cGMP analog. Results show that 8-Br-cAMP alone failed to significantly increase phosphorylation at Ser$_{157}$ or Ser$_{239}$, but PKI significantly increased phosphorylation at both sites induced by the analogs (Figs. 4A, 4B). Neither DT2 nor CALC had an effect on VASP phosphorylation in the presence of 8Br-cAMP. Additionally, when using 8Br-cAMP multiple kinase inhibition had no effect greater than that of PKI alone at either site (data not shown).

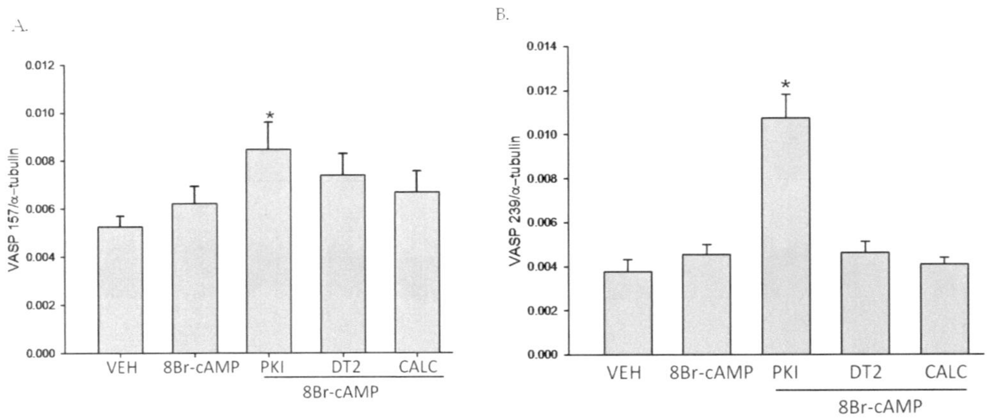

Fig. 4. (A) Effect of kinase inhibitors and 8Br-cAMP on VASPSer$_{157}$ phosphorylation. VSMCs were pretreated with PKI, DT2, CALC or vehicle for 30 min prior to stimulation with 8Br-cAMP (10uM) for 30 minutes (n=10). (B) Effect of kinase inhibitors and 8Br-cAMP on VASPSer$_{239}$ phosphorylation. VSMCs were pretreated with PKI, DT2, CALC or vehicle for 30 min prior to stimulation with 8Br-cAMP (10uM) for 30 minutes (n=10). Values are the means ± SE. * = different from vehicle (P<0.05).

In comparison, in these cells 8Br-cGMP alone failed to increase Ser$_{157}$ phosphorylation; however, co-treatment with PKI or CALC increased Ser$_{157}$ phosphorylation (Fig. 5A). On the other hand, 8Br-cGMP alone significantly increased phosphorylation at Ser$_{239}$ and a trend for potentiation by PKI was observed (p = 0.059; Fig. 5B).

Co-treatment with an inhibitor of PKG and PKC blunted the increase induced by 8-Br-cGMP at Ser$_{239}$ (Fig. 6).

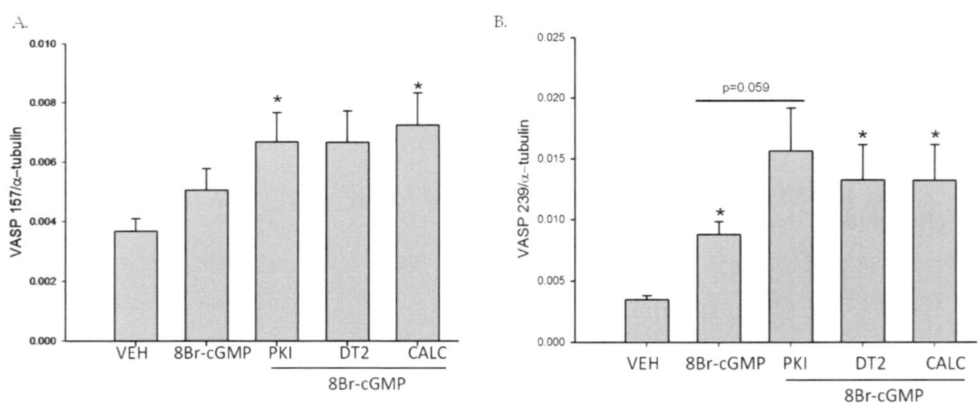

Fig. 5. (A) Effect of kinase inhibitors and 8Br-cGMP on VASPSer$_{157}$ phosphorylation. VSMCs were pretreated with PKI, DT2, CALC or vehicle for 30 min prior to stimulation with 8Br-cGMP (10uM) for 30 minutes (n=10). (B) Effect of kinase inhibitors and 8Br-cGMP on VASPSer$_{239}$ phosphorylation. VSMCs were pretreated with PKI, DT2, CALC or vehicle for 30 min prior to stimulation with 8Br-cGMP (10uM) for 30 minutes (n=10). Values are the means ± SE. * = different from vehicle (P<0.05).

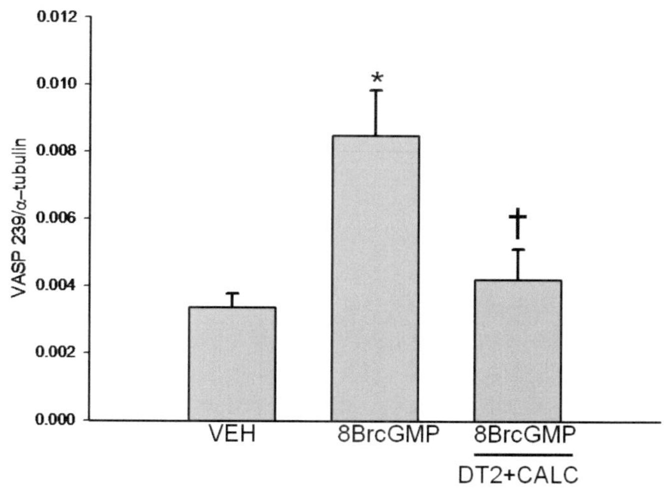

Fig. 6. Effect of CALC and DT2 on 8Br-cGMP-induced VASPSer$_{239}$ phosphorylation. VSMCs were pretreated with CALC and DT2 or vehicle for 30 min prior to stimulation with 8Br-cGMP for 30 minutes (n=8). Values are the means ± SE. * = different from vehicle ($P<0.05$); † = different from 8Br-cGMP ($P<0.05$).

3.4 Effect of BAY on Kinase activity

To determine if BAY increases kinase activity, VSMCs were pretreated with inhibitors of kinases prior to stimulation with BAY and selective kinase activity assays were performed. BAY increased the activity of PKA ~86% from (0.0278 ± 0.0223 to 0.0519 ± 0.0415). PKI decreased the activity back to control levels (0.0214 ± 0.0160). Similarly, BAY increased the activity of PKG ~55% from (0.0108 ± 0.0042 to 0.0167 ± 0.007) and DT2 decreased the activity back to control levels (0.0116 ± 0.003).

3.5 Effect of kinase inhibitors and BAY on Proliferation

Previously we concluded that BAY inhibits proliferation of both rat A7R5 (Mendelev *et al.*, 2009) and rat primary VSMCs (Joshi *et al.*, 2011). To determine the influence of kinases on BAY-induced growth suppression, three complementary assays were performed. In the MTT assay, a mitochondria reductase-dependent approach, cells were pretreated with kinase inhibitors prior to stimulation with BAY. BAY (non-significantly) decreased proliferation (~25%) after 24 hrs, and inhibition of either PKA or PKG significantly potentiated this effect (Fig. 7A). Using the BrdU assay, BAY significantly inhibited DNA replication after 48 hours, yet inhibition of any individual kinase had no further effect at this time point (Fig. 7B). Using hemocytometry, BAY reduced cell numbers after both 48 (Fig. 7C) and 72 hrs (Fig. 7D). Inhibition of PKA at 48 hrs (Fig. 7C) and either PKA or PKG at 72 hrs (Fig. 7D) potentiated this effect. On the other hand, a lower concentration of DT2 (1μM) had no effect at 72 hrs (Fig. 7E). The Neutral red and MTT assays done at 16 hrs demonstrate that BAY and all inhibitor combinations were not cytotoxic (data not shown).

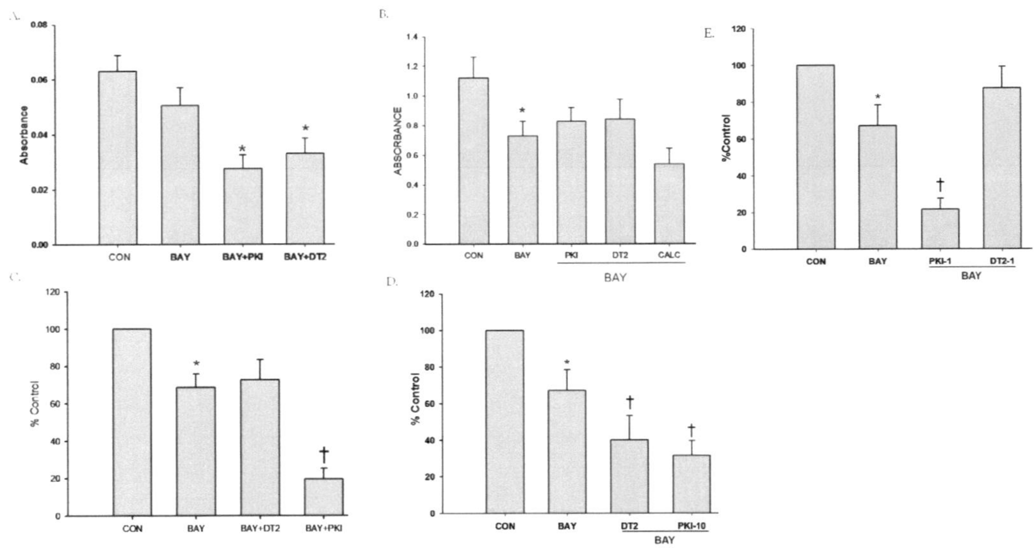

Fig. 7. (A) MTT assay to test the effect of PKI, DT2 and BAY on proliferation. VSMCs were pretreated with PKI, DT2 or vehicle for 30 min prior to stimulation with BAY (10uM) for 24 hours (n=9). (B) BrdU assay to test the effect of PKI, DT2, CALC and BAY on proliferation. VSMCs were pretreated with PKI, DT2, CALC or vehicle for 30 min prior to stimulation with BAY (10uM) for 48 hours (n=10). (C) Hemocytometry assay to test the effect of PKI, DT2 and BAY on proliferation. VSMCs were pretreated with PKI, DT2 or vehicle for 30 min prior to stimulation with BAY (10uM) for 48 hours (n=7). (D) Hemocytometry assay to test the effect of PKI, DT2 and BAY on proliferation. VSMCs were pretreated with PKI, DT2 or vehicle for 30 min prior to stimulation with BAY (10uM) after 72 hours (n=16). Values are the means ± SE. * = different from vehicle ($P<0.05$); † = different from vehicle and BAY ($P<0.01$). (E) Hemocytometry assay to test the effect of PKI, DT2 and BAY on proliferation. VSMCs were pretreated with PKI, DT2 or vehicle for 30 min prior to stimulation with BAY (10uM) for 72 hours (n=4). Values are the means ± SE. * = different from vehicle ($P<0.05$); † = different from vehicle and BAY ($P<0.05$).

3.6 Effect of kinase inhibitors and BAY on Migration

Previously we observed that BAY inhibits cell migration in rat primary VSMCs (Joshi *et al.*, 2011). In order to determine the role of kinases on this anti-migratory effect of BAY, VSMCs were pretreated with inhibitors of PKG, PKA or PKC prior to stimulation with BAY and migration was assessed using the wounding assay. BAY inhibited migration ~40% after 6 hours and co-incubation of BAY with DT-2 or CALC inhibited migration ~13% or ~17%, respectively (data not shown).

3.7 The effect of kinase inhibitors on BAY-induced VASP phosphorylation in high passaged rat primary VSMCs

In comparison to studies performed at low passage, in higher passage VSMCs (passage > 6) either PKI or two chemically dissimilar inhibitors of PKG, DT3 and KT5823, significantly increased phosphorylation of VASP at Ser_{157} but had no effect at Ser_{239} (Suppl. Figs. 1A and

1B, resp.). Notably, in these higher passage cells inhibition of PKC by CALC did not affect phosphorylation at either site versus that of BAY alone. In these cells, inhibition of multiple kinases had no effect on Ser_{157} but any combination of kinase inhibitors inhibited VASP phosphorylation at Ser_{239} (data not shown).

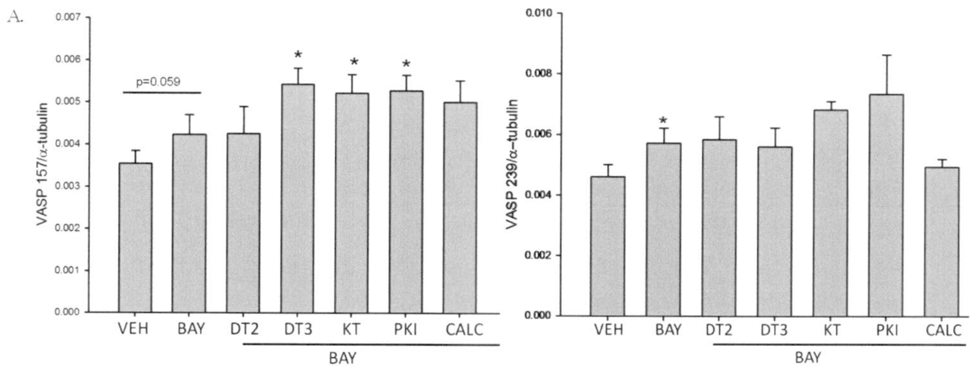

Suppl. Fig. 1. (A) Effect of kinase inhibitors on BAY 41-2272 (BAY)-induced $VASPSer_{157}$ phosphorylation. Higher passaged (p > 6) VSMCs were pretreated with PKI, DT2, DT3, KT, CALC or vehicle for 30 min prior to stimulation with BAY (10uM) for 60 minutes (n=8). Values are the means ± SE. * = different from BAY (P<0.05). (B) Effect of kinase inhibitors on BAY-induced $VASPSer_{239}$ phosphorylation. Higher passaged (p > 6) VSMCs were pretreated with PKI, DT2, DT3, KT, CALC or vehicle for 30 min prior to stimulation with BAY (10uM) for 60 minutes (n=9). Values are the means ± SE. * = different from vehicle (P<0.05).

3.8 Effect of kinase inhibitors on cyclic nucleotide analog-induced VASP phosphorylation

In higher passage VSMCs (p > 6), 8-Br-cGMP alone increased phosphorylation at Ser_{239} only (Suppl. Figs. 2A, 2B), and kinase inhibition was without effect at either site.

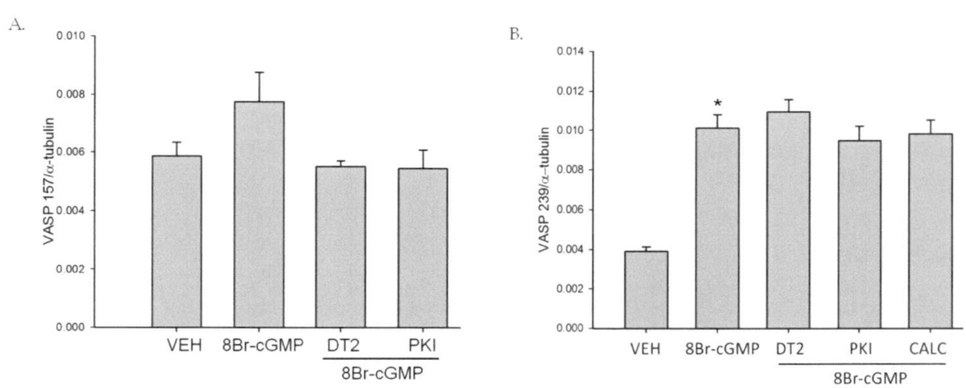

Suppl. Fig. 2. (A) Effect of kinase inhibitors and 8Br-cGMP on $VASPSer_{157}$ phosphorylation. VSMCs were pretreated with PKI, DT2 or vehicle for 30 min prior to stimulation with 8Br-cGMP (10uM) for 30 minutes (n=3). (B) Effect of kinase inhibitors and 8Br-cGMP on $VASPSer_{239}$ phosphorylation. VSMCs were pretreated with PKI, DT2, CALC or vehicle for

30 min prior to stimulation with 8Br-cGMP (10uM) for 30 minutes (n=21). Values are the means ± SE. * = different from vehicle ($P<0.05$).

8-Br-cAMP had no effect on site-specific VASP phosphorylation (Suppl. Fig 3A, 3B.). However, PKI further increased 8-Br-cAMP-induced phosphorylation at Ser_{239} (Suppl. Fig. 3B). In VSMCs p > 6 any combination of kinase inhibitors did not affect cAMP or cGMP-selective VASP phosphorylation (data not shown).

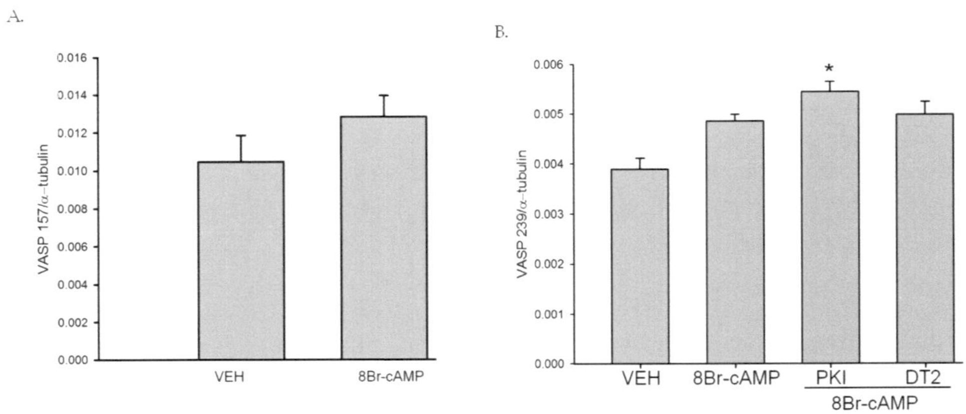

Suppl. Fig. 3. (A) Effect of 8Br-cAMP on $VASPSer_{157}$ phosphorylation. VSMCs were stimulated with 8Br-cAMP (10uM) for 30 minutes (n=15). (B) Effect of kinase inhibitors and 8Br-cAMP on $VASPSer_{239}$ phosphorylation. VSMCs were pretreated with PKI, DT2 or vehicle for 30 min prior to stimulation with 8Br-cAMP (10uM) for 30 minutes (n=20). Values are the means ± SE. * = different from vehicle ($P<0.05$).

3.9 Effect of BAY on Kinase expression

In untreated VSMCs, Western analysis shows that there is an increase in total PKA in passage 6 cells versus passage 2 cells, yet concentration-dependent stimulation with BAY did not alter the amount of total PKA in either cohort (Suppl. Fig. 4A). Interestingly, BAY (at 10 μM) largely increased the amount of phosphorylated PKA (at Ser_{338}) in p6 VSMCs but not in p2 cells (Suppl. Fig. 4A). In comparison, the amount of basal untreated PKG decreased with increasing passage, and stimulation with BAY had no additive effect (Suppl. Fig. 4B).

4. Discussion

The sGC signal transduction system plays an important regulatory role in cardiovascular disease due to the production of cyclic nucleotides and their impact on downstream effectors. VSMCs primarily regulate vascular tone and contraction and are important in the medial layer of arteries. In animal models of vascular injury, intimal and medial thickening is attributed to phenotypic switching of resident VSMCs and their proliferation and migration from the adventitia and media to the intimal space (Janssens *et al.*, 1998; Kawashima *et al.*, 2001; von der Thusen *et al.*, 2004). Therefore, augmentation of neointimal formation in iatrogenic response to therapeutic intervention compromises flow and increases the risk of a cardiac event. For that reason understanding the cellular mechanisms involved in vascular growth is of critical importance.

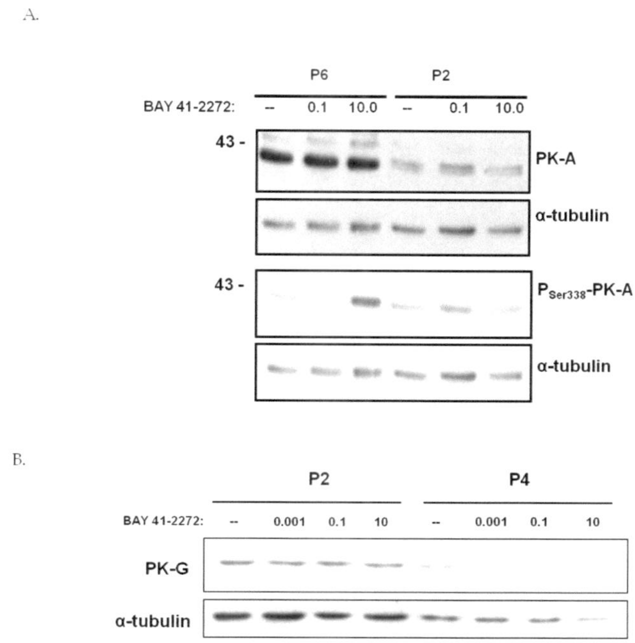

Suppl. Fig. 4. (A) Western analysis for Protein Kinase A (PKA) and phosphorylated-PKASer$_{338}$. VSMCs cytosolic preparations were incubated with antibodies generated against PKA and phospho-PKA. (B) Western analysis for Protein Kinase G (PKG). VSMCs cytosolic preparations were incubated with an antibody generated against PKG.

In this study we provide sound evidence that BAY, a synthetic stimulator of sGC, increases the activity of protein kinases and downstream effectors and that this system has the ability to control growth of VSM. Novel results show that activity from PKA, PKG and PKC has the capacity to increase phosphorylation of cellular proteins including VASP, a multi-functional cytoskeletal protein capable of regulating VSM cell proliferation and migration. Notably, the implication of multiple kinases in the regulation and activation of VASP provides evidence for potential crosstalk among otherwise distinct kinases within the cell. Functionally, strong evidence is presented linking cyclic nucleotide-mediated kinase activation and VASP phosphorylation with inhibition of VSM growth through modulation of proliferation and migration. BAY-induced site-specific VASP phosphorylation is regulated by kinases even though cross-talk exists among these (and perhaps other) kinases in VSM. These data demonstrate the importance of understanding cyclic nucleotide-directed kinases in regulating signaling proteins and their roles for inducing specific cellular responses.

VSMC proliferation, a critical mechanism in vessel remodeling, has been reported to be positively controlled by phosphorylation of VASP at Ser$_{157}$ in VSMCs (Chen *et al.*, 2004). VASP$_{Ser157}$ is the site preferentially phosphorylated by PKA, but evidence suggests that both PKG and PKC can also phosphorylate Ser$_{157}$ (Butt *et al.*, 1994; Chitaley *et al.*, 2004). In this study we report that inhibition of PKA alone has no effect on Ser_{157} phosphorylation of VASP induced by BAY (Fig 1A); however, in the presence of a PKG inhibitor phosphorylation at this site is increased (Fig 2A). On the other hand, $VASPSer_{239}$ is the site reported to be preferentially phosphorylated by PKG (Butt *et al.*, 1994), but PKA can also phosphorylate this site (Butt *et al.*, 1994). In this study, inhibition of PKG alone had no effect on VASP phosphorylation at Ser_{239} induced by BAY (Fig. 1B); however, in the presence of an inhibitor of PKC, BAY-induced phosphorylation was increased. These intriguing results demonstrate that there is crosstalk among these kinases in response to VASP phosphorylation induced by BAY: with respect to Ser_{157}, PKA and PKG are primarily involved, whereas at Ser_{239} the major kinases are PKA and PKC.

Cyclic nucleotide activity and localization in cells is regulated by PDEs which modulate the levels of cyclic nucleotide levels and maintains them at a steady state. Moreover, kinases have been shown to regulate PDE activity (Bender *et al.*, 2006). It has been reported that PKA phosphorylates and increases the activities of PDE3 and 4 (Bender *et al.*, 2006; Degerman *et al.*, 1997; Houslay *et al.*, 2007), while an increase in cGMP binding and phosphorylation by PKG increases the activity of PDE5 (Bender *et al.*, 2006; Rybalkin *et al.*, 2002). BAY has been shown to increase cyclic nucleotide content in VSMCs (Joshi *et al.*, 2011; Mendelev *et al.*, 2009); therefore, it is logical that PDE and kinase activity could potentially be modulated by BAY treatment. Preliminary data suggest that with respect to Ser_{157}, inhibition of PDE5 increased its phosphorylation while inhibition of PDE3 and PDE4 decreased its phosphorylation in response to BAY. Here we demonstrate that inhibition of PDE3 or PDE5 in the presence of an inhibitor of PKG increases the phosphorylation at this site while inhibition of PDE4 had no effect (Fig 3A). In the presence of PKA inhibition, Ser_{157} phosphorylation was increased and PDE inhibition had no additional effect (Fig 3B). Considering that PDE3 is a cGMP-inhibited PDE, following concomitant inhibition of both PDE5 and PKG, the resulting increase in cAMP and PKA is expected as an enhancement of BAY-induced signaling. The same observation is seen following simultaneous inhibition of PKG and PDE5, as there is an increase in cGMP which could then inhibit PDE3 and increase cAMP leading to additive phosphorylation at Ser_{157}. On the other hand, in the presence of an inhibitor of PKA and individual PDE inhibitors, no further phosphorylation is shown at Ser_{157} suggesting that this previously observed inhibitory effect is rescued.

At $VASP_{Ser239}$, preliminary data suggest that inhibition of PDE activity decreased phosphorylation at this site in response to BAY. Here we demonstrate that in the presence of an inhibitor of PKA or PKG, inhibition of PDEs no longer decreases Ser_{239} phosphorylation and in fact, inhibition of PDE5 with ZAP in the presence of PKI increased Ser_{239} phosphorylation (Fig 3C). These results support involvement of kinases in the regulation of PDEs. In this study we are inhibiting activation of PDEs via kinases as well as inhibiting the actual PDE, thereby increasing the cyclic nucleotide content of the cell and as a result increasing phosphorylation of VASP. This supports the argument that an increase in both cyclic nucleotides in a cell at the same time is inhibitory because two kinases are active simultaneously thereby activating opposing pathways. In addition, these findings confirm the theory of intracellular compartmentalization of second messengers and their effectors as described elsewhere (Di Benedetto *et al.*, 2008; Houslay *et al.*, 2007; Zaccolo *et al.*, 2002).

Activation of kinases in other pathways may negatively regulate phosphorylation of VASP. It is reported that PKG is the primary kinase responsible for phosphorylation of Ser_{239} (Butt *et al.*, 1994); in turn, PKG inhibition would hinder Ser_{239} phosphorylation and PKA would then become the primary kinase responsible for phosphorylation at this site. To further add to the complexity of these findings and to demonstrate the importance of kinase regulation and localization of proteins we demonstrate that inhibition of PDE5 in the presence of an inhibitor of PKA potentates the phosphorylation at Ser_{239}. This demonstrates that cGMP does indeed directly phosphorylate VASP at this site independent of cAMP/PKA signals (Fig. 3C).

BAY has been shown to increase the content of both cAMP and cGMP in VSMCs (Joshi *et al.*, 2011; Mendelev *et al.*, 2009); therefore, to directly assess its mechanisms of action and what kinases may be involved in site specific phosphorylation of VASP, cyclic nucleotide analogs which directly activate PKA and PKG were used. 8Br-cAMP is a direct activator of PKA, the kinase which directly phosphorylates Ser_{157} (Butt *et al.*, 1994). As one would expect addition of this analog should increase Ser_{157} phosphorylation of VASP due to the increase in PKA activity. We did not observe an increase with the analog alone (Fig. 4A); however, inhibition of PKA in the presence of the analog significantly increased VASP phosphorylation at this site (Fig. 4A). Interestingly, inhibition of PKG or PKC in the presence of 8Br-cAMP had no effect at Ser_{157}. Although PKG is reported to be the primary kinase responsible for phosphorylation of Ser_{239} there could potentially be crosstalk involving other kinases at this site. This observation was evidenced with Ser_{239} phosphorylation whereby inhibition of PKA increased phosphorylation in the presence of the cyclic nucleotide analog (Fig. 4B). These data could provide an argument against the selectivity of the analogs in activating kinases because 8Br-cAMP may be activating PKG or some other protein that in reality is phosphorylating VASP. 8Br-cGMP is a direct activator of PKG and alone has no effect on Ser_{157} but markedly increases Ser_{239} as expected (Fig 5A, 5B). However, inhibition of PKA and PKC increased phosphorylation at Ser_{157} in the presence of 8Br-cGMP (Fig 5A). Inhibition of PKA slightly increased Ser_{239} phosphorylation above that of 8Br-cGMP alone (Fig 5B), while inhibition of either PKG or PKC slightly but not significantly increases phosphorylation. We could infer here that PKA and PKG are competing for the same site; therefore inhibition of PKA increases the phosphorylation at any particular site. Alternatively, we could conclude that PKA negatively regulates phosphorylation of VASP because inhibition of PKA increases phosphorylation of both sites. In any event, these results demonstrate that cyclic nucleotide analogs and BAY use different mechanisms to phosphorylate VASP. This is demonstrated in the results shown here whereby inhibition of both PKG and PKC (Fig. 6) inhibit phosphorylation at Ser_{239}. We can conclude that both PKG and PKC are actively involved in phosphorylation of Ser_{239} in response to 8Br-cGMP. In comparison, multiple kinase inhibition had no effect on 8Br-cAMP in its phosphorylation of VASP. These data clearly demonstrate that phosphorylation and thus regulation of VASP at these sites is complex in that all three major protein kinases (PKA/G/C) are involved. Furthermore, BAY, originally characterized as a direct activator of sGC, equally increases both PKA and PKG activity in these cells (Fig. 7). Thus, it appears that PKA may be the more active kinase involved in the phosphorylative control of VASP in VSM.

Direct involvement of PKA in BAY-mediated reduction in proliferation is seen in both the MTT and hemocytometry results in this study, which suggest that PKA activity is necessary for proliferation of VSMCs (Fig 8). The involvement of PKA in proliferation was previously suggested by Chen and colleagues (Chen *et al.*, 2004) where they found that phosphorylation of VASP by PKA stimulates proliferation. The difference between our study and that of

Chen and associates is that our study uses rat primary VSMCs and phosphorylation of VASP was in response to a pharmacologic stimulus, whereas theirs was done in VASP knockout (KO) mice where wild type or mutant VASP was transduced and phosphorylation induced by NO. These results suggest that the cellular mechanism of inhibition of cell growth induced by BAY is precise and unique to the drug. Interestingly, in our study the effects of PKG on the ability of BAY to inhibit proliferation in the MTT and BrdU assays (Figs. 8A, 8B) is similar to that of PKA; however, according to hemocytometric analysis these effects are not apparent until a later timepoint of 72 hrs (Fig 8C).

In the study by Chen and others (Chen *et al.*, 2004), the investigators show that genetic KO of VASPSer$_{239}$ promotes migration. In our previous study (Joshi *et al.*, 2011), we showed that BAY phosphorylates Ser$_{239}$ and in turn inhibits migration, which appears specific as a mechanism of action since inhibition of PKG fully reversed this effect. In the current study, data support these previous findings. Interestingly, the effect that PKG has on migration is similar to what is seen with PKC. Considering the influence of 8Br-cGMP on VASP$_{Ser239}$ (Fig. 6), these data support the involvement of PKG and PKC with VASP phosphorylation at Ser$_{239}$ and their regulatory role in migration of VSMCs.

To summarize these functional data, with respect to proliferation VASPSer$_{157}$ and PKA appear to be more involved and for migration VASPSer$_{239}$ and PKG may be more involved in VSMCs. In an instance where one kinase is reduced or rendered dysfunctional, other kinases may compensate providing multiple kinase regulation of any particular functional event. Additionally, these data demonstrate that PDEs help to maintain tight regulation of cyclic nucleotides in these cells. Inhibition of PDEs increases the activity of proteins involved in phosphorylation of VASP, and inhibition or feedback regulation occurs because both cyclic nucleotides are high at the same time demonstrating the importance of compartmentalization of proteins. It is also possible that BAY increases the activity of some alternate pathways that may be involved with VASP phosphorylation in these cells.

In separate experiments, higher passaged cells were used to evaluate the influence of phenotypic switching in the response to cyclic nucleotide and kinase signaling. In these cells (p>6) treated with BAY, inhibition of PKA increased phosphorylation of VASP at Ser$_{157}$ but had no effect on Ser$_{239}$, while inhibition of PKG increased phosphorylation of Ser$_{157}$ but had no effect on Ser$_{239}$ (Suppl. Fig. 1). On the other hand, inhibition of any kinase in the presence of a cyclic nucleotide analog had no effect (Suppl. Figs 2-3), except in the presence of 8Br-cAMP at Ser$_{239}$ (suppl. Fig. 3B). We also show that the amounts of total PKA and phosphorylated PKA increase and that the amount of PKG protein decreases with passage (suppl. Figs. 5A, 5B). Thus, these signaling events and in particular kinase expression must be considered with elevated passage as a model of phenotypic changes associated with injury or disease.

5. Conclusion

In conclusion, in this study we demonstrate that: 1) BAY increases phosophorylation of VASP at both Ser$_{157}$ and Ser$_{239}$; 2) BAY increases both PKA and PKG activity; 3) BAY reduces proliferation and migration in VSMCs; 4) inhibition of PKA in response to BAY augments inhibition of proliferation of VSMCs yet has no effect on migration; 5) multiple kinases appear to be involved in regulation of VASP phosphorylation; 6) PKA appears to be the kinase that is most active in these cells in response to BAY, and 7) kinases are involved in regulation of PDEs that regulate VASP phosphorylation in VSMCs. These findings provide valuable details about the biochemical cell signaling associated with cyclic nucleotide second messengers in VSM and highlight BAY as a potential anti-growth modulator in VSM.

6. Acknowledgements

We would like to acknowledge Jonathan C. Fox and Patti Shaver for assistance with isolation and culture of rat primary vascular smooth muscle cells. This project was supported by Award Number R01HL081720 from the National Institutes of Health National Heart, Lung, and Blood Institute (NHLBI), by ARRA Award Number R01HL081720-03S2, and by Post-doctoral Research Supplement Award Number R01HL081720-05S1 from the NHLBI. This content is solely the responsibility of the authors and does not necessarily represent the official views of the NHLBI or the NIH.

7. References

Abel K, Mieskes G, Walter U (1995). Dephosphorylation of the focal adhesion protein VASP in vitro and in intact human platelets. *FEBS Lett* 370(3): 184-188.

Bender AT, Beavo JA (2006). Cyclic nucleotide phosphodiesterases: molecular regulation to clinical use. *Pharmacol Rev* 58(3): 488-520.

Benz PM, Blume C, Seifert S, Wilhelm S, Waschke J, Schuh K, *et al.* (2009). Differential VASP phosphorylation controls remodeling of the actin cytoskeleton. *J Cell Sci* 122(Pt 21): 3954-3965.

Blume C, Benz PM, Walter U, Ha J, Kemp BE, Renne T (2007). AMP-activated protein kinase impairs endothelial actin cytoskeleton assembly by phosphorylating vasodilator-stimulated phosphoprotein. *J Biol Chem* 282(7): 4601-4612.

Boerrigter G, Burnett JC, Jr. (2007). Nitric oxide-independent stimulation of soluble guanylate cyclase with BAY 41-2272 in cardiovascular disease. *Cardiovasc Drug Rev* 25(1): 30-45.

Bononi A, Agnoletto C, De Marchi E, Marchi S, Patergnani S, Bonora M, *et al.* (2011). Protein kinases and phosphatases in the control of cell fate. *Enzyme Res* 2011: 329098.

Butt E, Abel K, Krieger M, Palm D, Hoppe V, Hoppe J, *et al.* (1994). cAMP- and cGMP-dependent protein kinase phosphorylation sites of the focal adhesion vasodilator-stimulated phosphoprotein (VASP) in vitro and in intact human platelets. *J Biol Chem* 269(20): 14509-14517.

Chen H, Kovar J, Sissons S, Cox K, Matter W, Chadwell F, *et al.* (2005). A cell-based immunocytochemical assay for monitoring kinase signaling pathways and drug efficacy. *Anal Biochem* 338(1): 136-142.

Chen L, Daum G, Chitaley K, Coats SA, Bowen-Pope DF, Eigenthaler M, *et al.* (2004). Vasodilator-stimulated phosphoprotein regulates proliferation and growth inhibition by nitric oxide in vascular smooth muscle cells. *Arterioscler Thromb Vasc Biol* 24(8): 1403-1408.

Chitaley K, Chen L, Galler A, Walter U, Daum G, Clowes AW (2004). Vasodilator-stimulated phosphoprotein is a substrate for protein kinase C. *FEBS Lett* 556(1-3): 211-215.

Christodoulides N, Durante W, Kroll MH, Schafer AI (1995). Vascular smooth muscle cell heme oxygenases generate guanylyl cyclase-stimulatory carbon monoxide. *Circulation* 91(9): 2306-2309.

Cohen P, Frame S (2001). The renaissance of GSK3. *Nat Rev Mol Cell Biol* 2(10): 769-776.

Cohen S, Fleischmann R (2010). Kinase inhibitors: a new approach to rheumatoid arthritis treatment. *Curr Opin Rheumatol* 22(3): 330-335.

Degerman E, Belfrage P, Manganiello VC (1997). Structure, localization, and regulation of cGMP-inhibited phosphodiesterase (PDE3). *J Biol Chem* 272(11): 6823-6826.

Di Benedetto G, Zoccarato A, Lissandron V, Terrin A, Li X, Houslay MD, *et al.* (2008). Protein kinase A type I and type II define distinct intracellular signaling compartments. *Circ Res* 103(8): 836-844.

Fry DW, Kraker AJ, McMichael A, Ambroso LA, Nelson JM, Leopold WR, *et al.* (1994). A specific inhibitor of the epidermal growth factor receptor tyrosine kinase. *Science* 265(5175): 1093-1095.

Gibbons GH, Dzau VJ (1994). The emerging concept of vascular remodeling. *N Engl J Med* 330(20): 1431-1438.

Hanks SK, Quinn AM (1991). Protein kinase catalytic domain sequence database: identification of conserved features of primary structure and classification of family members. *Methods Enzymol* 200: 38-62.

Heidenreich KA, Toledo SP, Kenner KA (1991). Regulation of protein phosphorylation by insulin and insulin-like growth factors in cultured fetal neurons. *Adv Exp Med Biol* 293: 379-384.

Herity NA, Ward MR, Lo S, Yeung AC (1999). Review: Clinical aspects of vascular remodeling. *J Cardiovasc Electrophysiol* 10(7): 1016-1024.

Houslay MD, Baillie GS, Maurice DH (2007). cAMP-Specific phosphodiesterase-4 enzymes in the cardiovascular system: a molecular toolbox for generating compartmentalized cAMP signaling. *Circ Res* 100(7): 950-966.

Hug H, Sarre TF (1993). Protein kinase C isoenzymes: divergence in signal transduction? *Biochem J* 291 (Pt 2): 329-343.

Ignarro LJ (1991). Heme-dependent activation of guanylate cyclase by nitric oxide: a novel signal transduction mechanism. *Blood Vessels* 28(1-3): 67-73.

Ignarro LJ, Harbison RG, Wood KS, Kadowitz PJ (1986). Activation of purified soluble guanylate cyclase by endothelium-derived relaxing factor from intrapulmonary artery and vein: stimulation by acetylcholine, bradykinin and arachidonic acid. *J Pharmacol Exp Ther* 237(3): 893-900.

Ignarro LJ, Kadowitz PJ (1985). The pharmacological and physiological role of cyclic GMP in vascular smooth muscle relaxation. *Annu Rev Pharmacol Toxicol* 25: 171-191.

Janssens S, Flaherty D, Nong Z, Varenne O, van Pelt N, Haustermans C, *et al.* (1998). Human endothelial nitric oxide synthase gene transfer inhibits vascular smooth muscle cell proliferation and neointima formation after balloon injury in rats. *Circulation* 97(13): 1274-1281.

Joshi CN, Martin DN, Fox JC, Mendelev NN, Brown TA, Tulis DA (2011). The Soluble Guanylate Cyclase Stimulator BAY41-2272 Inhibits Vascular Smooth Muscle Growth through the PKA and PKG Pathways. *J Pharmacol Exp Ther* 339: 394-402.

Karin M (2005). Inflammation-activated protein kinases as targets for drug development. *Proc Am Thorac Soc* 2(4): 386-390; discussion 394-385.

Kawashima S, Yamashita T, Ozaki M, Ohashi Y, Azumi H, Inoue N, *et al.* (2001). Endothelial NO synthase overexpression inhibits lesion formation in mouse model of vascular remodeling. *Arterioscler Thromb Vasc Biol* 21(2): 201-207.

Krause M, Bear JE, Loureiro JJ, Gertler FB (2002). The Ena/VASP enigma. *J Cell Sci* 115(Pt 24): 4721-4726.

Kwiatkowski AV, Gertler FB, Loureiro JJ (2003). Function and regulation of Ena/VASP proteins. *Trends Cell Biol* 13(7): 386-392.

Lissandron V, Zaccolo M (2006). Compartmentalized cAMP/PKA signalling regulates cardiac excitation-contraction coupling. *J Muscle Res Cell Motil* 27(5-7): 399-403.

Liu XM, Peyton KJ, Mendelev NN, Wang H, Tulis DA, Durante W (2009). YC-1 stimulates the expression of gaseous monoxide-generating enzymes in vascular smooth muscle cells. *Mol Pharmacol* 75(1): 208-217.

Lucas KA, Pitari GM, Kazerounian S, Ruiz-Stewart I, Park J, Schulz S, *et al.* (2000). Guanylyl cyclases and signaling by cyclic GMP. *Pharmacol Rev* 52(3): 375-414.

Manning G, Whyte DB, Martinez R, Hunter T, Sudarsanam S (2002). The protein kinase complement of the human genome. *Science* 298(5600): 1912-1934.

Mayers RM, Leighton B, Kilgour E (2005). PDH kinase inhibitors: a novel therapy for Type II diabetes? *Biochem Soc Trans* 33(Pt 2): 367-370.

Mendelev NN, Williams VS, Tulis DA (2009). Antigrowth properties of BAY 41-2272 in vascular smooth muscle cells. *J Cardiovasc Pharmacol* 53(2): 121-131.

Mitchison TJ, Cramer LP (1996). Actin-based cell motility and cell locomotion. *Cell* 84(3): 371-379.

Mullershausen F, Russwurm M, Friebe A, Koesling D (2004). Inhibition of phosphodiesterase type 5 by the activator of nitric oxide-sensitive guanylyl cyclase BAY 41-2272. *Circulation* 109 (14): 1711-1713.

Ozaki M, Kawashima S, Yamashita T, Hirase T, Namiki M, Inoue N, *et al.* (2002). Overexpression of endothelial nitric oxide synthase accelerates atherosclerotic lesion formation in apoE-deficient mice. *J Clin Invest* 110(3): 331-340.

Pawson T, Nash P (2000). Protein-protein interactions define specificity in signal transduction. *Genes Dev* 14(9): 1027-1047.

Reinhard M, Jarchau T, Walter U (2001). Actin-based motility: stop and go with Ena/VASP proteins. *Trends Biochem Sci* 26(4): 243-249.

Roger S, Badier-Commander C, Paysant J, Cordi A, Verbeuren TJ, Feletou M (2010). The anti-aggregating effect of BAY 41-2272, a stimulator of soluble guanylyl cyclase, requires the presence of nitric oxide. *Br J Pharmacol* 161(5): 1044-1058.

Rudic RD, Shesely EG, Maeda N, Smithies O, Segal SS, Sessa WC (1998). Direct evidence for the importance of endothelium-derived nitric oxide in vascular remodeling. *J Clin Invest* 101(4): 731-736.

Rybalkin SD, Rybalkina IG, Feil R, Hofmann F, Beavo JA (2002). Regulation of cGMP-specific phosphodiesterase (PDE5) phosphorylation in smooth muscle cells. *J Biol Chem* 277(5): 3310-3317.Scott JD (1991). Cyclic nucleotide-dependent protein kinases. *Pharmacol Ther* 50(1): 123-145.

Shi Y, Pieniek M, Fard A, O'Brien J, Mannion JD, Zalewski A (1996). Adventitial remodeling after coronary arterial injury. *Circulation* 93(2): 340-348.

Siow RC, Sato H, Mann GE (1999). Heme oxygenase-carbon monoxide signalling pathway in atherosclerosis: anti-atherogenic actions of bilirubin and carbon monoxide? *Cardiovasc Res* 41(2): 385-394.

Stasch JP, Becker EM, Alonso-Alija C, Apeler H, Dembowsky K, Feurer A, *et al.* (2001). NO-independent regulatory site on soluble guanylate cyclase. *Nature* 410(6825): 212-215.

Stasch JP, Hobbs AJ (2009). NO-independent, haem-dependent soluble guanylate cyclase stimulators. *Handb Exp Pharmacol*(191): 277-308.

Taylor SS, Buechler JA, Yonemoto W (1990). cAMP-dependent protein kinase: framework for a diverse family of regulatory enzymes. *Annu Rev Biochem* 59: 971-1005.

Taylor SS, Knighton DR, Zheng J, Ten Eyck LF, Sowadski JM (1992). cAMP-dependent protein kinase and the protein kinase family. *Faraday Discuss*(93): 143-152.

Thomson DM, Ascione MP, Grange J, Nelson C, Hansen MD (2011). Phosphorylation of VASP by AMPK alters actin binding and occurs at a novel site. *Biochem Biophys Res Commun.*

Tulis DA, Bohl Masters KS, Lipke EA, Schiesser RL, Evans AJ, Peyton KJ, *et al.* (2002). YC-1-mediated vascular protection through inhibition of smooth muscle cell proliferation and platelet function. *Biochem Biophys Res Commun* 291(4): 1014-1021.

Tulis DA, Durante W, Liu X, Evans AJ, Peyton KJ, Schafer AI (2001a). Adenovirus-mediated heme oxygenase-1 gene delivery inhibits injury-induced vascular neointima formation. *Circulation* 104(22): 2710-2715.

Tulis DA, Durante W, Peyton KJ, Evans AJ, Schafer AI (2001b). Heme oxygenase-1 attenuates vascular remodeling following balloon injury in rat carotid arteries. *Atherosclerosis* 155(1): 113-122.

Tulis DA, Keswani AN, Peyton KJ, Wang H, Schafer AI, Durante W (2005). Local administration of carbon monoxide inhibits neointima formation in balloon injured rat carotid arteries. *Cell Mol Biol (Noisy-le-grand)* 51(5): 441-446.

Vasioukhin V, Bauer C, Yin M, Fuchs E (2000). Directed actin polymerization is the driving force for epithelial cell-cell adhesion. *Cell* 100(2): 209-219.

von der Thusen JH, Fekkes ML, Passier R, van Zonneveld AJ, Mainfroid V, van Berkel TJ, *et al.* (2004). Adenoviral transfer of endothelial nitric oxide synthase attenuates lesion formation in a novel murine model of postangioplasty restenosis. *Arterioscler Thromb Vasc Biol* 24(2): 357-362.

Xiang YK (2011). Compartmentalization of beta-adrenergic signals in cardiomyocytes. *Circ Res* 109(2): 231-244.

Zaccolo M, Pozzan T (2002). Discrete microdomains with high concentration of cAMP in stimulated rat neonatal cardiac myocytes. *Science* 295(5560): 1711-1715.

Zhuang S, Nguyen GT, Chen Y, Gudi T, Eigenthaler M, Jarchau T, *et al.* (2004). Vasodilator-stimulated phosphoprotein activation of serum-response element-dependent transcription occurs downstream of RhoA and is inhibited by cGMP-dependent protein kinase phosphorylation. *J Biol Chem* 279(11): 10397-10407.

7

The Physiological Relationship of Endothelial Protein Kinase G with Endothelial Nitric Oxide Synthase

Theresa A. John[1] and J. Usha Raj[2]
[1]*Department of Pharmacology, Lagos State University College of Medicine, Lagos,*
[2]*Department of Pediatrics, College of Medicine of the University of Illinois at Chicago, Chicago, Illinois,*
[1]*Nigeria*
[2]*USA*

1. Introduction

Furchgott and his co-workers reported the vascular actions of an unknown biological mediator that was named endothelium-derived relaxing factor (EDRF).[1] Nitric oxide (NO) was soon discovered as a vasodilator that had identical properties with EDRF.[2,3,4] NO produced by the endothelium diffuses to the adjacent vascular smooth muscle cells causing them to relax. This function is important in many vascular beds such as the coronary circulation.[5, 6]

Nitric oxide synthases produce nitric oxide. In the endothelium, constitutive endothelial nitric oxide synthase (eNOS) produces nitric oxide (NO) under normal conditions and in response to chemical and physical stimuli such as shear stress. [6, 7, 8, 9] NO produced from endothelial cells or from platelets serves protective functions such as inhibiting platelet aggregation and adhesion and inhibiting leukocyte adhesion.[10, 11] Nitric oxide produced by constitutive neuronal nitric oxide synthase (nNOS) is a neuromediator in the central nervous system[12, 13] and in peripheral non adrenergic-non cholinergic nerves[14, 15, 16] where is mediates many physiological functions.

The second messenger or "molecular receptor" activated by NO to mediate these physiologic functions is soluble guanylate cyclase (GC).[17, 18] In the vasculature, this pathway has been clearly outlined to involve intracellular increase in the concentration of cyclic guanosine monophosphate (cGMP),[19, 20] activation of protein kinase G (PKG, cyclic GMP-dependent protein kinase),[21] and modulation of activities of cyclic nucleotide phosphodiesterases (PDE, 2, 3, and 5). In vascular smooth muscle cells the net effect is decreased phosphorylation of the myosin light chain, reduction of cytosolic Ca,$^{2+}$ and relaxation of smooth muscle cells. PKG plays a key role in vasodilation. PKG catalyzes the phosphorylation of certain proteins. It also activates myosin light chain phosphatase[22] causing dephosphorylation of mysosin light chain and vasorelaxation.

A third isoform of the enzyme is inducible nitric oxide synthase (iNOS) which is induced by immunologic stimuli and activated in pathologic conditions. It produces cytotoxic amounts of NO for host defense. [23, 24, 25, 26] All three terms: endothelial, neuronal, and inducible are traditional and presently exceptions have been identified in the variety of NOS enzymes.

Many cardiovascular diseases and pathologic conditions are related to abnormalities and irregularities of the NOS function and NO signaling. For example nNOS overproduction of NO is involved in some forms of epilepsy and iNOS overproduction of NO causes septic shock. Chronic diseases linked to deranged NO production include atherosclerosis, hypertension, cerebral and coronary vasospasm, and erectile dysfunction. Therefore, NO delivery and NO control are therapeutic targets.[27, 28, 29] Inhaled NO is especially useful for neonates with persistent pulmonary hypertension.[30,31] Nitric oxide donors are drugs used to enhance or substitute NO availability. They include organic nitrates and nitrites and have been used to lower blood pressure and for ischemic heart disease. Control of endogenous production of NO is dependent on our knowledge of the enzyme, NOS, that produces NO. Many chemicals have been used to compete with the amino acid L-arginine substrate from which NOS synthesizes NO. Some are experimental tools used to study NOS function, e.g. asymmetric dimethyl-l-arginine. NOS inhibitors may be useful in septic shock and neurodegenerative diseases.[6] However, we know very little about the life cycle of the enzyme nitric oxide synthase and how it is physiologically regulated: synthesized, deactivated, recycled or replenished in basal conditions in normal endothelial cells. Our recent findings of the relationship between protein kinase G and endothelial nitric oxide synthase open new thoughts on the regulation of nitric oxide synthase. These findings are discussed in this chapter.

2. Protein kinase G isoforms and their distribution in endothelial cells

Protein kinase G (PKG) is a serine/threonine-specific protein kinase. It is found in vascular smooth muscle cells (SMCs) and there it mediates the effect of endothelial NO on vascular SMCs to cause relaxation. Recently, we showed the abundance of PKG in quiescent ovine neonatal lung microvascular endothelial cells (LMVECs) and investigations were carried out to find out if NO within endothelial cells mediates any physiological function through endothelial PKG. A feed-back role of endothelial PKG on activated endothelial NOS appears to be a physiologic regulation of NO production.

PKG is likely to have an important physiologic role in endothelial cells because PKG isoforms are present in normal endothelial cells in distinct patterns of distribution (Figure 1).[32]

3. Physical intracellular relationship of NOS and PKG: Colocalization of PKG 1 beta and serine-116-phosphorylated endothelial nitric oxide synthase

Recently, using specific epitope fluorescent immunolabeling of monolayers and confocal microscopy, NOS was shown to colocalize with PKG in neonatal ovine LMVECs.[32] The confocal imaging of colocalized serine-116-phosphorylated NOS with PKG 1 beta in quiescent MVECs indicated a spatial relationship between the two molecules in intracellular domains (Figure 2). Thus we suspect a role for PKG (1 beta) in normal NOS metabolism in endothelial cells.

Till present, the importance of PKG in endothelial NO signaling has been focused on smooth muscle cell PKG. NO produced by the endothelium freely diffuses across biological

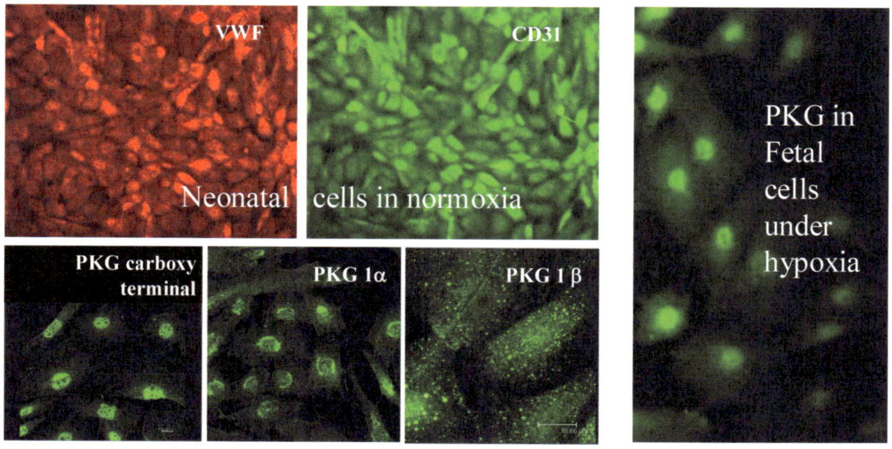

Fig. 1. Fluorescence immunohistochemistry expression of characteristic endothelial markers and some protein kinase G (PKG) isoforms in primary cultures of lamb lung microvascular endothelial cells (LMVECs). The figure shows newborn lamb (all left panels) and term fetal cells (right panel). The upper left panels show 100% of cells in the monolayer stained for both von Willebrand Factor (vWF; red alexa 568 staining) and cluster differentiating type antigen 31 (CD31, green alexa 488 staining), characterizing them as endothelial cells. The lower left panels show alexa 488-conjugated secondary antibodies attached to respective primary antibodies. The PKG carboxy terminal is expressed in the cytosol and more intensely in the perinuclear region of LMVEC; PKG 1α staining is localized to the Golgi and endoplasmic reticulum regions of the cells; and PKG 1β staining is localized to vesicular structures distributed all over the cell.

membranes and stimulates soluble guanylate cyclase in smooth muscle cells leading to increased intracellular cGMP, activation of PKG, and smooth muscle cell relaxation. The rest of the evidence in this chapter presents an emerging role of an equivalent NO-cyclic GMP-PKG signaling pathway in the endothelial cell that appears to produce a negative regulation of NOS (possibly in addition to endothelial relaxation).

4. Effects of protein kinase G modulator drugs on endothelial intracellular nitric oxide production

Basal levels of NO production have been difficult to determine because NO has a very short half-life of 5s and stimulated NO release is only about 500±20 nmol/s.[33] Many previous studies have measured stimulated NO production using agonists such as carbachol and calcium ionophore and these investigations studied factors altering such NOS function. The experimental tool, diaminofluorescein with a difluoro modification (DAF-FM) forms a relatively photo-stable, water soluble adduct, a fluorescent benzotriazole, with NO[34] at concentrations as low as 3 nM and can trap NO as it is produced. Importantly, low doses of DAF can pick up minute differences between groups with minimal diaminofluorescein (DAF) autofluorescence.[35, 36] The more important novelty about DAF is probably the opportunity it affords us to measure and monitor basal NO as it is produced by living cells. DAF is thus an interesting tool for studying basal NOS function [37] which we need to understand in order to better regulate nitric oxide production therapeutically. We detected

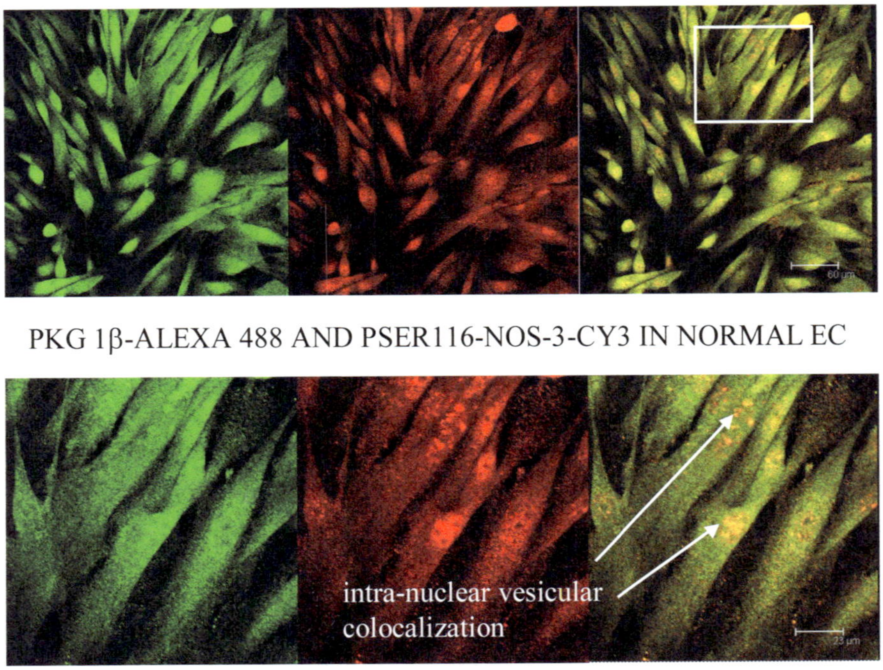

Fig. 2. PKG1beta and serine-116-phosphorylated NOS moieties colocalize in the cytosol and nuclear region vesicles. Quadruplet monolayers of untreated passage 3 neonatal ovine lung microvascular endothelial cells were fixed with 4% paraformaldehyde, permeabilized with 0.2 % triton X-100, blocked with 20% bovine serum albumin and then incubated with 1: 200 dilutions of goat anti PKG 1beta and rabbit anti serine116-phosphorylated NOS primary antibodies for 2h at room temperature. They were counterstained with 1: 125 dilutions of alexa 488-conjugated donkey anti goat and cy-3 conjugated F(ab)2 fragment specific sheep anti-rabbit secondary antibodies. Pictures were taken with a Leica DMIRE2 confocal inverted microscope and Leica v2.61 confocal software. The PKG 1beta and pSer116-NOS markers colocalized in the cytosol and intensely in vesicles in the nuclear region.

nitric oxide produced by isolated LMVECs under basal conditions by DAF fluorescence and it was inhibited by general NOS inhibitors NG-nitro-L-arginine (L-NNA) or N-nitro-L-arginine methyl ester (L-NAME). Here we summarize a number of published experiments that show that PKG activation negatively regulates activated NOS.[32, 38, 39, 40, 41]

Effects of PKG activation on real time basal and stimulated NO production are shown in Figure 3 and a dose dependency of the inhibitory effect of PKG on basal NO production is shown in Figure 4. The inhibitory effect doubled from 20 nM to 2 micromolar 8-bromo cyclic guanosine monophosphate (8-Br-cGMP) in Figure 4 indicating that this inhibitory mechanism is sensitive to low physiological levels of 8-Br-cGMP. The inhibition of NOS activity by 8-Br-cGMP was prevented by the peptide YGRKKRRQRRRPPLRKKKKKH (DT2), a potent PKG inhibitor, indicating that 8-Br-cGMP inhibits NOS through PKG (Figure

4). To make such observations, it was necessary to first subtract the large proportion of extraneous fluorescence. Basal NO production is at a very low level compared to stimulated production (Figure 3). The expansion of the fluorescence axis by extraneous fluorescence can minimize or obliterate observation of the effect of drugs on basal NOS activity.

NOS activity can be activated by such stimuli as cholinergic receptor agonists, shear stress, bradykinin receptor agonists, and calcium ionophore.[42, 43] Previously, factors modulating NOS were studied largely by investigating stimulated NOS activity, NOS activity *in vitro* and in transfection-modified cells using mainly the citrulline assay. Such activation and the interaction of NOS and its binding partners may be modulated by proteins such as heat shock protein 90 (HSP90),[44] nitric oxide synthase traffic inducer (NOSTRIN),[45] and eNOS interacting protein (NOSIP),[46] as well as kinase action by Akt (protein kinase B, PKB) and PKA (protein kinase A).[47] Basal endothelial nitric oxide production has been difficult to measure and study with the citrulline assay. The sensitivity and ease of DAF[48] is a valuable tool that has enabled us to study more closely this narrow but important margin of biological activity (Figure 3). Possibly, derangement of basal NOS activity rather than alteration of factors involved in potentiated NOS activity may be the underlying factor in some NOS-related vascular diseases because pathology is often a sustained deviation from the basal or balanced state.

The data showing a physiological inhibitory role of PKG on basal NOS activity may be of clinical significance. The importance of this is not clear but it seems that activated endothelial NOS is not turned off until endothelial PKG is activated, thus ensuring a relationship between endothelial NO generation and corresponding vascular smooth muscle PKG function. In Figure 5 the effects of various PKG activator and inhibitor drugs also show that nitric oxide production in endothelial cells may be negatively regulated by protein kinase G feedback. In Figure 6 and Figure 7, the use of NOS inhibition shows that the protein kinase G effect appears to be specifically on nitric oxide synthase function.

Lack of basal PKG activity produced by the antagonist, guanosine-3′, 5′-cyclic monophosphorothionate 8-bromo Rp isomer, increases basal NOS activity and the effect of the cGMP antagonist is inhibited by NOS inhibitor L-NNA (Figure 6). In Figure 6, after the 5 min time point, the effect of the competitive inhibitor of NOS, LNNA becomes equilibrated and the PKG inhibition can be seen to sustain a cumulative active pool of NOS.

In another experiment using an extended time-course of up to 360 min, L-NNA also enhanced the effect of PKG activator 8-Br-cGMP but reversed the effect of another PKG inhibitor Rp-8-Br-PET-cGMPS on NOS function (Figure 7). The graphs obtained had correlation values of $R^2 > 0.9$ showing a real relationship between fluorescence changes and time of reaction with drugs. The slopes of the graphs all fall within a narrow range of 7.5 – 8.19 showing lack of variation and indicating parallel shifts. This suggests a molecular relationship between NOS and PKG modulators based on receptor activity.

These results suggest that the effects of PKG are directly on active NOS function. In summary, basal (and activated) NOS activity is regulated by the downstream activation of PKG by NO-stimulated cGMP production in endothelial cells. It appears that when NOS is activated, the NO produced diffuses to the smooth muscle but some of the NO triggers an endothelial PKG sensitive signaling that results in inhibition of the activated NOS within the endothelial cell.

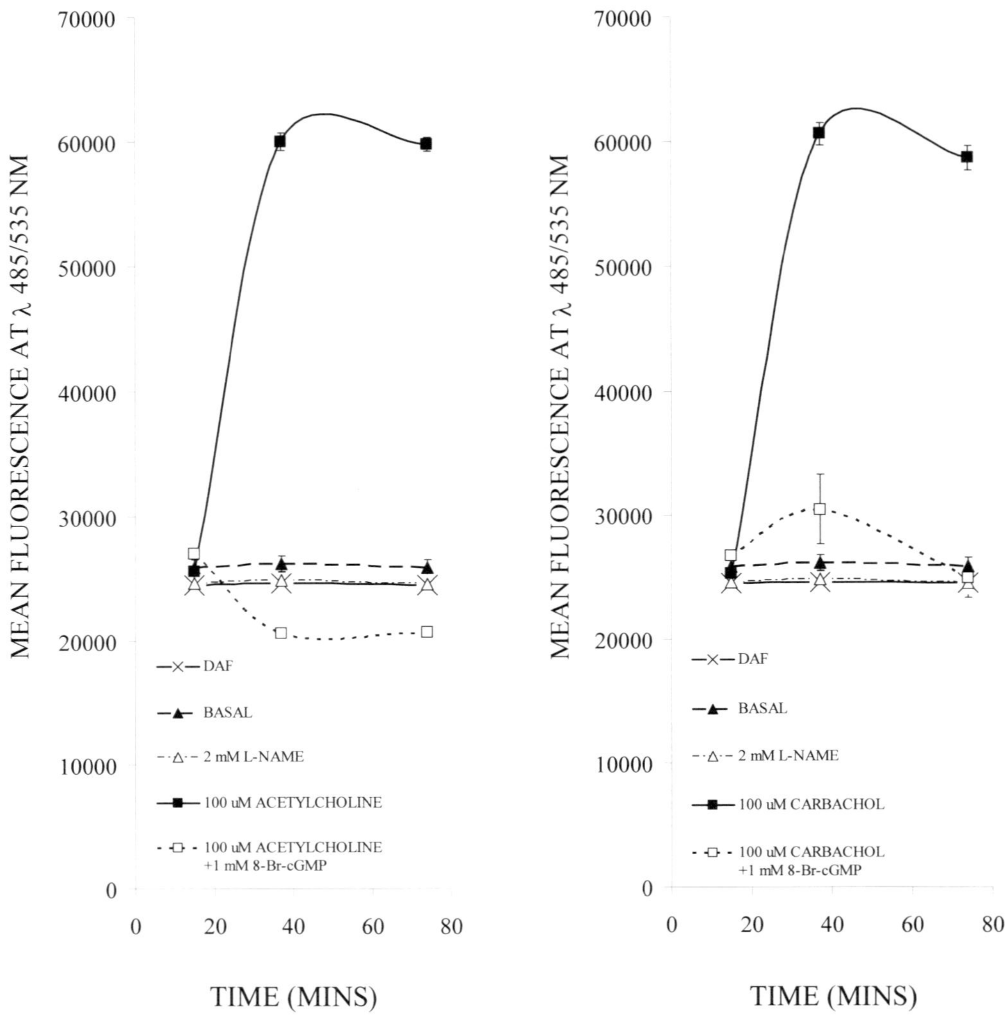

Fig. 3.[38] Exogenously administered 8-Br-cGMP attenuates stimulated nitric oxide synthase function in ovine LMVECs. Confluent monolayers of sibling cells of passage 4 were used. A set of 12 wells with no cells was used to measure DAF autofluorescence in the reaction mixture over the time-course of the experiment. To determine the level of extraneous fluorescence in the experimental groups, a group of 12 microcultures were treated with NOS inhibitor 2 mM L-NAME 1 h before DAF. In all the other groups, drugs (100 microM acetyl choline or 100 microM carbachol followed by 1 mM 8-Br-cGMP) were administered just before a mixture of 1 microM DAF FM and 2.4 microM DAF-FM diacetate. The basal NO group received only the fluorophore. Cells were measured at intervals, starting at 15 min after DAF, for total NO (in intracellular and cell bathing medium) that combined with DAF to produce fluorescence. All graphs shown are from means of raw fluorescence ± SE of 8 or more microcultures per point from a single experiment. Figure shows a representative of 2 similar experiments. 8-Br-cGMP antagonized the stimulating effects of both acetylcholine and carbachol on nitric oxide production.

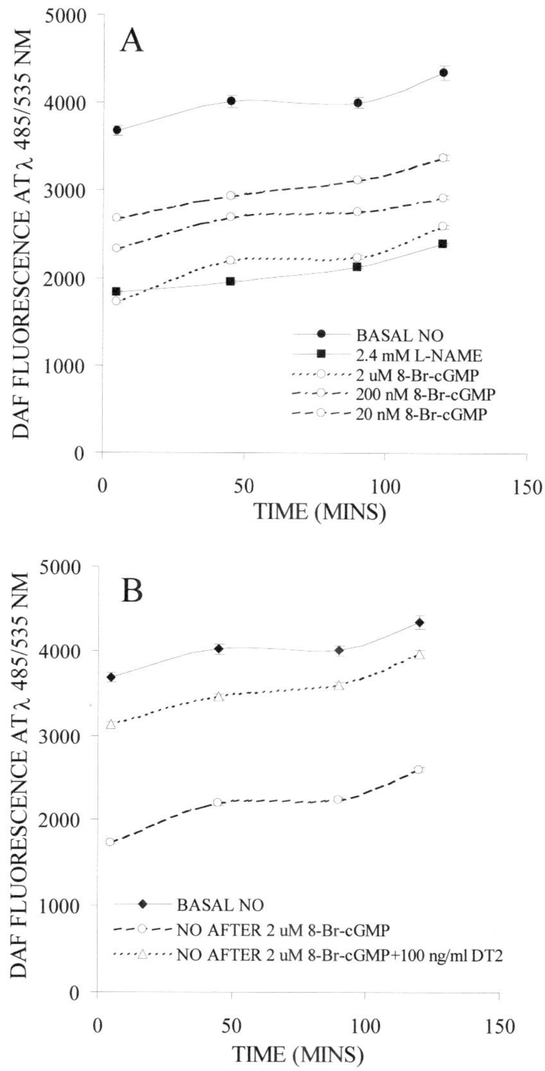

Fig. 4.38 Exogenously administered 8-Br-cGMP attenuates basal nitric oxide synthase function in ovine LMVECs. Confluent monolayers of sibling cells of passage 2 were used in 96 well clusters. They were treated with NOS inhibitor 2.4 mM L-NAME to determine the level of extraneous fluorescence in the experimental groups, with 20 nM - 2 microM 8-Br-cGMP to observe its effect on basal nitric oxide production, with PKG inhibitor 100 ng/ml DT2 to determine if 8-Br-cGMP acts through PKG. DT2 or L-NAME were administered 1 h before DAF. 8-Br-cGMP was administered 5 min before DAF. A mixture of 0.5 microM DAF FM and 0.8 microM DAF-FM diacetate was used to measure the NO in the cell bathing medium and intracellular NO together. The basal NO group received only the fluorophore. All graphs shown are from means of corrected fluorescence ± SE of 8 or 12 microcultures per point from a single experiment. Final fluorescence measurements are raw means minus corresponding means of DAF autofluorescence. The figure is a representative of 3 experiments.

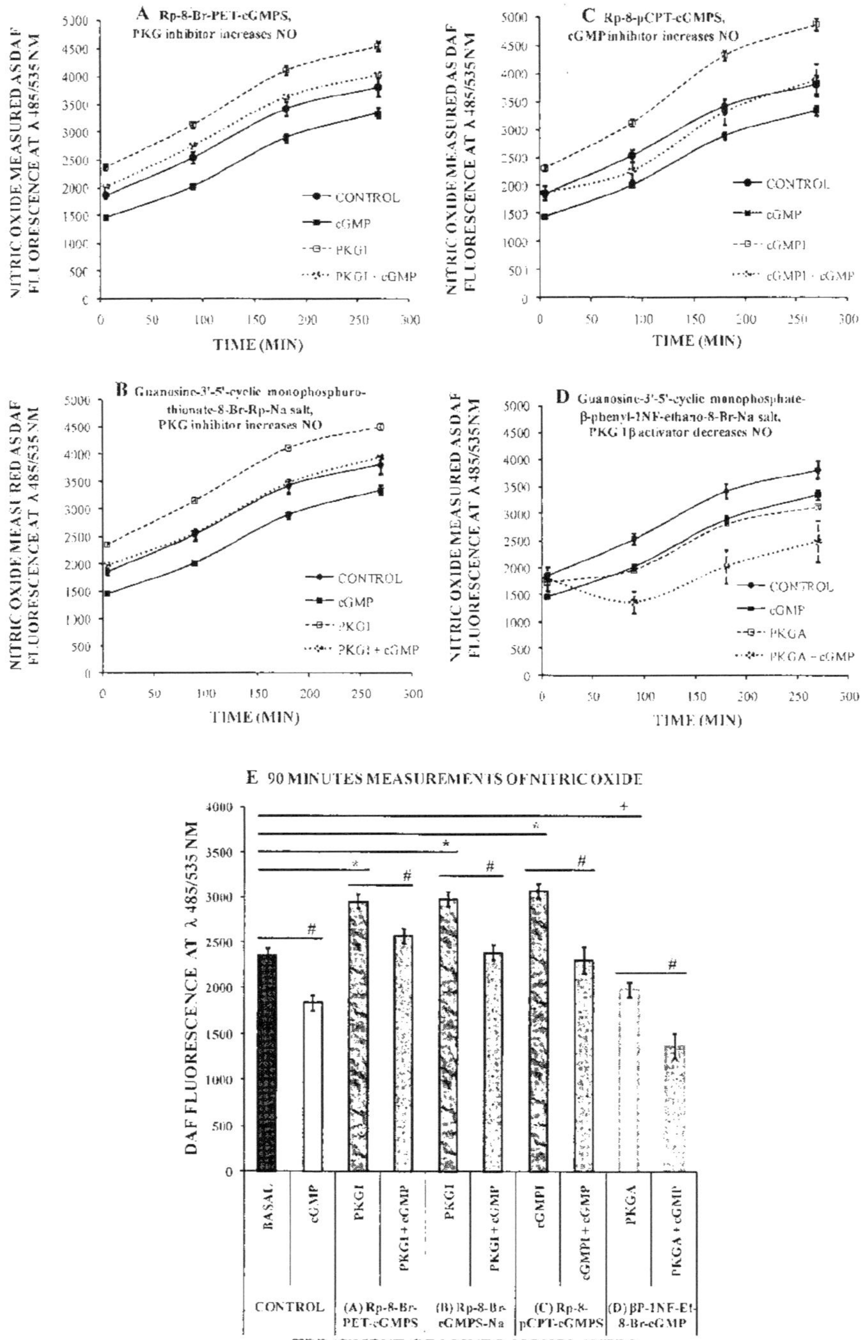

Fig. 5.[39] Timecourse of the inhibition of nitric oxide production by protein kinase G: modifications by PKG inhibitor and activator drugs. Nitric oxide measurements were done using 4,5-diamino-fluorescein derivative, DAF-FM, (0.5 µM DAF-FM and 0.8 µM DAF FM

diacetate) added to stably confluent microcultures of LMVECs in 96-well clusters. To show the basic effect of PKG activation, cells were incubated with 100 μM 8-Br-cGMP. Other experimental groups were preincubated with a modulator drug: PKG inhibitors 100 μM Rp-8-Br-PET-cGMPS or 50 nM guanosine 3′-5′-cyclic mono-phosphoro thionate-8-Br-Rp isomer Na salt; cGMP inhibitor: 4 μM Rp-8-pCPTcGMPS; or PKG activator: 100 μM guanosine-3′-5′-cyclic monophosphate-β-phenyl-1NF-ethano-8-bromo sodium salt. Inhibitors were added 30 min before DAF and activators were added with DAF. Total NO produced after 5, 90, 180 and 270 min of incubation with DAF at 37 °C was measured using a Victor 1420 multilabel counter (Waltham, Massachusetts) with a fluorescein setting (λex/em 485/535 nm). Using the 90 min time point, the data plotted as mean ± SE show that 8-Br-cGMP reduces NO production in controls and in all the tests with cGMP/PKG inhibitor drugs as well as in tests with the PKG activator ($p<0.05\#$ in all cases). The cGMP/PKG inhibitor drugs (A), (B) and (C) significantly increased NO production ($p<0.05^*$ in all cases) and the PKG activator guanosine-3′-5′-cyclic monophosphate-β-phenyl-1NF-ethano-8-bromo sodium salt (D) significantly decreased NO production ($p<0.05+$) (lower figure).

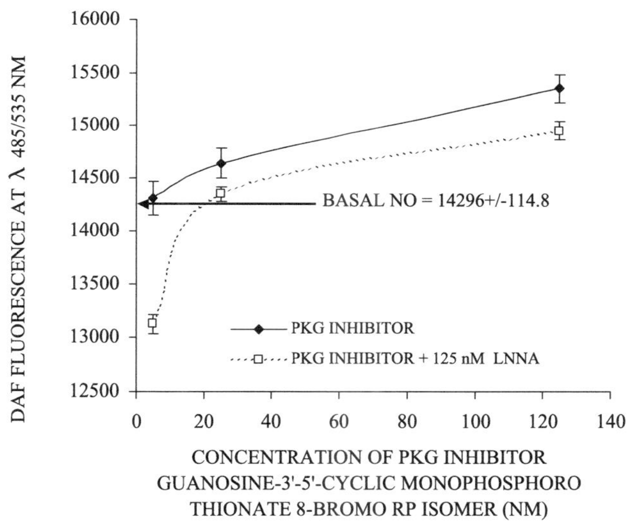

Fig. 6.[38] Basal nitric oxide production in ovine LMVECs is enhanced by lack of basal endothelial PKG activity and this effect is attenuated by nitric oxide synthase inhibitor LNNA. Confluent monolayers of sibling cells of passage 4 were used in groups of N=12 for basal and N=6 for test groups of microcultures. Cells were treated with different concentrations of cGMP antagonist guanosine-3′, 5′-cyclic monophosphorothionate 8-bromo Rp isomer (for PKG inhibition) added 30 min before DAF or with the same concentrations of the cGMP antagonist plus 125 nM L-NNA previously added 2h before DAF. A mixture of 0.5 microM DAF FM and 1 microM DAF-FM diacetate was used to measure total intracellular and cell bathing medium NO that combined with DAF to produce fluorescence. Measurements were made 15 min after DAF was applied. The arrow shows the level of basal NO at 15 min in an identical third group not treated with drugs. All graphs shown are from means ± SE. Figure shows a representative of 2 similar experiments. The PKG inhibition dose-dependently increased NO measured. The inhibition of this effect by L-NNA shows it is through NOS.

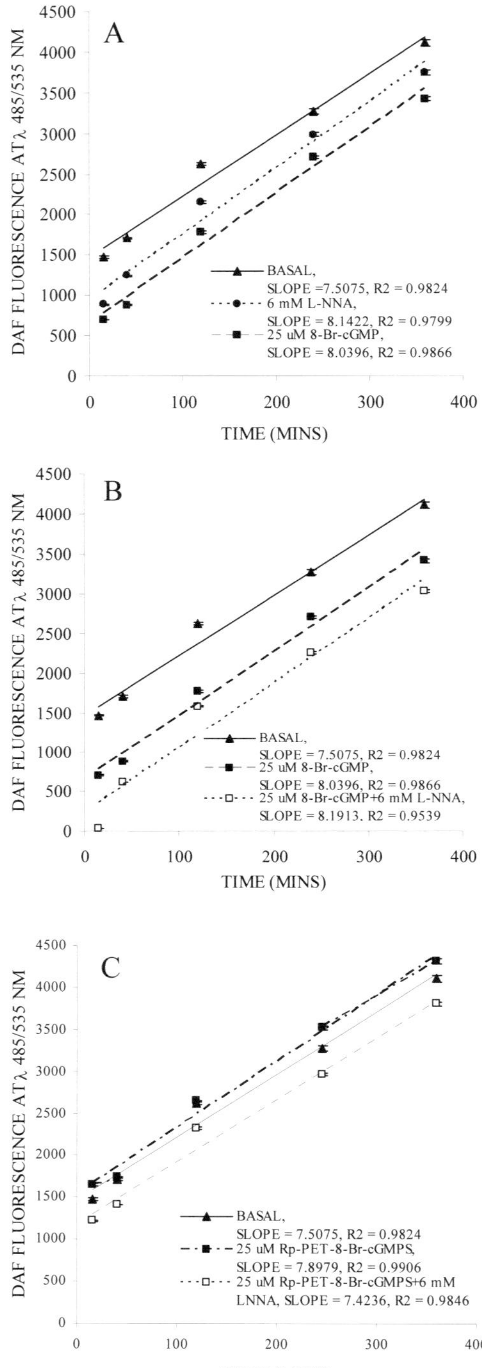

Fig. 7.[38] Specificity of PKG effects on NOS function: NOS inhibitor LNNA enhances the effect of PKG activator 8-Br-cGMP on NOS function and reverses the effect of PKG inhibitor

Rp-8-Br-PET-cGMPS on NOS function. Confluent monolayers of sibling cells of passage 4 were used. Inhibitors, 6 mM L-NNA and 25 microM Rp-8-Br-PET-cGMPS were administered 2h and 1h respectively before DAF and the PKG activator, 25 microM 8-Br-cGMP < 5 min before DAF. A mixture of 0.5 microM DAF FM and 0.8 microM DAF-FM diacetate was used to measure total intracellular and cell bathing medium NO that combined with DAF to produce fluorescence. All graphs shown are means ± SE of 12 for basal and 6 for tests microcultures per point. Figure shows a representative of 2 similar experiments. A: Inhibition by L-NNA or 8-Br-cGMP produced a parallel shift of the basal NOS activity graph downwards. B: Synergy of the effects of L-NNA and 8-Br-cGMP co-administered. C: The effect of PKG inhibition produces a parallel shift of basal NOS activity graph upwards and this shift is antagonized by L-NNA.

5. Relationship of PKG activation and phosphorylation of NOS at various sites

Constitutive nitric oxide synthase in endothelial cells is localized to caveolae [49, 50] where it docks into the intracellular domain 4 of the bradykinin B2 receptor.[51] The structural protein of caveolae, caveolin-1, also binds to NOS keeping it inactive[52] Activation of NOS leading to its dissociation from the complex is calcium dependent.[9, 52] A further activation on serine 1177/1179 is produced by kinase activity.[42] Several kinases are known to phosphorylate NOS,[53] regulating its activity negatively or positively under experimental conditions. Apart from constitutive caveolin-1, other negative regulators of NOS are NOSIP[54] and NOSTRIN.[55] Both interfere with the association of NOS with caveolae and cause its redistribution from the plasma membrane to intercellular compartments with a decrease in nitric oxide (NO) production.

Three positive regulators of NOS have been identified. The protein kinase AKt (PKB) phosphorylates NOS on serine 1177/1179, enhancing NOS activation.[56] Protein kinase A also phosphorylates NOS to increase its activity.[47] HSP90 is a molecular scaffold that facilitates the interaction of kinases and substrates including NOS. It facilitates the dissociation of NOS from caveolae in response to calcium-calmodulin.[44, 57] An emerging role of protein kinase G in NOS regulation is interesting because not only is it a regular downstream component of the physiologic NO signaling pathway that could give feedback regulation, but also because experimentally, its involvement in NOS phosphorylation can be related to its regulation of NOS function. Presently, PKG is unnoticed as a physiological regulator of NOS.[58]

We recently published fluorescence activated cell sorter (FACS) analysis of the relationship between PKG and endothelial NOS (Figures 8-10).[39] In summary, 5 microM 8-Br-cGMP in <5 min caused an increase in N-terminal labeling of NOS and a decrease in both C-terminal and serine 1177 labeling of NOS. 8-Br-cGMP appeared to increase PKG 1α and to decrease PKG 1β labeling. Changes in other phosphorylation sites were less consistent but overall mean channel fluorescence increased from 19.92 to 217.36 for serine 116 and decreased from 329.27 to 254.03 for threonine 495 phosphorylation. Data indicated that PKG caused both molecular and phosphorylation changes in NOS. Thus an implicated role of PKG and serine 116 phosphorylation site on NOS may be a future focus for studying the termination of activated NOS function and the vesicular processing of the spent enzyme. Several evidences from confocal microscopy further develop this.

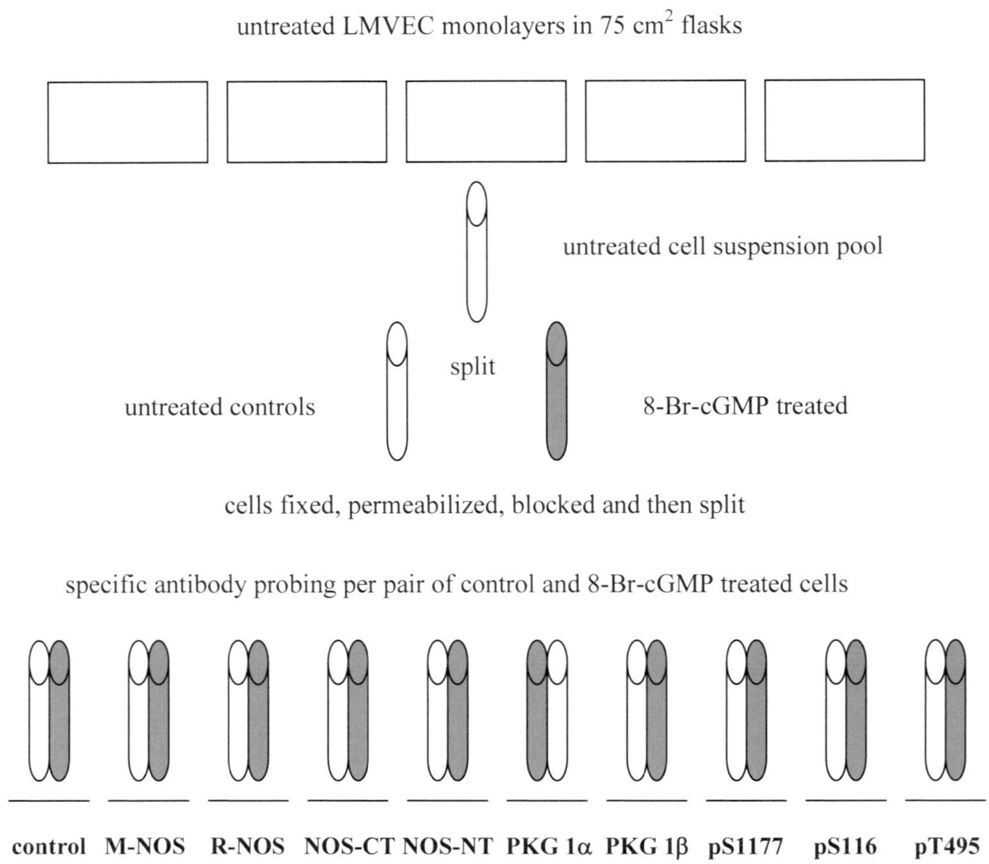

Fig. 8.39 Schema of experimental protocol for fluorescence activated cell sorter analysis of the effect of cGMP on antibody labeling of NOS regions and protein kinase G isoforms. Sibling cells were pooled and equally distributed for controls and tests.

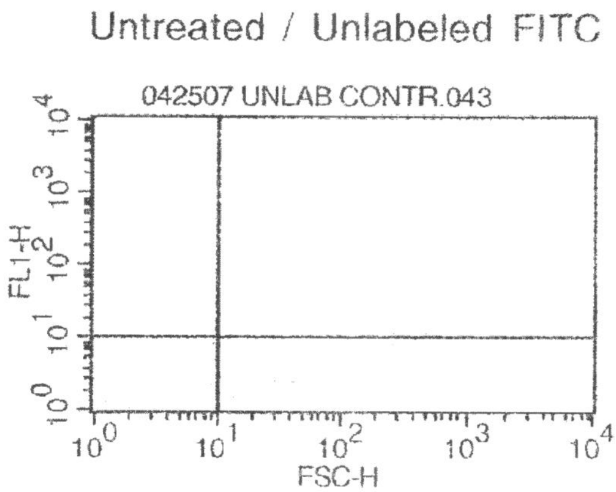

A

B

FACS ANALYSIS	1ST		2ND		3RD	
	CONTROL	8-Br-cGMP	CONTROL	8-Br-cGMP	CONTROL	8-Br-cGMP
NOS-NT	1.78	4.71	0.25	0.52	0.07	1.43
NOS-CT	4.42	1.84	1.41	0.92	1.15	0.85
pS1177-NOS	2.28	1.39	2.05	0.83	0.75	0.62
pS116-NOS	2.32	1.86	0.36	0.41	0.22	0.5
pT495-NOS	2.44	1.84	1.04	1.75	0.68	1.16

C

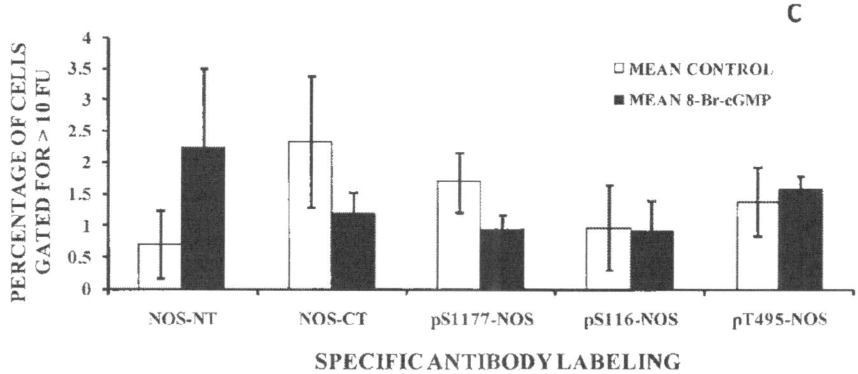

Fig. 9.[39] Effect of PKG activation on percentage of cells with high fluorescence intensity for various immunolabeling. Cell suspensions in phosphate buffered saline (PBS) containing 5% antibiotic/antimycotic mixture were split equally into control and 5 µM 8-Br-cGMP treatment groups before fixation with 4% paraformaldehyde. Cell were permeabilized with methanol, blocked with 20 mg/ml bovine serum albumin (BSA) and then labeled with 1: 100 primary antibodies against NOS N-terminus, NOS C-terminus, or serine 1177, serine 116, or threonine 495 phosphorylated NOS for 2h. They were counterlabelled with FITC conjugated secondary antibodies and ran through a BD FACSCallibur. Cells with positive signal of fluorescence were analysed using a FITC (fluorescein) filter (FL1). Upper figure shows gating of cells for 10 fluorescence units (y-axis) and 10 forward scatter units (x-axis). Middle table shows actual Cell Quest read out values for % gated. The mean values of the three determinations were plotted in the lower figure.

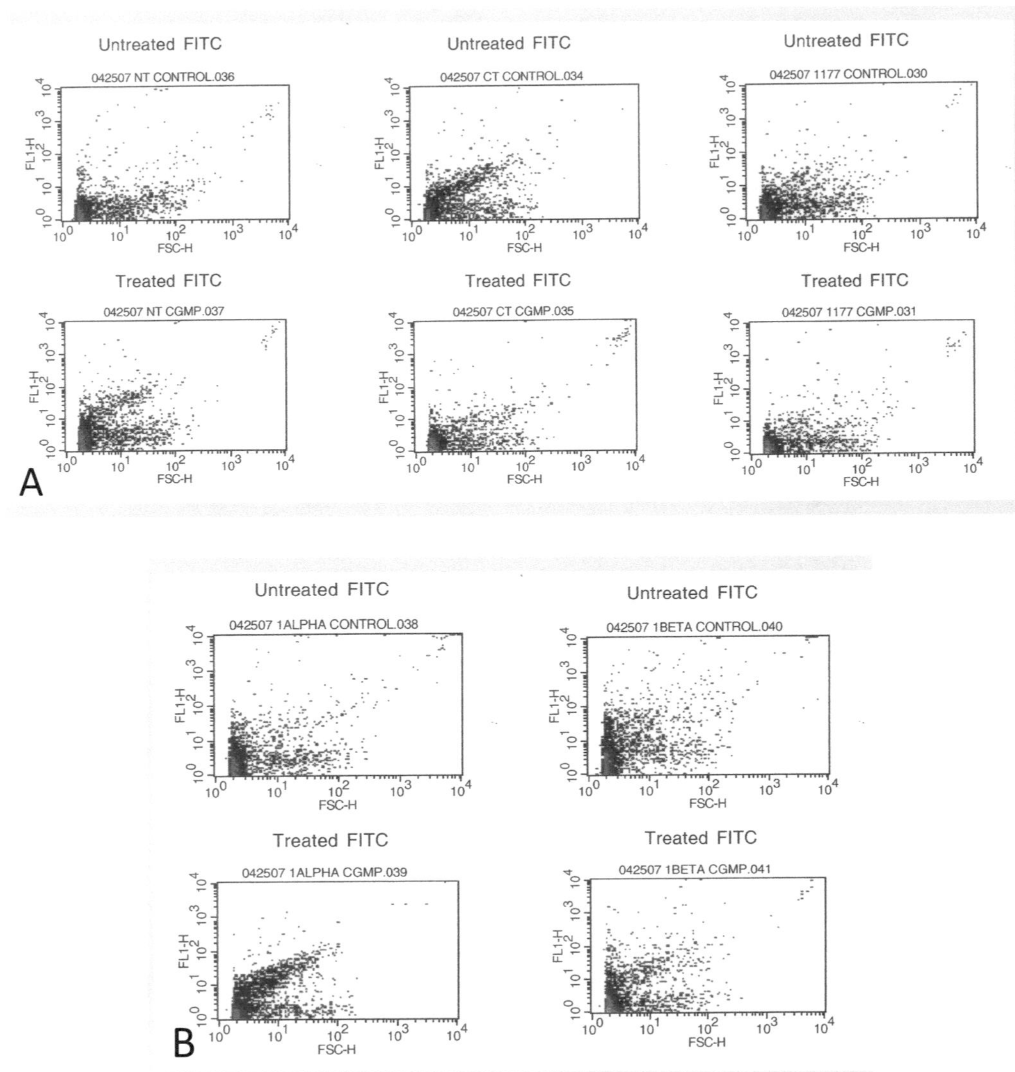

Fig. 10.[39] A. FACS analysis of the effect of PKG activation on NOS N-terminal and C-terminal specific rabbit antibody labeling and pSer1177 NOS specific antibody labeling. Cells were immunolabeled as described for Figure 9. The fluorescence intensities of sample populations were captured with a Cell Quest software. Upper channels show control cells and lower channels show PKG activator 5 μM 8-Br-cGMP treated cells. Left panels were labeled with a rabbit anti N-terminal of NOS antibody. Middle panels were labeled with a rabbit anti C-terminus of NOS antibody. Right panels were labeled with a rabbit anti serine 1177 phosphorylated NOS antibody. The fluorescence intensity range (y-axis) was increased by PKG activation for N-terminal labeling and decreased by PKG activation for C-terminal labeling. The fluorescence intensity range was decreased by PKG activation for serine 1177 phosphorylation. B. FACS analysis of the effect of PKG activation on antibody labeling of protein kinase G isoforms. Upper channels show control cells and lower channels show PKG

activator 5 μM 8-BrcGMP treated cells. Left panels were labeled with a goat anti PKG 1α antibody. Right panels were labeled with a goat anti PKG 1β antibody. PKG activation increased total number of cells with >10^1 fluorescence units (FU) (y-axis) labeling for PKG 1α and decreased the same labeling for PKG 1β.

6. PKG phosphorylated nitric oxide synthase in endothelial cells

The colocalization of serine 116-phosphorylated NOS with PKG suggests that PKG phosphorylates NOS at serine 116 but this has not been directly demonstrated. If this is true, serine 116 phosphorylation of NOS may be connected with the process of terminating the action of NOS after synthesizing NO. The functional data showing modulation of NO production by cGMP and the FACS data showing the effect of cGMP on C- and N-terminal expressions of NOS and phosphorylation expressions of NOS add up to suggest that serine-116 phosphorylated NOS is a spent form of the enzyme. Figure 11 from our published data[32, 41] indicates that serine116-phosphorylated NOS is not like nascent NOS that is known to be associated with the plasmalemma and golgi. Serine 116 phosphorylated NOS is found in the cytosol and in endoplasmic reticulum (ER) and nuclear site vesicles. Figure 12[41] clearly shows that this form of NOS is not associated with the plasma membrane. On the other hand, unphosphorylated NOS colocalizes with caveolin in the plasmalemma and in the Golgi (Figure 13).[41]

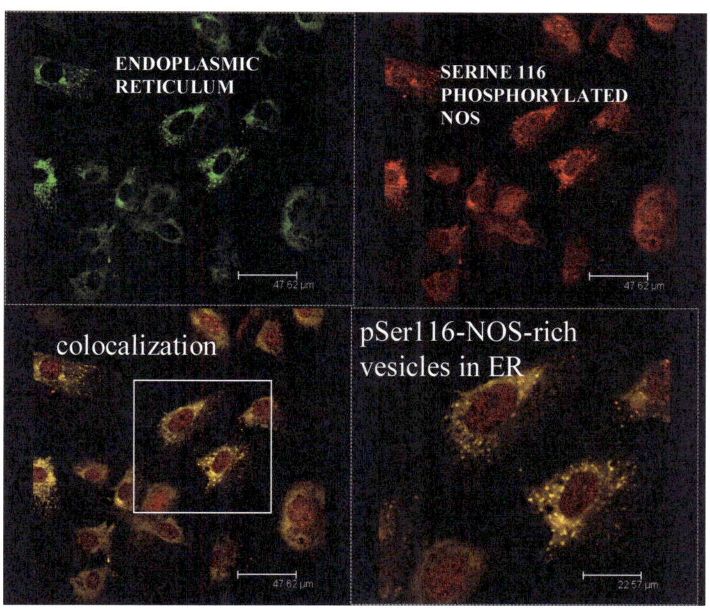

Fig. 11. Serine 116-phosphorylated NOS is found in vesicular structures in endothelial cell endoplasmic reticulum and nuclear regions. Triple monolayers of untreated passage 3 neonatal ovine lung microvascular endothelial cells were fixed with 4% paraformaldehyde, permeabilized with 0.2% triton X-100, blocked with 20% bovine serum albumin and then incubated with mouse anti protein disulfide isomerase (1: 200) (an endoplasmic reticulum (ER)-specific marker) and rabbit anti serine116-phosphorylated NOS (1: 200) primary antibodies for 2h at room temperature. They were counterstained with 1: 125 dilutions of F(ab)2 fragment specific sheep anti mouse conjugated to green fluorescein isothiocyanate

(FITC) and sheep anti rabbit conjugated to red cy-3 secondary antibodies. Pictures were taken with a DMIRE2 confocal inverted microscope and Leica v2.61 confocal software. Insert shows that a serine 116 phosphorylated NOS moiety passes through the ER and nuclear regions of quiescent cells in certain processing vesicles indicating a NOS metabolic pathway.

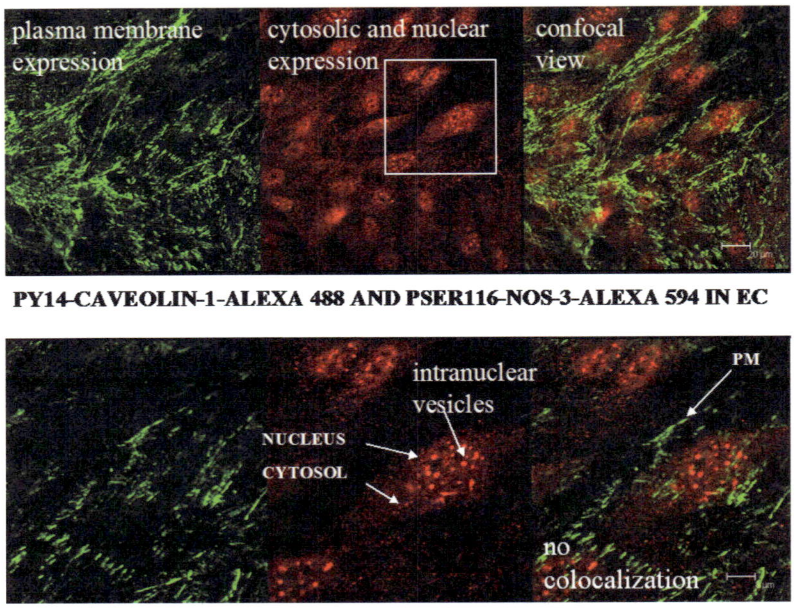

Fig. 12. Plasmalemma associated tyrosine14-phosphorylated caveolin-1 and serine 116-phosphorylated NOS are differentially located in endothelial cells. Triple monolayers of untreated passage 2 neonatal ovine lung microvascular endothelial cells were fixed with 4% paraformaldehyde, permeabilized with 0.2% triton X-100, blocked with 20% bovine serum albumin and then incubated with mouse anti tyrosine14-phosphorylated caveolin-1 (1: 200) and rabbit anti serine116-phosphorylated NOS (1: 200) primary antibodies for 2h at room temperature. They were counterstained with 1: 125 dilutions of F(ab')2 fragment specific goat secondary antibodies against mouse conjugated to green alexa 488 and against rabbit conjugated to red alexa 594. Pictures of a field representative of all three monolayers are shown taken with a DMIRE2 confocal inverted microscope and Leica v2.61 confocal software. Upper panel shows separation of pY14-caveolin-1 marker distinctively in plasma membrane and pSer116-NOS marker distinctively in the cytosol and nuclear region. Insert from upper panel shown in lower panel depicts the intense distribution of the pSer116-NOS moiety in nuclear region vesicles.

Caveolin-1 is a regulator of NOS that has been well studied. From published work, immunoprecipitation of NOS from bovine lung microvascular endothelial cells resulted in co-precipitation of caveolin-1.[49] Caveolin-1 is well known to be abundant in the membrane region.[59,60] The binding of caveolin-1 to NOS has been shown to negatively regulate NOS activity. [51,52] NOS is bound to caveolin-1 before being activated by extracellular signals and PKG does not appear to be involved in such inhibition of NOS at the plasma membrane. In the evidence presented here, PKG activation decreases the immunodetection of serine 1177 phosphorylated moiety of NOS showing that it may be involved in deactivation of activated NOS (Figure 14).

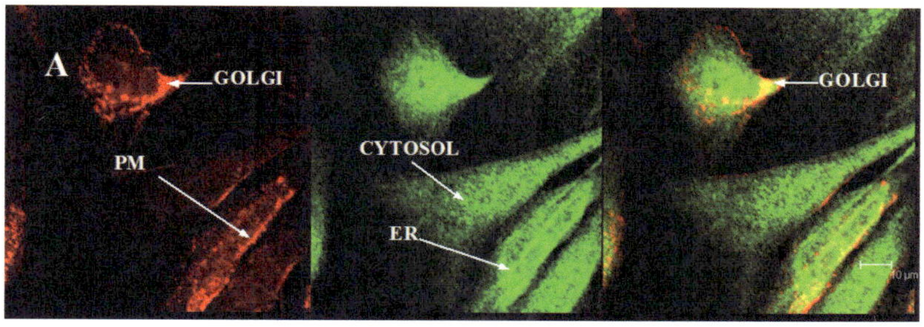

CAVEOLIN-1-ALEXA 594 AND NOS-3-ALEXA 488 IN NORMAL EC

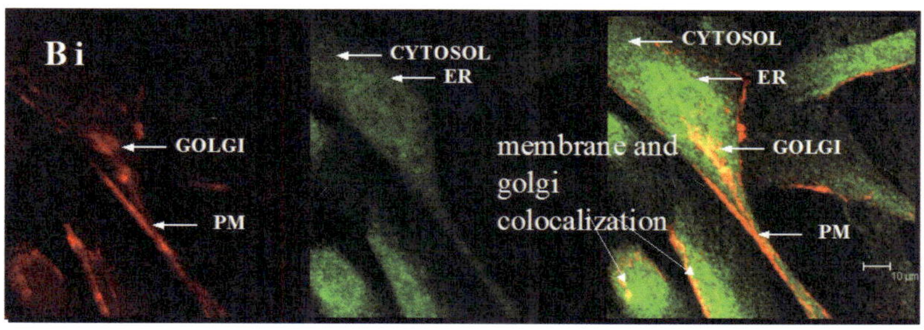

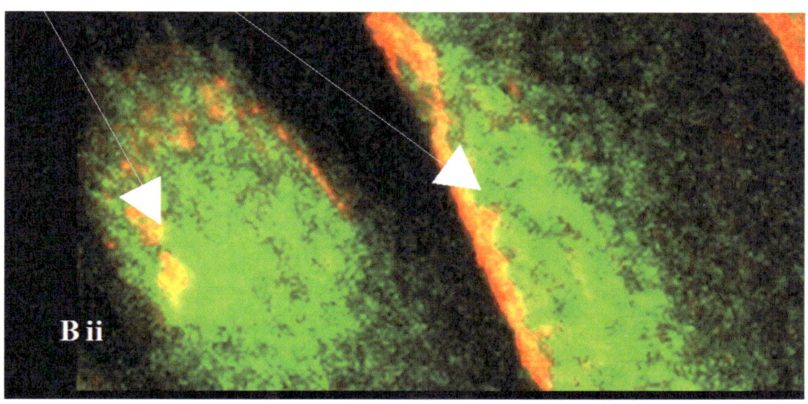

MEMBRANE AND GOLGI COLOCALIZATION OF CAVEOLIN-1 AND NOS

Fig. 13. Caveolin-1 and endothelial NOS (NOS-3) colocalize in the plasmalemma and Golgi in untreated endothelial cells. Triple monolayers of passage 2 neonatal ovine lung microvascular endothelial cells were fixed with 4% paraformaldehyde, permeabilized with 0.2% triton X-100, blocked with 20% bovine serum albumin and then incubated with affinity purified rabbit anti caveolin-1 (1: 200) and monoclonal anti NOS(1: 200) primary antibodies overnight at 4C. They were counterstained with 1: 125 dilutions of F(ab')2 fragment specific goat secondary antibodies against rabbit conjugated to red alexa 594 and against mouse conjugated to green alexa 488. Pictures were taken with a DMIRE2 confocal inverted microscope and Leica v2.61 confocal software. Arrows show areas of caveolin-1 and NOS co-localization seen in random fields from two monolayers (A and B i) with en-largement of some membrane portion of B to show numerous minute membranous colocalization foci (yellow, B ii) (scale shows 10 micro meters).

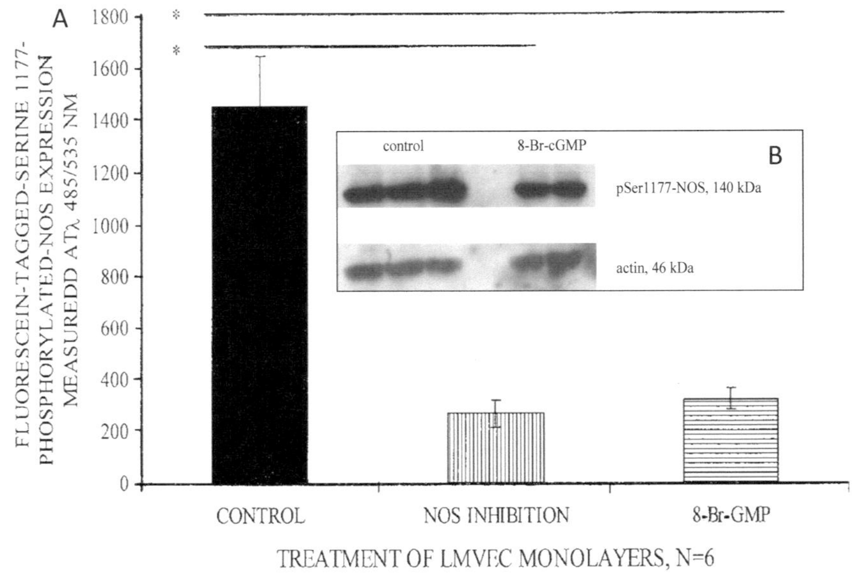

Fig. 14.[39] A: PKG activation decreases pSer1177-NOS immunofluorescence. To quantitatively determine the effect of PKG activation on pSer1177-NOS expression, stably confluent monolayers of LMVECs grown in 96-well clusters were treated with NOS inhibitor 1 mM L-NNA 1h before or PKG activator 20 µM 8-Br-cGMP (Calbiochem, CA) 10 min before fixation. Monolayers fixed with 4% paraformaldehyde were permeabilized with 0.2 % triton X-100 in PBS for 2 min, blocked with Chemicon® blocking reagent by 30 min incubation, and then incubated with rabbit anti-pSer1177-NOS primary antibody (Santa Cruz) overnight at 4C. The cells were counter-labeled with an antirabbit fluorescein (FITC)-tagged secondary antibody (Sigma, MO). The fluorescence was read with a Victor 1420 multilabel counter with a fluorescein setting (λex/em 485/535 nm). Both L-NNA and 8-Br-cGMP significantly* decreased fluorescence produced by the presence of fluorescein-tagged serine 1177 phosphorylation of NOS ($p<0.01$ in both). Figure is a representative of two studies. Insert, B, shows similar results of cGMP decreasing serine 1177 phosphorylated NOS expression in western blots.

The process of regulation of NOS after production of nitric oxide is not yet delineated[61, 62, 63, ,42] and may be governed by subcellular translocation involving the Golgi network.[64,65, 66, 67] The nucleus has not been considered as playing a prominent role in the metabolism of NOS. We localized serine-116- phosphorylated NOS (pSer116-NOS) in distinct vesicles in the cell nuclei as well as in the endoplasmic reticulum. At both sites, we found pSer116-NOS colocalized with protein kinase G1β. Thus, PKG appears to be directly involved in inactivation of NOS after NO production and to be chaperoned with spent NOS.

7. The implication of increased intracellular serine116-NOS detection after exogenous cGMP administration to cultured cells

Both in-cell fluorescence and Western blots detections suggest that serine-116-phosphorylated NOS is normally found in the endothelial cells and can be increased by exogenous cGMP administration (Figures 15 and 16). It appears that there is a basal mechanism of removal of this phosphorylated NOS and administration of excess cGMP overwhelms this mechanism causing an accumulation. The mechanism does not seem to be by destruction of the serine-116 terminal (FACS analysis) but may be by internalization and translocation in subcellular compartments. Vesicles are implicated as the use of filipin, which disrupts lipid membranes, disturbed the effects of cGMP on NOS (Figure 17).[32]

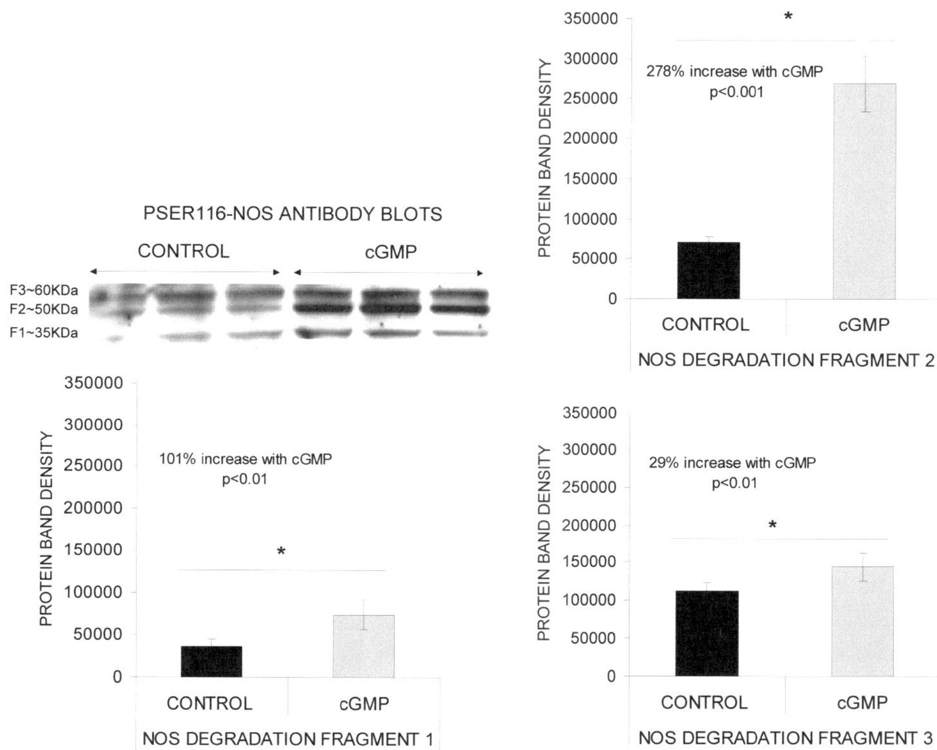

Fig. 15.[40] PKG activation increases anti-serine116-NOS antibody labeling of NOS N-terminal fractions in normal endothelial cells. LMVECs from a single suspension were equally seeded in 100mm diameter dishes. Five-day-confluent triplicate monolayers were used for controls and identical triplicate monolayers of sibling cells were used as tests treated with 300 nM 8-Br-cGMP (Calbiochem, CA) for 15 min before routine immunoblot analysis. After electrophoresis and transfer of proteins to a membrane, the membrane was blocked with 5% non-fat milk and then incubated with a primary antibody against phosphorylated serine-116 region of NOS (1: 200 concentration, Sigma, St. Louis) at 4C for 24h with gentle rocking. Protein bands were visualized by horseradish peroxidase conjugated secondary antibodies (Amersham, Buckinghamshire, UK) using Super Signal® West Pico Chemiluminescent Substrate kit (Pierce, Rockford, IL) to prepare the developing solution. Membranes were exposed to Blue Lite Autorad Films (ISC BioExpress, Kaysville, UT), and the films were processed in a HOPE developer. The film developed from western blotting showed that serine-116 phosphorylated NOS can be normally found in control cells as enzyme fragments and that these fragments are increased by PKG activation.

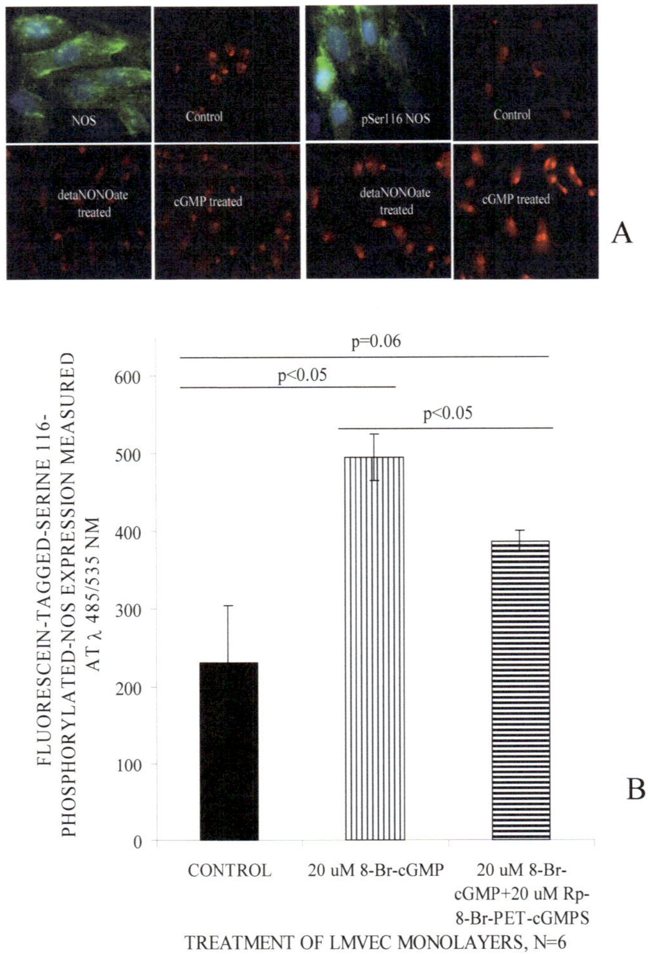

Fig. 16.[32] Protein kinase G (PKG) activation decreases immunofluorescence of nitric oxide synthase (NOS) and increases that of the serine 116 phosphorylated analogue, pSer116-NOS, in lung microvascular endothelial cells (LMVECs). A: The PKG-modulating drugs were incubated with cells 30 min before fixation. Cells were processed using routine fluorescence immunohistochemistry with specific primary antibodies and species corresponding secondary antibodies. Left upper panels show that eNOS in control cells is found in the plasmalemma and cytosol as green Alexa 488-tagged NOS and also in the nucleus as red cy3-tagged NOS; the lower panels show that both 170 micromol/L diethylenetriamine NONOate (DETANONOate) and 3 micromol/L 8-Br-cGMP decrease NOS-cy3 expression. Right upper panels show that pSer116-NOS in control cells is found in the cytosol as green Alexa 488-tagged pSer116-NOS and also in the nucleus as red cy3-tagged NOS; the lower panels show that both 170 micromol/L DETANONOate and 3 micromol/L 8-Br-cGMP increased pSer116-NOS-cy3 expression in the cytosol and nucleus. Figures are representative of two experiments. B: The graph shows quantitative modification of serine 116 immunofluorescence by 8-Br-cGMP. Stably confluent monolayers of LMVEC (n = 3) were treated with the PKG activator 20 micro mol/L 8-Br-cGMP with/without 20

micromol/L PKG inhibitor Rp-phenyl-1,N2-etheno-8-bromoguanosine-3',5'-cyclic monophosphorothionate (Rp-8-bromo-PET-cGMPS) 10 min before fixation with 4% paraformaldehyde. The inhibitors were given 1 h before the activator. Fluorescence immunohistochemistry was performed with fluorescein-tagged secondary antibody. The fluorescence was read with a Victor 1420 multilabel counter (Perkin-Elmer, Waltham, MA, USA) with a fluorescein setting (λex/em 485/535 nm). Fluorescence intensity was increased by 8-Br-cGMP ($p < 0.05$) and the effect was attenuated by Rp-8-Br-PET-cGMPS ($p < 0.05$). The figure is a representative of two experiments.

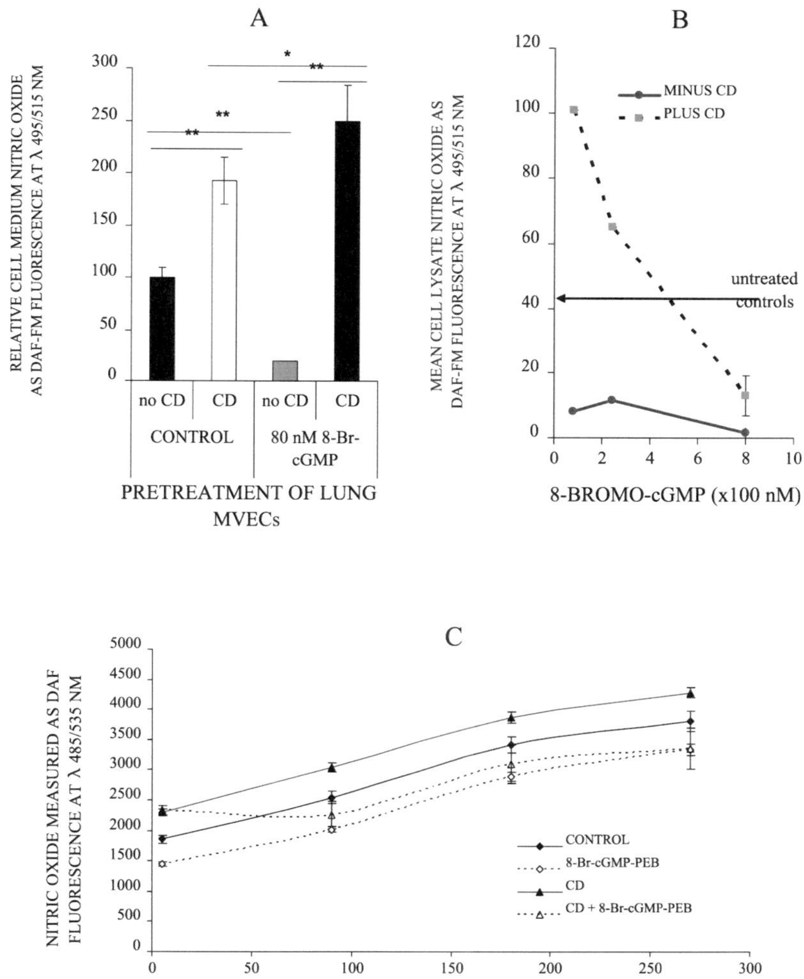

Fig. 17.[32] Lipid vesicles are involved in protein kinase G (PKG) interaction with nitric oxide synthase (NOS). Confluent monolayers were treated with cholesterol extractor 2 micromol/L cyclodextrin (CD) for 2 h to disrupt lipid vesicles and then incubated with various concentrations of 8-Br-cGMP and a combination of 0.5 micromol/L DAF-FM and cell-permeable 0.8 micromol/L DAF-FM diacetate, plus 4 microg/mL calcium ionophore to stimulate nitric oxide (NO) production. Nitric oxide produced was measured after 15 min of

incubation with DAF. Control cells were similarly treated with the exclusion of only CD. (A) Cell medium NO measurements from cells treated with CD showed greater NO production. In the absence of CD, 8-Br-cGMP caused a decrease in NO production. Cells with disrupted lipid lacked 8-Br-cGMP-inhibition of NOS. (B) Cell lysate NO measurements increased for various concentrations of 8-Br-cGMP in cells treated with CD compared with cells not treated with CD. Lipid disruption with CD raised NO production significantly ($p < 0.01$ for each 8-Br-cGMP concentration; $n = 6$). The effect of lipid disruption diminished with increasing concentrations of 8-Br-cGMP. (C) Time-course of the effect of CD on basal NO production. Cells were used in groups of 12 wells each for various controls and tests. All test groups were pretreated before adding DAF with 100 mmol/L PKG activator guanosine-3'-5'-cyclic monophosphonate-β-phenyl-1NF-ethano-8-Br Na salt (8-Br-cGMP-PEB; 10 min before) without or with 20 mmol/L CD (given 2 h before) or with CD alone. Controls were included. Using paired t-tests, the differences at the 270 min time-point between both the controls and 8-Br-cGMP-PEB groups and the controls and CD groups were significant at $p < 0.05$.

8. Hypothetical comparison of the negative regulatory roles of caveolin-1 and PKG on endothelial NOS - a review of literature.

Figure 18 and Table 1 compare the role of the well-studied caveolin-1 and the emerging role of PKG in inhibiting NOS. Both are endogenous components of cellular function under physiologic conditions. Caveolin-1 has been known as a negative regulator of NOS for many years but there is lack of significant clinical usage derived from this in terms of drug and molecular targeting and endogenous NO regulation. The suggested feedback role of PKG in the NOS metabolic cycle needs further studies for possible clinical applications. One aspect may be to target the N-terminal moiety (serine 116 region) that might be involve in transcription and *de novo* synthesis of NOS enzyme. This may be manipulated for utilization in vascular, neuronal, and other functional disorders involving NO.

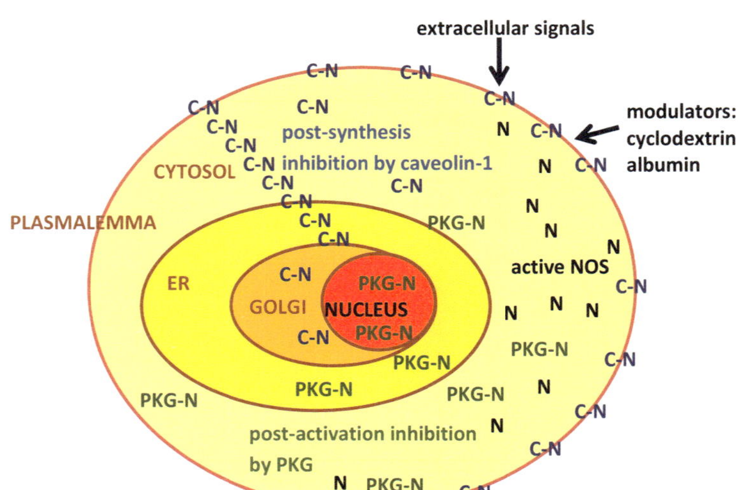

Fig. 18. Hypothetical schema of differential subcellular co-localization of caveolin-1 or PKG1beta as inhibitors of NOS in endothelial cells. Caveolin-1 co-localizes with and is

known to inhibit nascent NOS produced in the Golgi or supplied to the plasma membrane for extracellular signals. The figure depicts caveolin-1 bound to NOS in the Golgi network and plasmalemma as Figure 13 indicates and as previous workers have shown.[67, 68, 69, 62, 63 ,64] The figure also depicts that PKG inhibits NOS after it is activated in the cytosol and is involved in the recycling of spent NOS as Figure 2 indicates. By some mechanism PKG colocalizes with NOS or with a NOS fragment in the nuclear region. Key: N = free nitric oxide synthase; C-N = hypothetical caveolin-1-NOS complex; PKG-N = hypothetical protein kinase G-NOS complex.

COMPARISON	CAVEOLIN-1	PKG
Molecular effect on NOS	post-translational inhibition[62, 63, 52]	post activity inhibition[32, 38]
Speculation on the form of NOS affected	nascent (membrane bound) NOS[64, 65, 66]	active NOS after NO production, or spent NOS[39, 40, 41]
Subcellular co-localization with NOS	plasma membrane, caveolae, golgi[50, 67, 68, 69]	Cytosol, nucleus, endoplasmic reticulum[32, 41]
Biological result	makes NOS available for extracellular signals[49, 52, 56, 43, 68]	pro-translocation phosphorylation for enzyme degradation and recycling[32, 40, 41]
Activators	calcium-calmodulin[53, 68, 70, 71] albumin[72, 73] membrane lipid disruption[72, 32] bradykinin, acetyl choline[66]	cGMP analogues[38, 39]
Effect of activators	increased NO production[72, 73]	decreased NO production[32, 38, 39]
Inhibitors	hypoxia[72, 74]	cGMP analogues[38, 39]
Effect of inhibitors	decreased NO production[72]	increased NO production[38, 39]

Table 1. Comparison of the negative regulatory roles of constitutive caveolin-1 and constitutive PKG on nitric oxide synthase in the endothelium.

Protein Kinase G inhibits activated NOS by a feedback mechanism. Its actions appear to involve phosphorylation changes as well as molecular changes on the NOS molecule (Figure 19) that determine the fate of the spent enzyme. The spent enzyme, from pictorial indications, appears to enter a vesicular system possibly as a degraded N-terminal form. The role of this cycle and the therapeutic possibilities linked to it are yet to be expanded.

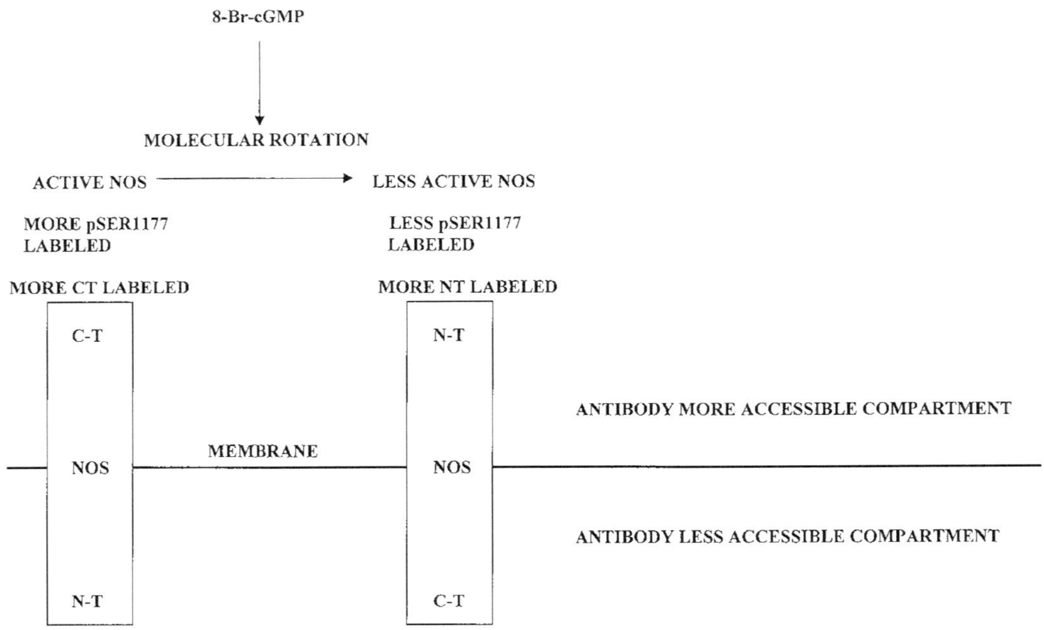

Fig. 19. A hypothetical depiction of molecular and phosphorylation effects of PKG on NOS.

9. Conclusions

- Protein kinase G isoforms are normally found in endothelial cells
- NO signaling activates guanylate cyclase and leads to activation of PKG in endothelial cells
- PKG activation via NO-GC-cGMP signaling in endothelial cells appears to exert a feedback on active NOS.
- Serine116-phosphorylated NOS appears to be spent NOS and to be chaperoned with PKG 1beta
- Serine116-phosphorylated NOS moiety enters a vesicular system including the nucleus.
- Serine116-phosphorylated NOS moiety may be an important messenger for replenishing the cell with nascent NOS enzyme.

10. Acknowledgements

The data presented here were taken from five papers published by the following journals: Clinical and Experimental Pharmacology and Physiology; The Anatomical Record; African

Journal of Medicine and Medical Sciences; Nigerian Journal of Physiology; and African Health Sciences. The authors appreciate their permissions for using the published data. Most of the work reported here on protein kinase G was done at the Los Angeles Biomedical Research Institute at Harbor UCLA Medical Center (LABiomed), Torrance, California. Grant funding and resources were provided by the laboratory of Prof. J. Usha Raj at the Division of Neonatology, Department of Pediatrics, LABiomed, Torrance, California, USA. The publication of the present Chapter was sponsored by Lagos State University College of Medicine, Lagos, Nigeria.

11. References

[1] Furchgott RF, Vanhoutte PM. Endothelium-derived relaxing and contracting factors. FASEB J 1989; Jul;3(9):2007-18.

[2] Palmer RM, Ferrige AG, Moncada S. Nitric oxide release accounts for the biological activity of endothelium-derived relaxing factor. Nature 1987; Jun 11-17;327(6122):524-6.

[3] Ignarro LJ, Buga GM, Wood KS, Byrns RE, Chaudhuri G. Endothelium-derived relaxing factor produced and released from artery and vein is nitric oxide. Proc Natl Acad Sci USA 1987; Dec;84(24):9265-9.

[4] Furchgott RF, Jothianandan D. Endothelium-Dependent and -Independent Vasodilation Involving Cyclic GMP: Relaxation Induced by Nitric Oxide, Carbon Monoxide and Light. Blood Vessels 1991;28:52-61 (Now J Vascular Research).

[5] Arnal JF, Dinh-Xuan AT, Pueyo M, Darblade B, Rami J. Endothelium-derived nitric oxide and vascular physiology and pathology. Cell Mol Life Sci 1999; 55(8-9):1078-87. Review.

[6] Hobbs AJ, Higgs A, Moncada S. Inhibition of nitric oxide synthase as a potential therapeutic target. Anna. Rev. Pharmacol. Tooxicol, 1999, 39:1337-1356.

[7] Moncada S, Palmer RM, Higgs EA. Biosynthesis of nitric oxide from L-arginine. A pathway for the regulation of cell function and communication. Biochem Pharmacol. 1989 Jun 1;38(11):1709-15.

[8] Vanhoutte PM. Endothelium and control of vascular function. State of the Art lecture. Hypertension. 1989 Jun;13(6 Pt 2):658-67. Review.

[9] Michel T, Feron O. Nitric oxide synthases. Which, where, how, and why? J. Clin. Invest. 1997; 100: 2146–52.

[10] Radomski MW, Palmer RM, Moncada S. An L-arginine/nitric oxide pathway present in human platelets regulates aggregation. Proc Natl Acad Sci U S A. 1990 Jul;87(13):5193-7.

[11] Moncada S, Higgs A. The L-arginine-nitric oxide pathway. N Engl J Med. 1993 Dec 30;329(27):2002-12.

[12] Garthwaite J. Glutamate, nitric oxide and cell-cell signalling in the nervous system. Trends Neurosci. 1991 Feb;14(2):60-7. Review

[13] Snyder SH, Bredt DS. Biological roles of nitric oxide. Sci Am. 1992 May;266(5):68-71, 74-7.

[14] Gillespie JS, Liu XR, Martin W. The effects of L-arginine and NG-monomethyl L-arginine on the response of the rat anococcygeus muscle to NANC nerve stimulation. Br J Pharmacol. 1989 Dec;98(4):1080-2.
[15] Rand MJ. Nitrergic transmission: nitric oxide as a mediator of non-adrenergic, non-cholinergic neuro-effector transmission. Clin Exp Pharmacol Physiol. 1992 Mar;19(3):147-69. Review
[16] Toda N. Regulation of blood pressure by nitroxidergic nerve. J Diabetes Complications. 1995 Oct-Dec;9(4):200-2. Review
[17] Arnold WP, Mittal CK, Katsuki S, Murad F. Nitric oxide activates guanylate cyclase and increases guanosine 3':5'-cyclic monophosphate levels in various tissue preparations. Proc Natl Acad Sci U S A. 1977 Aug;74(8):3203-7.
[18] Ignarro LJ. Heme-dependent activation of guanylate cyclase by nitric oxide: a novel signal transduction mechanism. Blood Vessels. 1991;28(1-3):67-73. Review
[19] Gruetter CA, Barry BK, McNamara DB, Gruetter DY, Kadowitz PJ, Ignarro L. Relaxation of bovine coronary artery and activation of coronary arterial guanylate cyclase by nitric oxide, nitroprusside and a carcinogenic nitrosoamine. J Cyclic Nucleotide Res. 1979;5(3):211-24.
[20] Murad F. The nitric oxide-cyclic GMP signal transduction system for intracellular and intercellular communication. Recent Prog Horm Res. 1994;49:239-48. Review.
[21] Gao Y, Dhanakoti S, Tolsa J-F and Raj JU. Role of protein kinse G in nitric oxide- and cGMP-induced relaxation of newborn ovine pulmonary veins. J App physiol 1999; 87: 993-998.
[22] Surks HK, Mochizuki N, Kasai Y, Georgescu SP, Tang KM, Ito M, Lincoln TM, Mendelsohn ME. Regulation of myosin phosphatase by a specific interaction with cGMP- dependent protein kinase Ialpha. Science. 1999 Nov 19;286(5444):1583-7.
[23] Marletta MA, Yoon PS, Iyengar R, Leaf CD, Wishnok JS. Macrophage oxidation of L-arginine to nitrite and nitrate: nitric oxide is an intermediate. Biochemistry. 1988 Nov 29;27(24):8706-11.
[24] Hibbs JB Jr, Taintor RR, Vavrin Z, Rachlin EM. Nitric oxide: a cytotoxic activated macrophage effector molecule. Biochem Biophys Res Commun. 1988 Nov 30;157(1):87-94. Erratum in: Biochem Biophys Res Commun 1989 Jan 31;158(2):624.
[25] Stuehr DJ, Nathan CF. Nitric oxide. A macrophage product responsible for cytostasis and respiratory inhibition in tumor target cells. J Exp Med. 1989 May 1;169(5):1543-55.
[26] Nussler AK, Billiar TR. Inflammation, immunoregulation, and inducible nitric oxide synthase. J Leukoc Biol. 1993 Aug;54(2):171-8. Review.
[27] Nathan C. The moving frontier in nitric oxide-dependent signaling. Sci STKE. 2004 Nov 2;2004(257):pe52. Review.
[28] Isenović E, Soskić S, Düngen HD, Dobutovic B, Elvis T, Simone I, Marche P. Regulation of endothelial nitric oxide synthase in pathophysiological conditions. Cardiovasc Hematol Disord Drug Targets. 2011 Oct 26. [Epub ahead of print]
[29] Ignarro LJ, Napoli C. Novel features of nitric oxide, endothelial nitric oxide synthase, and atherosclerosis. Curr Diab Rep. 2005 Feb;5(1):17-23. Review.

[30] Hwang SJ, Lee KH, Hwang JH, Choi CW, Shim JW, Chang YS, Park WS. Factors affecting the response to inhaled nitric oxide therapy in persistent pulmonary hypertension of the newborn infants. Yonsei Med J. 2004 Feb 29;45(1):49-55

[31] Mourani PM, Ivy DD, Gao D, Abman SH. Pulmonary vascular effects of inhaled nitric oxide and oxygen tension in bronchopulmonary dysplasia. Am J Respir Crit Care Med. 2004 Nov 1;170(9):1006-13. Epub 2004 Jun 7.

[32] John TA, Ibe BO, Raj JU. Regulation of endothelial nitric oxide synthase in neonatal lamb lung microvascular endothelial cells: role of protein kinase G, lipid structures and serine 116 phosphorylation. Clinical and Experimental Pharmacology and Physiology 2008; Feb;35(2):148-58. Epub Sep 24 2007.

[33] Brovkovych V, Stolarczyk E, Oman J, Tomboulian P, Malinski T. Direct electrochemical measurement of nitric oxide in vascular endothelium. J Pharmaceutical and Biochemical Analysis 1999; 19: 135-143.

[34] Kojima H, Sakurai K, Kikuchi K, Kawahara S, Kirino Y, Nagoshi H, Hirata Y, Nagano T. Development of a fluorescent indicator for nitric oxide based on the fluorescein chromophore. Chem Pharm Bull (Tokyo). 1998 Feb; 46(2): 373-5.

[35] Rathel TR, Leikert JF, Vollmar AM, Dirsch VM. Application of 4,5-diaminofluorescein to reliably measure nitric oxide released from endothelial cells in vitro. Biol Proced Online. 2003; 5: 136-142. Epub 2003 Jun 15.

[36] Leikert JF, Rathel TR, Muller C, Vollmar AM, Dirsch VM. Reliable in vitro measurement of nitric oxide released from endothelial cells using low concentrations of the fluorescent probe 4,5-diaminofluorescein. FEBS Lett. 2001 Oct 5; 506(2): 131-4.

[37] Lopez-Figueroa MO, Caamano C, Marin R, Guerra B, Alonso R, Morano MI, Akil H, Watson SJ. Characterization of basal nitric oxide production in living cells. Biochim Biophys Acta. 2001 Sep 26; 1540(3): 253-64.

[38] John TA. Protein kinase G inhibits basal and stimulated nitric oxide synthase activity in neonatal ovine lung microvascular endothelial cells. Af J Med med Sci 2010; 39: 89-98.

[39] John TA, Raj JU. A fluorescence activated cell sorter analysis of the relationship between protein kinase G and endothelial nitric oxide synthase. Anatomical Record 2010; 293(10): 1755-1765.

[40] John TA. Protein kinase G may exert pro-degradation inhibition on nitric oxide synthase. Nigerian Journal of Physiological Sciences 2011; 26(1): 1-6.

[41] John TA, Raj JU. Inhibitors caveolin-1 and protein kinase G show differential subcellular colocalization with nitric oxide synthase. Af Health Sc. 11(4): 526-534.

[42] Mount PF, Kemp BE, Power DA. Regulation of endothelial and myocardial NO synthesis by multi-site eNOS phosphorylation. J Mol Cell Cardiol. 2007 Feb;42(2):271-9. Epub 2006 Jul 12. Review

[43] Fleming I, Busse R. Molecular mechanisms involved in the regulation of the endothelial nitric oxide synthase. Am J Physiol Regul Integr Comp Physiol. 2003 Jan;284(1):R1-12. Review.

[44] Gratton JP, Fontana J, O'Connor DS, Garcia-Cardena G, McCabe TJ, Sessa WC. Reconstitution of an endothelial nitric-oxide synthase (eNOS), hsp90, and caveolin-

1 complex in vitro. Evidence that hsp90 facilitates calmodulin stimulated displacement of eNOS from caveolin-1. J Biol Chem 2000; 275: 22268-22272.
[45] Schilling K, Opitz N, Wiesenthal A, Oess S, Tikkanen R, Muller-Esterl W, Icking A. Translocation of endothelial nitric-oxide synthase involves a ternary complex with caveolin-1 and NOSTRIN. Mol Biol Cell. 2006 Sep; 17(9):3870-80. Epub 2006 Jun 28.
[46] Schleicher M, Brundin F, Gross S, Muller-Esterl W, Oess S. Cell cycle-regulated inactivation of endothelial NO synthase through NOSIP-dependent targeting to the cytoskeleton. Mol Cell Biol. 2005 Sep; 25(18):8251-8.
[47] Boo YC, Sorescu G, Boyd N, Shiojima I, Walsh K, Du J, Jo H. Shear stress stimulates phosphorylation of endothelial nitric-oxide synthase at Ser1179 by Akt-independent mechanisms: role of protein kinase A. J Biol Chem. 2002 Feb 1;277(5):3388-96. Epub 2001 Nov 29.
[48] Leikert JF, Räthel TR, Wohlfart P, Cheynier V, Vollmar AM, Dirsch VM. Red wine polyphenols enhance endothelial nitric oxide synthase expression and subsequent nitric oxide release from endothelial cells. Circulation. 2002 Sep 24;106(13):1614-7.
[49] García-Cardeña G, Oh P, Liu J, Schnitzer JE, Sessa WC. Targeting of nitric oxide synthase to endothelial cell caveolae via palmitoylation: implications for nitric oxide signaling. Proc Natl Acad Sci USA 1996; 25;93(13): 6448-53.
[50] Shaul PW, Smart EJ, Robinson LJ, German Z, Yuhanna IS, Ying Y, Anderson RG, Michel T. Acylation targets emdothelial nitric-oxide synthase to plasmalemmal caveolae. J Biol Chem 1996; 271(11):6518-22.
[51] Ju H, Venema VJ, Marrero MB, Venema RC. Inhibitory interactions of the bradykinin B2 receptor with endothelial nitric-oxide synthase. J Biol Chem 1998; 273(37): 24025-24029.
[52] Feron O, Saldana F, Michel J B, Michel T. The endothelial nitric-oxide synthase-caveolin regulatory cycle. J Biol Chem 1998; 273: 3125-3128.
[53] Bredt DS, Ferris CD, Snyder SH. Nitric oxide synthase regulatory sites. Phosphorylation by cyclic AMP-dependent protein kinase, protein kinase C, and calcium/calmodulin protein kinase; identification of flavin and calmodulin binding sites. J Biol Chem. 1992 Jun 5;267(16):10976-81.
[54] Dedio J, Konig P, Wohlfart P, Schroeder C, Kummer W, Muller-Esterl W. NOSIP, a novel modulator of endothelial nitric oxide synthase activity. FASEB J 2001; 15: 79-89.
[55] Zimmermann K, Opitz N, Dedio J, Renne C, Muller-Esterl W, Oess S. NOSTRIN: a protein modulating nitric oxide release and subcellular distribution of endothelial nitric oxide synthase. Proc Natl Acad Sci USA 2002; 99: 17167-17172.
[56] Fulton D, Gratton J-P, Sessa WC. Post-translation control of endothelial nitric oxide synthase: why isn't calcium/calmodulin enough? JPET 2001; 299(3): 818-824.
[57] Fleming I, Busse R. Signal transduction of eNOS activation. Cardiovascular Res.1999 Aug 15; 43(3): 532-541.
[58] Rafikov R, Fonseca FV, Kumar S, Pardo D, Darragh C, Elms S, Fulton D, Black SM. eNOS activation and NO function: structural motifs responsible for the posttranslational control of endothelial nitric oxide synthase activity. J Endocrinol. 2011 Sep;210(3):271-84. Epub 2011 Jun 3. Review

[59] Rothberg KG, Heuser JE, Donzell WC, Ying Y-S, Glenney JR, Anderson RGW. Caveolin, a protein component of caveolae membrane coats. Cell 1992; 68:673-682.

[60] Das K, Lewis RY, Scherer PE, Lisanti MP. The membrane-spanning domains of caveolins-1 and -2 mediate the formation of caveolin hetero-oligomers. Implications for the assembly of caveolae membranes in vivo. J Biol Chem 1999; 274:18721-18728.

[61] Oess S, Icking A, Fulton D, Govers R, Müller-Esterl W. Subcellular targeting and trafficking of nitric oxide synthases. Biochem J. 2006 Jun 15;396(3):401-9. Review

[62] Garcia-Cardena G, Fan R, Stern DF, Liu J, Sessa WC. Endothelial nitric oxide synthase is regulated by tyrosine phosphorylation and interacts with caveolin-1. J Biol Chem 1996; 271(44):27237-40.

[63] Ju H, Zou R, Venema VJ, Venema RC. Direct interaction of endothelial nitric-oxide synthase and caveolin-1 inhibits synthase activity. J Biol Chem 1997; 272(30):18522-5.

[64] Feron O, Michel JB, Sase K, Michel T. Dynamic regulation of endothelial nitric oxide synthase: complementary roles of dual acylation and caveolin interactions. Biochemistry 1998; 37(1):193-200.

[65] Michel T. Targeting and translocation of endothelial nitric oxide synthase. Braz J Med Biol Res. 1999 Nov;32(11):1361-6. Review

[66] Prabhakar P, Thatte HS, Goetz RM, Cho MR, Golan DE, Michel T. Receptor-regulated translocation of endothelial nitric-oxide synthase. J Biol Chem 1998; 273(42): 27383-8.

[67] O'Brien AJ, Young HM, Povey JM, Furness JB. Nitric oxide synthase is localized predominantly in the Golgi apparatus and cytoplasmic vesicles of vascular endothelial cells. Histochem Cell Biol. 1995 Mar;103(3):221-5

[68] Fulton D, Babbitt R, Zoellner S, Fontana J, Acevedo L, McCabe TJ, Iwakiri Y, Sessa WC.Targeting of endothelial nitric-oxide synthase to the cytoplasmic face of the Golgi complex or plasma membrane regulates Akt- versus calcium-dependent mechanisms for nitric oxide release. J Biol Chem. 2004 Jul 16;279(29):30349-57. Epub 2004 May 10

[69] Zhang Q, Church JE, Jagnandan D, Catravas JD, Sessa WC, Fulton D. Functional relevance of Golgi- and plasma membrane-localized endothelial NO synthase in reconstituted endothelial cells. Arterioscler Thromb Vasc Biol 2006; 26(5):1015-21.Epub 2006 Mar 2.

[70] Michel JB, Feron O, Sacks D, Michel T. Reciprocal regulation of endothelial nitric-oxide synthase by Ca^{2+}-calmodulin and caveolin. J Biol Chem 1997; 272(25):15583-6.

[71] Forstermann U, Pollock JS, Schmidt HH, Heller M, Murad F. Calmodulin-dependent endothelium-derived relaxing factor/nitric oxide synthase activity is present in the particulate and cytosolic fractions of bovine aortic endothelial cells. Proc Natl Acad Sci USA 1991; 88: 1788-1792.

[72] John TA, Ibe BO, Raj JU. Oxygen alters caveolin-1 and nitric oxide synthase-3 functions in ovine fetal and neonatal lung microvascular endothelial cells. Am J Physiol Lung Cell Mol Physiol 2006; 291(5): L1079-93.

[73] Maniatis NA, Brovkovych V, Allen SE, John TA, Shajahan AN, Tiruppathi C, Vogel SM, Skidgel RA, Malik AB, Minshall RD. Novel mechanism of endothelial nitric oxide synthase activation mediated by caveolae internalization in endothelial cells. Circ Res. 2006 Oct 13;99(8):870-7. Epub 2006 Sep 14.

[74] Murata T, Sato K, Hori M, Ozaki H, Karaki H. Decreased endothelial nitric-oxide synthase (eNOS) activity resulting from abnormal interaction between eNOS and its regulatory proteins in hypoxia-induced pulmonary hypertension. J Biol Chem. 2002 Nov 15;277(46):44085-92. Epub 2002 Aug 15.

8

The Role of Tyrosine Kinases in the Pathogenesis and Treatment of Lung Disease

Jonathan Lam and Stewart J. Levine
*Laboratory of Asthma and Lung Inflammation,
Cardiovascular and Pulmonary Branch,
National Heart, Lung, and Blood Institute,
National Institutes of Health, Bethesda,
USA*

1. Introduction

Tyrosine kinases (TKs) are key regulators of signal transduction pathways that modulate essential biological processes, such as cellular differentiation, metabolism and proliferation, as well as, protein synthesis, cell cycle progression and apoptosis (Schlessinger, 2000). These enzymes modify protein function by transferring phosphate groups from adenosine triphosphate (ATP) or guanosine triphosphate (GTP) to free hydroxyl groups on tyrosine residues. Under physiological conditions, kinase activity is regulated by protein phosphatases that dephosphorylate and inactivate signaling pathways. However, these signaling pathways can become dysregulated in lung diseases, such as non-small cell lung cancer (NSCLC), asthma, chronic obstructive pulmonary disease (COPD) and idiopathic pulmonary fibrosis (IPF). This chapter will provide an overview of the pathways mediated by receptor and non-receptor tyrosine kinases, as well as their role in the pathogenesis of lung disease. Furthermore, the emerging role of TK inhibitors (TKIs) for the treatment of NSCLC, asthma, COPD and IPF will be discussed.

1.1 Receptor tyrosine kinases and non-receptor tyrosine kinases

Receptor tyrosine kinases (RTKs) play essential roles in growth factor and cytokine receptor signaling. Over 58 RTKs have been identified, which include epidermal growth factor receptors (EGFR), ephrin receptors, fibroblast growth factor receptors (FGFR), platelet derived growth factor receptors (PDGFR), RAR-related orphan receptors (ROR), vascular endothelial growth factor receptors (VEGFR), the hepatocyte growth factor receptor (MET) and the insulin receptor (Hubbard & Miller, 2007; Robinson, Wu, & Lin, 2000). RTKs are comprised of an extracellular ligand-binding domain, a transmembrane domain and an intracellular kinase domain (Robinson et al., 2000). Upon ligand binding, RTKs undergo conformational changes that induce homo- or hetero- dimerization and autophosphorylation of intracellular kinase domains at specific tyrosine residues. This activates downstream signaling pathways that recruit non-receptor tyrosine kinases (non-RTKs) and activate serine/threonine kinase signaling pathways.

The EGFR belongs to a family of RTKs that includes EGFR, ERBB2 (HER2), ERBB3 (HER3) and ERBB4 (HER4), which play key roles in cellular growth, proliferation and differentiation. All EGFR family members have intrinsic TK activity, except ERBB3 (Sibilia, Kroismayr, Lichtenberger, Natarajan, Hecking, & Holcmann, 2007). EGF receptors exist as inactive monomers that undergo conformational changes upon binding of EGF or transforming growth factor-α (TGF-α), which facilitates receptor homodimerization or heterodimerization (Fig. 1) (Garrett et al., 2002; Odaka, Kohda, Lax, Schlessinger, & Inagaki, 1997; Ogiso et al., 2002). This is followed by intermolecular autophosphorylation of key tyrosine residues in the activation loop of catalytic TK domains (Grandal & Madshus, 2008). The carboxy-terminal phosphotyrosine residues then recruit signaling molecules with Src homology 2 (SH2) and protein tyrosine-binding domains to activate downstream signaling pathways. Downstream EGF receptor signaling pathways include the mitogen-activated protein kinase (MAPK) pathway, phosphoinositol 3'-kinase (PI3K), phospholipase C (PLC) and Janus tyrosine kinases (JAK) and signal transducers and activators of transcription (STAT)

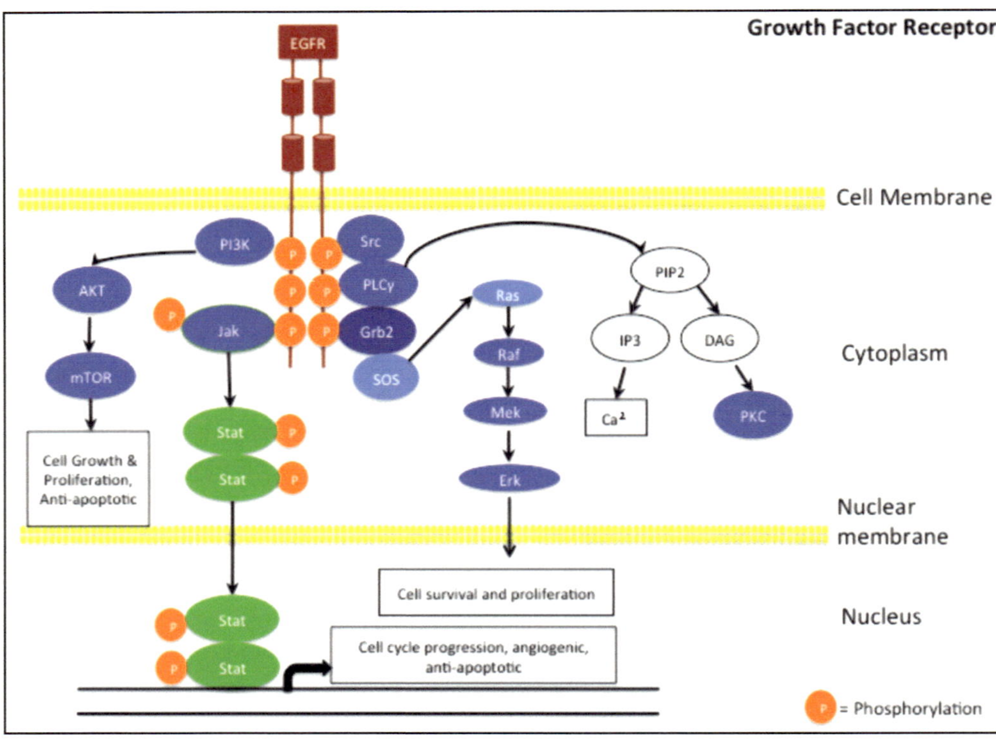

Fig. 1. Growth Factor Receptor Signaling Pathways. Upon ligand binding, tyrosine residues on the EGFR intracellular domain become autophosphorylated. This allows the receptor to interact with SH2 or PTB adaptor proteins (e.g., Grb2), which activates downstream signaling pathways, such as PLC-γ, PI3K-Akt, Ras-Raf-Mek-Eer and JAK-STATs, to modulate cell proliferation, migration, differentiation, and apoptosis. mTOR, mammalian target of rapamycin; Sos, son of sevenless; Grb2, growth factor receptor-bound protein 2; Mek, MAPK kinase; Erk, extracellular signal-regulated kinase. Adapted from reference (Ratushny, Astsaturov, Burtness, Golemis, & Silverman, 2009).

(Ciardiello & Tortora, 2008; Laskin & Sandler, 2004). Activation of the MAPK signaling pathway occurs via the small GTPase, Ras, which exchanges GDP for GTP and stimulates the Raf-Mek-Erk cascade, to modulate key cellular processes, including gene transcription, G1/S cell-cycle progression and cell proliferation (Ho & Laskin, 2009). The PLC pathway is activated via cleavage of phosphatidylinositol (4,5)-bisphosphate (PIP2) on the cell membrane to form second messenger molecules, inositol triphosphate (IP3) and diacylglyerol (DAG). IP3 triggers the release of intracellular calcium stores, while DAG activates protein kinase C (PKC) (Milano, De Rosa, Iaffaioli, & Caponigro, 2007). The PI3K pathway activates Akt and mammalian target of rapamycin (mTOR), which regulates gene transcription, cell proliferation and migration (Herbst, Heymach, & Lippman, 2008). The JAK-STAT pathway (JAK2, STAT1, STAT3 and STAT5) stimulates transcription of nuclear factors that promote inflammation, cell survival and oncogenesis (Quesnelle, Boehm, & Grandis, 2007; Silva, 2004).

In contrast, non-RTKs cannot bind ligands and do not possess a transmembrane domain, but are activated by cytoplasmic tyrosine kinases. There are 32 cytoplasmic non-RTKs that can be divided into nine sub-families, which include Src, Csk (cytoplasmic tyrosine kinase), Ack (activated p21CDC42 kinase, tyrosine kinase, non-receptor 2 (Tnk2)), Fak (focal adhesion kinase), Tec, Fes/Fer, Syk (spleen tyrosine kinase), Abl (Abelson murine leukemia viral oncogene homolog), and JAKs (Manning, Whyte, Martinez, Hunter, & Sudarsanam, 2002; Parsons & Parsons, 2004). These non-RTKs activate downstream signaling cascades and transcription factors to initiate gene transcription and thereby modify important cellular functions. As discussed below, cytokine receptors and immune receptors are examples of receptors that signal via non-RTKs.

1.2 Cytokine receptors

Cytokines are small signaling proteins that regulate key physiological cellular functions, such as inflammation, and innate and adaptive immune responses. Receptors for cytokines belonging to the Type I and Type II cytokine families, as well as members of the interleukin (IL)-6 and IL-11 families, signal via JAK non-RTKs. For example, Type I cytokine receptors, such as IL-4 and IL-13 receptors, are multimeric receptors that share the common cytokine receptor γ-subunit (γc), whereas the IL-3, IL-5 and granulocyte-macrophage colony stimulating factor (GM-CSF) receptors share the common cytokine receptor β-chain (β_c), (Pesu, Laurence, Kishore, Zwillich, Chan, & O'Shea, 2008). Upon ligand binding, receptor oligomerization and signal transduction occur via the interaction of specific JAK isoforms (JAK1, JAK2, JAK3 and TYK2) with the intracytoplasmic domains of cytokine receptors (Fig. 2) (Ghoreschi, Laurence, & O'Shea, 2009).

The Type I cytokines, prolactin, erythropoietin, thrombopoietin, GM-CSF, IL-3 and IL-5, signal through JAK2, while IL-6 family members (IL-6, IL-11), IL-10 family members (IL-10, IL-19, IL-20, IL-22) and the Type II cytokine, interferon (IFN)-γ, signal through JAK1 and JAK2 (Ghoreschi et al., 2009). In contrast, JAK3 only associates with receptors that share the common γ-chain (γc), such as receptors for IL-2, IL-4, IL-7, IL-9, IL-15 and IL-21 (Argetsinger et al., 1993; Bacon, McVicar, Ortaldo, Rees, O'Shea, & Johnston, 1995; Johnston et al., 1994; Muller et al., 1993; Parganas et al., 1998; Parham et al., 2002). The biological importance of JAK signaling pathways is illustrated by the congenital absence of JAK3, which causes a

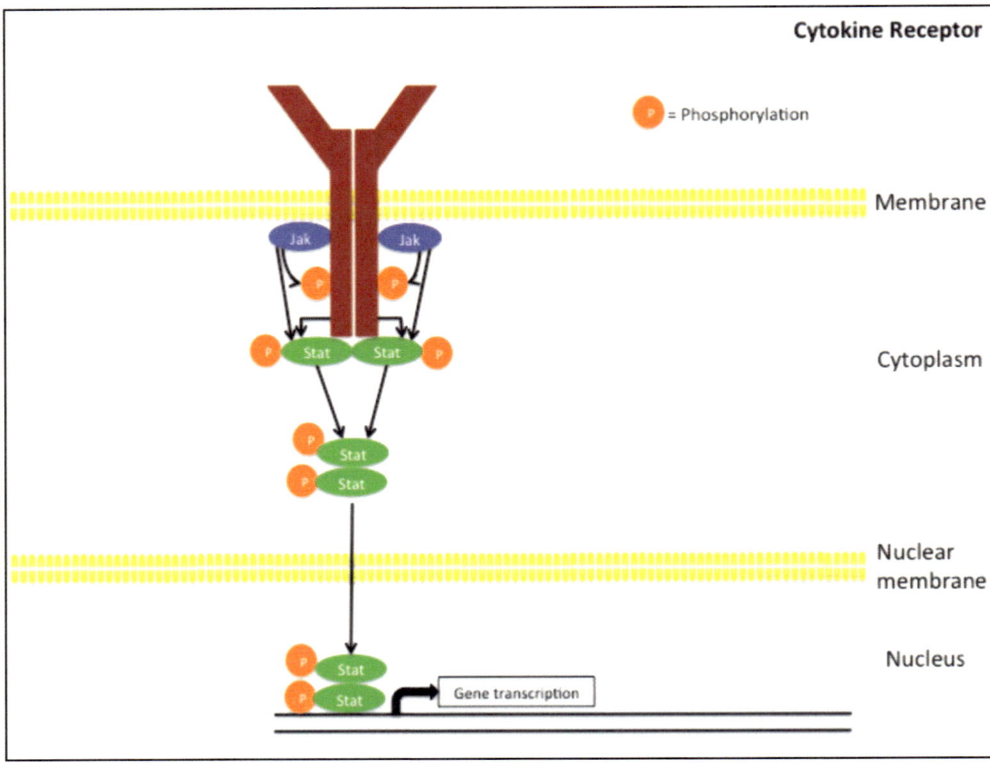

Fig. 2. Cytokine Receptor Signaling Pathways. Following ligand binding to cytokine receptors, receptor-associated JAKs become activated and phosphorylate specific receptor tyrosine residues on the receptor intracytoplasmic domain. This leads to the recruitment of STATs, which are then tyrosine-phosphorylated. Activated STATs dimerize, translocate to the nucleus, and bind to STAT binding sites in the promoter regions of target genes to induce gene transcription. Adapted from reference (Malaviya & Laskin, 2010).

primary immunodeficiency disorder termed T–B+NK– severe combined immunodeficiency disease. This demonstrates the key role for JAK3 in the development of T cells and natural killer cells (Macchi et al., 1995; Russell et al., 1995). JAK activation leads to tyrosine phosphorylation and the recruitment and activation of STAT transcription factors. There are seven STAT family members (STAT1–6 with two STAT5 proteins) that form homo- or hetero-dimers and translocate to the nucleus after being phosphorylated in the cytoplasm to regulate gene transcription (Darnell, 1997). For example, activation of the IL-4 receptor stimulates JAK1 to associate with the IL-4Rα chain and JAK3 to associate with the γc chain, which results in the activation of STAT6(Takeda et al., 1996). In contrast, IL-13 can interact with both the IL-4Rα chain and the IL-13Rα1 chain to activate JAK1, JAK2, Tyk2, and STAT6 (Pernis & Rothman, 2002; Schindler, 2002; Wills-Karp et al., 1998).

1.3 Immune receptors

The high affinity IgE receptor (FcεRI), the T cell receptor (TCR) and the B cell receptor (BCR) are immune receptors expressed by mast cells, T cells and B cells, respectively. These receptors are composed of antigen-binding subunits and signal transducing subunits that consist of one or more immunoreceptor tyrosine-based activation motifs (ITAMs), [Y(2X)I/L(6–8X)Y(2X)I/L] (Pitcher & van Oers, 2003). FcεRI, is a tetrameric protein ($\alpha\beta\gamma2$ chains) consisting of the IgE binding α chain, a signal-amplifying β chain and two disulfide-linked γ chains, which contain ITAMs for signal transduction (Fig. 3) (Novak, Kraft, & Bieber, 2001). The TCR consists of either αβ or γδ heterodimer subunits that recognize antigen presented within the context of Major Histocompatibility Complex (MHC) molecules (Fig. 4). The BCR consists of a membrane immunoglobulin isotype molecule (IgD, IgM, IgE or IgG) that binds antigen and Igαβ heterodimers, which contain ITAM sequences (Fig. 5) (Schamel & Reth, 2000). Upon tyrosine phosphorylation, ITAMs act as docking sites for non-RTKs and SH2 containing adaptor proteins or as phosphotyrosine-binding domains to activate signal transduction cascades.

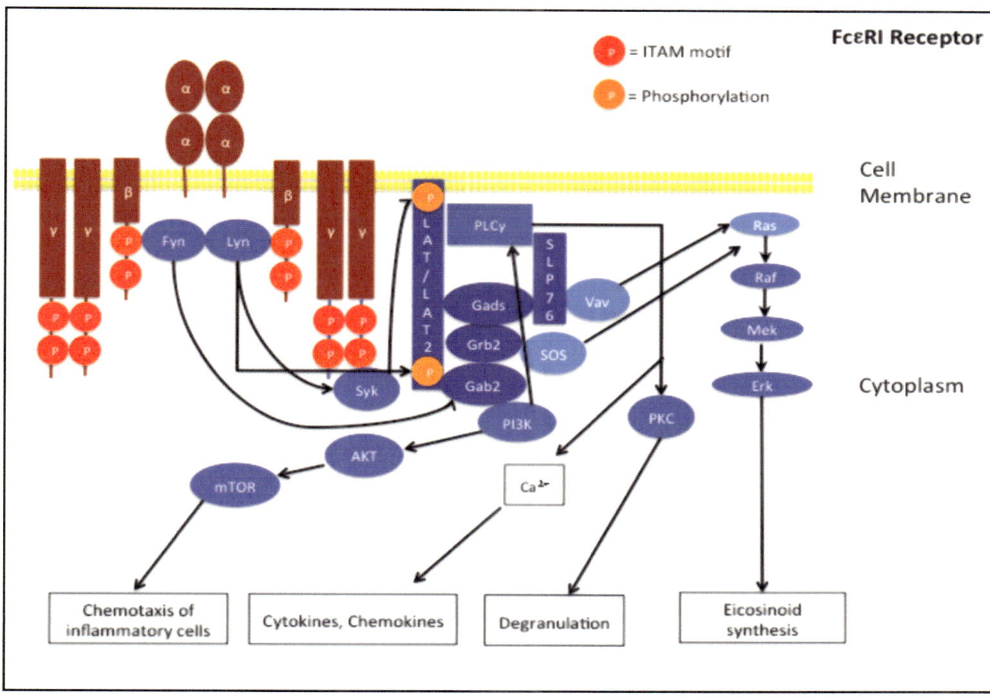

Fig. 3. FcεRI Receptor Signaling Pathways. Engagement of the high affinity FcεRI on mast cells and basophils by the Fc region of IgE results in phosphorylation of receptor ITAMs by Lyn and Fyn, thereby generating docking sites for these proteins, as well as for Syk. Binding to ITAMs up-regulates Fyn, Lyn and Syk kinase activity that phosphorylate downstream signaling molecules. Multi-protein complexes are then formed, which activate MAPK pathways and generate second messenger molecules, leading to cellular activation. Adapted from reference (Colgan & Hankel, 2010).

Signaling via immune receptors is initiated by Src-family kinases (SFKs) that phosphorylate ITAMs (Lowell, 2011). This recruits Syk to the SH2 domain of phosphorylated ITAM motifs and activates downstream signaling pathways. SFKs consist of nine members; Src, Fyn, Lck, Hck, Lyn, Fgr, Blk, Yes, and Yrk. Src is widely expressed, whereas the expression of

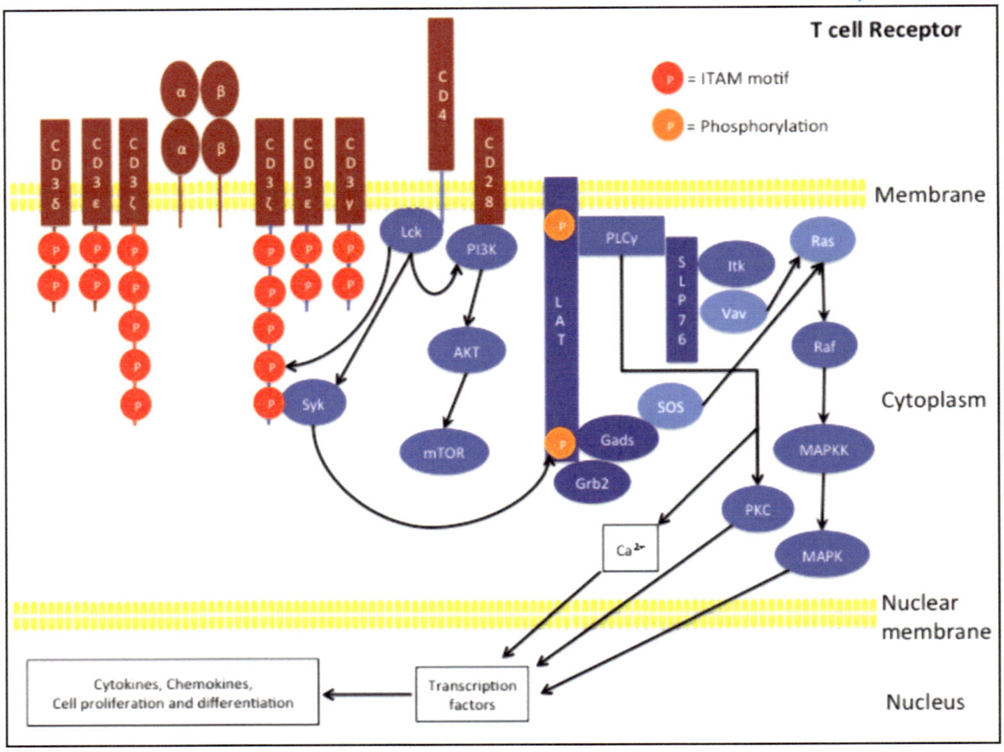

Fig. 4. T Cell Receptor (TCR) Signaling Pathways. TCR engagement by antigen bound to major histocompatibility complex proteins activates Lck, which phosphorylates ITAMs in the CD3 zeta chains to recruit and activate Syk. This leads to phosphorylation of downstream signaling molecules and formation of multi-protein complexes nucleated by the LAT adapter protein. Components of these complexes activate MAPK pathways, generate second messenger molecules and activate transcription factors that induce T cell proliferation and differentiation. Adapted from reference (Colgan & Hankel, 2010).

hematopoietic cell kinase (Hck) and lymphoid cell kinase (Lck) are more restricted (Thomas & Brugge, 1997). Src normally adopts an inactive conformation, such that Y416 in the activation loop is buried between the N-lobe and C-lobe of the kinase domain. This conformation is maintained by interactions between the SH3 domain and the linker region connecting the SH2 domain with the catalytic domain, as well as by binding of the SH2 domain to the C-terminal tail in response to phosphorylation of Y527. Activation occurs when the high affinity ligand binding disrupts the interactions between the SH2 and SH3 domains, thereby allowing unfolding and exposure of Y416 for autophosphorylation (Benati & Baldari, 2008). This recruits the SH2-containing Syk to phosphorylate the adaptor proteins, LAT and LAB (linker for activation of T and B cells, respectively), Grb2, Gab2, Gads, SLP-76 and SLP-65 (SH2 domain-containing leukocyte protein of 76kDa/65kDa), as well as the GTP exchange factors, Sos and Vav, which activate MAPK signaling (Colgan & Hankel, 2010). This activates the Tec family non-RTKs, Itk (inducible T cell kinase) and Btk (Bruton's tyrosine kinase), as well as PLCγ and PI3K. The Tec family can also activate PLCγ to hydrolyse PIP2 into IP3 and DAG, which in turn mobilizes intracellular calcium and

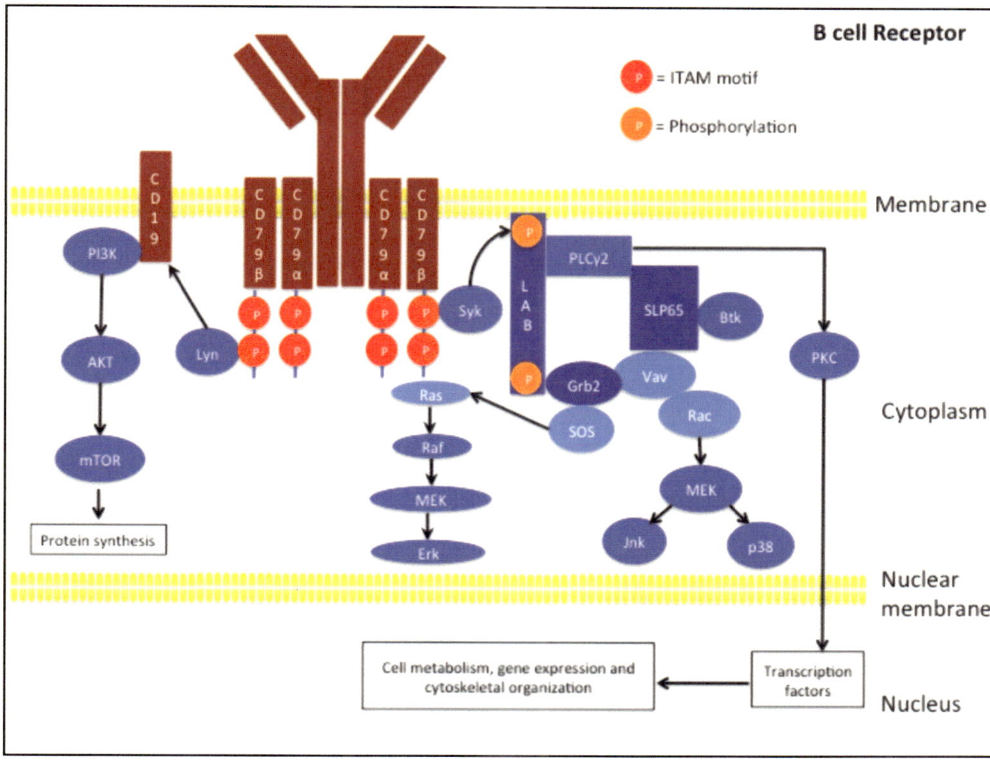

Fig. 5. B Cell Receptor Signaling Pathways. Aggregation of the B-cell antigen receptor (BCR) by antigen leads to rapid activation of the Src family kinases, Lyn, Blk, and Fyn, as well as the Syk and Btk. This initiates the formation of a 'signalosome' composed of the BCR, Src family kinases, adaptor proteins such as CD19, and signaling enzymes, such as PLCγ2, PI3K, and Vav. Signals emanating from the signalosome activate multiple signaling cascades that involve kinases, GTPases, and transcription factors, which modulate cell proliferation and differentiation, as well as antibody generation. Adapted from reference (Wong, 2005).

activates PKC isoforms β, ζ and θ (Miller & Berg, 2002). This is essential for activation of the transcription factor nuclear factor-κB (NF-κB), which mediates immune receptor signaling via CARMA1/BCL10/MALT1 and IκB kinases (Li, Rickert, & Karin, 2004).

Cross-linking of IgE bound to FcεRI by multivalent antigen activates Lyn, followed by activation of Syk, which phosphorylates LAT and SLP-76 (Kawakami & Galli, 2002). This activates Btk and induces mast cell degranulation and activation. The TCR heterodimer associates with CD3 molecules that contain ITAM sequences (Colgan & Hankel, 2010). During TCR activation, Lck and Fyn phosphorylate ITAMs on CD3, which recruits Syk to phosphorylate LAT and SLP-76. This activates Itk causing T cell activation (Nel, 2002; Samelson, 2002). Upon BCR activation, phosphorylation of ITAMs by Lyn, Fyn or Blk recruits Syk, which phosphorylates LAB and SLP-65 to activate Btk (Dal Porto, Gauld, Merrell, Mills, Pugh-Bernard, & Cambier, 2004). This leads to cellular proliferation and differentiation, which generates a population of antibody-secreting plasma and memory B cells.

2. The role of tyrosine kinases in non small cell lung cancer

Lung cancer is the leading cause of cancer-related death worldwide and accounted for an estimated 157,300 deaths in the US in 2010 (Jemal, Siegel, Xu, & Ward, 2010). Approximately 85% to 90% of all lung cancer cases are NSCLCs, such as adenocarcinoma, squamous cell carcinoma and large-cell carcinoma. Advanced-stage NSCLC is currently an incurable disease and less than 30% of patients with metastatic NSCLC respond to platinum-based chemotherapy. The EGFR is activated in more than half of the patients with NSCLC as a result of either protein over-expression, gene mutations or an increase in gene copy number (Balak et al., 2006; Chitale et al., 2009; Ding et al., 2008; Soh et al., 2009; Weir et al., 2007). EGFR mutations are usually heterozygous, with the mutant alleles showing gene amplification involving exons 18 to 21 that encode the kinase domain. Therefore, inhibitors of EGFR signaling have been developed as a novel approach for the treatment of NSCLC.

2.1 Use of EGFR inhibitors for the treatment of NSCLC

The EGFR inhibitors, erlotinib and gefitinib, were developed to attenuate excessive EGFR signaling as a novel approach for the treatment of NSCLC (Cataldo, Gibbons, Perez-Soler, & Quintas-Cardama, 2011). Erlotinib (Tarceva®) is currently approved by the U.S. Food and Drug Administration as a second line therapy for patients with locally advanced or metastatic NSCLC after failure of at least one prior chemotherapy regimen or as maintenance therapy for patients with NSCLC whose disease has not progressed after four cycles of platinum-based first-line chemotherapy (Genentech, 2011). In contrast, continued treatment with genfitinib (Iressa®) as monotherapy is limited to patients who are responding or have previously responded to gefinitib treatment for NSCLC after failure of both platinum-based and docetaxel chemotherapies (AstraZeneca, 2005). This restriction is based upon a large, placebo-controlled, randomized trial in patients with advanced NSCLC, which did not show an improvement in survival with gefitinib as second- or third-line treatment for NSCLC (Thatcher et al., 2005). Gefitinib may continue to be used in clinical trials under a new drug application.

These tyrosine kinase inhibitors (TKIs) are small molecules that bind orthosteric and/or allosteric sites to competitively inhibit ATP phosphorylation or irreversibly inhibit its activity (Noble, Endicott, & Johnson, 2004). Erlotinib and gefitinib are most effective in patients with EGFR mutations, such as L858R and G719S, which result in increased EGFR activation (Carey et al., 2006; Yun et al., 2007; Yun et al., 2008). Inhibition of EGFR mutants leads to increased cell death mediated by an apoptotic pathway that is dependent upon Bim, a Bcl-2 family member that is pro-apoptotic and regulated by Erk signaling (Costa et al., 2008; Cragg, Kuroda, Puthalakath, Huang, & Strasser, 2007; Deng et al., 2007; Gong et al., 2007). Although most patients tolerate erlotinib and gefitinib, some patients experience serious side effects such as diarrhea, rash, nausea and interstitial lung disease (Makris et al., 2007; Shah et al., 2005; Shepherd et al., 2005). Patients may have primary resistance due to EGFR drug-resistant mutations, such as EGFRvIII, which is a constitutively active form of EGFR caused by the deletion of exons 2 - 7 (Greulich et al., 2005; Inukai et al., 2006; Maheswaran et al., 2008; Prudkin, Tang, & Wistuba, 2009; Wu et al., 2008), mutations in downstream signaling pathways that co-occur with EGFR mutations, such as the PI3K catalytic subunit, PIK3CA (Kawano et al., 2006), or mutations that occur in other downstream genes, such as k-Ras and b-Raf (Brose et al., 2002; Davies et al., 2002; Linardou et al., 2008; Mok et al., 2009; Pao et al., 2005b; Wheeler, Dunn, & Harari, 2010; Zhang & Chang, 2008). Furthermore, despite an initial positive response to treatment, acquired resistance can develop following 6 to 12 months of therapy with erlotinib or gefitinib. Acquired resistance to erlotinib or gefintib can occur via several mechanisms, such as second-site mutations and amplification of the *MET* oncogene, which encodes a RTK for hepatocyte growth factor that activates a ERBB3 (HER3)-dependent PI3K/Akt pathway (Bean et al., 2007; Engelman et al., 2007; Wheeler et al., 2010; Zhang & Chang, 2008). For example, the T790M mutation occurs in 50% of EGFR-mutant tumors that develop acquired resistance to erlotinib or gefitinib via a mechanism that involves increased affinity for ATP binding (Fig. 6) (Kobayashi et al., 2005; Pao et al., 2005a; Wheeler et al., 2010; Yun et al., 2008).

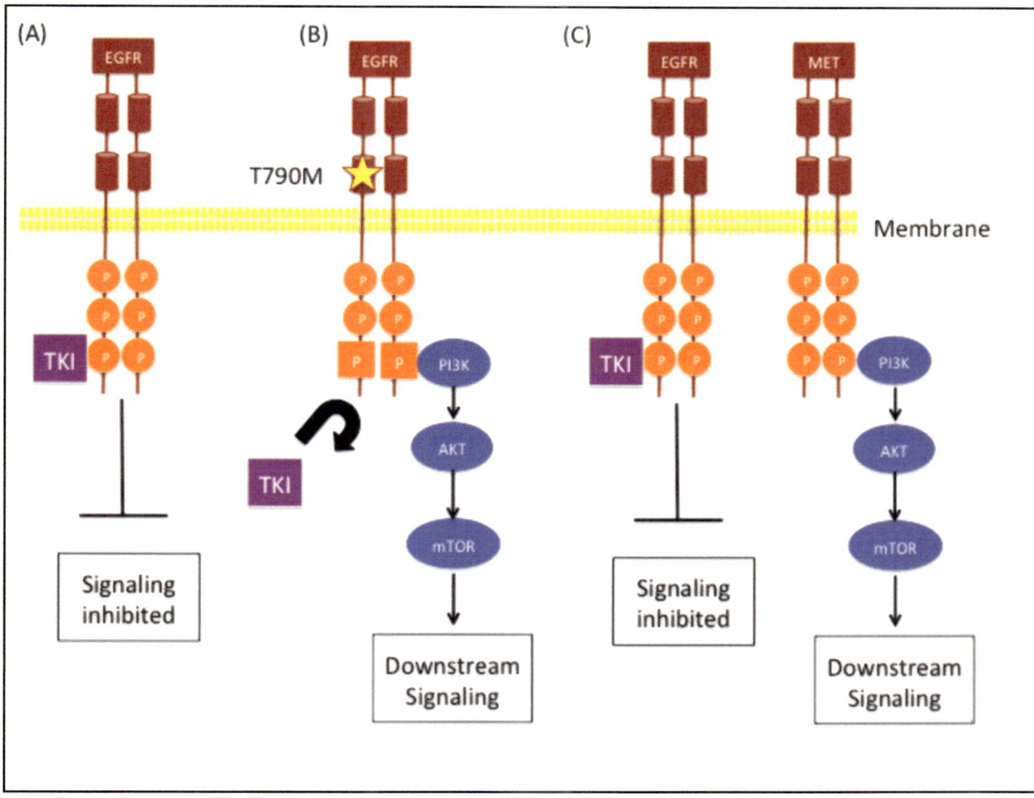

Fig. 6. Inhibition of EGFR Signaling by Tyrosine Kinase Inhibitors and Mechanisms of Resistance in Non-small Cell Lung Cancers. **Panel A.** Intracellular binding of TKIs, erlotinib or gefitinib, to the ATP-binding site of the tyrosine kinase domain of EGFR blocks kinase activity and inhibits downstream signaling pathways responsible for cellular proliferation. **Panel B.** The T790M mutation, detected in approximately 50% of patients who relapse while receiving an EGFR TKI, causes steric hindrance that prevents TKI binding and promotes constitutive activation of the mutated EGFR kinase. **Panel C.** Amplification of the *MET* oncogene activates downstream signaling through the PI3K–Akt pathway in an EGFR-independent fashion despite effective EGFR tyrosine kinase inhibition by TKI. Adapted from reference (Cataldo et al., 2011).

2.2 Strategies to overcome EGFR resistance

Various strategies have been developed to address TKI resistance, such as the development of second and third generation TKIs. The irreversible inhibitor, BIBW2992 (Afatinib; Boehringer Ingelheim), has been shown to have potent activity against EGFR and ERBB2 that overcomes resistance secondary to T790M EGFR mutations. BIBW2992 has been shown to reduce the survival of cancer cell lines and induce tumor regression in xenograft and transgenic lung cancer models (Li et al., 2008). Phase I clinical trials have demonstrated that daily oral BIBW2992 has durable anti-tumor activity and is safely tolerated (Yap et al., 2010). Combination therapy with BIBW2992 and an EGFR-specific antibody, cetuximab, further overcame T790M EGFR resistance in a murine model (Regales et al., 2009), which suggests that this approach may represent an alternative treatment option. Although combined inhibition of VEGFR and EGFR has been reported to reduce tumor growth in xenograft models of EGFR inhibitor resistance (Naumov et al., 2009), this approach did not improve survival in a phase III clinical trial of NSCLC when erlotinib was combined with bevacizumab, an antibody against VEGF (Herbst et al., 2011). Combination of erlotinib with sorafenib, a multi-TKI of VEGFR, c-Kit, PDGFR, b-Raf and c-Raf, had some therapeutic benefit in EGFR resistant patients with NSCLC as the median overall survival was prolonged to 8 months in the erlotinib/sorafenib group as compared to 4.5 months in the placebo/erlotinib group ($P = 0.019$) (Spigel et al., 2011). The same strategy might also be used to overcome mutations in pathways downstream of EGFR (Sos et al., 2010). For example, simultaneous targeting of EGFR and its downstream target, Akt, using BIBW2992 and rapamycin, reduced tumor size and protein phosphorylation in murine NSCLC models (Perera et al., 2009). Similar findings have been shown using erlotinib combined with rapamycin to treat NSCLC cell lines (A549, H1299, H1650 and H1975) *in vitro* (Nakachi et al., 2010). Furthermore, alternative TKIs, such as sorafenib and crizotinib, an inhibitor of anaplastic lymphoma kinase (Alk) and Met, might be used to overcome mutations in other EGFR-related genes, such as b-Raf (Brose et al., 2002; Davies et al., 2002), k-Ras (Pao et al., 2005b) and Alk (Soda et al., 2007)(Comoglio, Giordano, & Trusolino, 2008; Ou et al., 2011; Takezawa et al., 2009).

3. Therapeutic use of TKIs in inflammatory airway diseases

New treatments are needed for patients with severe asthma or COPD. While the majority of asthmatics can be adequately controlled with low-to-moderate doses of inhaled corticosteroids, approximately 5% to 10% of patients have severe disease that is refractory to standard treatments (Program, 2007; Wenzel & Busse, 2007). These individuals have persistent symptoms and recurrent disease exacerbations despite high-doses of inhaled corticosteroids plus long-acting β2-agonists or oral corticosteroids. New treatment options are needed as oral corticosteroids are associated with serious and potentially debilitating side effects, such as diabetes, hypertension, weight gain, impaired host defense, reduced bone density, cataracts, skin atrophy, and myopathy. Limited treatment options also exist for patients with COPD, which include inhaled and oral corticosteroids, inhaled β2-agonists, inhaled anti-cholinergics and phosphodiesterase inhibitors.

Signaling via RTKs and non-RTKs has been implicated in the pathogenesis of asthma and COPD, especially in patients with severe disease who are resistant to corticosteroid therapy (Adcock, Chung, Caramori, & Ito, 2006). Therefore, TKIs have been proposed as an alternative approach for the treatment of asthma and COPD. In addition, although TKIs are designed to target specific protein kinases, they frequently have activity towards additional targets, which may be valuable in treating diseases other than those for which the TKI was originally developed. For example, imatinib (Gleevec™) was originally designed as a Bcr-Abl inhibitor for the treatment of chronic myelogenous leukemia (CML), but was also found to inhibit c-Kit, PDGFRα/β, and CSF1R (Buchdunger et al., 2000; Druker et al., 1996; Heinrich, Griffith, Druker, Wait, Ott, & Zigler, 2000). Although these off-target effects may lead to unanticipated toxicities, it may also represent an opportunity to extend the utility of TKIs for the treatment of lung disease.

3.1 Asthma

Asthma is a common respiratory illness that afflicts roughly 300 million people worldwide (Braman, 2006). Asthma prevalence is highest in developed countries, including the UK (>15%), USA (~11%) and Australia (~15%), where it accounts for approximately 1% to 2% of the healthcare annual budget (Masoli, Fabian, Holt, & Beasley, 2004). There is a positive correlation between healthcare costs and asthma severity, as individuals with severe disease and frequent exacerbations represent approximately 5% to 10% of patients, but account for up to 50% of the costs (Godard, Chanez, Siraudin, Nicoloyannis, & Duru, 2002). Asthma is characterized by airway inflammation, airway remodeling, mucus hypersecretion and enhanced smooth muscle contractility (i.e., airway hyperreactivity) (Anderson, 2008; Barnes, 2008; Fanta, 2009; Holgate & Polosa, 2006; Rogers, 2004). Allergic airway inflammation is mediated by the recruitment of eosinophils, neutrophils, basophils, activated mast cells, and Th2-type CD4+ T cells that produce a characteristic set of cytokines, typified by IL-4, IL-5, IL-9, and IL-13 (Barnes, 2008). Structural changes that lead to airway wall remodeling include mucous cell metaplasia, airway smooth muscle cell hypertrophy and hyperplasia, epithelial cell proliferation, subepithelial fibrosis, basement membrane thickening, and vascular hyperplasia (Anderson, 2008; Barnes, 2008; Cohen et al., 2007; Fanta, 2009; Holgate & Polosa, 2006). Airway hyperreactivity (AHR) is an additional cardinal feature of asthma that results in increased airflow resistance and airflow obstruction (Fanta, 2009). Mucus hypersecretion also contributes to airway obstruction via mucus plugging in the airway (Del Donno, Bittesnich, Chetta, Olivieri, & Lopez-Vidriero, 2000; Rogers, 2007).

3.1.1 The potential role of RTK inhibitors for the treatment of asthma

Multiple lines of evidence support a role for RTKs in the pathogenesis of asthma. For example, increased tyrosine phosphorylation has been observed in the airway epithelium of patients with severe asthma who are resistant to corticosteroids (Hamilton et al., 2005). Similarly, PDGFR-β expression is increased in patients with severe asthma and may contribute to fibrotic airway remodeling responses (Lewis et al., 2005). Furthermore, a WNT/tenascin/PDGFR pathway has been implicated in the pathogenesis of airway smooth muscle hyperplasia and hypertrophy in a murine model of allergen-induced asthma (Cohen, Ihida-Stansbury, Lu, Panettieri, Jones, & Morrisey, 2009). Expression of c-Kit and its ligand stem cell factor (SCF) are also increased in the airways of asthmatic patients as compared to non-asthmatic control subjects (Al-Muhsen, Shablovsky, Olivenstein, Mazer, & Hamid, 2004; Bradding, Walls, & Holgate, 2006). Enhanced c-Kit signaling plays an important role in the activation of mast cells and dendritic cells, AHR, mucus hyperproduction, collagen deposition, and airway inflammation in asthma (Krishnamoorthy et al., 2008; Paniagua et al., 2006; Yu, Tsai, Tam, Jones, Zehnder, & Galli, 2006). In experimental models of asthma, SCF has been shown to mediate AHR, airway remodeling, and the production of pro-inflammatory cytokines (TNF, IL-5) and chemokines (CCL2, CCL5, CCL6, CCL17), and mucous cell metaplasia (Berlin, Hogaboam, & Lukacs, 2006; Berlin, Lincoln, Tomkinson, & Lukacs, 2004; Campbell, Hogaboam, Lincoln, & Lukacs, 1999; Dolgachev, Thomas, Berlin, & Lukacs, 2007). Lastly, VEGF signaling has been shown to induce an asthmatic phenotype in murine models (Lee et al., 2004).

These data support the concept of investigating the role of TKIs for the treatment of asthma. For example, imatinib has been shown to attenuate airway inflammation, AHR and fibrosis in murine models of allergic disease (Berlin & Lukacs, 2005). This was also demonstrated with sunitinib, a RTK inhibitor of VEGFR, PDGFR, c-Kit and fetal liver tyrosine kinase receptor 3 (FLT-3), which is used as an alternative treatment for imatinib-resistant CML (Huang, Liu, Du, Yao, & Yin, 2009). Furthermore, masitinib, a multi-target RTK inhibitor of c-Kit and PDGFR, has recently been shown to improve disease control in a clinical trial of severe, corticosteroid-resistant asthma (Humbert et al., 2009). Finally, EGFR activation plays a central role in the induction of mucin synthesis and mucous cell hyperplasia (Burgel & Nadel, 2004; Tamaoka et al., 2008). Gefinitib treatment inhibits EGFR and PI3K/Akt activation in ovalbumin (OVA) sensitized mice, which suggests that this approach might be beneficial for the treatment of asthma (Hur et al., 2007). House dust mite (HDM), a common aeroallergen that causes atopic asthma, has been shown to enhance EGFR signaling and epithelial-to-mesenchymal transition (EMT) in airway epithelial cells and thereby promote airway remodeling in asthma (Heijink, Postma, Noordhoek, Broekema, & Kapus, 2010). Furthermore, inhibition of EGFR signaling prevents TGF-β/HDM-induced EMT, which suggests that this approach might be utilized to attenuate airway remodeling in asthma.

3.1.2 Treatment of asthma with non-RTK inhibitors

3.1.2.1 Src kinase inhibitors

In human eosinophils, binding of eotaxin (CCL11) to CCR3 activates SFKs, such as Hck, Fgr and Lyn, as well as Syk, to enhance tyrosine phosphorylation, chemotaxis and respiratory burst. Cross-linking of FcεRI with multivalent antigen in mast cells and basophils activates Lyn with resultant calcium mobilization, actin polymerization, shape change, and secretion of pro-inflammatory cytokines, such as IL-6, IL-13 and TNF-α (Amoui, Draberova, Tolar, & Draber, 1997; Furumoto, Nunomura, Terada, Rivera, & Ra, 2004; Kepley, Wilson, & Oliver, 1998; Vonakis et al., 2005). Lyn deficient mice develop severe persistent asthma, which demonstrates that Lyn is a critical negative regulator of Th2 immunity (Beavitt et al., 2005). Furthermore, a Lyn blocking peptide and the Src-selective inhibitor, PP1, can attenuate eosinophil activation, differentiation and survival (Adachi, Stafford, Sur, & Alam, 1999; Lynch, Giembycz, Daniels, Barnes, & Lindsay, 2000). Lck is expressed primarily by T lymphocytes and plays an essential role in immune responses (Faith, Akdis, Akdis, Simon, & Blaser, 1997; Molina et al., 1992). Lck deficient mice have impaired thymocyte development, lack Th2 cells and are unable to mount antigen-dependent immune responses (Karnitz et al., 1992). Thus, Lck plays a crucial role in T-cell maturation and antigen-induced T-cell activation, which suggests that a Lck inhibitor could be utilized for the treatment of autoimmune and inflammatory diseases (Burchat et al., 2002). SFKs have also been reported to mediate angiotensin II-induced hyperresponsiveness of rat bronchial smooth muscle (Sakai et al., 2010).

3.1.2.2 Syk inhibitors

Syk, which is activated by integrins and oxidative stress, has an important role in signal transduction (Au-Yeung et al., 2009). Syk is expressed by all hematopoietic cells, as well as by fibroblasts, endothelial cells, hepatocytes and epithelium (Ulanova et al., 2005). Its role in modulating the function of non-hematopoietic cells, however, is uncertain. Mice with a spontaneous point mutation in Syk have altered TCR signaling that leads to the positive selection of autoimmune T cells and autoimmune arthritis (Sakaguchi et al., 2003). Interestingly, this was not observed in a patient with a Syk variant (Picard et al., 2009). Syk induces mast cell degranulation via FcεRI signaling, T and B lymphocyte development and activation, as well as IL-5- and GM-CSF-mediated eosinophil survival (Yanagi, Inatome, Takano, & Yamamura, 2001). The inhibition of Syk by antisense nucleotides blocks eosinophil survival and its deletion in mast cells prevents IgE-induced mast cell activation and AHR (Matsubara et al., 2006; Stenton et al., 2002). Thus, Syk inhibition has the potential to suppress mast cell-driven airway bronchoconstriction, hyperreactivity and inflammation. Consistent with this, mice with an inducible deletion of Syk or those treated with a pharmacological Syk inhibitor (BAY61-3606) displayed reduced OVA-induced eosinophilia (Wex et al., 2011). BAY61-3606 has also been reported to attenuate OVA-induced airway inflammation, lipid mediator release, cytokine synthesis and mast cell degranulation in rats (Yamamoto et al., 2003). Similarly, another Syk inhibitor, R406, inhibited airway hyperreactivity, pulmonary eosinophilia and goblet cell metaplasia in a murine model of OVA-induced asthma (Matsubara et al., 2006). The Syk-selective inhibitor, piceatannol, has also been shown to attenuate antigen-mediated bronchial contraction of guinea pig airways *in vitro* (Seow, Chue, & Wong, 2002). Anti-sense approaches to knockdown Syk expression have also suppressed airway inflammation in rat asthma models, as well as antigen-induced contraction of isolated tracheas (Stenton et al., 2002). Finally, the Syk inhibitor, R-112, has also been shown to improve symptoms of seasonal allergic rhinitis (Meltzer, Berkowitz, & Grossbard, 2005). An alternative compound, R-343, has completed phase I clinical trials for asthma and phase II trials are being planned (Norman, 2009).

3.1.2.3 Tec inhibitors

Inducible T cell kinase (Itk) plays an important role in T cell activation and differentiation, as well as in the differentiation and function of Th2-type cells, which modulate the pathogenesis of asthma. Consistent with this, Itk deficient mice have reduced antigen-specific T cell recruitment to the lung, decreased IL-5 and IL-13 production, reduced antigen-mediated T cell proliferative responses, as well as attenuated eosinophil infiltration and mucus production (Mueller & August, 2003). Reduced airway inflammation in Itk knockout mice is caused by an increase in $\gamma\delta$ T cells that attenuate mast cell responses (Felices, Yin, Kosaka, Kang, & Berg, 2009; Qi et al., 2009). Itk deficient mice also have reduced AHR (Ferrara, Mueller, Sahu, Ben-Jebria, & August, 2006). Itk deficiency is also associated with a decrease in IL-17A expression by Th17 cells that is mediated via the transcription factor, NFATc1 (nuclear factor of activated T cells) (Gomez-Rodriguez et al., 2009). Lastly, double Itk and Btk knockout mice have severely impaired FcεRI-dependent mast cell responses (Iyer et al., 2011). Pharmacological Itk inhibitors have been developed as potential new treatments for inflammatory diseases (Das et al., 2006). BMS-488516 and BMS-509744 inhibit Itk kinase activity by reducing TCR-induced functions, including PLCγ1 tyrosine phosphorylation, calcium mobilization, IL-2 secretion, and T-cell proliferation *in vitro*, as well as suppress IL-2 production in mice (Lin et al., 2004). BMS-509744 also significantly diminished lung inflammation in a mouse model of OVA-induced allergic asthma (Lin et al., 2004). Additional Itk inhibitors have been synthesized based upon the (4 or 5-aryl)pyrazolyl-indole scaffold that is selective for Itk (Velankar et al., 2010). Furthermore, irreversible inhibition of Btk with the highly selective inhibitor, PCI-32765, has been shown to block IgE-mediated activation of human basophils (MacGlashan, Honigberg, Smith, Buggy, & Schroeder, 2011).

3.1.2.4 JAK inhibitors

JAK inhibition represents another potential therapeutic approach for the treatment of asthma. JAK3 is expressed by many cells involved in the pathogenesis of asthma, such as mast cells, T cells, macrophages and dendritic cells. For example, RANTES induces rapid tyrosine phosphorylation of CCR5, activation of JAK2 and JAK3, and formation of STAT1 and STAT3 dimers in human T cells (Bacon, Szabo, Yssel, Bolen, & Schall, 1996; Wong & Fish, 1998; Wong et al., 2001). Stromal cell-derived factor-1a (SDF-1a) triggers CXCR4 receptor dimerization and activates JAK2, JAK3, and Syk, causing transendothelial migration of human T cell lines (Ticchioni et al., 2002; Vila-Coro, Rodriguez-Frade, Martin De Ana, Moreno-Ortiz, Martinez, & Mellado, 1999). Although JAK3 is primarily responsible for cytokine signaling, it also has other functional roles. For example, JAK3 plays a pivotal role in FcεRI-mediated mast cell responses. Consistent with this, the JAK3 inhibitor, CP-690550 attenuated pulmonary eosinophilia, as well as IL-13 and eotaxin production in an OVA-model of allergic airway disease, while targeting JAK3 with WHI-P97 attenuated both pulmonary eosinophilia and AHR (Kudlacz, Conklyn, Andresen, Whitney-Pickett, & Changelian, 2008; Malaviya et al., 2000; Malaviya, Zhu, Dibirdik, & Uckun, 1999). Similarly, the JAK3 inhibitor, WHI-P131 has been shown to inhibit IgE-mediated mast cell degranulation (Kudlacz et al., 2008; Malaviya et al., 2000; Malaviya et al., 1999). Furthermore, a pan-JAK inhibitor, pyridone 6 encapsulated in polyactic-coglycolic acid nanoparticles, has been shown to suppress Th2 inflammation and pulmonary eosinophilia, but not airway hyperreactivity in a murine OVA-challenge model (Matsunaga et al., 2011).

3.2 Chronic Obstructive Pulmonary Disease

COPD is a common lung disease, with an estimated prevalence of 210 million individuals in 2007. Furthermore, COPD is the fifth most common cause of mortality worldwide and accounted for approximately 3 million deaths in 2005 (Marwick & Chung, 2010). COPD-related deaths are predicted to rise by 30% in the next 10 years and will become the third leading cause of death worldwide by 2030. COPD is typically caused by the exposure of noxious particles or gases, such as those contained in cigarette smoke or fossil fuels found in environmental air pollution (Buist et al., 2007; Chung & Adcock, 2008; Mannino & Buist, 2007). COPD is manifested by the destruction of the lung parenchyma with resultant emphysema and inflammation. Patients with COPD have an increase in lung macrophages, neutrophils and CD8+ T cells (Hogg et al., 2004; Saetta, Turato, Maestrelli, Mapp, & Fabbri, 2001). This inflammatory process is orchestrated by chemokines, such as CCL2, CXCL1, CXCL8, and cytokines, such as TNF, IL-1β, IL-6, and IFN-γ (Barnes, 2008).

Mucous cell metaplasia is another important manifestation of COPD that results in mucus overproduction and airway obstruction in patients with chronic bronchitis. Consistent with this, expression of EGFRs and airway mucins are increased in airway epithelial cells of smokers, as compared with non-smokers (de Boer, Hau, van Schadewijk, Stolk, van Krieken, & Hiemstra, 2006; Takeyama et al., 2001). Signaling through EGFR involves dual oxidase 1 and the production of reactive oxygen species to activate TNF-alpha converting enzyme (TACE, ADAM17), which induces mucin production via a pathway involving increased TGF-α shedding and EGFR phosphorylation (Shao & Nadel, 2005). Binding of CCL20 to CCR6 has been shown to exaggerate EGFR-dependent MUC5AC production by human airway epithelial cells (Kim, Lewis, & Nadel, 2011). EGFR activation also participates in CXCL8 (IL-8) production by bronchial epithelial cells (Richter et al., 2002). Furthermore, EGFR signaling via STAT and Erk1/2 reduces expression of IL-8 and ICAM-1 by bronchial epithelial cells following rhinovirus infection, which is a common trigger of COPD exacerbations (Liu, Gualano, Hibbs, Anderson, & Bozinovski, 2008). The role of EGFR activation in airway mucin production and IL-8 expression suggests that EGFR blockade might be considered as a treatment approach for COPD. However, a clinical trial of an inhaled EGFR inhibitor (BIBW2948 BS) in 48 patients with COPD did not significantly decrease airway epithelial cell mucin stores and was also poorly tolerated (Woodruff et al., 2010). Thus, additional studies will be required to establish whether there is a role for EGFR inhibition in the treatment of COPD. Lastly, COPD patients may develop NSCLC due to altered EGFR signaling and aberrant methylation, which may have a unique genetic profile compared to NSCLC patients without COPD (Suzuki et al., 2010).

3.3 Idiopathic Pulmonary Fibrosis

Idiopathic pulmonary fibrosis (IPF) is a progressive lung disorder that is associated with a relentless deterioration in lung function and a median survival of only 2.5 to 3.5 years after diagnosis (King, Pardo, & Selman, 2011; Ley, Collard, & King, 2011). IPF has an estimated prevalence of 13 to 20 cases per 100,000 and has a higher predominance in men than women. In addition, the frequency of IPF increases with age, with disease typically occurring in patients greater than 50 years of age. The etiology of IPF may reflect a dysregulated airway epithelial cell repair response to repetitive injury, which causes apoptosis of epithelial and endothelial cells. This results in the release of pro-fibrotic mediators that mediate the migration and proliferation of mesenchymal cells, the formation of fibroblast/myofibroblast foci, alveolar destruction and excessive collagen deposition despite minimal inflammation. Various stimuli such as cigarette smoke, wood or metal dust, viral infection and gastro-esophageal reflux/aspiration can contribute to the development of IPF (Iwai, Mori, Yamada, Yamaguchi, & Hosoda, 1994; Raghu, Yang, Spada, Hayes, & Pellegrini, 2006; Steele et al., 2005; Tobin, Pope, Pellegrini, Emond, Sillery, & Raghu, 1998; Ueda et al., 1992). Genes that have been associated with the pathogenesis of IPF include surfactant protein C (Thomas et al., 2002), surfactant protein A2 (Nogee, Dunbar, Wert, Askin, Hamvas, & Whitsett, 2002; Wang et al., 2009), mucin 5B (*MUC5B*) (Seibold et al., 2011) and human telomerase reverse transcriptase (*TERT*) and telomerase RNA component (*TERC*) (Alder et al., 2008; Armanios et al., 2007; Cronkhite et al., 2008; Tsakiri et al., 2007).

At present, there is no effective treatment for IPF. Only lung transplantation has reported to prolong survival (King et al., 2011). The finding that fibrogenic growth factors, such as TGF-β, PDGF, FGF and VEGF signal via RTKs has lead to the concept that TKIs might be utilized as a novel treatment approach for IPF (Downey, 2011). Consistent with this, inhibition of pro-fibrotic signaling pathways mediated by TGF-α, PDGF, FGF, IGF and VEGF can prevent matrix deposition and fibroblast proliferation (Abdollahi et al., 2005; Chaudhary et al., 2007; Choi et al., 2009). Furthermore, the EGFR inhibitor, gefitinib has been reported to prevent bleomycin-induced IPF in murine models (Ishii, Fujimoto, & Fukuda, 2006; Wang et al., 2010), while gefitinib and erlotinib have been shown to reduce TGF-α-induced IPF in a murine model (Hardie et al., 2008). The role of TKIs for the treatment of IPF has been investigated in clinical trials. A recent multicenter, double-blind trial of imatinib in 119 patients with mild-to-moderate IPF did not improve survival or lung function (Daniels, Lasky, Limper, Mieras, Gabor, & Schroeder, 2010). In contrast, more promising results were reported by a recent 12-month, phase 2 clinical trial of 432 patients that investigated the efficacy of the TKI, BIBF 1120 (Indedanib), as compared to placebo (Richeldi et al., 2011). BIBF 1120 is a multi-receptor tyrosine kinase inhibitor that targets PDGFR, VEGFRs 1, 2 and 3 and FGFRs 1, 2 and 3. This study found a trend towards a reduction in decline of lung function, with fewer acute exacerbations and preserved quality of life. This result suggests that TKIs that target multiple fibrogenic pathways might represent an efficacious treatment approach for IPF (Downey, 2011). This could represent a major advance for the treatment of IPF patients.

4. Conclusion

Activation of RTKs and non-RTKs play key roles in the pathogenesis of NSCLC and lung diseases, such as asthma, COPD and IPF. Although the TKI, erlotinib, has advanced to clinical practice as a second-line treatment for NSCLC, TKIs have not yet been shown to be effective for the treatment of severe asthma or COPD. Recent results, however, have suggested that use of a TKI that inhibits multiple tyrosine kinase targets, might slow disease progression in IPF. If confirmed to be efficacious, TKI treatment could represent a major advance for the treatment of this progressive and fatal lung disease. Thus, the future is bright as continued advances in the development of TKIs may lead to novel treatment approaches for patients with severe lung diseases with high associated mortalities, such as IPF and NSCLC. Furthermore, personalizing TKI therapy to target specific RTKs and non-RTKs that mediate disease pathogenesis may allow for more accurate treatment and reduce unwanted off-target effects.

5. Acknowledgement

This work was funded by the Division of Intramural Research, National Heart, Lung, and Blood Institute, National Institutes of Health.

6. References

Abdollahi, A., Li, M., Ping, G., Plathow, C., Domhan, S., Kiessling, F., Lee, L. B., McMahon, G., Grone, H. J., Lipson, K. E., & Huber, P. E. 2005. Inhibition of platelet-derived growth factor signaling attenuates pulmonary fibrosis. *J Exp Med*, 201(6): 925-935.

Adachi, T., Stafford, S., Sur, S., & Alam, R. 1999. A novel Lyn-binding peptide inhibitor blocks eosinophil differentiation, survival, and airway eosinophilic inflammation. *J Immunol*, 163(2): 939-946.

Adcock, I. M., Chung, K. F., Caramori, G., & Ito, K. 2006. Kinase inhibitors and airway inflammation. *Eur J Pharmacol*, 533(1-3): 118-132.

Al-Muhsen, S. Z., Shablovsky, G., Olivenstein, R., Mazer, B., & Hamid, Q. 2004. The expression of stem cell factor and c-kit receptor in human asthmatic airways. *Clin Exp Allergy*, 34(6): 911-916.

Alder, J. K., Chen, J. J., Lancaster, L., Danoff, S., Su, S. C., Cogan, J. D., Vulto, I., Xie, M., Qi, X., Tuder, R. M., Phillips, J. A., 3rd, Lansdorp, P. M., Loyd, J. E., & Armanios, M. Y. 2008. Short telomeres are a risk factor for idiopathic pulmonary fibrosis. *Proceedings of the National Academy of Sciences of the United States of America*, 105(35): 13051-13056.

Amoui, M., Draberova, L., Tolar, P., & Draber, P. 1997. Direct interaction of Syk and Lyn protein tyrosine kinases in rat basophilic leukemia cells activated via type I Fc epsilon receptors. *Eur J Immunol*, 27(1): 321-328.

Anderson, G. P. 2008. Endotyping asthma: new insights into key pathogenic mechanisms in a complex, heterogeneous disease. *Lancet*, 372(9643): 1107-1119.

Argetsinger, L. S., Campbell, G. S., Yang, X., Witthuhn, B. A., Silvennoinen, O., Ihle, J. N., & Carter-Su, C. 1993. Identification of JAK2 as a growth hormone receptor-associated tyrosine kinase. *Cell*, 74(2): 237-244.

Armanios, M. Y., Chen, J. J., Cogan, J. D., Alder, J. K., Ingersoll, R. G., Markin, C., Lawson, W. E., Xie, M., Vulto, I., Phillips, J. A., 3rd, Lansdorp, P. M., Greider, C. W., & Loyd, J. E. 2007. Telomerase mutations in families with idiopathic pulmonary fibrosis. *The New England Journal of Medicine*, 356(13): 1317-1326.

AstraZeneca. 2005. Iressa® Gefitinib package insert.

Au-Yeung, B. B., Deindl, S., Hsu, L. Y., Palacios, E. H., Levin, S. E., Kuriyan, J., & Weiss, A. 2009. The structure, regulation, and function of ZAP-70. *Immunol Rev*, 228(1): 41-57.

Bacon, C. M., McVicar, D. W., Ortaldo, J. R., Rees, R. C., O'Shea, J. J., & Johnston, J. A. 1995. Interleukin 12 (IL-12) induces tyrosine phosphorylation of JAK2 and TYK2: differential use of Janus family tyrosine kinases by IL-2 and IL-12. *J Exp Med*, 181(1): 399-404.

Bacon, K. B., Szabo, M. C., Yssel, H., Bolen, J. B., & Schall, T. J. 1996. RANTES induces tyrosine kinase activity of stably complexed p125FAK and ZAP-70 in human T cells. *J Exp Med*, 184(3): 873-882.

Balak, M. N., Gong, Y., Riely, G. J., Somwar, R., Li, A. R., Zakowski, M. F., Chiang, A., Yang, G., Ouerfelli, O., Kris, M. G., Ladanyi, M., Miller, V. A., & Pao, W. 2006. Novel D761Y and common secondary T790M mutations in epidermal growth factor receptor-mutant lung adenocarcinomas with acquired resistance to kinase inhibitors. *Clin Cancer Res*, 12(21): 6494-6501.

Barnes, P. J. 2008. The cytokine network in asthma and chronic obstructive pulmonary disease. *J Clin Invest*, 118(11): 3546-3556.

Bean, J., Brennan, C., Shih, J. Y., Riely, G., Viale, A., Wang, L., Chitale, D., Motoi, N., Szoke, J., Broderick, S., Balak, M., Chang, W. C., Yu, C. J., Gazdar, A., Pass, H., Rusch, V., Gerald, W., Huang, S. F., Yang, P. C., Miller, V., Ladanyi, M., Yang, C. H., & Pao, W. 2007. MET amplification occurs with or without T790M mutations in EGFR mutant lung tumors with acquired resistance to gefitinib or erlotinib. *Proceedings of the National Academy of Sciences of the United States of America*, 104(52): 20932-20937.

Beavitt, S. J., Harder, K. W., Kemp, J. M., Jones, J., Quilici, C., Casagranda, F., Lam, E., Turner, D., Brennan, S., Sly, P. D., Tarlinton, D. M., Anderson, G. P., & Hibbs, M. L. 2005. Lyn-deficient mice develop severe, persistent asthma: Lyn is a critical negative regulator of Th2 immunity. *J Immunol*, 175(3): 1867-1875.

Benati, D., & Baldari, C. T. 2008. SRC family kinases as potential therapeutic targets for malignancies and immunological disorders. *Curr Med Chem*, 15(12): 1154-1165.

Berlin, A. A., Hogaboam, C. M., & Lukacs, N. W. 2006. Inhibition of SCF attenuates peribronchial remodeling in chronic cockroach allergen-induced asthma. *Lab Invest*, 86(6): 557-565.

Berlin, A. A., Lincoln, P., Tomkinson, A., & Lukacs, N. W. 2004. Inhibition of stem cell factor reduces pulmonary cytokine levels during allergic airway responses. *Clin Exp Immunol*, 136(1): 15-20.

Berlin, A. A., & Lukacs, N. W. 2005. Treatment of cockroach allergen asthma model with imatinib attenuates airway responses. *Am J Respir Crit Care Med*, 171(1): 35-39.

Bradding, P., Walls, A. F., & Holgate, S. T. 2006. The role of the mast cell in the pathophysiology of asthma. *J Allergy Clin Immunol*, 117(6): 1277-1284.

Braman, S. S. 2006. The global burden of asthma. *Chest*, 130(1 Suppl): 4S-12S.

Brose, M. S., Volpe, P., Feldman, M., Kumar, M., Rishi, I., Gerrero, R., Einhorn, E., Herlyn, M., Minna, J., Nicholson, A., Roth, J. A., Albelda, S. M., Davies, H., Cox, C., Brignell, G., Stephens, P., Futreal, P. A., Wooster, R., Stratton, M. R., & Weber, B. L. 2002. BRAF and RAS mutations in human lung cancer and melanoma. *Cancer Res*, 62(23): 6997-7000.

Buchdunger, E., Cioffi, C. L., Law, N., Stover, D., Ohno-Jones, S., Druker, B. J., & Lydon, N. B. 2000. Abl protein-tyrosine kinase inhibitor STI571 inhibits in vitro signal transduction mediated by c-kit and platelet-derived growth factor receptors. *The Journal of Pharmacology and Experimental Therapeutics*, 295(1): 139-145.

Buist, A. S., McBurnie, M. A., Vollmer, W. M., Gillespie, S., Burney, P., Mannino, D. M., Menezes, A. M., Sullivan, S. D., Lee, T. A., Weiss, K. B., Jensen, R. L., Marks, G. B., Gulsvik, A., & Nizankowska-Mogilnicka, E. 2007. International variation in the prevalence of COPD (the BOLD Study): a population-based prevalence study. *Lancet*, 370(9589): 741-750.

Burchat, A. F., Calderwood, D. J., Friedman, M. M., Hirst, G. C., Li, B., Rafferty, P., Ritter, K., & Skinner, B. S. 2002. Pyrazolo[3,4-d]pyrimidines containing an extended 3-substituent as potent inhibitors of Lck -- a selectivity insight. *Bioorganic & Medicinal Chemistry Letters*, 12(12): 1687-1690.

Burgel, P. R., & Nadel, J. A. 2004. Roles of epidermal growth factor receptor activation in epithelial cell repair and mucin production in airway epithelium. *Thorax*, 59(11): 992-996.

Campbell, E., Hogaboam, C., Lincoln, P., & Lukacs, N. W. 1999. Stem cell factor-induced airway hyperreactivity in allergic and normal mice. *Am J Pathol*, 154(4): 1259-1265.

Carey, K. D., Garton, A. J., Romero, M. S., Kahler, J., Thomson, S., Ross, S., Park, F., Haley, J. D., Gibson, N., & Sliwkowski, M. X. 2006. Kinetic analysis of epidermal growth factor receptor somatic mutant proteins shows increased sensitivity to the epidermal growth factor receptor tyrosine kinase inhibitor, erlotinib. *Cancer Res*, 66(16): 8163-8171.

Cataldo, V. D., Gibbons, D. L., Perez-Soler, R., & Quintas-Cardama, A. 2011. Treatment of non-small-cell lung cancer with erlotinib or gefitinib. *N Engl J Med*, 364(10): 947-955.

Chaudhary, N. I., Roth, G. J., Hilberg, F., Muller-Quernheim, J., Prasse, A., Zissel, G., Schnapp, A., & Park, J. E. 2007. Inhibition of PDGF, VEGF and FGF signalling attenuates fibrosis. *Eur Respir J*, 29(5): 976-985.

Chitale, D., Gong, Y., Taylor, B. S., Broderick, S., Brennan, C., Somwar, R., Golas, B., Wang, L., Motoi, N., Szoke, J., Reinersman, J. M., Major, J., Sander, C., Seshan, V. E., Zakowski, M. F., Rusch, V., Pao, W., Gerald, W., & Ladanyi, M. 2009. An integrated genomic analysis of lung cancer reveals loss of DUSP4 in EGFR-mutant tumors. *Oncogene*, 28(31): 2773-2783.

Choi, J. E., Lee, S. S., Sunde, D. A., Huizar, I., Haugk, K. L., Thannickal, V. J., Vittal, R., Plymate, S. R., & Schnapp, L. M. 2009. Insulin-like growth factor-I receptor blockade improves outcome in mouse model of lung injury. *Am J Respir Crit Care Med*, 179(3): 212-219.

Chung, K. F., & Adcock, I. M. 2008. Multifaceted mechanisms in COPD: inflammation, immunity, and tissue repair and destruction. *Eur Respir J*, 31(6): 1334-1356.

Ciardiello, F., & Tortora, G. 2008. EGFR antagonists in cancer treatment. *N Engl J Med*, 358(11): 1160-1174.

Cohen, E. D., Ihida-Stansbury, K., Lu, M. M., Panettieri, R. A., Jones, P. L., & Morrisey, E. E. 2009. Wnt signaling regulates smooth muscle precursor development in the mouse lung via a tenascin C/PDGFR pathway. *J Clin Invest*, 119(9): 2538-2549.

Cohen, L., E, X., Tarsi, J., Ramkumar, T., Horiuchi, T. K., Cochran, R., DeMartino, S., Schechtman, K. B., Hussain, I., Holtzman, M. J., & Castro, M. 2007. Epithelial cell proliferation contributes to airway remodeling in severe asthma. *Am J Respir Crit Care Med*, 176(2): 138-145.

Colgan, J. D., & Hankel, I. L. 2010. Signaling pathways critical for allergic airway inflammation. *Curr Opin Allergy Clin Immunol*, 10(1): 42-47.

Comoglio, P. M., Giordano, S., & Trusolino, L. 2008. Drug development of MET inhibitors: targeting oncogene addiction and expedience. *Nat Rev Drug Discov*, 7(6): 504-516.

Costa, D. B., Nguyen, K. S., Cho, B. C., Sequist, L. V., Jackman, D. M., Riely, G. J., Yeap, B. Y., Halmos, B., Kim, J. H., Janne, P. A., Huberman, M. S., Pao, W., Tenen, D. G., & Kobayashi, S. 2008. Effects of erlotinib in EGFR mutated non-small cell lung cancers with resistance to gefitinib. *Clin Cancer Res*, 14(21): 7060-7067.

Cragg, M. S., Kuroda, J., Puthalakath, H., Huang, D. C., & Strasser, A. 2007. Gefitinib-induced killing of NSCLC cell lines expressing mutant EGFR requires BIM and can be enhanced by BH3 mimetics. *PLoS Med*, 4(10): 1681-1689; discussion 1690.

Cronkhite, J. T., Xing, C., Raghu, G., Chin, K. M., Torres, F., Rosenblatt, R. L., & Garcia, C. K. 2008. Telomere shortening in familial and sporadic pulmonary fibrosis. *American Journal of Respiratory And Critical Care Medicine*, 178(7): 729-737.

Dal Porto, J. M., Gauld, S. B., Merrell, K. T., Mills, D., Pugh-Bernard, A. E., & Cambier, J. 2004. B cell antigen receptor signaling 101. *Mol Immunol*, 41(6-7): 599-613.

Daniels, C. E., Lasky, J. A., Limper, A. H., Mieras, K., Gabor, E., & Schroeder, D. R. 2010. Imatinib treatment for idiopathic pulmonary fibrosis: Randomized placebo-controlled trial results. *American journal of respiratory and critical care medicine*, 181(6): 604-610.

Darnell, J. E., Jr. 1997. STATs and gene regulation. *Science*, 277(5332): 1630-1635.

Das, J., Furch, J. A., Liu, C., Moquin, R. V., Lin, J., Spergel, S. H., McIntyre, K. W., Shuster, D. J., O'Day, K. D., Penhallow, B., Hung, C. Y., Doweyko, A. M., Kamath, A., Zhang, H., Marathe, P., Kanner, S. B., Lin, T. A., Dodd, J. H., Barrish, J. C., & Wityak, J.

2006. Discovery and SAR of 2-amino-5-(thioaryl)thiazoles as potent and selective Itk inhibitors. *Bioorganic & Medicinal Chemistry Letters*, 16(14): 3706-3712.

Davies, H., Bignell, G. R., Cox, C., Stephens, P., Edkins, S., Clegg, S., Teague, J., Woffendin, H., Garnett, M. J., Bottomley, W., Davis, N., Dicks, E., Ewing, R., Floyd, Y., Gray, K., Hall, S., Hawes, R., Hughes, J., Kosmidou, V., Menzies, A., Mould, C., Parker, A., Stevens, C., Watt, S., Hooper, S., Wilson, R., Jayatilake, H., Gusterson, B. A., Cooper, C., Shipley, J., Hargrave, D., Pritchard-Jones, K., Maitland, N., Chenevix-Trench, G., Riggins, G. J., Bigner, D. D., Palmieri, G., Cossu, A., Flanagan, A., Nicholson, A., Ho, J. W., Leung, S. Y., Yuen, S. T., Weber, B. L., Seigler, H. F., Darrow, T. L., Paterson, H., Marais, R., Marshall, C. J., Wooster, R., Stratton, M. R., & Futreal, P. A. 2002. Mutations of the BRAF gene in human cancer. *Nature*, 417(6892): 949-954.

de Boer, W. I., Hau, C. M., van Schadewijk, A., Stolk, J., van Krieken, J. H., & Hiemstra, P. S. 2006. Expression of epidermal growth factors and their receptors in the bronchial epithelium of subjects with chronic obstructive pulmonary disease. *Am J Clin Pathol*, 125(2): 184-192.

Del Donno, M., Bittesnich, D., Chetta, A., Olivieri, D., & Lopez-Vidriero, M. T. 2000. The effect of inflammation on mucociliary clearance in asthma: an overview. *Chest*, 118(4): 1142-1149.

Deng, J., Shimamura, T., Perera, S., Carlson, N. E., Cai, D., Shapiro, G. I., Wong, K. K., & Letai, A. 2007. Proapoptotic BH3-only BCL-2 family protein BIM connects death signaling from epidermal growth factor receptor inhibition to the mitochondrion. *Cancer Res*, 67(24): 11867-11875.

Ding, L., Getz, G., Wheeler, D. A., Mardis, E. R., McLellan, M. D., Cibulskis, K., Sougnez, C., Greulich, H., Muzny, D. M., Morgan, M. B., Fulton, L., Fulton, R. S., Zhang, Q., Wendl, M. C., Lawrence, M. S., Larson, D. E., Chen, K., Dooling, D. J., Sabo, A., Hawes, A. C., Shen, H., Jhangiani, S. N., Lewis, L. R., Hall, O., Zhu, Y., Mathew, T., Ren, Y., Yao, J., Scherer, S. E., Clerc, K., Metcalf, G. A., Ng, B., Milosavljevic, A., Gonzalez-Garay, M. L., Osborne, J. R., Meyer, R., Shi, X., Tang, Y., Koboldt, D. C., Lin, L., Abbott, R., Miner, T. L., Pohl, C., Fewell, G., Haipek, C., Schmidt, H., Dunford-Shore, B. H., Kraja, A., Crosby, S. D., Sawyer, C. S., Vickery, T., Sander, S., Robinson, J., Winckler, W., Baldwin, J., Chirieac, L. R., Dutt, A., Fennell, T., Hanna, M., Johnson, B. E., Onofrio, R. C., Thomas, R. K., Tonon, G., Weir, B. A., Zhao, X., Ziaugra, L., Zody, M. C., Giordano, T., Orringer, M. B., Roth, J. A., Spitz, M. R., Wistuba, II, Ozenberger, B., Good, P. J., Chang, A. C., Beer, D. G., Watson, M. A., Ladanyi, M., Broderick, S., Yoshizawa, A., Travis, W. D., Pao, W., Province, M. A., Weinstock, G. M., Varmus, H. E., Gabriel, S. B., Lander, E. S., Gibbs, R. A., Meyerson, M., & Wilson, R. K. 2008. Somatic mutations affect key pathways in lung adenocarcinoma. *Nature*, 455(7216): 1069-1075.

Dolgachev, V., Thomas, M., Berlin, A., & Lukacs, N. W. 2007. Stem cell factor-mediated activation pathways promote murine eosinophil CCL6 production and survival. *J Leukoc Biol*.

Downey, G. P. 2011. Resolving the scar of pulmonary fibrosis. *N Engl J Med*, 365(12): 1140-1141.

Druker, B. J., Tamura, S., Buchdunger, E., Ohno, S., Segal, G. M., Fanning, S., Zimmermann, J., & Lydon, N. B. 1996. Effects of a selective inhibitor of the Abl tyrosine kinase on the growth of Bcr-Abl positive cells. *Nat Med*, 2(5): 561-566.

Engelman, J. A., Zejnullahu, K., Mitsudomi, T., Song, Y., Hyland, C., Park, J. O., Lindeman, N., Gale, C. M., Zhao, X., Christensen, J., Kosaka, T., Holmes, A. J., Rogers, A. M., Cappuzzo, F., Mok, T., Lee, C., Johnson, B. E., Cantley, L. C., & Janne, P. A. 2007. MET amplification leads to gefitinib resistance in lung cancer by activating ERBB3 signaling. *Science*, 316(5827): 1039-1043.

Faith, A., Akdis, C. A., Akdis, M., Simon, H. U., & Blaser, K. 1997. Defective TCR stimulation in anergized type 2 T helper cells correlates with abrogated p56(lck) and ZAP-70 tyrosine kinase activities. *J Immunol*, 159(1): 53-60.

Fanta, C. H. 2009. Asthma. *N Engl J Med*, 360(10): 1002-1014.

Felices, M., Yin, C. C., Kosaka, Y., Kang, J., & Berg, L. J. 2009. Tec kinase Itk in gammadeltaT cells is pivotal for controlling IgE production in vivo. *Proc Natl Acad Sci U S A*, 106(20): 8308-8313.

Ferrara, T. J., Mueller, C., Sahu, N., Ben-Jebria, A., & August, A. 2006. Reduced airway hyperresponsiveness and tracheal responses during allergic asthma in mice lacking tyrosine kinase inducible T-cell kinase. *J Allergy Clin Immunol*, 117(4): 780-786.

Furumoto, Y., Nunomura, S., Terada, T., Rivera, J., & Ra, C. 2004. The FcepsilonRIbeta immunoreceptor tyrosine-based activation motif exerts inhibitory control on MAPK and IkappaB kinase phosphorylation and mast cell cytokine production. *J Biol Chem*, 279(47): 49177-49187.

Garrett, T. P., McKern, N. M., Lou, M., Elleman, T. C., Adams, T. E., Lovrecz, G. O., Zhu, H. J., Walker, F., Frenkel, M. J., Hoyne, P. A., Jorissen, R. N., Nice, E. C., Burgess, A. W., & Ward, C. W. 2002. Crystal structure of a truncated epidermal growth factor receptor extracellular domain bound to transforming growth factor alpha. *Cell*, 110(6): 763-773.

Genentech. 2011. Tarceva® erlotinib package insert.

Ghoreschi, K., Laurence, A., & O'Shea, J. J. 2009. Janus kinases in immune cell signaling. *Immunol Rev*, 228(1): 273-287.

Godard, P., Chanez, P., Siraudin, L., Nicoloyannis, N., & Duru, G. 2002. Costs of asthma are correlated with severity: a 1-yr prospective study. *Eur Respir J*, 19(1): 61-67.

Gomez-Rodriguez, J., Sahu, N., Handon, R., Davidson, T. S., Anderson, S. M., Kirby, M. R., August, A., & Schwartzberg, P. L. 2009. Differential expression of interleukin-17A and -17F is coupled to T cell receptor signaling via inducible T cell kinase. *Immunity*, 31(4): 587-597.

Gong, Y., Somwar, R., Politi, K., Balak, M., Chmielecki, J., Jiang, X., & Pao, W. 2007. Induction of BIM is essential for apoptosis triggered by EGFR kinase inhibitors in mutant EGFR-dependent lung adenocarcinomas. *PLoS Med*, 4(10): e294.

Grandal, M. V., & Madshus, I. H. 2008. Epidermal growth factor receptor and cancer: control of oncogenic signalling by endocytosis. *J Cell Mol Med*, 12(5A): 1527-1534.

Greulich, H., Chen, T. H., Feng, W., Janne, P. A., Alvarez, J. V., Zappaterra, M., Bulmer, S. E., Frank, D. A., Hahn, W. C., Sellers, W. R., & Meyerson, M. 2005. Oncogenic transformation by inhibitor-sensitive and -resistant EGFR mutants. *PLoS Med*, 2(11): e313.

Hamilton, L. M., Puddicombe, S. M., Dearman, R. J., Kimber, I., Sandstrom, T., Wallin, A., Howarth, P. H., Holgate, S. T., Wilson, S. J., & Davies, D. E. 2005. Altered protein tyrosine phosphorylation in asthmatic bronchial epithelium. *Eur Respir J*, 25(6): 978-985.

Hardie, W. D., Davidson, C., Ikegami, M., Leikauf, G. D., Le Cras, T. D., Prestridge, A., Whitsett, J. A., & Korfhagen, T. R. 2008. EGF receptor tyrosine kinase inhibitors diminish transforming growth factor-alpha-induced pulmonary fibrosis. *Am J Physiol Lung Cell Mol Physiol*, 294(6): L1217-1225.

Heijink, I. H., Postma, D. S., Noordhoek, J. A., Broekema, M., & Kapus, A. 2010. House dust mite-promoted epithelial-to-mesenchymal transition in human bronchial epithelium. *Am J Respir Cell Mol Biol*, 42(1): 69-79.

Heinrich, M. C., Griffith, D. J., Druker, B. J., Wait, C. L., Ott, K. A., & Zigler, A. J. 2000. Inhibition of c-kit receptor tyrosine kinase activity by STI 571, a selective tyrosine kinase inhibitor. *Blood*, 96(3): 925-932.

Herbst, R. S., Ansari, R., Bustin, F., Flynn, P., Hart, L., Otterson, G. A., Vlahovic, G., Soh, C. H., O'Connor, P., & Hainsworth, J. 2011. Efficacy of bevacizumab plus erlotinib versus erlotinib alone in advanced non-small-cell lung cancer after failure of standard first-line chemotherapy (BeTa): a double-blind, placebo-controlled, phase 3 trial. *Lancet*, 377(9780): 1846-1854.

Herbst, R. S., Heymach, J. V., & Lippman, S. M. 2008. Lung cancer. *N Engl J Med*, 359(13): 1367-1380.

Ho, C., & Laskin, J. 2009. EGFR-directed therapies to treat non-small-cell lung cancer. *Expert Opin Investig Drugs*, 18(8): 1133-1145.

Hogg, J. C., Chu, F., Utokaparch, S., Woods, R., Elliott, W. M., Buzatu, L., Cherniack, R. M., Rogers, R. M., Sciurba, F. C., Coxson, H. O., & Pare, P. D. 2004. The nature of small-airway obstruction in chronic obstructive pulmonary disease. *N Engl J Med*, 350(26): 2645-2653.

Holgate, S. T., & Polosa, R. 2006. The mechanisms, diagnosis, and management of severe asthma in adults. *Lancet*, 368(9537): 780-793.

Huang, M., Liu, X., Du, Q., Yao, X., & Yin, K. S. 2009. Inhibitory effects of sunitinib on ovalbumin-induced chronic experimental asthma in mice. *Chin Med J (Engl)*, 122(9): 1061-1066.

Hubbard, S. R., & Miller, W. T. 2007. Receptor tyrosine kinases: mechanisms of activation and signaling. *Curr Opin Cell Biol*, 19(2): 117-123.

Humbert, M., de Blay, F., Garcia, G., Prud'homme, A., Leroyer, C., Magnan, A., Tunon-de-Lara, J. M., Pison, C., Aubier, M., Charpin, D., Vachier, I., Purohit, A., Gineste, P., Bader, T., Moussy, A., Hermine, O., & Chanez, P. 2009. Masitinib, a c-kit/PDGF receptor tyrosine kinase inhibitor, improves disease control in severe corticosteroid-dependent asthmatics. *Allergy*, 64(8): 1194-1201.

Hur, G. Y., Lee, S. Y., Lee, S. H., Kim, S. J., Lee, K. J., Jung, J. Y., Lee, E. J., Kang, E. H., Jung, K. H., Kim, J. H., Shin, C., Shim, J. J., In, K. H., Kang, K. H., & Yoo, S. H. 2007. Potential use of an anticancer drug gefinitib, an EGFR inhibitor, on allergic airway inflammation. *Exp Mol Med*, 39(3): 367-375.

Inukai, M., Toyooka, S., Ito, S., Asano, H., Ichihara, S., Soh, J., Suehisa, H., Ouchida, M., Aoe, K., Aoe, M., Kiura, K., Shimizu, N., & Date, H. 2006. Presence of epidermal growth factor receptor gene T790M mutation as a minor clone in non-small cell lung cancer. *Cancer Res*, 66(16): 7854-7858.

Ishii, Y., Fujimoto, S., & Fukuda, T. 2006. Gefitinib prevents bleomycin-induced lung fibrosis in mice. *Am J Respir Crit Care Med*, 174(5): 550-556.

Iwai, K., Mori, T., Yamada, N., Yamaguchi, M., & Hosoda, Y. 1994. Idiopathic pulmonary fibrosis. Epidemiologic approaches to occupational exposure. *American Journal of Respiratory and Critical Care Medicine*, 150(3): 670-675.

Iyer, A. S., Morales, J. L., Huang, W., Ojo, F., Ning, G., Wills, E., Baines, J. D., & August, A. 2011. Absence of Tec family kinases interleukin-2 inducible T cell kinase (Itk) and Bruton's tyrosine kinase (Btk) severely impairs Fc epsilonRI-dependent mast cell responses. *J Biol Chem*, 286(11): 9503-9513.

Jemal, A., Siegel, R., Xu, J., & Ward, E. 2010. Cancer statistics, 2010. *CA Cancer J Clin*, 60(5): 277-300.

Johnston, J. A., Kawamura, M., Kirken, R. A., Chen, Y. Q., Blake, T. B., Shibuya, K., Ortaldo, J. R., McVicar, D. W., & O'Shea, J. J. 1994. Phosphorylation and activation of the Jak-3 Janus kinase in response to interleukin-2. *Nature*, 370(6485): 151-153.

Karnitz, L., Sutor, S. L., Torigoe, T., Reed, J. C., Bell, M. P., McKean, D. J., Leibson, P. J., & Abraham, R. T. 1992. Effects of p56lck deficiency on the growth and cytolytic effector function of an interleukin-2-dependent cytotoxic T-cell line. *Mol Cell Biol*, 12(10): 4521-4530.

Kawakami, T., & Galli, S. J. 2002. Regulation of mast-cell and basophil function and survival by IgE. *Nature reviews. Immunology*, 2(10): 773-786.

Kawano, O., Sasaki, H., Endo, K., Suzuki, E., Haneda, H., Yukiue, H., Kobayashi, Y., Yano, M., & Fujii, Y. 2006. PIK3CA mutation status in Japanese lung cancer patients. *Lung Cancer*, 54(2): 209-215.

Kepley, C. L., Wilson, B. S., & Oliver, J. M. 1998. Identification of the Fc epsilonRI-activated tyrosine kinases Lyn, Syk, and Zap-70 in human basophils. *The Journal of Allergy and Clinical Immunology*, 102(2): 304-315.

Kim, S., Lewis, C., & Nadel, J. A. 2011. CCL20/CCR6 feedback exaggerates epidermal growth factor receptor-dependent MUC5AC mucin production in human airway epithelial (NCI-H292) cells. *J Immunol*, 186(6): 3392-3400.

King, T. E., Jr., Pardo, A., & Selman, M. 2011. Idiopathic pulmonary fibrosis. *Lancet*.

Kobayashi, S., Boggon, T. J., Dayaram, T., Janne, P. A., Kocher, O., Meyerson, M., Johnson, B. E., Eck, M. J., Tenen, D. G., & Halmos, B. 2005. EGFR mutation and resistance of non-small-cell lung cancer to gefitinib. *N Engl J Med*, 352(8): 786-792.

Krishnamoorthy, N., Oriss, T. B., Paglia, M., Fei, M., Yarlagadda, M., Vanhaesebroeck, B., Ray, A., & Ray, P. 2008. Activation of c-Kit in dendritic cells regulates T helper cell differentiation and allergic asthma. *Nat Med*, 14(5): 565-573.

Kudlacz, E., Conklyn, M., Andresen, C., Whitney-Pickett, C., & Changelian, P. 2008. The JAK-3 inhibitor CP-690550 is a potent anti-inflammatory agent in a murine model of pulmonary eosinophilia. *Eur J Pharmacol*, 582(1-3): 154-161.

Laskin, J. J., & Sandler, A. B. 2004. Epidermal growth factor receptor: a promising target in solid tumours. *Cancer Treat Rev*, 30(1): 1-17.

Lee, C. G., Link, H., Baluk, P., Homer, R. J., Chapoval, S., Bhandari, V., Kang, M. J., Cohn, L., Kim, Y. K., McDonald, D. M., & Elias, J. A. 2004. Vascular endothelial growth factor (VEGF) induces remodeling and enhances TH2-mediated sensitization and inflammation in the lung. *Nat Med*, 10(10): 1095-1103.

Lewis, C. C., Chu, H. W., Westcott, J. Y., Tucker, A., Langmack, E. L., Sutherland, E. R., & Kraft, M. 2005. Airway fibroblasts exhibit a synthetic phenotype in severe asthma. *J Allergy Clin Immunol*, 115(3): 534-540.

Ley, B., Collard, H. R., & King, T. E., Jr. 2011. Clinical course and prediction of survival in idiopathic pulmonary fibrosis. *Am J Respir Crit Care Med*, 183(4): 431-440.

Li, D., Ambrogio, L., Shimamura, T., Kubo, S., Takahashi, M., Chirieac, L. R., Padera, R. F., Shapiro, G. I., Baum, A., Himmelsbach, F., Rettig, W. J., Meyerson, M., Solca, F., Greulich, H., & Wong, K. K. 2008. BIBW2992, an irreversible EGFR/HER2 inhibitor highly effective in preclinical lung cancer models. *Oncogene*, 27(34): 4702-4711.

Li, Z. W., Rickert, R. C., & Karin, M. 2004. Genetic dissection of antigen receptor induced-NF-kappaB activation. *Mol Immunol*, 41(6-7): 701-714.

Lin, T. A., McIntyre, K. W., Das, J., Liu, C., O'Day, K. D., Penhallow, B., Hung, C. Y., Whitney, G. S., Shuster, D. J., Yang, X., Townsend, R., Postelnek, J., Spergel, S. H., Lin, J., Moquin, R. V., Furch, J. A., Kamath, A. V., Zhang, H., Marathe, P. H., Perez-Villar, J. J., Doweyko, A., Killar, L., Dodd, J. H., Barrish, J. C., Wityak, J., & Kanner, S. B. 2004. Selective Itk inhibitors block T-cell activation and murine lung inflammation. *Biochemistry*, 43(34): 11056-11062.

Linardou, H., Dahabreh, I. J., Kanaloupiti, D., Siannis, F., Bafaloukos, D., Kosmidis, P., Papadimitriou, C. A., & Murray, S. 2008. Assessment of somatic k-RAS mutations as a mechanism associated with resistance to EGFR-targeted agents: a systematic review and meta-analysis of studies in advanced non-small-cell lung cancer and metastatic colorectal cancer. *Lancet Oncol*, 9(10): 962-972.

Liu, K., Gualano, R. C., Hibbs, M. L., Anderson, G. P., & Bozinovski, S. 2008. Epidermal growth factor receptor signaling to Erk1/2 and STATs control the intensity of the epithelial inflammatory responses to rhinovirus infection. *J Biol Chem*, 283(15): 9977-9985.

Lowell, C. A. 2011. Src-family and Syk kinases in activating and inhibitory pathways in innate immune cells: signaling cross talk. *Cold Spring Harb Perspect Biol*, 3(3).

Lynch, O. T., Giembycz, M. A., Daniels, I., Barnes, P. J., & Lindsay, M. A. 2000. Pleiotropic role of lyn kinase in leukotriene B(4)-induced eosinophil activation. *Blood*, 95(11): 3541-3547.

Macchi, P., Villa, A., Giliani, S., Sacco, M. G., Frattini, A., Porta, F., Ugazio, A. G., Johnston, J. A., Candotti, F., O'Shea, J. J., & et al. 1995. Mutations of Jak-3 gene in patients with autosomal severe combined immune deficiency (SCID). *Nature*, 377(6544): 65-68.

MacGlashan, D., Jr., Honigberg, L. A., Smith, A., Buggy, J., & Schroeder, J. T. 2011. Inhibition of IgE-mediated secretion from human basophils with a highly selective Bruton's tyrosine kinase, Btk, inhibitor. *Int Immunopharmacol*, 11(4): 475-479.

Maheswaran, S., Sequist, L. V., Nagrath, S., Ulkus, L., Brannigan, B., Collura, C. V., Inserra, E., Diederichs, S., Iafrate, A. J., Bell, D. W., Digumarthy, S., Muzikansky, A., Irimia, D., Settleman, J., Tompkins, R. G., Lynch, T. J., Toner, M., & Haber, D. A. 2008. Detection of mutations in EGFR in circulating lung-cancer cells. *N Engl J Med*, 359(4): 366-377.

Makris, D., Scherpereel, A., Copin, M. C., Colin, G., Brun, L., Lafitte, J. J., & Marquette, C. H. 2007. Fatal interstitial lung disease associated with oral erlotinib therapy for lung cancer. *BMC Cancer*, 7: 150.

Malaviya, R., Chen, C. L., Navara, C., Liu, X. P., Keenan, M., Waurzyniak, B., & Uckun, F. M. 2000. Treatment of allergic asthma by targeting janus kinase 3-dependent leukotriene synthesis in mast cells with 4-(3', 5'-dibromo-4'-hydroxyphenyl)amino-6,7-dimethoxyquinazoline (WHI-P97). *J Pharmacol Exp Ther*, 295(3): 912-926.

Malaviya, R., & Laskin, D. L. 2010. Janus kinase-3 dependent inflammatory responses in allergic asthma. *Int Immunopharmacol*, 10(8): 829-836.

Malaviya, R., Zhu, D., Dibirdik, I., & Uckun, F. M. 1999. Targeting Janus kinase 3 in mast cells prevents immediate hypersensitivity reactions and anaphylaxis. *J Biol Chem*, 274(38): 27028-27038.

Manning, G., Whyte, D. B., Martinez, R., Hunter, T., & Sudarsanam, S. 2002. The protein kinase complement of the human genome. *Science*, 298(5600): 1912-1934.

Mannino, D. M., & Buist, A. S. 2007. Global burden of COPD: risk factors, prevalence, and future trends. *Lancet*, 370(9589): 765-773.

Marwick, J. A., & Chung, K. F. 2010. Glucocorticoid insensitivity as a future target of therapy for chronic obstructive pulmonary disease. *Int J Chron Obstruct Pulmon Dis*, 5: 297-309.

Masoli, M., Fabian, D., Holt, S., & Beasley, R. 2004. The global burden of asthma: executive summary of the GINA Dissemination Committee report. *Allergy*, 59(5): 469-478.

Matsubara, S., Li, G., Takeda, K., Loader, J. E., Pine, P., Masuda, E. S., Miyahara, N., Miyahara, S., Lucas, J. J., Dakhama, A., & Gelfand, E. W. 2006. Inhibition of spleen tyrosine kinase prevents mast cell activation and airway hyperresponsiveness. *Am J Respir Crit Care Med*, 173(1): 56-63.

Matsunaga, Y., Inoue, H., Fukuyama, S., Yoshida, H., Moriwaki, A., Matsumoto, T., Matsumoto, K., Asai, Y., Kubo, M., Yoshimura, A., & Nakanishi, Y. 2011. Effects of a Janus kinase inhibitor, pyridone 6, on airway responses in a murine model of asthma. *Biochem Biophys Res Commun*, 404(1): 261-267.

Meltzer, E. O., Berkowitz, R. B., & Grossbard, E. B. 2005. An intranasal Syk-kinase inhibitor (R112) improves the symptoms of seasonal allergic rhinitis in a park environment. *J Allergy Clin Immunol*, 115(4): 791-796.

Milano, A., De Rosa, V., Iaffaioli, R. V., & Caponigro, F. 2007. Downstream intracellular effectors of epidermal growth factor receptor as targets for anticancer therapy. *Expert Opin Ther Targets*, 11(6): 771-782.

Miller, A. T., & Berg, L. J. 2002. New insights into the regulation and functions of Tec family tyrosine kinases in the immune system. *Curr Opin Immunol*, 14(3): 331-340.

Mok, T. S., Wu, Y. L., Thongprasert, S., Yang, C. H., Chu, D. T., Saijo, N., Sunpaweravong, P., Han, B., Margono, B., Ichinose, Y., Nishiwaki, Y., Ohe, Y., Yang, J. J., Chewaskulyong, B., Jiang, H., Duffield, E. L., Watkins, C. L., Armour, A. A., & Fukuoka, M. 2009. Gefitinib or carboplatin-paclitaxel in pulmonary adenocarcinoma. *N Engl J Med*, 361(10): 947-957.

Molina, T. J., Kishihara, K., Siderovski, D. P., van Ewijk, W., Narendran, A., Timms, E., Wakeham, A., Paige, C. J., Hartmann, K. U., Veillette, A., & et al. 1992. Profound block in thymocyte development in mice lacking p56lck. *Nature*, 357(6374): 161-164.

Mueller, C., & August, A. 2003. Attenuation of immunological symptoms of allergic asthma in mice lacking the tyrosine kinase ITK. *J Immunol*, 170(10): 5056-5063.

Muller, M., Briscoe, J., Laxton, C., Guschin, D., Ziemiecki, A., Silvennoinen, O., Harpur, A. G., Barbieri, G., Witthuhn, B. A., Schindler, C., & et al. 1993. The protein tyrosine kinase JAK1 complements defects in interferon-alpha/beta and -gamma signal transduction. *Nature*, 366(6451): 129-135.

Nakachi, I., Naoki, K., Soejima, K., Kawada, I., Watanabe, H., Yasuda, H., Nakayama, S., Yoda, S., Satomi, R., Ikemura, S., Terai, H., Sato, T., & Ishizaka, A. 2010. The combination of multiple receptor tyrosine kinase inhibitor and mammalian target of rapamycin inhibitor overcomes erlotinib resistance in lung cancer cell lines through c-Met inhibition. *Mol Cancer Res*, 8(8): 1142-1151.

Naumov, G. N., Nilsson, M. B., Cascone, T., Briggs, A., Straume, O., Akslen, L. A., Lifshits, E., Byers, L. A., Xu, L., Wu, H. K., Janne, P., Kobayashi, S., Halmos, B., Tenen, D., Tang, X. M., Engelman, J., Yeap, B., Folkman, J., Johnson, B. E., & Heymach, J. V. 2009. Combined vascular endothelial growth factor receptor and epidermal growth factor receptor (EGFR) blockade inhibits tumor growth in xenograft models of EGFR inhibitor resistance. *Clin Cancer Res*, 15(10): 3484-3494.

Nel, A. E. 2002. T-cell activation through the antigen receptor. Part 1: signaling components, signaling pathways, and signal integration at the T-cell antigen receptor synapse. *J Allergy Clin Immunol*, 109(5): 758-770.

Noble, M. E., Endicott, J. A., & Johnson, L. N. 2004. Protein kinase inhibitors: insights into drug design from structure. *Science*, 303(5665): 1800-1805.

Nogee, L. M., Dunbar, A. E., 3rd, Wert, S., Askin, F., Hamvas, A., & Whitsett, J. A. 2002. Mutations in the surfactant protein C gene associated with interstitial lung disease. *Chest*, 121(3 Suppl): 20S-21S.

Norman, P. 2009. A novel Syk kinase inhibitor suitable for inhalation: R-343(?)--WO-2009031011. *Expert Opin Ther Pat*, 19(10): 1469-1472.

Novak, N., Kraft, S., & Bieber, T. 2001. IgE receptors. *Curr Opin Immunol*, 13(6): 721-726.

Odaka, M., Kohda, D., Lax, I., Schlessinger, J., & Inagaki, F. 1997. Ligand-binding enhances the affinity of dimerization of the extracellular domain of the epidermal growth factor receptor. *J Biochem*, 122(1): 116-121.

Ogiso, H., Ishitani, R., Nureki, O., Fukai, S., Yamanaka, M., Kim, J. H., Saito, K., Sakamoto, A., Inoue, M., Shirouzu, M., & Yokoyama, S. 2002. Crystal structure of the complex of human epidermal growth factor and receptor extracellular domains. *Cell*, 110(6): 775-787.

Ou, S. H., Kwak, E. L., Siwak-Tapp, C., Dy, J., Bergethon, K., Clark, J. W., Camidge, D. R., Solomon, B. J., Maki, R. G., Bang, Y. J., Kim, D. W., Christensen, J., Tan, W., Wilner, K. D., Salgia, R., & Iafrate, A. J. 2011. Activity of crizotinib (PF02341066), a dual mesenchymal-epithelial transition (MET) and anaplastic lymphoma kinase (ALK) inhibitor, in a non-small cell lung cancer patient with de novo MET amplification. *J Thorac Oncol*, 6(5): 942-946.

Paniagua, R. T., Sharpe, O., Ho, P. P., Chan, S. M., Chang, A., Higgins, J. P., Tomooka, B. H., Thomas, F. M., Song, J. J., Goodman, S. B., Lee, D. M., Genovese, M. C., Utz, P. J., Steinman, L., & Robinson, W. H. 2006. Selective tyrosine kinase inhibition by imatinib mesylate for the treatment of autoimmune arthritis. *J Clin Invest*, 116(10): 2633-2642.

Pao, W., Miller, V. A., Politi, K. A., Riely, G. J., Somwar, R., Zakowski, M. F., Kris, M. G., & Varmus, H. 2005a. Acquired resistance of lung adenocarcinomas to gefitinib or erlotinib is associated with a second mutation in the EGFR kinase domain. *PLoS Med*, 2(3): e73.

Pao, W., Wang, T. Y., Riely, G. J., Miller, V. A., Pan, Q., Ladanyi, M., Zakowski, M. F., Heelan, R. T., Kris, M. G., & Varmus, H. E. 2005b. KRAS mutations and primary resistance of lung adenocarcinomas to gefitinib or erlotinib. *PLoS Med*, 2(1): e17.

Parganas, E., Wang, D., Stravopodis, D., Topham, D. J., Marine, J. C., Teglund, S., Vanin, E. F., Bodner, S., Colamonici, O. R., van Deursen, J. M., Grosveld, G., & Ihle, J. N. 1998. Jak2 is essential for signaling through a variety of cytokine receptors. *Cell*, 93(3): 385-395.

Parham, C., Chirica, M., Timans, J., Vaisberg, E., Travis, M., Cheung, J., Pflanz, S., Zhang, R., Singh, K. P., Vega, F., To, W., Wagner, J., O'Farrell, A. M., McClanahan, T., Zurawski, S., Hannum, C., Gorman, D., Rennick, D. M., Kastelein, R. A., de Waal

Malefyt, R., & Moore, K. W. 2002. A receptor for the heterodimeric cytokine IL-23 is composed of IL-12Rbeta1 and a novel cytokine receptor subunit, IL-23R. *J Immunol*, 168(11): 5699-5708.
Parsons, S. J., & Parsons, J. T. 2004. Src family kinases, key regulators of signal transduction. *Oncogene*, 23(48): 7906-7909.
Perera, S. A., Li, D., Shimamura, T., Raso, M. G., Ji, H., Chen, L., Borgman, C. L., Zaghlul, S., Brandstetter, K. A., Kubo, S., Takahashi, M., Chirieac, L. R., Padera, R. F., Bronson, R. T., Shapiro, G. I., Greulich, H., Meyerson, M., Guertler, U., Chesa, P. G., Solca, F., Wistuba, II, & Wong, K. K. 2009. HER2YVMA drives rapid development of adenosquamous lung tumors in mice that are sensitive to BIBW2992 and rapamycin combination therapy. *Proc Natl Acad Sci U S A*, 106(2): 474-479.
Pernis, A. B., & Rothman, P. B. 2002. JAK-STAT signaling in asthma. *J Clin Invest*, 109(10): 1279-1283.
Pesu, M., Laurence, A., Kishore, N., Zwillich, S. H., Chan, G., & O'Shea, J. J. 2008. Therapeutic targeting of Janus kinases. *Immunol Rev*, 223: 132-142.
Picard, C., Dogniaux, S., Chemin, K., Maciorowski, Z., Lim, A., Mazerolles, F., Rieux-Laucat, F., Stolzenberg, M. C., Debre, M., Magny, J. P., Le Deist, F., Fischer, A., & Hivroz, C. 2009. Hypomorphic mutation of ZAP70 in human results in a late onset immunodeficiency and no autoimmunity. *Eur J Immunol*, 39(7): 1966-1976.
Pitcher, L. A., & van Oers, N. S. 2003. T-cell receptor signal transmission: who gives an ITAM? *Trends Immunol*, 24(10): 554-560.
Program, N. A. E. a. P. 2007. *Expert Panel Report 3: Guidelines for the Diagnosis and Managment of Asthma*: National Heart, Lung, and Blood Institute, National Institutes of Health, U.S. Department of Health and Human Services.
Prudkin, L., Tang, X., & Wistuba, II. 2009. Germ-line and somatic presentations of the EGFR T790M mutation in lung cancer. *J Thorac Oncol*, 4(1): 139-141.
Qi, Q., Xia, M., Hu, J., Hicks, E., Iyer, A., Xiong, N., & August, A. 2009. Enhanced development of CD4+ gammadelta T cells in the absence of Itk results in elevated IgE production. *Blood*, 114(3): 564-571.
Quesnelle, K. M., Boehm, A. L., & Grandis, J. R. 2007. STAT-mediated EGFR signaling in cancer. *J Cell Biochem*, 102(2): 311-319.
Raghu, G., Yang, S. T., Spada, C., Hayes, J., & Pellegrini, C. A. 2006. Sole treatment of acid gastroesophageal reflux in idiopathic pulmonary fibrosis: a case series. *Chest*, 129(3): 794-800.
Ratushny, V., Astsaturov, I., Burtness, B. A., Golemis, E. A., & Silverman, J. S. 2009. Targeting EGFR resistance networks in head and neck cancer. *Cell Signal*, 21(8): 1255-1268.
Regales, L., Gong, Y., Shen, R., de Stanchina, E., Vivanco, I., Goel, A., Koutcher, J. A., Spassova, M., Ouerfelli, O., Mellinghoff, I. K., Zakowski, M. F., Politi, K. A., & Pao, W. 2009. Dual targeting of EGFR can overcome a major drug resistance mutation in mouse models of EGFR mutant lung cancer. *J Clin Invest*, 119(10): 3000-3010.
Richeldi, L., Costabel, U., Selman, M., Kim, D. S., Hansell, D. M., Nicholson, A. G., Brown, K. K., Flaherty, K. R., Noble, P. W., Raghu, G., Brun, M., Gupta, A., Juhel, N., Kluglich, M., & du Bois, R. M. 2011. Efficacy of a tyrosine kinase inhibitor in idiopathic pulmonary fibrosis. *N Engl J Med*, 365(12): 1079-1087.
Richter, A., O'Donnell, R. A., Powell, R. M., Sanders, M. W., Holgate, S. T., Djukanovic, R., & Davies, D. E. 2002. Autocrine ligands for the epidermal growth factor receptor

mediate interleukin-8 release from bronchial epithelial cells in response to cigarette smoke. *Am J Respir Cell Mol Biol*, 27(1): 85-90.

Robinson, D. R., Wu, Y. M., & Lin, S. F. 2000. The protein tyrosine kinase family of the human genome. *Oncogene*, 19(49): 5548-5557.

Rogers, D. F. 2004. Airway mucus hypersecretion in asthma: an undervalued pathology? *Curr Opin Pharmacol*, 4(3): 241-250.

Rogers, D. F. 2007. Physiology of airway mucus secretion and pathophysiology of hypersecretion. *Respir Care*, 52(9): 1134-1146; discussion 1146-1139.

Russell, S. M., Tayebi, N., Nakajima, H., Riedy, M. C., Roberts, J. L., Aman, M. J., Migone, T. S., Noguchi, M., Markert, M. L., Buckley, R. H., O'Shea, J. J., & Leonard, W. J. 1995. Mutation of Jak3 in a patient with SCID: essential role of Jak3 in lymphoid development. *Science*, 270(5237): 797-800.

Saetta, M., Turato, G., Maestrelli, P., Mapp, C. E., & Fabbri, L. M. 2001. Cellular and structural bases of chronic obstructive pulmonary disease. *Am J Respir Crit Care Med*, 163(6): 1304-1309.

Sakaguchi, N., Takahashi, T., Hata, H., Nomura, T., Tagami, T., Yamazaki, S., Sakihama, T., Matsutani, T., Negishi, I., Nakatsuru, S., & Sakaguchi, S. 2003. Altered thymic T-cell selection due to a mutation of the ZAP-70 gene causes autoimmune arthritis in mice. *Nature*, 426(6965): 454-460.

Sakai, H., Nishimura, A., Watanabe, Y., Nishizawa, Y., Hashimoto, Y., Chiba, Y., & Misawa, M. 2010. Involvement of Src family kinase activation in angiotensin II-induced hyperresponsiveness of rat bronchial smooth muscle. *Peptides*, 31(12): 2216-2221.

Samelson, L. E. 2002. Signal transduction mediated by the T cell antigen receptor: the role of adapter proteins. *Annual Review of Immunology*, 20: 371-394.

Schamel, W. W., & Reth, M. 2000. Monomeric and oligomeric complexes of the B cell antigen receptor. *Immunity*, 13(1): 5-14.

Schindler, C. W. 2002. Series introduction. JAK-STAT signaling in human disease. *J Clin Invest*, 109(9): 1133-1137.

Schlessinger, J. 2000. Cell signaling by receptor tyrosine kinases. *Cell*, 103(2): 211-225.

Seibold, M. A., Wise, A. L., Speer, M. C., Steele, M. P., Brown, K. K., Loyd, J. E., Fingerlin, T. E., Zhang, W., Gudmundsson, G., Groshong, S. D., Evans, C. M., Garantziotis, S., Adler, K. B., Dickey, B. F., du Bois, R. M., Yang, I. V., Herron, A., Kervitsky, D., Talbert, J. L., Markin, C., Park, J., Crews, A. L., Slifer, S. H., Auerbach, S., Roy, M. G., Lin, J., Hennessy, C. E., Schwarz, M. I., & Schwartz, D. A. 2011. A common MUC5B promoter polymorphism and pulmonary fibrosis. *N Engl J Med*, 364(16): 1503-1512.

Seow, C. J., Chue, S. C., & Wong, W. S. 2002. Piceatannol, a Syk-selective tyrosine kinase inhibitor, attenuated antigen challenge of guinea pig airways in vitro. *Eur J Pharmacol*, 443(1-3): 189-196.

Shah, N. T., Kris, M. G., Pao, W., Tyson, L. B., Pizzo, B. M., Heinemann, M. H., Ben-Porat, L., Sachs, D. L., Heelan, R. T., & Miller, V. A. 2005. Practical management of patients with non-small-cell lung cancer treated with gefitinib. *J Clin Oncol*, 23(1): 165-174.

Shao, M. X., & Nadel, J. A. 2005. Dual oxidase 1-dependent MUC5AC mucin expression in cultured human airway epithelial cells. *Proc Natl Acad Sci U S A*, 102(3): 767-772.

Shepherd, F. A., Rodrigues Pereira, J., Ciuleanu, T., Tan, E. H., Hirsh, V., Thongprasert, S., Campos, D., Maoleekoonpiroj, S., Smylie, M., Martins, R., van Kooten, M., Dediu, M., Findlay, B., Tu, D., Johnston, D., Bezjak, A., Clark, G., Santabarbara, P., &

Seymour, L. 2005. Erlotinib in previously treated non-small-cell lung cancer. *N Engl J Med*, 353(2): 123-132.

Sibilia, M., Kroismayr, R., Lichtenberger, B. M., Natarajan, A., Hecking, M., & Holcmann, M. 2007. The epidermal growth factor receptor: from development to tumorigenesis. *Differentiation*, 75(9): 770-787.

Silva, C. M. 2004. Role of STATs as downstream signal transducers in Src family kinase-mediated tumorigenesis. *Oncogene*, 23(48): 8017-8023.

Soda, M., Choi, Y. L., Enomoto, M., Takada, S., Yamashita, Y., Ishikawa, S., Fujiwara, S., Watanabe, H., Kurashina, K., Hatanaka, H., Bando, M., Ohno, S., Ishikawa, Y., Aburatani, H., Niki, T., Sohara, Y., Sugiyama, Y., & Mano, H. 2007. Identification of the transforming EML4-ALK fusion gene in non-small-cell lung cancer. *Nature*, 448(7153): 561-566.

Soh, J., Okumura, N., Lockwood, W. W., Yamamoto, H., Shigematsu, H., Zhang, W., Chari, R., Shames, D. S., Tang, X., MacAulay, C., Varella-Garcia, M., Vooder, T., Wistuba, II, Lam, S., Brekken, R., Toyooka, S., Minna, J. D., Lam, W. L., & Gazdar, A. F. 2009. Oncogene mutations, copy number gains and mutant allele specific imbalance (MASI) frequently occur together in tumor cells. *PLoS One*, 4(10): e7464.

Sos, M. L., Rode, H. B., Heynck, S., Peifer, M., Fischer, F., Kluter, S., Pawar, V. G., Reuter, C., Heuckmann, J. M., Weiss, J., Ruddigkeit, L., Rabiller, M., Koker, M., Simard, J. R., Getlik, M., Yuza, Y., Chen, T. H., Greulich, H., Thomas, R. K., & Rauh, D. 2010. Chemogenomic profiling provides insights into the limited activity of irreversible EGFR Inhibitors in tumor cells expressing the T790M EGFR resistance mutation. *Cancer Res*, 70(3): 868-874.

Spigel, D. R., Burris, H. A., 3rd, Greco, F. A., Shipley, D. L., Friedman, E. K., Waterhouse, D. M., Whorf, R. C., Mitchell, R. B., Daniel, D. B., Zangmeister, J., Bass, J. D., & Hainsworth, J. D. 2011. Randomized, double-blind, placebo-controlled, phase II trial of sorafenib and erlotinib or erlotinib alone in previously treated advanced non-small-cell lung cancer. *J Clin Oncol*, 29(18): 2582-2589.

Steele, M. P., Speer, M. C., Loyd, J. E., Brown, K. K., Herron, A., Slifer, S. H., Burch, L. H., Wahidi, M. M., Phillips, J. A., 3rd, Sporn, T. A., McAdams, H. P., Schwarz, M. I., & Schwartz, D. A. 2005. Clinical and pathologic features of familial interstitial pneumonia. *American Journal of Respiratory and Critical Care Medicine*, 172(9): 1146-1152.

Stenton, G. R., Ulanova, M., Dery, R. E., Merani, S., Kim, M. K., Gilchrist, M., Puttagunta, L., Musat-Marcu, S., James, D., Schreiber, A. D., & Befus, A. D. 2002. Inhibition of allergic inflammation in the airways using aerosolized antisense to Syk kinase. *J Immunol*, 169(2): 1028-1036.

Suzuki, M., Wada, H., Yoshino, M., Tian, L., Shigematsu, H., Suzuki, H., Alaa, M., Tamura, H., Fujiwara, T., Nagato, K., Motohashi, S., Moriya, Y., Hoshino, H., Yoshida, S., Shibuya, K., Hiroshima, K., Nakatani, Y., & Yoshino, I. 2010. Molecular characterization of chronic obstructive pulmonary disease-related non-small cell lung cancer through aberrant methylation and alterations of EGFR signaling. *Ann Surg Oncol*, 17(3): 878-888.

Takeda, K., Tanaka, T., Shi, W., Matsumoto, M., Minami, M., Kashiwamura, S., Nakanishi, K., Yoshida, N., Kishimoto, T., & Akira, S. 1996. Essential role of Stat6 in IL-4 signalling. *Nature*, 380(6575): 627-630.

Takeyama, K., Jung, B., Shim, J. J., Burgel, P. R., Dao-Pick, T., Ueki, I. F., Protin, U., Kroschel, P., & Nadel, J. A. 2001. Activation of epidermal growth factor receptors is responsible for mucin synthesis induced by cigarette smoke. *Am J Physiol Lung Cell Mol Physiol*, 280(1): L165-172.

Takezawa, K., Okamoto, I., Yonesaka, K., Hatashita, E., Yamada, Y., Fukuoka, M., & Nakagawa, K. 2009. Sorafenib inhibits non-small cell lung cancer cell growth by targeting B-RAF in KRAS wild-type cells and C-RAF in KRAS mutant cells. *Cancer Res*, 69(16): 6515-6521.

Tamaoka, M., Hassan, M., McGovern, T., Ramos-Barbon, D., Jo, T., Yoshizawa, Y., Tolloczko, B., Hamid, Q., & Martin, J. G. 2008. The epidermal growth factor receptor mediates allergic airway remodelling in the rat. *Eur Respir J*, 32(5): 1213-1223.

Thatcher, N., Chang, A., Parikh, P., Rodrigues Pereira, J., Ciuleanu, T., von Pawel, J., Thongprasert, S., Tan, E. H., Pemberton, K., Archer, V., & Carroll, K. 2005. Gefitinib plus best supportive care in previously treated patients with refractory advanced non-small-cell lung cancer: results from a randomised, placebo-controlled, multicentre study (Iressa Survival Evaluation in Lung Cancer). *Lancet*, 366(9496): 1527-1537.

Thomas, A. Q., Lane, K., Phillips, J., 3rd, Prince, M., Markin, C., Speer, M., Schwartz, D. A., Gaddipati, R., Marney, A., Johnson, J., Roberts, R., Haines, J., Stahlman, M., & Loyd, J. E. 2002. Heterozygosity for a surfactant protein C gene mutation associated with usual interstitial pneumonitis and cellular nonspecific interstitial pneumonitis in one kindred. *American Journal of Respiratory and Critical Care Medicine*, 165(9): 1322-1328.

Thomas, S. M., & Brugge, J. S. 1997. Cellular functions regulated by Src family kinases. *Annu Rev Cell Dev Biol*, 13: 513-609.

Ticchioni, M., Charvet, C., Noraz, N., Lamy, L., Steinberg, M., Bernard, A., & Deckert, M. 2002. Signaling through ZAP-70 is required for CXCL12-mediated T-cell transendothelial migration. *Blood*, 99(9): 3111-3118.

Tobin, R. W., Pope, C. E., 2nd, Pellegrini, C. A., Emond, M. J., Sillery, J., & Raghu, G. 1998. Increased prevalence of gastroesophageal reflux in patients with idiopathic pulmonary fibrosis. *American Journal of Respiratory and Critical Care Medicine*, 158(6): 1804-1808.

Tsakiri, K. D., Cronkhite, J. T., Kuan, P. J., Xing, C., Raghu, G., Weissler, J. C., Rosenblatt, R. L., Shay, J. W., & Garcia, C. K. 2007. Adult-onset pulmonary fibrosis caused by mutations in telomerase. *Proc Natl Acad Sci U S A*, 104(18): 7552-7557.

Ueda, T., Ohta, K., Suzuki, N., Yamaguchi, M., Hirai, K., Horiuchi, T., Watanabe, J., Miyamoto, T., & Ito, K. 1992. Idiopathic pulmonary fibrosis and high prevalence of serum antibodies to hepatitis C virus. *The American Review of Respiratory Disease*, 146(1): 266-268.

Ulanova, M., Puttagunta, L., Marcet-Palacios, M., Duszyk, M., Steinhoff, U., Duta, F., Kim, M. K., Indik, Z. K., Schreiber, A. D., & Befus, A. D. 2005. Syk tyrosine kinase participates in beta1-integrin signaling and inflammatory responses in airway epithelial cells. *Am J Physiol Lung Cell Mol Physiol*, 288(3): L497-507.

Velankar, A. D., Quintini, G., Prabhu, A., Weber, A., Hunaeus, G., Voland, B., Wuest, M., Orjeda, C., Harel, D., Varghese, S., Gore, V., Patil, M., Gayke, D., Herdemann, M., Heit, I., & Zaliani, A. 2010. Synthesis and biological evaluation of novel (4 or 5-aryl)pyrazolyl-indoles as inhibitors of interleukin-2 inducible T-cell kinase (ITK). *Bioorg Med Chem*, 18(12): 4547-4559.

Vila-Coro, A. J., Rodriguez-Frade, J. M., Martin De Ana, A., Moreno-Ortiz, M. C., Martinez, A. C., & Mellado, M. 1999. The chemokine SDF-1alpha triggers CXCR4 receptor dimerization and activates the JAK/STAT pathway. *FASEB J*, 13(13): 1699-1710.

Vonakis, B. M., Gibbons, S. P., Jr., Rotte, M. J., Brothers, E. A., Kim, S. C., Chichester, K., & MacDonald, S. M. 2005. Regulation of rat basophilic leukemia-2H3 mast cell secretion by a constitutive Lyn kinase interaction with the high affinity IgE receptor (Fc epsilon RI). *J Immunol*, 175(7): 4543-4554.

Wang, P., Tian, Q., Liang, Z. X., Yang, Z., Xu, S. F., Sun, J. P., & Chen, L. A. 2010. Gefitinib attenuates murine pulmonary fibrosis induced by bleomycin. *Chin Med J (Engl)*, 123(16): 2259-2264.

Wang, Y., Kuan, P. J., Xing, C., Cronkhite, J. T., Torres, F., Rosenblatt, R. L., DiMaio, J. M., Kinch, L. N., Grishin, N. V., & Garcia, C. K. 2009. Genetic defects in surfactant protein A2 are associated with pulmonary fibrosis and lung cancer. *American Journal of Human Genetics*, 84(1): 52-59.

Weir, B. A., Woo, M. S., Getz, G., Perner, S., Ding, L., Beroukhim, R., Lin, W. M., Province, M. A., Kraja, A., Johnson, L. A., Shah, K., Sato, M., Thomas, R. K., Barletta, J. A., Borecki, I. B., Broderick, S., Chang, A. C., Chiang, D. Y., Chirieac, L. R., Cho, J., Fujii, Y., Gazdar, A. F., Giordano, T., Greulich, H., Hanna, M., Johnson, B. E., Kris, M. G., Lash, A., Lin, L., Lindeman, N., Mardis, E. R., McPherson, J. D., Minna, J. D., Morgan, M. B., Nadel, M., Orringer, M. B., Osborne, J. R., Ozenberger, B., Ramos, A. H., Robinson, J., Roth, J. A., Rusch, V., Sasaki, H., Shepherd, F., Sougnez, C., Spitz, M. R., Tsao, M. S., Twomey, D., Verhaak, R. G., Weinstock, G. M., Wheeler, D. A., Winckler, W., Yoshizawa, A., Yu, S., Zakowski, M. F., Zhang, Q., Beer, D. G., Wistuba, II, Watson, M. A., Garraway, L. A., Ladanyi, M., Travis, W. D., Pao, W., Rubin, M. A., Gabriel, S. B., Gibbs, R. A., Varmus, H. E., Wilson, R. K., Lander, E. S., & Meyerson, M. 2007. Characterizing the cancer genome in lung adenocarcinoma. *Nature*, 450(7171): 893-898.

Wenzel, S. E., & Busse, W. W. 2007. Severe asthma: lessons from the Severe Asthma Research Program. *J Allergy Clin Immunol*, 119(1): 14-21; quiz 22-13.

Wex, E., Bouyssou, T., Duechs, M. J., Erb, K. J., Gantner, F., Sanderson, M. P., Schnapp, A., Stierstorfer, B. E., & Wollin, L. 2011. Induced Syk deletion leads to suppressed allergic responses but has no effect on neutrophil or monocyte migration in vivo. *Eur J Immunol*.

Wheeler, D. L., Dunn, E. F., & Harari, P. M. 2010. Understanding resistance to EGFR inhibitors-impact on future treatment strategies. *Nat Rev Clin Oncol*, 7(9): 493-507.

Wills-Karp, M., Luyimbazi, J., Xu, X., Schofield, B., Neben, T. Y., Karp, C. L., & Donaldson, D. D. 1998. Interleukin-13: central mediator of allergic asthma. *Science*, 282(5397): 2258-2261.

Wong, M., & Fish, E. N. 1998. RANTES and MIP-1alpha activate stats in T cells. *J Biol Chem*, 273(1): 309-314.

Wong, M., Uddin, S., Majchrzak, B., Huynh, T., Proudfoot, A. E., Platanias, L. C., & Fish, E. N. 2001. Rantes activates Jak2 and Jak3 to regulate engagement of multiple signaling pathways in T cells. *J Biol Chem*, 276(14): 11427-11431.

Wong, W. S. 2005. Inhibitors of the tyrosine kinase signaling cascade for asthma. *Curr Opin Pharmacol*, 5(3): 264-271.

Woodruff, P. G., Wolff, M., Hohlfeld, J. M., Krug, N., Dransfield, M. T., Sutherland, E. R., Criner, G. J., Kim, V., Prasse, A., Nivens, M. C., Tetzlaff, K., Heilker, R., & Fahy, J.

V. 2010. Safety and efficacy of an inhaled epidermal growth factor receptor inhibitor (BIBW 2948 BS) in chronic obstructive pulmonary disease. *Am J Respir Crit Care Med*, 181(5): 438-445.

Wu, J. Y., Wu, S. G., Yang, C. H., Gow, C. H., Chang, Y. L., Yu, C. J., Shih, J. Y., & Yang, P. C. 2008. Lung cancer with epidermal growth factor receptor exon 20 mutations is associated with poor gefitinib treatment response. *Clin Cancer Res*, 14(15): 4877-4882.

Yamamoto, N., Takeshita, K., Shichijo, M., Kokubo, T., Sato, M., Nakashima, K., Ishimori, M., Nagai, H., Li, Y. F., Yura, T., & Bacon, K. B. 2003. The orally available spleen tyrosine kinase inhibitor 2-[7-(3,4-dimethoxyphenyl)-imidazo[1,2-c]pyrimidin-5-ylamino]nicotinamide dihydrochloride (BAY 61-3606) blocks antigen-induced airway inflammation in rodents. *J Pharmacol Exp Ther*, 306(3): 1174-1181.

Yanagi, S., Inatome, R., Takano, T., & Yamamura, H. 2001. Syk expression and novel function in a wide variety of tissues. *Biochem Biophys Res Commun*, 288(3): 495-498.

Yap, T. A., Vidal, L., Adam, J., Stephens, P., Spicer, J., Shaw, H., Ang, J., Temple, G., Bell, S., Shahidi, M., Uttenreuther-Fischer, M., Stopfer, P., Futreal, A., Calvert, H., de Bono, J. S., & Plummer, R. 2010. Phase I trial of the irreversible EGFR and HER2 kinase inhibitor BIBW 2992 in patients with advanced solid tumors. *Journal of Clinical Oncology: Official Journal of the American Society of Clinical Oncology*, 28(25): 3965-3972.

Yu, M., Tsai, M., Tam, S. Y., Jones, C., Zehnder, J., & Galli, S. J. 2006. Mast cells can promote the development of multiple features of chronic asthma in mice. *J Clin Invest*, 116(6): 1633-1641.

Yun, C. H., Boggon, T. J., Li, Y., Woo, M. S., Greulich, H., Meyerson, M., & Eck, M. J. 2007. Structures of lung cancer-derived EGFR mutants and inhibitor complexes: mechanism of activation and insights into differential inhibitor sensitivity. *Cancer Cell*, 11(3): 217-227.

Yun, C. H., Mengwasser, K. E., Toms, A. V., Woo, M. S., Greulich, H., Wong, K. K., Meyerson, M., & Eck, M. J. 2008. The T790M mutation in EGFR kinase causes drug resistance by increasing the affinity for ATP. *Proc Natl Acad Sci U S A*, 105(6): 2070-2075.

Zhang, X., & Chang, A. 2008. Molecular predictors of EGFR-TKI sensitivity in advanced non-small cell lung cancer. *Int J Med Sci*, 5(4): 209-217.

9

Myotonic Dystrophy Protein Kinase: Structure, Function and Its Possible Role in the Pathogenesis of Myotonic Dystrophy Type 1

Jonathan J. Magaña[1], Rocío Suárez-Sánchez[1], Norberto Leyva-García[1], Bulmaro Cisneros[2] and Oscar Hernández-Hernández[1]

[1]*Department of Genetics, National Rehabilitation Institute, Mexico City,*
[2]*Department of Genetics and Molecular Biology, Center for Research and Advanced Studies of the National Polytechnic Institute (CINVESTAV-IPN), Mexico City, Mexico*

1. Introduction

Myotonic dystrophy protein kinase (DMPK) is a member of the AGC super family of serine/threonine protein kinases (Caenepeel et al., 2004; Manning et al., 2002). The *DMPK* human gene encodes several alternative spliced protein products believed to be involved in remodeling of the actin cytoskeleton, mitochondrial dynamics, ion homeostasis and nuclear envelope stability. DMPK and its isoforms are mainly expressed in skeletal, heart and smooth muscle, and brain, the main targets of myotonic dystrophy type 1 (DM1) (Groenen et al., 2000; Ueda et al., 2000). DM1 is the most common form of muscular dystrophy in adults with a frequency of 1 in 8,000 individuals worldwide. It is a multisystem dominantly inherited disorder characterized by myotonia, progressive muscular weakness and wasting, cardiac defects, cataracts and frontal balding, as well as several central nervous system (CNS) manifestations (Harper et al., 2002). The disease is caused by the expansion of an unstable $(CTG)_n$ repeat in the 3′-untranslated region (3′-UTR) of the *DMPK* gene (Brook et al., 1992; Fu et al., 1992; Mahadevan, M. et al., 1992). In healthy population the CTG tract is polymorphic with alleles ranging from 5 to 37 in length. Individuals carrying the DM1 premutation, a tract between 38 and 49 CTG repeats, generally are asymptomatic but are at risk of transmitting a pathological expanded mutation. In contrast, a CTG expansion between 50 and 4000 CTG repeats results in DM1 disease. Affected families show the phenomenon of anticipation; longer expansions correlate with an earlier age of onset and more severe course in subsequent generations. Based on their clinical presentation DM1 is classified into four subtypes: late onset, classic DM1, childhood onset and congenital DM1 (CDM) (Harley et al., 1992). An RNA-mediated dominant gain-of-function is currently accepted as the pathogenic mechanism to explain features of the DM1. *DMPK* toxic transcripts accumulate as nuclear foci (Davis et al., 1997; Taneja et al., 1995), interfering with the activity of RNA-interacting proteins and altering RNA metabolism, notably the splicing programme (Ranum & Day, 2004). Although reduced DMPK protein levels in DM1 tissues are well described (Fu et al., 1993; Hofmann-Radvanyi et al., 1993; Krahe et al., 1995; Novelli et al., 1993), the specific role of DMPK in the pathogenesis of the disease is not completely understood. Studies focused on the biology of this protein had received little attention, as protein product of *DMPK* gene seemed to have no direct role in the disease. Identification of novel DMPK targets, and comparative differences of DMPK isoforms have started to emerge (Groenen et al., 2000; Kaliman & Llagostera, 2008; Mulders et al., 2011; Oude Ophuis et al., 2009a; van Herpen et al., 2005). This chapter offers the current knowledge about the structure, expression patterns and function of DMPK and its isoforms. A better understanding of DMPK biology will enable us to speculate how this protein could be involved in the pathophysiology of DM1.

2. Myotonic Dystrophy Protein Kinase gene

2.1 Localization and structure of the *DMPK* gene

The *DMPK* human gene is located on the long (q) arm of chromosome 19 at position 13.3 covering 13 kb in length. The gene is composed of 15 exons coding a full-length protein of 692 amino acids with a composite domain structure. The *DMPK* gene is transcribed in the telomere-to-centromere orientation, giving rise to a set of primary transcripts that are subject to extensive alternative splicing (Figure 1). The $(CTG)_n$ repeat lies within the 3'-UTR of the gene, in the exon 15 downstream of the translation stop signal, and approximately 500 bp upstream of the poly (A) signal (Groenen et al., 2000; Jansen et al., 1992; Mahadevan, M.S. et al., 1993).

Characterization of the promoter region has showed that *DMPK* gene contains a conserved GC box to which the transcription factor Sp1 binds. Four E-box elements within the first intron mediate interactions with upstream promoter elements to up-regulate its transcription in myoblast (Storbeck et al., 1998). The $(CTG)_n$ repeat tract is located in a gene-dense region; in fact the expansion overlaps with the promoter region of the downstream neighboring *SIX5* gene. *SIX5* expression is needed for eye development in the fruit fly and regulates distal limb muscle development in the mouse (Otten & Tapscott, 1995; Ranum & Day, 2004; Wang et al., 1994). In DM1 an anti-sense *DMPK* transcript is originated from *SIX5* regulatory region. This transcript extends into an insulator element on the 3'UTR of *DMPK* regulating the environmental chromatin structure of the region (Cho et al., 2005).

2.2 Alternative splicing and DMPK isoforms

Initial bioinformatic analysis of the *DMPK* gene, and subsequent expression studies have demonstrated that six major *DMPK* isoforms can arise in both humans and mice as a result of extensive alternative splicing. In the full-length polypeptide, four distinct domains can be distinguished: a leucine-rich N terminus comprised by 40 amino acid that are shared by all DMPK isoforms, a serine/threonine protein kinase domain specified by exons 2 to 8, an α-helical coiled-coil region encoded by the exons 9 to 12, the VSGGG motif and a the C terminus. The presence of VSGGG motif and the nature of the C-terminus, the most important distinction among DMPK isoforms, are both determined by alternative splicing (Groenen et al., 2000; Jansen et al., 1992; Mahadevan, M.S. et al., 1993) (Figure 1).

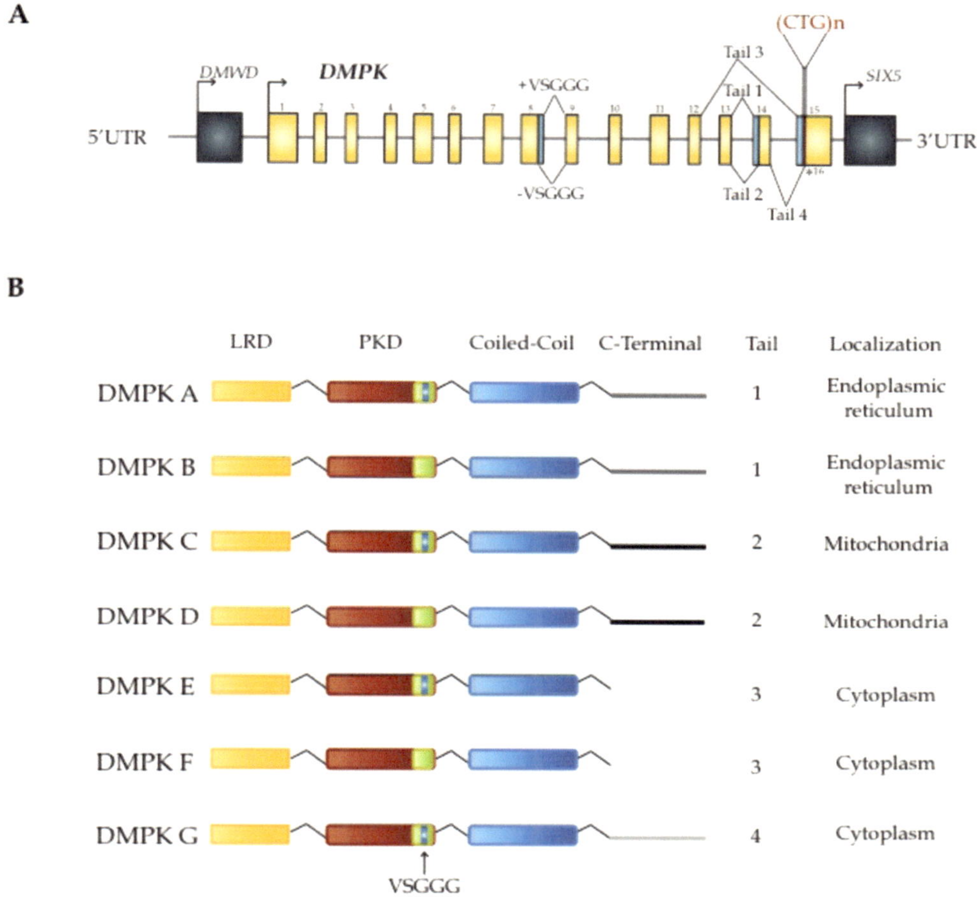

Fig. 1. The human *DMPK* gene and its splicing isoforms. (A) *DMPK* gene is composed of 15 exons, represented by rectangles. The $(CTG)_n$ expansion mutation over exon 15 is shown. Exons that are subject to alternative splicing are also indicated. (B) Schematic representation of the protein domain organization of the DMPK isoforms. All isoforms have a leucine-rich N-terminus domain (LRD), a serine/threonine protein kinase domain (PDK), a VSGGG motif, and an α-helical coiled-coil domain. C-terminal tail is specific to each isoform and determines their subcellular localization.

It can be distinguished DMPK isoforms with (A to D isoforms) or without a C-terminal extension (DMPK E and F). The alternative splice of the first four nucleotides of exon 14 produces two different open reading frames (ORFs) encoding proteins with C-terminal tail 1 or 2. An additional C-terminal with tail 3 is originated by the direct joining of exons 12 and 15 sequences, due to the splicing of exons 13-14. In this case, an ORF with stop codon at the beginning of exon 15 is generated. Altogether, the six distinct isoforms constitutes the DMPK profile in different cell types and tissues.

As predicted, the alternative splicing of *DMPK* pre-mRNA determines the structure and function of *DMPK* products. For instance, the inclusion of exon 14 generates ~ 70-kDa proteins with a hydrophobic C terminus carrying the tail 1 (isoforms A and B), whereas exclusion of the first four nucleotides from this exon results in the production of ~ 70-kDa proteins with less hydrophobic C terminus carrying the tail 2 (isoforms C and D). The presence of both tail 1 and tail 2 confers to the protein the capacity to anchor into membranes of endoplasmic reticulum and mitochondria. In contrast, isoforms with tail 3 as C terminus (isoforms E and F) adopt a cytosolic localization (van Herpen et al., 2005; Wansink et al., 2003). Interestingly, a tissue-specific expression pattern of DMPK isoforms have been showed in brain and smooth, cardiac and skeletal muscle.

There is a cryptic splice acceptor site in exon 15 of the human *DMPK* gene that defines the inclusion of an additional exon (exon 16). The fusion between exons 14 and 16 results in removal of the 5′ part of exon 15, including the $(CUG)_n$ repeat, generating the so-called *DMPK G* isoform. This transcript encodes a ~ 69-kDa protein with a unique C-terminal tail (tail 4). Because of the absence of the $(CUG)_n$ repeat, DMPK *G* transcripts can freely exit the nucleus, which might result in efficient DMPK G expression in DM1 cells (Tiscornia & Mahadevan, 2000; Wansink et al., 2003).

3. The DMPK family of proteins

DMPK belongs to the AGC super family of related serine/threonine protein kinases (Caenepeel et al., 2004; Manning et al., 2002). DMPK shows higher homology with myotonic dystrophy kinase-related Cdc42-binding kinases (MRCKs), comprise by MRCKα, MRCKβ and MRCKγ proteins (Figure 2). Regarding to the DMPK kinase domain, there is a 60% of sequence identity between DMPK and MRCKs (Leung et al., 1998; Ng et al., 2004). DMPK is closely related to the Rho-associated kinases I and II (ROCK1/ROCK2), with 45 and 43% identity respectively (Riento & Ridley, 2003), as well as to citron kinase (CRIK), with about 41% sequence identity (Madaule et al., 1998). DMPK is also related in a lower degree to the NDR family (consisting of NDR1, NDR2) and Lat's proteins (Lat 1 or Lat 2), and more distantly, to PKA, PKB (Akt), and PKC, the archetypes of the AGC group (Caenepeel et al., 2004; Manning et al., 2002).

3.1 Structural domains of DMPK

All DMPK family members are composed of the four following domains: an N terminal leucine-rich domain, a kinase domain, an α-helical coiled-coil region, and the C-terminal domain (Groenen et al., 2000; Mahadevan, M.S. et al., 1993) (Figure 1). DMPK and other family members have important differences among them, mainly in their size and overall protein domain composition. MRCKs, ROCKs and CRIK contain a number of additional domains with distinct functions, such as the GTPase binding domain, the pleckstrin homology domain (PH), the cysteine-rich domain and the citron homology domain (CNH), as shown in the Figure 2 (Riento & Ridley, 2003; Zhao & Manser, 2005). Both DMPK and ROCKs are able to form dimmers. Structural studies on DMPK, ROCKs, and MRCK have shown that dimerization depends on the regions N- and C-terminal immediately adjacent to the kinase domain (Doran et al., 2004; Garcia et al., 2006; Jacobs et al., 2006; Tan et al., 2001). DMPK is regulated by activation loop phosphorylation, dimerization and trans-autophosphorylation (Elkins et al., 2009). Although specific sites have not yet been identified, the residues Ser234, Thr240 and Thr403 conserved in DMPK and MRCKs, are probably implicated in DMPK activation (Tan et al., 2001; Wansink et al., 2003). In the following sections, the structural domains that comprise DMPK are described in detail.

3.1.1 Lucine-Rich Domain

In DMPK, the Leucine-Rich Domain (LRD) encodes a peptide of approximately 70 amino acids, in which, almost every fourth amino acid is a leucine. This domain constitutes a highly non-polar amino-terminal region that is constant in all DMPK isoforms, and presents 93% of sequence identity between human and mouse forms. This high degree of interspecies conservation suggests a potential role for this domain in DMPK activity (van der Ven et al., 1993; Waring et al., 1996). A segment of this region conforms the leucine zipper motif, which is conserved in all DMPK family members. Although the MRCKα and ROCK-II leucine zipper motifs mediate the dimerization, or even oligomerization, of GTPase-regulated kinases, the DMPK leucine zipper motif seems not to be involved in multimerization (Bush et al., 2000; Zhang & Epstein, 2003). Then, the specific role played by the DMPK N-terminal leucine-rich domain remains to be revealed. In many cases, protein kinase activity is modulated by domains flanking the catalytic domain (Manning et al., 2002). Therefore, it is thought that the leucine-rich domain may regulate the DMPK kinase activity, due to its proximity to the serine/threonine protein domain. The leucine-rich domain might link DMPK to signaling modules, or alternatively may help DMPK to localize to specific subcellular compartments (Manning et al., 2002). Further studies are necessary to clarify the functional activity of this domain.

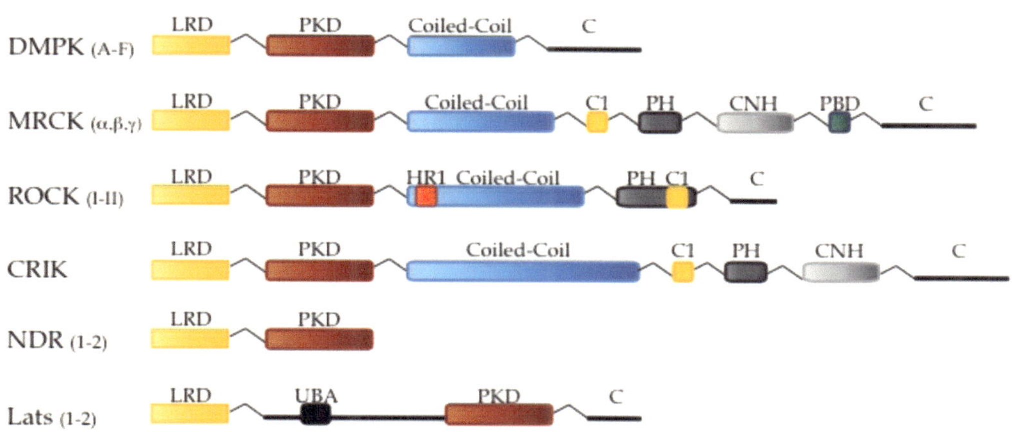

Fig. 2. Domain organization of DMPK and five major AGC proteins closely related. All proteins present a N-terminal leucine-rich domain (LRD), a protein kinase domain (PKD) and a C-terminal domain (C); most of them present also a coiled coil domain. AGC proteins have some domains that are distinctive of certain members, such as protein kinase C conserved region 1 domain or cysteine-rich domain (C1), pleckstrin homology domain (PH), citron homology domain (CNH), P21 Rho binding domain (PBD), Rho effector domain (HR1) and ubiquitin-associated domain (UBA).

3.1.2 Protein kinase domain

The catalytic domain between residues 70 and 349 carries out the kinase function. This domain transfers the gamma phosphate from nucleoside triphosphate (ATP) to the hydroxyl group of serine or threonine residues on the target protein, which results in a conformational change that affects protein function. Serine/Threonine protein kinase domain has eleven major conserved subdomains, separated by regions of lower conservation (Hanks & Hunter, 1995). In the active conformation, the kinase domain folds in a structure of two lobes: an N-terminal lobe formed by β sheets with a single α helices (αB and αC helices) and a larger C-terminal lobe containing mainly α helices and the activation loop (Huse & Kuriyan, 2002; Krupa et al., 2004; Nolen et al., 2004). Activation loop phosphorylation and autophosphorylation of DMPK induce a conformational change that stabilizes the kinase domain in an open conformation permissive for substrate binding. The ATP binding site is situated in the cleft at the interface of the N-terminal and C-terminal lobes. This highly conserved region creates the phosphate-binding loop site (P-loop), that is a glycine-rich GXGXΦG consensus sequence (with Φ usually being phenylalanine or tyrosine) (Hanks & Hunter, 1995). DMPK is also known as a Lys/Arg-directed kinase. An invariant lysine (Lys100) in the ATP binding region makes contact with the α– and β-phosphates of ATP, promoting in coordination with the P-loop the phosphotransfer from ATP (Wansink et al., 2003).

It is generally accepted that, in the nonphosphorylated state, activation loop folds into the active site blocking the binding of ATP and peptide substrate (Adams, 2003; Hanks & Hunter, 1995; Johnson et al., 1996). However, mass spectrometry of crystallized recombinant DMPK (unphosphorylated) has showed that activation loop occupies a well-ordered conformation that does not impede access to the nucleotide or substrates binding sites, resembling an active conformation (Elkins et al., 2009). A conserved structure in DMPK and ROCK1, the αEF/αF loop, interact with the activation loop to stabilize this active conformation. Also important for catalysis is the position of helix αC in close proximity to the active site. This helix is highly structured, unlike the structures of several inactive kinases, and formation of a bridge between αC and β3 (residues Glu119 and Lys100 of DMPK) is an indicator of the active kinase conformation. This suggests that DMPK could be active without phosphorylation, however there is still the question of what the structural impact of phosphorylation at the conserved phosphorylation sites (Ser234, Thr240 and Thr403) would be (Elkins et al., 2009).

3.1.3 The VSGGG Motif

The VSGGG motif is unique among DMPK-related kinases. As previously mentioned, the presence of this pentapeptide is defined by alternative splicing, and is found in A, C, E and G DMPK isoforms (Figure 2) (Wansink et al., 2003). It has been shown that VSGGG motif modulates DMPK autophosphorylation activity, an event that is usually required to regulate activity of many protein kinases (Johnson et al., 1996; van Herpen et al., 2005). The presence of Serine 379 in the VSGGG motif has a marked effect on the conformational state of the protein but it is not clear whether the presence of the motif increases the kinase activity by stabilization of the active conformation, or if the motif makes a turn-motif serine/threonine more available for autophosphorylation. Interestingly, it has been showed that DMPK isoforms lacking the VSGGG motif (B, D, and F) are also phosphorylated in this region.

Adjacent sequences the VSGGG motif may be used as phosphoracceptors (residues Thr375 or Thr379) or a different conformational regulation is possibly taking place.

3.1.4 Coiled-coil region

The coiled-coil domain is a common structural motif, formed by approximately 5% of all amino acids in proteins (Wolf et al., 1997), including transcription factors, signaling enzymes and motor proteins. Typically, it consists of two to five α-helices wrapped around each other into a left-handed helix to form a supercoil. Whereas regular α-helices go through 3.6 residues for each complete turn, the distortion imposed upon each helix within a left-handed coiled-coil in DMPK lowers this value to around 3.5 residues. Thus a heptad repeat (seven residues) occurs every two turns of the helix (Landschulz et al., 1988; Lupas, 1996).

The most commonly observed type of coiled-coil is left-handed, with anywhere from two (in designed coiled-coils) to 200 of these repeats in a protein (Burkhard et al., 2001; Kohn et al., 1997). The heptad repeat is commonly denoted $(a\text{-}b\text{-}c\text{-}d\text{-}e\text{-}f\text{-}g)_n$ in one helix, and $(a'\text{-}b'\text{-}c'\text{-}d'\text{-}e'\text{-}f'\text{-}g')_n$ in the other one (Lupas, 1996; Mason & Arndt, 2004). In this model, a and d are typically nonpolar core residues found at the interface of the two helices, whereas e and g are solvent exposed polar residues that give specificity between the two helices through electrostatic interactions. Interactions between a-a′ and d-d′ helices drive the "hydrophobic collapse" and formation of the supercoil, whereas e-g′ and g-e′ ionic interactions determine the specificity of the interhelical interaction (Mason & Arndt, 2004).

In DMPK the coiled-coil domain consist of only ~9 heptad repeats, containing approximately 65 amino acids. The DMPK coiled-coil domain is smaller than that of other AGC family members (Figure 2). DMPK isoforms could form high-molecular-weight complexes, due to homo- and heteromultimerization of up to 10 DMPK molecules (Bush et al., 2000; Zhang & Epstein, 2003). This multimerization is mediated by coiled-coil interactions, and occurs independently of alternatively spliced protein segments or DMPK activity (van Herpen et al., 2005). When a coiled-coil heptad repeat is mutated to glycines, the coiled-coil mutants are still capable of autophosphorylation and transphosphorylation, but the rates of their kinase activities are significantly lowered. Furthermore, it has been shown that the coiled-coil domain is important for the cellular localization of DMPK mitochondrial isoforms (van Herpen et al., 2005). In summary, the coiled-coil domain is involved in DMPK multimerization, substrate binding, kinase activity and subcellular localization.

3.1.5 C-terminal domain

The C-terminal domain appears to play an autoinhibitory role as C-terminally truncated DMPK activity is increased by 3-fold for MBP and 10-fold for myosin phosphatase target subunit (MYPT1). This inhibitory activity is mapped in the residues 550 to 629, which has also been implicated in oligomerization (Bush et al., 2000; Wansink et al., 2003).

Interestingly, alternative splicing affecting the C-terminal domain appears to determine the subcellular localization of DMPK isoforms (Groenen et al., 2000). DMPK isoforms with long C-terminal domain (DMPK A to D) anchor into membranes of the endoplasmic reticulum or mitochondria, whereas isoforms with no C-terminal extension (DMPK E and F) adopt a cytosolic localization.

The structure of DMPK illustrates that the most varied part of AGC proteins is the long C-terminal domain, compared with the structure of the closely related ROCKs, MRCKs and CRIK. Long C-terminal domain containing 96 and 97 amino acids, are named as tail 1 and tail 2, respectively. These tails are characterized by unique structures without homology to any other protein and by a single C-terminal hydrophobic stretch. The presence of tail 1 anchored DMPK A and DMPK B into membranes of the endoplasmic reticulum, whereas tail 2 is responsible for DMPK association with the mitochondrial outer membrane (van Herpen et al., 2005; Wansink et al., 2003) (Figure 2). Interestingly, presence of the coiled-coil region in DMPK C and DMPK D is essential for tail 2 to target isoforms to the mitochondrial membrane. Not much is known about the relationship between tail anchor structure and membrane specificity, and its avidity for lipid composition. A specific role of the signal recognition particle for targeting of similar proteins to the ER has been reported (Abell et al., 2004), but detailed insight on the mechanism of membrane insertion is currently lacking. Data from various studies suggest that shared structural elements in the C terminal reflect the only critical features implicated in membrane selectivity. For instance, the fact that amino acids in the DMPK A tail are all neutral is consistent with this observation, and may determine binding strength. Taking these data together, we can conclude that the strong hydrophobic character of the DMPK A isoform constitutes the endoplasmic reticulum targeting signal. This is compatible with other studies showing that artificially engineered stretches of hydrophobic amino acids of different lengths can be sufficient for endoplasmic reticulum targeting (Whitley et al., 1996; Yang et al., 1997). On the contrary, targeting of DMPK C to the mitochondrial outer membrane seems to be more elaborate, and additional structural information is necessary to make conclusions.

On the other hand, as showed figure 2, DMPK E and DMPK F isoforms do not contain the C-terminal domain. They present only two amino acids long segment, named tail 3, without any targeting information (Mulders et al., 2011; Whitley et al., 1996; Yang et al., 1997). Although the coiled-coil domain confers by itself no targeting properties, as evidenced from the cytosolic localization of DMPK E, the amino acid information in this domain seems to be a critical component for the correct association of DMPK C to the mitochondrial outer membrane. Finally, the minor human isoform DMPK G, which shows a cytosolic localization, has a C-Terminal domain with a function still unknown. Some authors have named this C-terminal domain as tail 4 (Wansink et al., 2003). Further studies are necessary to known in depth the impact of the C-terminal domain on DMPK isoforms cellular localization.

4. Expression pattern of DMPK

4.1 Tissue distribution of DMPK

Expression data based on mRNA analysis (*in-situ* hybridization, northern blot and RT-PCR) have showed a wide expression profile of *DMPK* in mouse and human tissue samples. *DMPK* mRNA is highly expressed in skeletal, cardiac, and smooth muscle and at lesser extent in different brain areas. Others tissues and organs that express *DMPK* mRNA include bone, testis, eye, skin, thymus, lung and liver. In contrast, no expression of *DMPK* transcript is found in ovary, pancreas and kidney (Jansen et al., 1992; O'Cochlain et al., 2004; Sarkar et al., 2004). The production of DMPK antibodies by using peptides or recombinant DMPK protein as immunogen has made possible the detection of DMPK at protein level (Lam et al.,

2000; Pham et al., 1998; Ueda et al., 2000). DMPK antibodies recognize protein products varying in size from ~42- to 84-kDa, being the smaller proteins (42- to 50-kDa) not real products of the DMPK gene (Lam et al., 2000; van der Ven et al., 1993). In general, mRNA and DMPK protein expression appear strongly correlated. Highest levels of *DMPK* gene products are found in smooth muscle (Jansen et al., 1992; O'Cochlain et al., 2004; Sarkar et al., 2004), heart and skeletal muscle (Groenen et al., 2000; Lam et al., 2000; O'Cochlain et al., 2004; Sarkar et al., 2004). DMPK at protein level has also been identified in several brain regions (van der Ven et al., 1993) both in neurons and glia. Interestingly astrocytes from brain cortex show higher DMPK levels compared with neurons (Oude Ophuis et al., 2009a). DMPK transcripts and protein products are also detected in fetal eyes, adult retina, corneal epithelia and optic nerve (Winchester et al., 1999).

Expression of DMPK during development has also been explored. DMPK expression is significantly increases in skeletal muscle between 9 and 16 weeks of human muscle development. The large increase in protein accumulation correlates with the formation of second-generation muscle fibers, and the major period of muscle formation. Increase in DMPK expression is also observed during *in vitro* myogenic differentiation (Furling et al., 2003). In other studies, DMPK protein expression has been revealed in eye, skin and intestine chick embryos as well as in heart and skeletal muscle, specifically in postmitotic myocytes, where it activates myotubes formation (Harmon et al., 2008). In brain and spinal cord of rat, DMPK begins to be expressed after birth and increases gradually to peak at postnatal day 21 in many regions. After that and proceeding to the adult stage, DMPK expression becomes more restricted to certain cell groups like spinal motor neurons (Balasubramanyam et al., 1998). Immunoreactive neurons have been observed in the early fetal frontal cortex and cerebellar granule cell layer before 29 weeks of gestation (Endo et al., 2000).

Regarding the DMPK isoforms expression pattern, it has been showed that long membrane-anchored isoforms are expressed in heart, diaphragm and skeletal muscle while short cytosolic isoforms are highly expressed in bladder and stomach. Interestingly, both long- and short-isoforms are expressed in diverse brain regions (Oude Ophuis et al., 2009a). Other reports have demonstrated that the four different ~74-kDa full-length DMPK isoforms are present in heart, skeletal muscle and brain, in contrast to the two C-terminally truncated ~68-kDa isoforms that are restricted to smooth muscle and to a lesser extent at heart (Groenen et al., 2000).

4.2 Differential subcellular localization

Detailed analyses of the subcellular localization of DMPK isoforms carrying distinct C termini have been performed. By applying electron and confocal microscopy and immunodetection approaches after overexpressing individual isoforms in several cell types, it has been demonstrated that unique sequences arrangement in the C terminus tail of DMPK control the specificity of anchoring into intracellular membranes. DMPK E and F mouse isoforms carrying a truncated tail present an evident cytosolic localization. In contrast, DMPK A, B, C and D mouse isoforms carrying a hydrophobic tail are associated specifically with either the endoplasmic reticulum or the mitochondrial outer membrane. In agreement with this, the corresponding human DMPK A and C proteins localized to mitochondria. Furthermore, the expression of mouse and human DMPK A, but not DMPK C isoforms in mammalian cells caused clustering of ER or mitochondria. That means that DMPK has a possible role in organelle distribution and dynamics (van Herpen et al., 2005; Wansink et al., 2003). Subcellular localization of the DMPK protein during cardiac myocytes differentiation has been analyzed in chicken and mouse cardiac myoblast, and in C2C12 cells transfected with a vector expressing the DMPK fused to the reporter protein GFP. The analysis revealed a shift in the DMPK subcellular localization during myogenesis, from the perinuclear region to the cellular membrane, suggesting the possibility that DMPK performs different roles during this process (Reddy et al., 1996). Electron microscopy analyses in adult rat spinal motor neurons reveal a localization of DMPK in the endoplasmic reticulum and dendritic microtubules, suggesting a function in membrane trafficking and secretion within neurons associated with cognition, memory, and motor control (Balasubramanyam et al., 1998). In addition, multiple subcellular localizations have been assigned to DMPK, including neuromuscular and myotendinous junctions (Berul et al., 1999; Berul et al., 2000; Shimokawa et al., 1997) and terminal cisternae and intercalated discs of skeletal muscle sarcoplasmic reticulum (Benders et al., 1997; Pall et al., 2003; Saba et al., 1999). Furthermore, DMPK localization to gap junctions of the intercalated discs has been reported in the heart and Purkinje fibers (Kaliman et al., 2005; Mounsey et al., 2000b).

5. DMPK function

5.1 Transgenic and knockout mice

In order to elucidate the DMPK function and its possible involvement in DM1 pathogenesis, several transgenic and knockout mice had been created. DMPK totally deficient ($DMPK^{-/-}$) mice develop a late-onset and progressive skeletal myopathy showing variation in muscle fiber size, increased fiber degeneration and fibrosis with ultra structural changes, and a 50% decrease in force generation. Inconsistent and minor changes in head and neck muscle fibers in adult mouse are also shown (Jansen et al., 1996; Reddy et al., 1996). In addition, the $DMPK^{-/-}$ mouse shows cardiac conductions defects, which include first-, second-, and third-degree atrioventricular (A–V) block in His-Purkinje regions. This effect demonstrates that conduction system is specifically compromised; supporting the idea that loss of DMPK plays a significant role in the cardiac DM1 phenotype (Berul et al., 1999; Berul et al., 2000). DMPK deficit also cause an enhanced basal contractility of single cardiomyocytes with associated increase in intracellular Ca^{2+}, suggesting that DMPK has a modulator role in the control of intracellular Ca^{2+} concentration (Benders et al., 1997; Pall et al., 2003). In addition, hypophosphorylation of phospholamban (PLN), a muscle-specific sarcoplasmic reticulum Ca^{2+}-ATPase (SERCA2a) inhibitor, causes a deregulation of Ca^{2+} uptake in sarcoplasmic reticulum vesicles of ventricular homogenates from $DMPK^{-/-}$ mice (Kaliman et al., 2005). An effect on Na^{+} current amplitude in $DMPK^{-/-}$ myocytes because of reduced channel number is also observed (Mounsey et al., 2000b; Reddy et al., 2002). Supporting this result, multiple late re-opening of Na^{+} channels is observed when DMPK is absent in both cardiac and skeletal muscles. This is reflected in a plateau of non-inactivating macroscopic Na^{+} current and prolongation of cardiac action potentials (Lee et al., 2003). Altogether these findings suggest the existence of a regulatory DMPK pathway for cardiac contractility, and provide a suitable molecular mechanism for DM1 heart pathology. A common feature in DM1 is insulin resistance, which is a major factor in the development of type 2 diabetes (Biddinger & Kahn, 2006). $DMPK^{-/-}$ mice exhibit impaired insulin signalling in muscle tissues but not in adipocytes and liver. These mice also display metabolic derangements such as abnormal glucose tolerance, reduced glucose uptake and impaired insulin-dependent GLUT4 trafficking in muscle. Altogether, these data indicate that reduced DMPK expression may directly influence the onset of insulin resistant in DM1 patients and suggest that DMPK could represent a gene of susceptibility to type 2-diabetes (Llagostera et al., 2007).

A transgenic murine line overexpressing the complete human DMPK gene (Tg26-hDMPK) has been generated. In this animal model, skeletal muscle shows myopathy with myotonic discharges coupled with deficit in sarcolemmal chloride channels. Deficient exercise endurance, fiber degeneration and smooth muscle tone deficit with systemic hypotension are also present in the Tg26-hDMPK mouse. Heart from these mice develops cardiomyopathic remodeling and propensity for dysrhythmia, characterized by overt intracellular calcium overload that promotes nuclear translocation of transcription factors responsible for maladaptive gene reprogramming. The cumulative stress induced by permanent overexpression of DMPK gene products increases the risk of some distinctive muscle traits of DM1 (O'Cochlain et al., 2004). Additional overexpression studies using different transgenic lines show enhanced neonatal mortality in addition to hypertrophic cardiomyopathy (Jansen et al., 1996). In DM1, membrane inclusions accumulate in lens fiber cells producing cataracts. Overexpression of DMPK with enzymatic activity in epithelial lens cells led to apoptotic-like blebbing of plasma membrane and reorganization of the actin cytoskeleton. This may be relevant to the removal of membrane organelles, a necessary process during normal lens differentiation (Jin et al., 2000).

5.2 DMPK function in muscle and brain tissues

The function of DMPK protein is not completely understood yet, however, several reports have described the participation of this protein in different cellular processes notably in both muscle and brain tissues.

DMPK has been described as an important factor during myogenesis. The differentiation process begins with the exit of the proliferating myoblast from the cell cycle and activation of the differentiation program throughout the expression of myogenic factors, such as MyoD and myogenin. Additionally, morphological changes are needed for myotube formation. Overexpression of DMPK in BC3H1 cells (Bush et al., 1996) or its depletion in C2C12 cells (Harmon et al., 2008) have demonstrated the role of DMPK as a regulator of myogenin expression during muscle differentiation. Overexpression of DMPK in chick cardiac and skeletal myocytes, and C2C12 cells provoked rounding cells that are brought to apoptosis, which suggest a pro-apoptotic role for DMPK. In addition, depletion of DMPK in C2C12 cells prevented cell fusion, a necessary step for myotube formation (Harmon et al., 2008). As mentioned previously, DMPK knockout transgenic mice have been employed to study the function of DMPK. Disruption of normal muscle function both in muscle and heart is attributed to altered DMPK expression (Jansen et al., 1996; Reddy et al., 1996). CDM, the most severe form of DM1, is associated with several developmental defects that include delayed muscle development. Myoblasts from CDM embryos have a reduced capacity to differentiate, and interestingly, produce less than 50% of DMPK protein levels, implying a specific role for DMPK in embryonic myocyte development (Furling et al., 2001; Furling et al., 2003). Overall, these data suggest that DMPK is crucially involved in myogenesis.

Ion homeostasis is a determinant mechanism regulated by membrane permeability to cations (Na^+, Ca^{2+}, K^+) or anions (Cl^-). Transmembrane ion flux is the predominant factor in controlling the excitation–contraction coupling mechanism. Alteration in the function of voltage-dependent L-type Ca^{2+} (the dihidropiridine receptor DHPR) and Na^+ channels has been shown in DMPK mice models (Benders et al., 1997), which evidences the importance of DMPK in the maintenance of this homeostasis. That is consistent with the fact that DHPR is a target of DMPK and that its activity is altered in DMPK$^{-/-}$ cells (Benders et al., 1997). Reduction in Na^+ current amplitude in DMPK$^{-/-}$ mice muscle has also been demonstrated (Mounsey et al., 2000b; Reddy et al., 2002). Furthermore, activity and membrane localization of chloride channel phospholemman (PLM), which regulates Cl^- currents, is also modulated by DMPK (Mounsey et al., 2000a). All these data suggest that DMPK is involved in regulating the initial events of excitation–contraction coupling in skeletal muscle, and controlled then, ion homeostasis. In the other hand, proteins involved indirect or directly with the cytoskeleton (MYPT1 and SRF) and nuclear envelope (Lamin A/C) integrity were reported as DMPK partners (Harmon et al., 2011; Iyer et al., 2003; Muranyi et al., 2001), suggesting the participation of DMPK in these cellular processes.

One of the most common DM1 alterations, the atrioventricular conduction block, was shown to be dependent on DMPK levels in a mechanism not yet understood (Berul et al., 2000; Saba et al., 1999). In addition, DMPK$^{-/-}$ mice with cardiac contractility deregulation exhibit an impaired SERCA2a function. This ATPase is regulated by PLN protein, which is phosphorylated by DMPK (Kaliman et al., 2005). Furthermore, cardiac dysfunction, hypertrophic cardiomyopathy, propensity for dysrhythmia and enhanced neonatal mortality has been reported in the DMPK mice (O'Cochlain et al., 2004). Altogether, these findings delineate a feasible molecular mechanism for DM1 heart pathology.

Participation of DMPK in DM1 brain physiology has also been reported. Mis-localization of DMPK occurs in DM1, and even at higher extent in the CDM (Endo et al., 2000). In the CNS an involvement of DMPK in important physiological processes has been suggested, including membrane trafficking and neuro-secretion. DMPK is expressed in a subpopulation of neurons associated with cognition, memory, and motor control. It is localized in the endoplasmic reticulum and dendritic microtubules within adult spinal motor neurons (Balasubramanyam et al., 1998). DMPK has been also associated with intracellular communication functions, due to its synaptic localization in cerebellum, hippocampus, midbrain and medulla (Whiting et al., 1995). DMPK might contribute to synaptic plasticity and cognitive function via cytoskeletal remodeling. Long-term potentiation (LTP) is a use-dependent form of synaptic plasticity that contributes to the cellular basis of memory storage and other cognitive functions. Changes in the actin cytoskeleton are important for mechanisms underlying LTP including changes in synaptic and dendritic spine shapes (Engert & Bonhoeffer, 1999; Fischer et al., 2000; Kim & Lisman, 1999). Phosphorylation of myosin phosphatase by DMPK promotes an increase in the phosphorylation state of myosin, which supports the assembly and contractility of the actin cytoskeleton (Muranyi et al., 2001). Thus, deregulation of DMPK expression could influences directly on actin cytoskeleton remodeling (Jin et al., 2000). These data together with abnormal LTP measures in a DMPK null mice (Schulz et al., 2003) suggest that DMPK have a key role in synaptic plasticity in such a way that DMPK depletion might contribute to the cognitive dysfunction associated with CDM, which is characterized by strong motor function and brain disability.

5.3 DMPK-protein interactions and function

DMPK protein is a serine/threonine kinase with preference for substrates with an arginine residue upstream of the phosphoacceptor site (serine or threonine), followed by a hydrophobic residue (usually leucine or valine), and another arginine (Bush et al., 2000). The consensus recognition motif on substrates -RxxS/TL/VR- is similar to that of PKC and CaMKII. DMPK has the capability to form multimeric complexes among individual molecules, as revealed by gel-filtration experiments on cell lysates expressing DMPK A, C, E or F isoforms (van Herpen et al., 2006). As remarked previously, coiled-coil domain mediates multimerization, which in turn modulates DMPK binding to substrate and phosphorylation. Multimerization also influences subcellular targeting properties of individual DMPK isoforms (van Herpen et al., 2006; Zhang & Epstein, 2003). On the other hand, DMPK targets are involved in important cell functions notably ion homeostasis and actin cytoskeleton remodeling (Table 1).

PROTEIN	*NORMAL FUNCTION*	*DM1 ASSOCIATED PHENOTYPE*	*REFERENCES*
PLN	Ion (Ca^{2+}) homeostasis	Cardiac contractility alterations	(Kaliman et al., 2005)
PLM	Ion (Cl^{-}) homeostasis	Myotonia	(Mounsey et al., 2000b) (Mounsey et al., 2000a)
DHPR	Ion (Ca^{2+}) homeostasis	Deregulation of excitation-contraction coupling	(Benders et al., 1997) (Timchenko, L. et al., 1995)
MYPT	Calcium sensitization Regulation of the cytoskeletal structure	Calcium desensitization Cytoskeleton rearrangement defects	(Muranyi et al., 2001) (Mulders et al., 2011)
SRF	Alpha-actin gene transcription regulator	Synaptic plasticity abnormalities Cataracts	(Iyer et al., 2003) (Schulz et al., 2003) (Jin et al., 2000)
Lamin A/C	Nuclear envelope integrity	Muscle wasting	(Harmon et al., 2011)
CUG-BP	Alternative splicing regulator	Splicing alteration	(Roberts et al., 1997)
ATP synthase a and β Trifunctional enzyme β	Mitochondrial proteins	Mitochondrial morphology (?)	(Forner et al., 2010) (Oude Ophuis et al., 2009b)

Table 1. Partners and phosphorylation targets of DMPK implicated in the DM1 phenotype.

Cardiac contractility dysfunction is an alteration commonly present in DM1 patients and DMPK-/- mouse model. Cardiac contraction and relaxation cycle is an important cell process regulated by intracellular calcium levels. In cardiac sarcoplasmic reticulum, SERCA2a regulates the cytoplasmic levels of Ca^{2+}, and SERCA2a in turn is tightly controlled by the sarcoplasmic reticulum membrane protein PLN. Hence, nonhosphorylated PLN inhibits

SERCA2a function, whereas phosphorylated PLN reverse such inhibition (Frank et al., 2003; MacLennan & Kranias, 2003). It has been reported that PLN is a target of PKA kinase and CaMKII. However, Kaliman and co-workers reported that DMPK colocalizes and interacts with PLN promoting its phosphorylation. In fact, PLN protein is found under a nonphosphorylated state in DMPK$^{-/-}$ mice, where Ca^{2+} uptake is highly reduced. Disruption of the interaction between PLN and DMPK and/or loss of the PLN activity regulatory mechanism, due to DMPK downregulation, might explain at least in part the cardiac pathology observed in DM1 patients (Kaliman et al., 2005).

Interestingly, the splicing regulator CUGBP, member of the CELF family, is also a DMPK target. CUGBP was described as a CUG-binding protein with high avidity to single-strand RNA containing CUG repeats (Ladd et al., 2001). It has been shown in DM1 patients and DMPK knockout mice that phosphorylation and intracellular localization of CUGBP is regulated by DMPK. By using immunoprecipitation and kinase activity assays, the interaction between CUGBP-DMPK and the subsequent CUGBP phosphorylation were demonstrated (Roberts et al., 1997). DMPK phosphorylates CUGBP thereby decreasing its nuclear levels, which is consistent with the augment of CUGBP nuclear levels found in cultured cells from DM1 patients (Roberts et al., 1997). A plausible interpretation of these findings is that DMPK downregulation results in increased levels of the nonphosphorylated form of CUGBP, which might cause an increase in the nuclear levels of CUGBP, thereby adversely affecting the processing of pre mRNAs. Several reports have described alteration in the RNA processing of CUGBP targets genes in DM1 patients, including human cardiac troponin T (cTNT) (Philips et al., 1998), transcription factor CCAAT/ enhancer-binding protein b (C/EBPb) (Timchenko, N.A. et al., 2001), and insulin receptor (IR) (Dansithong et al., 2005; Savkur et al., 2001). On the other hand, an increase in the steady-state levels of CUGBP was revealed in DM1 tissues. This increase is due to an augmentation in the stability of the protein, which is mediated by PKC phosphorylation (Kuyumcu-Martinez et al., 2007; Timchenko, N.A. et al., 2001). Further studies are necessary to identify specific serine/threonine residues on CUGBP that could be phosphorylated in a PKC-or DMPK-dependent manner. The complete understanding of the two pathways concurring in CUGBP phosphorylation would help to better know DM1 pathogenesis.

PLM is a major membrane substrate for PKA and PKC. In oocytes from *Xenopus,* PLM induces hyperpolarization-activated non-inactivating chloride current ($I_{Cl(PLM)}$) (Moorman et al., 1992; Palmer et al., 1991). By *in vitro* assays Mounsey and co-workers showed that PLM is a substrate for DMPK, and that co-expression of both PLM and DMPK reduces $I_{Cl(PLM)}$ by about half (Mounsey et al., 2000b). Site-directed mutagenesis has demonstrated that reduced $I_{Cl(PLM)}$ is provoked by decreased PLM expression at the plasma membrane, as a consequence of its phosphorylation by DMPK. Furthermore, electrophysiological studies have evidenced reduced muscle chloride conductance and membrane depolarization in myotonic muscles. On the other hand, mutations in CLCN1 have been reported in the myotonia congenital disease condition. As myotonia can be induced by reducing chloride conductance, at the plasma membrane, the exploration of PLM phosphorylation state and intracellular localization in the mouse model Tg2–hDMPK mouse, which shows reduced chloride channel expression (O'Cochlain et al., 2004), would provide insights into the development of DM1-asscociated myotonia (Mounsey et al., 2000a).

Another target for DMPK is the MYPT1, a component of the myosin phosphatase (MP). MP phosphatase activity is inhibited by phosphorylation mediated by Rho-kinase (Hartshorne et al., 1998), and such inhibition increases the levels of myosin light chain phosphorylation, which in turn leads to Ca^{2+} sensitization in smooth muscle (Hartshorne et al., 1998). At subcellular level, phosphorylation of MYPT1 by DMPK produces cytoskeletal rearrangements in non-muscle cells (Muranyi et al., 2001), whereas *in-vitro* DMPK can phosphorylate MYTP1 in the same way that Rho-kinase, suggesting a regulatory role for DMPK in myosin function (Muranyi et al., 2001).

It has been described that deregulation of ion influxes is a key factor for development of the DM1 phenotype. Particularly, Ca^{2+} conductance abnormalities may contribute to hyperexcitability of sarcoplasmic membrane in DM1. DHPR, the voltage gating L-type Ca^{2+} channel, shows an altered activity in differentiated myotubes derived from DMPK$^{-/-}$ mice (Benders et al., 1997). By applying *in-vitro* approaches it was demonstrated that DMPK phosphorylates the β-subunit of DHPR in a serine residue. These data demonstrate that DHPR is substrate for DMPK, and suggest that DMPK is involved in the initial events of excitation–contraction coupling in skeletal muscle, a mechanism regulated by ion homeostasis (Timchenko, L. et al., 1995). DMPK has also been associated with SRF, the serum response transcription factor. SRF is a phosphoprotein that regulates skeletal and cardiac alpha-actin gene transcription. DMPK phosphorylates the Thr159 residue localized in the α 1 coil of the DNA-binding domain on a MADS box. By using site-directed mutagenesis it has demonstrated that phosphorylation on Thr159 regulates cardiac α-actin expression at promoter level (Iyer et al., 2003).

By using high-resolution mass spectrometry, a new group of proteins associated with DMPK has been reported (Forner et al., 2010). Authors identified 15 putative partners of DMPK grouped as follows: mitochondrial proteins (ATP synthase subunit α, ATP synthase subunit β, trifunctional enzyme subunit β), contractile and myofibrillar proteins (myosin heavy chain 7, myosin regulatory light chain 2-skeletal muscle isoform, myosin light chain 1-skeletal muscle isoform, myosin light chain 3), heat shock proteins (HSP) (α-crystallin B chain, heat shock protein β-1, heat shock protein β-6) and other proteins (adenylate kinase isoenzyme 1, carbonic anhydrase 3, peroxiredoxin-1, peroxiredoxin-2, similar to glyceraldehyde-3-phosphate dehydrogenase). To study in detail some of these DMPK interactions, these authors used immunoprecipitation and pull-down assays, and showed direct interaction of DMPK with αB-crystallin/HSPB5 and HSP25/HSPB1. These findings suggest that DMPK isoforms might modulate some HSPs functions at diverse subcellular localizations (i.e. mitochondrial membrane rather than cytoplasm) or under different physiological conditions (i.e. oxidative stress opposed to heat production) (Forner et al., 2010).

Finally, the participation of DMPK in the integrity of the nuclear envelope (NE) has recently been disclosed. NE is an important actor that determines the nuclear structure and regulates gene expression. Key structural elements of NE are the nuclear type-A and type-B lamins. Lamin A/C and lamin B1 proteins interact with proteins of the inner nuclear membrane and chromatin (Hetzer, 2010). Mutations in inner NE proteins are the genetic cause of Emery-Dreiffus muscular dystrophy (EDMD) and limb girdle muscular dystrophy 1b (LGMD1B) (Roux & Burke, 2007). Thus, NE defects may represent a common mechanism of muscle wasting in muscular dystrophies. DMPK localizes to the NE forming a protein complex with lamin A/C, and importantly, overexpression or depletion of DMPK disrupts the nuclear envelope localization of both lamin A/C and lamin B1, and causes ultimately nuclear fragmentation. Thus, DMPK expression seems to be critical to maintain NE stability. As nuclear instability is a common mechanism of muscle wasting in muscular dystrophies (Harmon et al., 2011), DMPK misexpression may be an important contributor to skeletal muscle wasting in DM1.

6. DMPK contribution to DM1 phenotype

6.1 Mechanisms of pathogenesis in DM1

DM1 is a multisystemic neuromuscular disease characterized by a wide range of clinical manifestations, including myotonia, muscle weakness and wasting, cardiac abnormalities, cognitive and behavioural alterations, and lens opacities (Harper et al., 2002). Cumulative experimental evidence has demonstrated that several DM1 manifestations are the result of the nuclear accumulation of CUG repeat-containing transcripts, which interfere with the alternative splicing, and gene expression programs (Ranum & Day, 2004). Besides the RNA gain-of-function mechanism, additional pathways might contribute to DM1, including chromatin rearrangements at the DM1 locus, interference RNA pathways and DMPK haploinsufficiency (Figure 3) (Sicot et al., 2011).

RNA gain-of-function mechanism: Mutated *DMPK* RNA expression results in decreased muscle-blind (MBNL) and increased CUGBP/Elav-like family member 1 (CELF1) activities. CUG repeat-containing transcripts form alternative secondary RNA structures that bind to and sequester MBNL in nuclear foci altering its normal distribution in the nucleoplasm (Jiang et al., 2004; Lin et al., 2006; Mankodi et al., 2005). In addition, CELF1 overexpression due to PKC phosphorylation has been described in DM1 cells (Kuyumcu-Martinez et al., 2007). Deregulation of the splicing transcription factors MBNL and CELF1 disrupts a tightly regulated developmental program, leading to altered expression of embryonic splicing isoforms in adult tissues (Osborne & Thornton, 2006; Ranum & Day, 2004). Muscle specific chloride channel (*ClC1*), insulin receptor (IR), cardiac troponin T (*cTNT*), myotubularin related protein 1 (*MTMR1*), microtubule binding protein Tau (*MAPT*), and NMDA glutamate receptor (*NR1*), are examples of genes affected at the alternative splicing level in DM1 (Du et al., 2010; Ranum & Day, 2004).

Leaching of transcription factors mechanism: It has been reported that mutant *DMPK* RNA interferes with the function of selected transcription factors, depleting them from the active chromatin by an RNA leaching mechanism. In fact, altered expression of *CLCN1* by leaching of Sp-family transcription factors was found in DM1 cells (Ebralidze et al., 2004). It is predicted that massive derangement at transcription level, due to leaching of transcript factors, is present in DM1 cells.

Chromatin rearrangements at the DM1 locus, and interference RNA (iRNA) pathways mechanism: *DMPK* gene is transcribed into sense and anti-sense transcripts (Cho et al., 2005). iRNA pathways might be activated by processing of dsRNA structures that could be form due to folding of CUG-containing transcripts into hairpin structures, or due to complementary hybridization between complementary sense and anti-sense DMPK transcripts. MicroRNA (miRNA) deregulation appears to be an additional mechanism

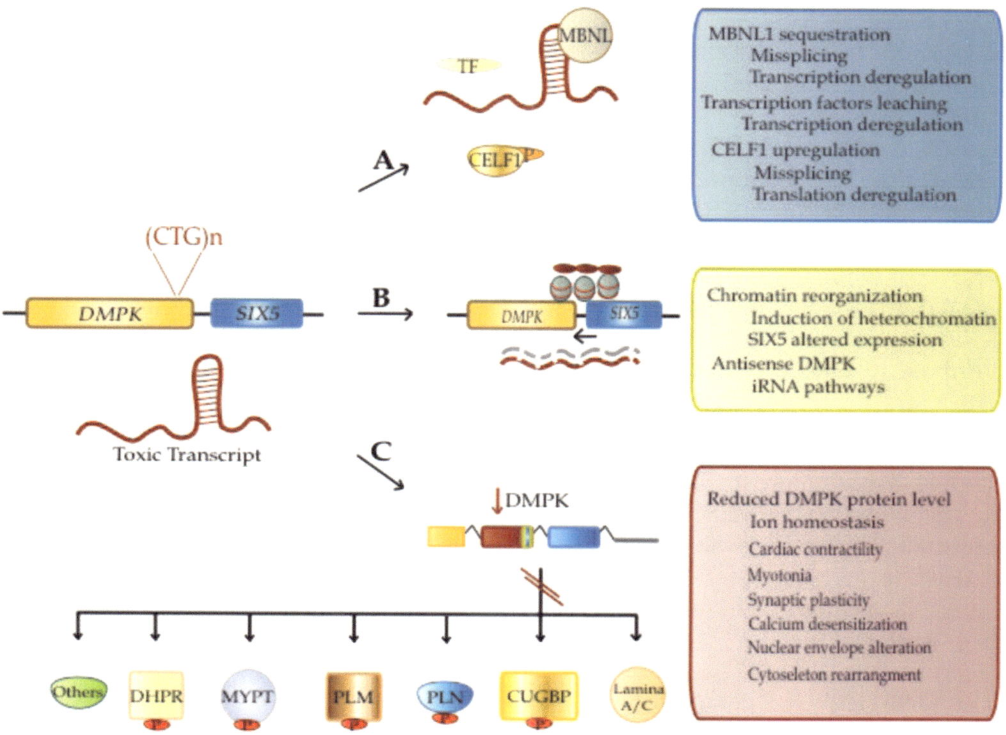

Fig. 3. Mechanisms of pathogenesis in DM1. Toxic CUG-containing transcripts form secondary structures and accumulate in the nucleus of DM1 cells, causing the multysistemic effects of DM1 throughout RNA gain-of-function mechanism (A), chromatin rearrangements at the DM1 locus (B) and DMPK haploinsuffiency (C). See text for details.

involved in DM1 biogenesis; in fact, the miRNAs pathway was found to be altered in DM1 skeletal muscle and heart. Interestingly, mi-R1 downregulation is mediated by MBNL1 depletion in cardiac DM1 tissue, which results in misprocessing of pre-miR-1 (Perbellini et al., 2011; Rau et al., 2011). On the other hand, processing of anti-sense transcript might be involved in chromatin ultrastructure regulation. It is known that anti-sense *DMPK* transcript extends into an insulator element located between *DMPK* and *SIX5* genes (Cho et al., 2005). CTCF-binding sites have been described on each side of the CTG repeat within the DMPK gene, restricting the extent of anti-sense RNA expression. Remarkable, the expanded mutant allele impairs CTCF binding, and induces spread of heterochromatin and regional CpG methylation in CDM cells. It is hypothesised that barrier function of the insulator is missing in DM1 due to disruption of CTCF binding, inducing higher DMPK expression in late embryogenesis, as a consequence of a high SIX5 enhancer activity, thereby contributing to the earlier disease onset in CDM (Sicot et al., 2011).

6.2 The possible role of DMPK and it isoforms in the pathogenesis of DM1

Although debated, many studies indicate that CTG expansion causes a decrease in the amount of DMPK protein by ~ 50%, with greatly reduced expression from the mutant allele and normal expression from the unaffected allele, a phenomenon known as haploinsufficiency (Bhagavati et al., 1996; Furling et al., 2001; Narang et al., 2000). Mice model generated to study the effects of DMPK deficiency have revealed an important contribution of DMPK to the development of myopathy and cardiac abnormalities (Berul et al., 1999; Jansen et al., 1996; Reddy et al., 1996). CDM is an entity of the disease that evidences the importance of DMPK expression; because, delayed muscle development and reduced capacity for muscle differentiation correlate with a decrease of more than 50% in DMPK levels. As stated above, it was recently reported that DMPK is a structural component of the nuclear envelope, forming a protein complex with lamin A/C (Harmon et al., 2011). In fact, alteration of DMPK expression disrupts the localization of lamins A/C and B1, causing nuclear fragmentation. Thus, deficient expression of DMPK in DM1 cells might contribute to nuclear instability, a common mechanism of muscle wasting in muscular dystrophies.

In skeletal muscle, the study of DMPK individual isoforms has provides insights into their potential contribution to the DM1 phenotype. In DMPK $^{-/-}$ mice myoblasts, exogenous expression of the cytosolic DMPK E isoform results in the presence of stress fibers, significant changes in myoblast polarity and delay myogenesis progression (Mulders et al., 2011). It is known that DMPK gene is expressed early in muscle development, but alternative splicing down-regulates DMPK E expression during myogenesis (Jansen et al., 1996; Oude Ophuis et al., 2009a). In contrast, membrane-anchored DMPK splice isoforms remain equally abundant between myoblast and myotubes, indicating that expression of these isoforms is not differentially regulated during myogenesis. Stress fibers formation induced by DMPK E is enhanced by myosin II light chain (MLC2) phosphorylation (Mulders et al., 2011). The phosphorylation status of MLC2 is dependent on myosin phosphatase (PP1) activity, which is inactivated by phosphorylation of its subunit MYPT1, a DMPK substrate. Hence, MCL2 phosphorylation levels are determined by the interplay between kinase and phosphatase activities, in which DMPK is involved. Exogenous expression of DMPK E causes also changes in cell shape and cell motility, which may be due to the abnormally rigid cytoskeletal organization caused by the high content of phosphorylated MLC2, which ultimately may impact myogenesis.

As mentioned previously, brain dysfunction is prominently implicated in DM1. Mental manifestations in CDM and adult-onset DM1 include mental retardation, cognitive and emotional disturbances, and hypersomnia (de Leon & Cisneros, 2008; Meola & Sansone, 2007). DMPK transcripts aggregated in foci were observed in brain of postmortem DM1 tissue (Jiang et al., 2004), suggesting that the RNA-gain of-function mechanism is involved in the development of DM1-asociated brain dysfunction. With respect to DMPK, it was recently demonstrated that both long and short DMPK isoforms are expressed across many brain regions, including brainstem, olfactory bulb, and striatum. In addition, DMPK protein was identified in astrocytes and U373 glioblastoma cells (Oude Ophuis et al., 2009a). Therefore, involvement of DMPK isoforms in DM1-mediated mental retardation can not be ruled out.

Finally, overexpression of human DMPK A in various cell lines causes abnormal mitochondrial morphology, mitochondrial clustering in the perinuclear region, loss of the mitochondrial membrane potential, increased autophagy activity, mitochondrial fragmentation, and eventually apoptosis (Oude Ophuis et al., 2009b). Interestingly, an abnormal mitochondrial morphology and mitochondrial dysfunction has been described in DM1 patients (Siciliano et al., 2001; Ueda et al., 1999). Furthermore, ATP synthase subunit α, ATP synthase subunit β, and trifunctional enzyme subunit β, three mitochondrial important components, have been described as possible partners of DMPK (Forner et al., 2010). Supporting the participation of DMPK isoforms in the DM1 phenotype and mitochondria morphology, overexpression of tail-anchored human DMPK isoforms in heart and skeletal muscles of Tg26-hDMPK mice causes accumulation of mitochondria in the subsarcolemmal space, disorganization of the mitochondrial cristae structure, and importantly, the appearance of DM1-like symptoms, including reduced workload tolerance, atrophy, cardiomyopathy, and myotonic myopathy (O'Cochlain et al., 2004). Oude Ophuis and cols. propose that expression and activity of DMPK isoforms should be tightly controlled, both in time and place. Thus, any alteration in the balance of DMPK isoforms expression due to changes in mRNA synthesis or pre-mRNA splicing, might contribute to the DM1 pathophysiology.

In addition, DMPK deficiency could be involved in specific process of DM1 biogenesis, according to its interaction with different protein partners and targets (Table 1).

7. Conclusion

DM1 is a neuromuscular multisystemic disease caused by the expansion of the CTG repeats in the 3'UTR of the *DMPK* gene. *DMPK* primary transcript gives rise to 6 isoforms with activity of serine/threonine kinase that differ each other in their capacity to anchor organelle membranes. DMPK is expressed mainly in heart, skeletal and smooth muscles, and brain, the most compromised tissues in DM1. Current knowledge indicates that clinical manifestations of DM1 are not due to a unique molecular mechanism; instead, it appears that different mechanisms operate in the development of the disease. In addition to the toxic RNA gain-of-function, the best-described mechanism, chromatin rearrangements of the DM1 locus, leaching of transcription factors, interference RNA pathways and altered expression of DMPK protein should be considered as contributors to DM1 biogenesis. Importantly, several partners and targets of DMPK have been reported, suggesting the involvement of DMPK in the specific cellular pathways affected in DM1. Although the physiological function of DMPK and its isoforms is not yet fully understood, growing body of experimental evidence strongly suggests a role for DMPK in DM1 pathophysiology. Future studies in different groups of DM1 patients (CDM versus classic DM1), as well as in animal and cells models with DMPK deficiency are required to define the participation of DMPK in DM1 biogenesis.

8. Acknowledgements

This work was supported by a Grant (PIFUTP08-164) from Science and Technology Institute of Mexico City (ICyTDF).

9. References

Abell, B.M.; Pool, M.R.; Schlenker, O.; Sinning, I. & High, S. (2004). Signal recognition particle mediates post-translational targeting in eukaryotes. *EMBO J*, Vol. 23, No. 14, (Jul 21 2004), pp. 2755-64, ISSN 0261-4189

Adams, J.A. (2003). Activation loop phosphorylation and catalysis in protein kinases: is there functional evidence for the autoinhibitor model? *Biochemistry*, Vol. 42, No. 3, (Jan 28 2003), pp. 601-7, ISSN 0006-2960

Balasubramanyam, A.; Iyer, D.; Stringer, J.L.; Beaulieu, C.; Potvin, A.; Neumeyer, A.M.; Avruch, J. & Epstein, H.F. (1998). Developmental changes in expression of myotonic dystrophy protein kinase in the rat central nervous system. *J Comp Neurol*, Vol. 394, No. 3, (May 11 1998), pp. 309-25, ISSN 0021-9967

Benders, A.A.; Groenen, P.J.; Oerlemans, F.T.; Veerkamp, J.H. & Wieringa, B. (1997). Myotonic dystrophy protein kinase is involved in the modulation of the Ca2+ homeostasis in skeletal muscle cells. *J Clin Invest*, Vol. 100, No. 6, (Sep 15 1997), pp. 1440-7, ISSN 0021-9738

Berul, C.I.; Maguire, C.T.; Aronovitz, M.J.; Greenwood, J.; Miller, C.; Gehrmann, J.; Housman, D.; Mendelsohn, M.E. & Reddy, S. (1999). DMPK dosage alterations result in atrioventricular conduction abnormalities in a mouse myotonic dystrophy model. *J Clin Invest*, Vol. 103, No. 4, (Feb 1999), pp. R1-7, ISSN 0021-9738

Berul, C.I.; Maguire, C.T.; Gehrmann, J. & Reddy, S. (2000). Progressive atrioventricular conduction block in a mouse myotonic dystrophy model. *J Interv Card Electrophysiol*, Vol. 4, No. 2, (Jun 2000), pp. 351-8, ISSN 1383-875X

Bhagavati, S.; Ghatpande, A. & Leung, B. (1996). Normal levels of DM RNA and myotonin protein kinase in skeletal muscle from adult myotonic dystrophy (DM) patients. *Biochim Biophys Acta*, Vol. 1317, No. 3, (Dec 16 1996), pp. 155-7, ISSN 0006-3002

Biddinger, S.B. & Kahn, C.R. (2006). From mice to men: insights into the insulin resistance syndromes. *Annu Rev Physiol*, Vol. 68, No. 2006), pp. 123-58, ISSN 0066-4278

Brook, J.D.; McCurrach, M.E.; Harley, H.G.; Buckler, A.J.; Church, D.; Aburatani, H.; Hunter, K.; Stanton, V.P.; Thirion, J.P.; Hudson, T. & et al. (1992). Molecular basis of myotonic dystrophy: expansion of a trinucleotide (CTG) repeat at the 3' end of a transcript encoding a protein kinase family member. *Cell*, Vol. 69, No. 2, (Apr 17 1992), pp. 385, ISSN 0092-8674

Burkhard, P.; Stetefeld, J. & Strelkov, S.V. (2001). Coiled coils: a highly versatile protein folding motif. *Trends Cell Biol*, Vol. 11, No. 2, (Feb 2001), pp. 82-8, ISSN 0962-8924

Bush, E.W.; Taft, C.S.; Meixell, G.E. & Perryman, M.B. (1996). Overexpression of myotonic dystrophy kinase in BC3H1 cells induces the skeletal muscle phenotype. *J Biol Chem*, Vol. 271, No. 1, (Jan 5 1996), pp. 548-52, ISSN 0021-9258

Bush, E.W.; Helmke, S.M.; Birnbaum, R.A. & Perryman, M.B. (2000). Myotonic dystrophy protein kinase domains mediate localization, oligomerization, novel catalytic activity, and autoinhibition. *Biochemistry*, Vol. 39, No. 29, (Jul 25 2000), pp. 8480-90, ISSN 0006-2960

Caenepeel, S.; Charydczak, G.; Sudarsanam, S.; Hunter, T. & Manning, G. (2004). The mouse kinome: discovery and comparative genomics of all mouse protein kinases. *Proc Natl Acad Sci U S A*, Vol. 101, No. 32, (Aug 10 2004), pp. 11707-12, ISSN 0027-8424

Cho, D.H.; Thienes, C.P.; Mahoney, S.E.; Analau, E.; Filippova, G.N. & Tapscott, S.J. (2005). Antisense transcription and heterochromatin at the DM1 CTG repeats are constrained by CTCF. *Mol Cell*, Vol. 20, No. 3, (Nov 11 2005), pp. 483-9, ISSN 1097-2765

Dansithong, W.; Paul, S.; Comai, L. & Reddy, S. (2005). MBNL1 is the primary determinant of focus formation and aberrant insulin receptor splicing in DM1. *J Biol Chem*, Vol. 280, No. 7, (Feb 18 2005), pp. 5773-80, ISSN 0021-9258

Davis, B.M.; McCurrach, M.E.; Taneja, K.L.; Singer, R.H. & Housman, D.E. (1997). Expansion of a CUG trinucleotide repeat in the 3' untranslated region of myotonic dystrophy protein kinase transcripts results in nuclear retention of transcripts. *Proc Natl Acad Sci U S A*, Vol. 94, No. 14, (Jul 8 1997), pp. 7388-93, ISSN 0027-8424

de Leon, M.B. & Cisneros, B. (2008). Myotonic dystrophy 1 in the nervous system: from the clinic to molecular mechanisms. *J Neurosci Res*, Vol. 86, No. 1, (Jan 2008), pp. 18-26, ISSN 0360-4012

Doran, J.D.; Liu, X.; Taslimi, P.; Saadat, A. & Fox, T. (2004). New insights into the structure-function relationships of Rho-associated kinase: a thermodynamic and hydrodynamic study of the dimer-to-monomer transition and its kinetic implications. *Biochem J*, Vol. 384, No. Pt 2, (Dec 1 2004), pp. 255-62, ISSN 1470-8728

Du, H.; Cline, M.S.; Osborne, R.J.; Tuttle, D.L.; Clark, T.A.; Donohue, J.P.; Hall, M.P.; Shiue, L.; Swanson, M.S.; Thornton, C.A. & Ares, M., Jr. (2010). Aberrant alternative splicing and extracellular matrix gene expression in mouse models of myotonic dystrophy. *Nat Struct Mol Biol*, Vol. 17, No. 2, (Feb 2010), pp. 187-93, ISSN 1545-9985

Ebralidze, A.; Wang, Y.; Petkova, V.; Ebralidse, K. & Junghans, R.P. (2004). RNA leaching of transcription factors disrupts transcription in myotonic dystrophy. *Science*, Vol. 303, No. 5656, (Jan 16 2004), pp. 383-7, ISSN 1095-9203

Elkins, J.M.; Amos, A.; Niesen, F.H.; Pike, A.C.; Fedorov, O. & Knapp, S. (2009). Structure of dystrophia myotonica protein kinase. *Protein Sci*, Vol. 18, No. 4, (Apr 2009), pp. 782-91, ISSN 1469-896X

Endo, A.; Motonaga, K.; Arahata, K.; Harada, K.; Yamada, T. & Takashima, S. (2000). Developmental expression of myotonic dystrophy protein kinase in brain and its relevance to clinical phenotype. *Acta Neuropathol*, Vol. 100, No. 5, (Nov 2000), pp. 513-20, ISSN 0001-6322

Engert, F. & Bonhoeffer, T. (1999). Dendritic spine changes associated with hippocampal long-term synaptic plasticity. *Nature*, Vol. 399, No. 6731, (May 6 1999), pp. 66-70, ISSN 0028-0836

Fischer, M.; Kaech, S.; Wagner, U.; Brinkhaus, H. & Matus, A. (2000). Glutamate receptors regulate actin-based plasticity in dendritic spines. *Nat Neurosci*, Vol. 3, No. 9, (Sep 2000), pp. 887-94, ISSN 1097-6256

Forner, F.; Furlan, S. & Salvatori, S. (2010). Mass spectrometry analysis of complexes formed by myotonic dystrophy protein kinase (DMPK). *Biochim Biophys Acta*, Vol. 1804, No. 6, (Jun 2010), pp. 1334-41, ISSN 0006-3002

Frank, K.F.; Bolck, B.; Erdmann, E. & Schwinger, R.H. (2003). Sarcoplasmic reticulum Ca2+-ATPase modulates cardiac contraction and relaxation. *Cardiovasc Res*, Vol. 57, No. 1, (Jan 2003), pp. 20-7, ISSN 0008-6363

Fu, Y.H.; Pizzuti, A.; Fenwick, R.G., Jr.; King, J.; Rajnarayan, S.; Dunne, P.W.; Dubel, J.; Nasser, G.A.; Ashizawa, T.; de Jong, P. & et al. (1992). An unstable triplet repeat in a gene related to myotonic muscular dystrophy. *Science*, Vol. 255, No. 5049, (Mar 6 1992), pp. 1256-8, ISSN 0036-8075

Fu, Y.H.; Friedman, D.L.; Richards, S.; Pearlman, J.A.; Gibbs, R.A.; Pizzuti, A.; Ashizawa, T.; Perryman, M.B.; Scarlato, G.; Fenwick, R.G., Jr. & et al. (1993). Decreased expression of myotonin-protein kinase messenger RNA and protein in adult form of myotonic dystrophy. *Science*, Vol. 260, No. 5105, (Apr 9 1993), pp. 235-8, ISSN 0036-8075

Furling, D.; Lemieux, D.; Taneja, K. & Puymirat, J. (2001). Decreased levels of myotonic dystrophy protein kinase (DMPK) and delayed differentiation in human myotonic dystrophy myoblasts. *Neuromuscul Disord*, Vol. 11, No. 8, (Nov 2001), pp. 728-35, ISSN 0960-8966

Furling, D.; Lam le, T.; Agbulut, O.; Butler-Browne, G.S. & Morris, G.E. (2003). Changes in myotonic dystrophy protein kinase levels and muscle development in congenital myotonic dystrophy. *Am J Pathol*, Vol. 162, No. 3, (Mar 2003), pp. 1001-9, ISSN 0002-9440

Garcia, P.; Ucurum, Z.; Bucher, R.; Svergun, D.I.; Huber, T.; Lustig, A.; Konarev, P.V.; Marino, M. & Mayans, O. (2006). Molecular insights into the self-assembly mechanism of dystrophia myotonica kinase. *FASEB J*, Vol. 20, No. 8, (Jun 2006), pp. 1142-51, ISSN 1530-6860

Groenen, P.J.; Wansink, D.G.; Coerwinkel, M.; van den Broek, W.; Jansen, G. & Wieringa, B. (2000). Constitutive and regulated modes of splicing produce six major myotonic dystrophy protein kinase (DMPK) isoforms with distinct properties. *Hum Mol Genet*, Vol. 9, No. 4, (Mar 1 2000), pp. 605-16, ISSN 0964-6906

Hanks, S.K. & Hunter, T. (1995). Protein kinases 6. The eukaryotic protein kinase superfamily: kinase (catalytic) domain structure and classification. *FASEB J*, Vol. 9, No. 8, (May 1995), pp. 576-96, ISSN 0892-6638

Harley, H.G.; Rundle, S.A.; Reardon, W.; Myring, J.; Crow, S.; Brook, J.D.; Harper, P.S. & Shaw, D.J. (1992). Unstable DNA sequence in myotonic dystrophy. *Lancet*, Vol. 339, No. 8802, (May 9 1992), pp. 1125-8, ISSN 0140-6736

Harmon, E.B.; Harmon, M.L.; Larsen, T.D.; Paulson, A.F. & Perryman, M.B. (2008). Myotonic dystrophy protein kinase is expressed in embryonic myocytes and is required for myotube formation. *Dev Dyn*, Vol. 237, No. 9, (Sep 2008), pp. 2353-66, ISSN 1058-8388

Harmon, E.B.; Harmon, M.L.; Larsen, T.D.; Yang, J.; Glasford, J.W. & Perryman, M.B. (2011). Myotonic dystrophy protein kinase is critical for nuclear envelope integrity. *J Biol Chem*, Vol. No. (Sep 26 2011), pp. ISSN 1083-351X

Harper, P.S.; van Engelen, B.G.; Eymard, B.; Rogers, M. & Wilcox, D. (2002). 99th ENMC international workshop: myotonic dystrophy: present management, future therapy. 9-11 November 2001, Naarden, The Netherlands. *Neuromuscul Disord*, Vol. 12, No. 6, (Aug 2002), pp. 596-9, ISSN 0960-8966

Hartshorne, D.J.; Ito, M. & Erdodi, F. (1998). Myosin light chain phosphatase: subunit composition, interactions and regulation. *J Muscle Res Cell Motil*, Vol. 19, No. 4, (May 1998), pp. 325-41, ISSN 0142-4319

Hetzer, M.W. (2010). The nuclear envelope. *Cold Spring Harb Perspect Biol*, Vol. 2, No. 3, (Mar 2010), pp. a000539, ISSN 1943-0264

Hofmann-Radvanyi, H.; Lavedan, C.; Rabes, J.P.; Savoy, D.; Duros, C.; Johnson, K. & Junien, C. (1993). Myotonic dystrophy: absence of CTG enlarged transcript in congenital forms, and low expression of the normal allele. *Hum Mol Genet*, Vol. 2, No. 8, (Aug 1993), pp. 1263-6, ISSN 0964-6906

Huse, M. & Kuriyan, J. (2002). The conformational plasticity of protein kinases. *Cell*, Vol. 109, No. 3, (May 3 2002), pp. 275-82, ISSN 0092-8674

Iyer, D.; Belaguli, N.; Fluck, M.; Rowan, B.G.; Wei, L.; Weigel, N.L.; Booth, F.W.; Epstein, H.F.; Schwartz, R.J. & Balasubramanyam, A. (2003). Novel phosphorylation target in the serum response factor MADS box regulates alpha-actin transcription. *Biochemistry*, Vol. 42, No. 24, (Jun 24 2003), pp. 7477-86, ISSN 0006-2960

Jacobs, M.; Hayakawa, K.; Swenson, L.; Bellon, S.; Fleming, M.; Taslimi, P. & Doran, J. (2006). The structure of dimeric ROCK I reveals the mechanism for ligand selectivity. *J Biol Chem*, Vol. 281, No. 1, (Jan 6 2006), pp. 260-8, ISSN 0021-9258

Jansen, G.; Mahadevan, M.; Amemiya, C.; Wormskamp, N.; Segers, B.; Hendriks, W.; O'Hoy, K.; Baird, S.; Sabourin, L.; Lennon, G. & et al. (1992). Characterization of the myotonic dystrophy region predicts multiple protein isoform-encoding mRNAs. *Nat Genet*, Vol. 1, No. 4, (Jul 1992), pp. 261-6, ISSN 1061-4036

Jansen, G.; Groenen, P.J.; Bachner, D.; Jap, P.H.; Coerwinkel, M.; Oerlemans, F.; van den Broek, W.; Gohlsch, B.; Pette, D.; Plomp, J.J.; Molenaar, P.C.; Nederhoff, M.G.; van Echteld, C.J.; Dekker, M.; Berns, A.; Hameister, H. & Wieringa, B. (1996). Abnormal myotonic dystrophy protein kinase levels produce only mild myopathy in mice. *Nat Genet*, Vol. 13, No. 3, (Jul 1996), pp. 316-24, ISSN 1061-4036

Jiang, H.; Mankodi, A.; Swanson, M.S.; Moxley, R.T. & Thornton, C.A. (2004). Myotonic dystrophy type 1 is associated with nuclear foci of mutant RNA, sequestration of muscleblind proteins and deregulated alternative splicing in neurons. *Hum Mol Genet*, Vol. 13, No. 24, (Dec 15 2004), pp. 3079-88, ISSN 0964-6906

Jin, S.; Shimizu, M.; Balasubramanyam, A. & Epstein, H.F. (2000). Myotonic dystrophy protein kinase (DMPK) induces actin cytoskeletal reorganization and apoptotic-like blebbing in lens cells. *Cell Motil Cytoskeleton*, Vol. 45, No. 2, (Feb 2000), pp. 133-48, ISSN 0886-1544

Johnson, L.N.; Noble, M.E. & Owen, D.J. (1996). Active and inactive protein kinases: structural basis for regulation. *Cell*, Vol. 85, No. 2, (Apr 19 1996), pp. 149-58, ISSN 0092-8674

Kaliman, P.; Catalucci, D.; Lam, J.T.; Kondo, R.; Gutierrez, J.C.; Reddy, S.; Palacin, M.; Zorzano, A.; Chien, K.R. & Ruiz-Lozano, P. (2005). Myotonic dystrophy protein kinase phosphorylates phospholamban and regulates calcium uptake in cardiomyocyte sarcoplasmic reticulum. *J Biol Chem*, Vol. 280, No. 9, (Mar 4 2005), pp. 8016-21, ISSN 0021-9258

Kaliman, P. & Llagostera, E. (2008). Myotonic dystrophy protein kinase (DMPK) and its role in the pathogenesis of myotonic dystrophy 1. *Cell Signal*, Vol. 20, No. 11, (Nov 2008), pp. 1935-41, ISSN 0898-6568

Kim, C.H. & Lisman, J.E. (1999). A role of actin filament in synaptic transmission and long-term potentiation. *J Neurosci*, Vol. 19, No. 11, (Jun 1 1999), pp. 4314-24, ISSN 1529-2401

Kohn, W.D.; Mant, C.T. & Hodges, R.S. (1997). Alpha-helical protein assembly motifs. *J Biol Chem*, Vol. 272, No. 5, (Jan 31 1997), pp. 2583-6, ISSN 0021-9258

Krahe, R.; Ashizawa, T.; Abbruzzese, C.; Roeder, E.; Carango, P.; Giacanelli, M.; Funanage, V.L. & Siciliano, M.J. (1995). Effect of myotonic dystrophy trinucleotide repeat expansion on DMPK transcription and processing. *Genomics*, Vol. 28, No. 1, (Jul 1 1995), pp. 1-14, ISSN 0888-7543

Krupa, A.; Preethi, G. & Srinivasan, N. (2004). Structural modes of stabilization of permissive phosphorylation sites in protein kinases: distinct strategies in Ser/Thr and Tyr kinases. *J Mol Biol*, Vol. 339, No. 5, (Jun 18 2004), pp. 1025-39, ISSN 0022-2836

Kuyumcu-Martinez, N.M.; Wang, G.S. & Cooper, T.A. (2007). Increased steady-state levels of CUGBP1 in myotonic dystrophy 1 are due to PKC-mediated hyperphosphorylation. *Mol Cell*, Vol. 28, No. 1, (Oct 12 2007), pp. 68-78, ISSN 1097-2765

Ladd, A.N.; Charlet, N. & Cooper, T.A. (2001). The CELF family of RNA binding proteins is implicated in cell-specific and developmentally regulated alternative splicing. *Mol Cell Biol*, Vol. 21, No. 4, (Feb 2001), pp. 1285-96, ISSN 0270-7306

Lam, L.T.; Pham, Y.C.; Nguyen, T.M. & Morris, G.E. (2000). Characterization of a monoclonal antibody panel shows that the myotonic dystrophy protein kinase, DMPK, is expressed almost exclusively in muscle and heart. *Hum Mol Genet*, Vol. 9, No. 14, (Sep 1 2000), pp. 2167-73, ISSN 0964-6906

Landschulz, W.H.; Johnson, P.F. & McKnight, S.L. (1988). The leucine zipper: a hypothetical structure common to a new class of DNA binding proteins. *Science*, Vol. 240, No. 4860, (Jun 24 1988), pp. 1759-64, ISSN 0036-8075

Lee, H.C.; Patel, M.K.; Mistry, D.J.; Wang, Q.; Reddy, S.; Moorman, J.R. & Mounsey, J.P. (2003). Abnormal Na channel gating in murine cardiac myocytes deficient in myotonic dystrophy protein kinase. *Physiol Genomics*, Vol. 12, No. 2, (Jan 15 2003), pp. 147-57, ISSN 1531-2267

Leung, T.; Chen, X.Q.; Tan, I.; Manser, E. & Lim, L. (1998). Myotonic dystrophy kinase-related Cdc42-binding kinase acts as a Cdc42 effector in promoting cytoskeletal reorganization. *Mol Cell Biol*, Vol. 18, No. 1, (Jan 1998), pp. 130-40, ISSN 0270-7306

Lin, X.; Miller, J.W.; Mankodi, A.; Kanadia, R.N.; Yuan, Y.; Moxley, R.T.; Swanson, M.S. & Thornton, C.A. (2006). Failure of MBNL1-dependent post-natal splicing transitions in myotonic dystrophy. *Hum Mol Genet*, Vol. 15, No. 13, (Jul 1 2006), pp. 2087-97, ISSN 0964-6906

Lupas, A. (1996). Coiled coils: new structures and new functions. *Trends Biochem Sci*, Vol. 21, No. 10, (Oct 1996), pp. 375-82, ISSN 0968-0004

Llagostera, E.; Catalucci, D.; Marti, L.; Liesa, M.; Camps, M.; Ciaraldi, T.P.; Kondo, R.; Reddy, S.; Dillmann, W.H.; Palacin, M.; Zorzano, A.; Ruiz-Lozano, P.; Gomis, R. & Kaliman, P. (2007). Role of myotonic dystrophy protein kinase (DMPK) in glucose homeostasis and muscle insulin action. *PLoS One*, Vol. 2, No. 11, 2007), pp. e1134, ISSN 1932-6203

MacLennan, D.H. & Kranias, E.G. (2003). Phospholamban: a crucial regulator of cardiac contractility. *Nat Rev Mol Cell Biol*, Vol. 4, No. 7, (Jul 2003), pp. 566-77, ISSN 1471-0072

Madaule, P.; Eda, M.; Watanabe, N.; Fujisawa, K.; Matsuoka, T.; Bito, H.; Ishizaki, T. & Narumiya, S. (1998). Role of citron kinase as a target of the small GTPase Rho in cytokinesis. *Nature*, Vol. 394, No. 6692, (Jul 30 1998), pp. 491-4, ISSN 0028-0836

Mahadevan, M.; Tsilfidis, C.; Sabourin, L.; Shutler, G.; Amemiya, C.; Jansen, G.; Neville, C.; Narang, M.; Barcelo, J.; O'Hoy, K. & et al. (1992). Myotonic dystrophy mutation: an unstable CTG repeat in the 3' untranslated region of the gene. *Science*, Vol. 255, No. 5049, (Mar 6 1992), pp. 1253-5, ISSN 0036-8075

Mahadevan, M.S.; Amemiya, C.; Jansen, G.; Sabourin, L.; Baird, S.; Neville, C.E.; Wormskamp, N.; Segers, B.; Batzer, M.; Lamerdin, J. & et al. (1993). Structure and genomic sequence of the myotonic dystrophy (DM kinase) gene. *Hum Mol Genet*, Vol. 2, No. 3, (Mar 1993), pp. 299-304, ISSN 0964-6906

Mankodi, A.; Lin, X.; Blaxall, B.C.; Swanson, M.S. & Thornton, C.A. (2005). Nuclear RNA foci in the heart in myotonic dystrophy. *Circ Res*, Vol. 97, No. 11, (Nov 25 2005), pp. 1152-5, ISSN 1524-4571

Manning, G.; Whyte, D.B.; Martinez, R.; Hunter, T. & Sudarsanam, S. (2002). The protein kinase complement of the human genome. *Science*, Vol. 298, No. 5600, (Dec 6 2002), pp. 1912-34, ISSN 1095-9203

Mason, J.M. & Arndt, K.M. (2004). Coiled coil domains: stability, specificity, and biological implications. *Chembiochem*, Vol. 5, No. 2, (Feb 6 2004), pp. 170-6, ISSN 1439-4227

Meola, G. & Sansone, V. (2007). Cerebral involvement in myotonic dystrophies. *Muscle Nerve*, Vol. 36, No. 3, (Sep 2007), pp. 294-306, ISSN 0148-639X

Moorman, J.R.; Palmer, C.J.; John, J.E., 3rd; Durieux, M.E. & Jones, L.R. (1992). Phospholemman expression induces a hyperpolarization-activated chloride current in Xenopus oocytes. *J Biol Chem*, Vol. 267, No. 21, (Jul 25 1992), pp. 14551-4, ISSN 0021-9258

Mounsey, J.P.; John, J.E., 3rd; Helmke, S.M.; Bush, E.W.; Gilbert, J.; Roses, A.D.; Perryman, M.B.; Jones, L.R. & Moorman, J.R. (2000a). Phospholemman is a substrate for myotonic dystrophy protein kinase. *J Biol Chem*, Vol. 275, No. 30, (Jul 28 2000a), pp. 23362-7, ISSN 0021-9258

Mounsey, J.P.; Mistry, D.J.; Ai, C.W.; Reddy, S. & Moorman, J.R. (2000b). Skeletal muscle sodium channel gating in mice deficient in myotonic dystrophy protein kinase. *Hum Mol Genet*, Vol. 9, No. 15, (Sep 22 2000b), pp. 2313-20, ISSN 0964-6906

Mulders, S.A.; van Horssen, R.; Gerrits, L.; Bennink, M.B.; Pluk, H.; de Boer-van Huizen, R.T.; Croes, H.J.; Wijers, M.; van de Loo, F.A.; Fransen, J.; Wieringa, B. & Wansink, D.G. (2011). Abnormal actomyosin assembly in proliferating and differentiating myoblasts upon expression of a cytosolic DMPK isoform. *Biochim Biophys Acta*, Vol. 1813, No. 5, (May 2011), pp. 867-77, ISSN 0006-3002

Muranyi, A.; Zhang, R.; Liu, F.; Hirano, K.; Ito, M.; Epstein, H.F. & Hartshorne, D.J. (2001). Myotonic dystrophy protein kinase phosphorylates the myosin phosphatase targeting subunit and inhibits myosin phosphatase activity. *FEBS Lett*, Vol. 493, No. 2-3, (Mar 30 2001), pp. 80-4, ISSN 0014-5793

Narang, M.A.; Waring, J.D.; Sabourin, L.A. & Korneluk, R.G. (2000). Myotonic dystrophy (DM) protein kinase levels in congenital and adult DM patients. *Eur J Hum Genet*, Vol. 8, No. 7, (Jul 2000), pp. 507-12, ISSN 1018-4813

Ng, Y.; Tan, I.; Lim, L. & Leung, T. (2004). Expression of the human myotonic dystrophy kinase-related Cdc42-binding kinase gamma is regulated by promoter DNA methylation and Sp1 binding. *J Biol Chem*, Vol. 279, No. 33, (Aug 13 2004), pp. 34156-64, ISSN 0021-9258

Nolen, B.; Taylor, S. & Ghosh, G. (2004). Regulation of protein kinases; controlling activity through activation segment conformation. *Mol Cell*, Vol. 15, No. 5, (Sep 10 2004), pp. 661-75, ISSN 1097-2765

Novelli, G.; Gennarelli, M.; Zelano, G.; Pizzuti, A.; Fattorini, C.; Caskey, C.T. & Dallapiccola, B. (1993). Failure in detecting mRNA transcripts from the mutated allele in myotonic dystrophy muscle. *Biochem Mol Biol Int*, Vol. 29, No. 2, (Feb 1993), pp. 291-7, ISSN 1039-9712

O'Cochlain, D.F.; Perez-Terzic, C.; Reyes, S.; Kane, G.C.; Behfar, A.; Hodgson, D.M.; Strommen, J.A.; Liu, X.K.; van den Broek, W.; Wansink, D.G.; Wieringa, B. & Terzic, A. (2004). Transgenic overexpression of human DMPK accumulates into hypertrophic cardiomyopathy, myotonic myopathy and hypotension traits of myotonic dystrophy. *Hum Mol Genet*, Vol. 13, No. 20, (Oct 15 2004), pp. 2505-18, ISSN 0964-6906

Osborne, R.J. & Thornton, C.A. (2006). RNA-dominant diseases. *Hum Mol Genet*, Vol. 15 Spec No 2, No. (Oct 15 2006), pp. R162-9, ISSN 0964-6906

Otten, A.D. & Tapscott, S.J. (1995). Triplet repeat expansion in myotonic dystrophy alters the adjacent chromatin structure. *Proc Natl Acad Sci U S A*, Vol. 92, No. 12, (Jun 6 1995), pp. 5465-9, ISSN 0027-8424

Oude Ophuis, R.J.; Mulders, S.A.; van Herpen, R.E.; van de Vorstenbosch, R.; Wieringa, B. & Wansink, D.G. (2009a). DMPK protein isoforms are differentially expressed in myogenic and neural cell lineages. *Muscle Nerve*, Vol. 40, No. 4, (Oct 2009a), pp. 545-55, ISSN 0148-639X

Oude Ophuis, R.J.; Wijers, M.; Bennink, M.B.; van de Loo, F.A.; Fransen, J.A.; Wieringa, B. & Wansink, D.G. (2009b). A tail-anchored myotonic dystrophy protein kinase isoform induces perinuclear clustering of mitochondria, autophagy, and apoptosis. *PLoS One*, Vol. 4, No. 11, 2009b), pp. e8024, ISSN 1932-6203

Palmer, C.J.; Scott, B.T. & Jones, L.R. (1991). Purification and complete sequence determination of the major plasma membrane substrate for cAMP-dependent protein kinase and protein kinase C in myocardium. *J Biol Chem*, Vol. 266, No. 17, (Jun 15 1991), pp. 11126-30, ISSN 0021-9258

Pall, G.S.; Johnson, K.J. & Smith, G.L. (2003). Abnormal contractile activity and calcium cycling in cardiac myocytes isolated from DMPK knockout mice. *Physiol Genomics*, Vol. 13, No. 2, (Apr 16 2003), pp. 139-46, ISSN 1531-2267

Perbellini, R.; Greco, S.; Sarra-Ferraris, G.; Cardani, R.; Capogrossi, M.C.; Meola, G. & Martelli, F. (2011). Dysregulation and cellular mislocalization of specific miRNAs in myotonic dystrophy type 1. *Neuromuscul Disord*, Vol. 21, No. 2, (Feb 2011), pp. 81-8, ISSN 1873-2364

Pham, Y.C.; Man, N.; Lam, L.T. & Morris, G.E. (1998). Localization of myotonic dystrophy protein kinase in human and rabbit tissues using a new panel of monoclonal antibodies. *Hum Mol Genet*, Vol. 7, No. 12, (Nov 1998), pp. 1957-65, ISSN 0964-6906

Philips, A.V.; Timchenko, L.T. & Cooper, T.A. (1998). Disruption of splicing regulated by a CUG-binding protein in myotonic dystrophy. *Science*, Vol. 280, No. 5364, (May 1 1998), pp. 737-41, ISSN 0036-8075

Ranum, L.P. & Day, J.W. (2004). Myotonic dystrophy: RNA pathogenesis comes into focus. *Am J Hum Genet*, Vol. 74, No. 5, (May 2004), pp. 793-804, ISSN 0002-9297

Rau, F.; Freyermuth, F.; Fugier, C.; Villemin, J.P.; Fischer, M.C.; Jost, B.; Dembele, D.; Gourdon, G.; Nicole, A.; Duboc, D.; Wahbi, K.; Day, J.W.; Fujimura, H.; Takahashi, M.P.; Auboeuf, D.; Dreumont, N.; Furling, D. & Charlet-Berguerand, N. (2011). Misregulation of miR-1 processing is associated with heart defects in myotonic dystrophy. *Nat Struct Mol Biol*, Vol. 18, No. 7, (Jul 2011), pp. 840-5, ISSN 1545-9985

Reddy, S.; Smith, D.B.; Rich, M.M.; Leferovich, J.M.; Reilly, P.; Davis, B.M.; Tran, K.; Rayburn, H.; Bronson, R.; Cros, D.; Balice-Gordon, R.J. & Housman, D. (1996). Mice lacking the myotonic dystrophy protein kinase develop a late onset progressive myopathy. *Nat Genet*, Vol. 13, No. 3, (Jul 1996), pp. 325-35, ISSN 1061-4036

Reddy, S.; Mistry, D.J.; Wang, Q.C.; Geddis, L.M.; Kutchai, H.C.; Moorman, J.R. & Mounsey, J.P. (2002). Effects of age and gene dose on skeletal muscle sodium channel gating in mice deficient in myotonic dystrophy protein kinase. *Muscle Nerve*, Vol. 25, No. 6, (Jun 2002), pp. 850-7, ISSN 0148-639X

Riento, K. & Ridley, A.J. (2003). Rocks: multifunctional kinases in cell behaviour. *Nat Rev Mol Cell Biol*, Vol. 4, No. 6, (Jun 2003), pp. 446-56, ISSN 1471-0072

Roberts, R.; Timchenko, N.A.; Miller, J.W.; Reddy, S.; Caskey, C.T.; Swanson, M.S. & Timchenko, L.T. (1997). Altered phosphorylation and intracellular distribution of a (CUG)n triplet repeat RNA-binding protein in patients with myotonic dystrophy and in myotonin protein kinase knockout mice. *Proc Natl Acad Sci U S A*, Vol. 94, No. 24, (Nov 25 1997), pp. 13221-6, ISSN 0027-8424

Roux, K.J. & Burke, B. (2007). Nuclear envelope defects in muscular dystrophy. *Biochim Biophys Acta*, Vol. 1772, No. 2, (Feb 2007), pp. 118-27, ISSN 0006-3002

Saba, S.; Vanderbrink, B.A.; Luciano, B.; Aronovitz, M.J.; Berul, C.I.; Reddy, S.; Housman, D.; Mendelsohn, M.E.; Estes, N.A., 3rd & Wang, P.J. (1999). Localization of the sites of conduction abnormalities in a mouse model of myotonic dystrophy. *J Cardiovasc Electrophysiol*, Vol. 10, No. 9, (Sep 1999), pp. 1214-20, ISSN 1045-3873

Sarkar, P.S.; Han, J. & Reddy, S. (2004). In situ hybridization analysis of Dmpk mRNA in adult mouse tissues. *Neuromuscul Disord*, Vol. 14, No. 8-9, (Sep 2004), pp. 497-506, ISSN 0960-8966

Savkur, R.S.; Philips, A.V. & Cooper, T.A. (2001). Aberrant regulation of insulin receptor alternative splicing is associated with insulin resistance in myotonic dystrophy. *Nat Genet*, Vol. 29, No. 1, (Sep 2001), pp. 40-7, ISSN 1061-4036

Schulz, P.E.; McIntosh, A.D.; Kasten, M.R.; Wieringa, B. & Epstein, H.F. (2003). A role for myotonic dystrophy protein kinase in synaptic plasticity. *J Neurophysiol*, Vol. 89, No. 3, (Mar 2003), pp. 1177-86, ISSN 0022-3077

Shimokawa, M.; Ishiura, S.; Kameda, N.; Yamamoto, M.; Sasagawa, N.; Saitoh, N.; Sorimachi, H.; Ueda, H.; Ohno, S.; Suzuki, K. & Kobayashi, T. (1997). Novel isoform of myotonin protein kinase: gene product of myotonic dystrophy is localized in the sarcoplasmic reticulum of skeletal muscle. *Am J Pathol*, Vol. 150, No. 4, (Apr 1997), pp. 1285-95, ISSN 0002-9440

Siciliano, G.; Mancuso, M.; Tedeschi, D.; Manca, M.L.; Renna, M.R.; Lombardi, V.; Rocchi, A.; Martelli, F. & Murri, L. (2001). Coenzyme Q10, exercise lactate and CTG trinucleotide expansion in myotonic dystrophy. *Brain Res Bull*, Vol. 56, No. 3-4, (Oct-Nov 1 2001), pp. 405-10, ISSN 0361-9230

Sicot, G.; Gourdon, G. & Gomes-Pereira, M. (2011). Myotonic dystrophy, when simple repeats reveal complex pathogenic entities: new findings and future challenges. *Hum Mol Genet*, Vol. 20, No. R2, (Oct 15 2011), pp. R116-23, ISSN 1460-2083

Storbeck, C.J.; Sabourin, L.A.; Waring, J.D. & Korneluk, R.G. (1998). Definition of regulatory sequence elements in the promoter region and the first intron of the myotonic dystrophy protein kinase gene. *J Biol Chem*, Vol. 273, No. 15, (Apr 10 1998), pp. 9139-47, ISSN 0021-9258

Tan, I.; Seow, K.T.; Lim, L. & Leung, T. (2001). Intermolecular and intramolecular interactions regulate catalytic activity of myotonic dystrophy kinase-related Cdc42-binding kinase alpha. *Mol Cell Biol*, Vol. 21, No. 8, (Apr 2001), pp. 2767-78, ISSN 0270-7306

Taneja, K.L.; McCurrach, M.; Schalling, M.; Housman, D. & Singer, R.H. (1995). Foci of trinucleotide repeat transcripts in nuclei of myotonic dystrophy cells and tissues. *J Cell Biol*, Vol. 128, No. 6, (Mar 1995), pp. 995-1002, ISSN 0021-9525

Timchenko, L.; Nastainczyk, W.; Schneider, T.; Patel, B.; Hofmann, F. & Caskey, C.T. (1995). Full-length myotonin protein kinase (72 kDa) displays serine kinase activity. *Proc Natl Acad Sci U S A*, Vol. 92, No. 12, (Jun 6 1995), pp. 5366-70, ISSN 0027-8424

Timchenko, N.A.; Cai, Z.J.; Welm, A.L.; Reddy, S.; Ashizawa, T. & Timchenko, L.T. (2001). RNA CUG repeats sequester CUGBP1 and alter protein levels and activity of CUGBP1. *J Biol Chem*, Vol. 276, No. 11, (Mar 16 2001), pp. 7820-6, ISSN 0021-9258

Tiscornia, G. & Mahadevan, M.S. (2000). Myotonic dystrophy: the role of the CUG triplet repeats in splicing of a novel DMPK exon and altered cytoplasmic DMPK mRNA isoform ratios. *Mol Cell*, Vol. 5, No. 6, (Jun 2000), pp. 959-67, ISSN 1097-2765

Ueda, H.; Shimokawa, M.; Yamamoto, M.; Kameda, N.; Mizusawa, H.; Baba, T.; Terada, N.; Fujii, Y.; Ohno, S.; Ishiura, S. & Kobayashi, T. (1999). Decreased expression of myotonic dystrophy protein kinase and disorganization of sarcoplasmic reticulum in skeletal muscle of myotonic dystrophy. *J Neurol Sci*, Vol. 162, No. 1, (Jan 1 1999), pp. 38-50, ISSN 0022-510X

Ueda, H.; Ohno, S. & Kobayashi, T. (2000). Myotonic dystrophy and myotonic dystrophy protein kinase. *Prog Histochem Cytochem*, Vol. 35, No. 3, 2000), pp. 187-251, ISSN 0079-6336 van der Ven, P.F.; Jansen, G.; van Kuppevelt, T.H.; Perryman, M.B.; Lupa, M.; Dunne, P.W.; ter Laak, H.J.; Jap, P.H.; Veerkamp, J.H.; Epstein, H.F. & et al. (1993). Myotonic dystrophy kinase is a component of neuromuscular junctions. *Hum Mol Genet*, Vol. 2, No. 11, (Nov 1993), pp. 1889-94, ISSN 0964-6906

van Herpen, R.E.; Oude Ophuis, R.J.; Wijers, M.; Bennink, M.B.; van de Loo, F.A.; Fransen, J.; Wieringa, B. & Wansink, D.G. (2005). Divergent mitochondrial and endoplasmic reticulum association of DMPK splice isoforms depends on unique sequence arrangements in tail anchors. *Mol Cell Biol*, Vol. 25, No. 4, (Feb 2005), pp. 1402-14, ISSN 0270-7306

van Herpen, R.E.; Tjeertes, J.V.; Mulders, S.A.; Oude Ophuis, R.J.; Wieringa, B. & Wansink, D.G. (2006). Coiled-coil interactions modulate multimerization, mitochondrial binding and kinase activity of myotonic dystrophy protein kinase splice isoforms. *FEBS J*, Vol. 273, No. 6, (Mar 2006), pp. 1124-36, ISSN 1742-464X

Wang, Y.H.; Amirhaeri, S.; Kang, S.; Wells, R.D. & Griffith, J.D. (1994). Preferential nucleosome assembly at DNA triplet repeats from the myotonic dystrophy gene. *Science*, Vol. 265, No. 5172, (Jul 29 1994), pp. 669-71, ISSN 0036-8075

Wansink, D.G.; van Herpen, R.E.; Coerwinkel-Driessen, M.M.; Groenen, P.J.; Hemmings, B.A. & Wieringa, B. (2003). Alternative splicing controls myotonic dystrophy protein kinase structure, enzymatic activity, and subcellular localization. *Mol Cell Biol*, Vol. 23, No. 16, (Aug 2003), pp. 5489-501, ISSN 0270-7306

Waring, J.D.; Haq, R.; Tamai, K.; Sabourin, L.A.; Ikeda, J.E. & Korneluk, R.G. (1996). Investigation of myotonic dystrophy kinase isoform translocation and membrane association. *J Biol Chem*, Vol. 271, No. 25, (Jun 21 1996), pp. 15187-93, ISSN 0021-9258

Whiting, E.J.; Waring, J.D.; Tamai, K.; Somerville, M.J.; Hincke, M.; Staines, W.A.; Ikeda, J.E. & Korneluk, R.G. (1995). Characterization of myotonic dystrophy kinase (DMK) protein in human and rodent muscle and central nervous tissue. *Hum Mol Genet*, Vol. 4, No. 6, (Jun 1995), pp. 1063-72, ISSN 0964-6906

Whitley, P.; Grahn, E.; Kutay, U.; Rapoport, T.A. & von Heijne, G. (1996). A 12-residue-long polyleucine tail is sufficient to anchor synaptobrevin to the endoplasmic reticulum membrane. *J Biol Chem*, Vol. 271, No. 13, (Mar 29 1996), pp. 7583-6, ISSN 0021-9258

Winchester, C.L.; Ferrier, R.K.; Sermoni, A.; Clark, B.J. & Johnson, K.J. (1999). Characterization of the expression of DMPK and SIX5 in the human eye and implications for pathogenesis in myotonic dystrophy. *Hum Mol Genet*, Vol. 8, No. 3, (Mar 1999), pp. 481-92, ISSN 0964-6906

Wolf, E.; Kim, P.S. & Berger, B. (1997). MultiCoil: a program for predicting two- and three-stranded coiled coils. *Protein Sci*, Vol. 6, No. 6, (Jun 1997), pp. 1179-89, ISSN 0961-8368

Yang, M.; Ellenberg, J.; Bonifacino, J.S. & Weissman, A.M. (1997). The transmembrane domain of a carboxyl-terminal anchored protein determines localization to the endoplasmic reticulum. *J Biol Chem*, Vol. 272, No. 3, (Jan 17 1997), pp. 1970-5, ISSN 0021-9258

Zhang, R. & Epstein, H.F. (2003). Homodimerization through coiled-coil regions enhances activity of the myotonic dystrophy protein kinase. *FEBS Lett*, Vol. 546, No. 2-3, (Jul 10 2003), pp. 281-7, ISSN 0014-5793

Zhao, Z.S. & Manser, E. (2005). PAK and other Rho-associated kinases--effectors with surprisingly diverse mechanisms of regulation. *Biochem J*, Vol. 386, No. Pt 2, (Mar 1 2005), pp. 201-14, ISSN 1470-8728

10

Protein Kinases in the Pathogenesis of Muscle Wasting

Fabio Penna[1], Domiziana Costamagna[1], Andrea Camperi[1], Maurizio Muscaritoli[2], Francesco M. Baccino[1] and Paola Costelli[1]
[1]*Department of Experimental Medicine and Oncology, University of Torino,*
[2]*Department of Clinical Medicine, 'Sapienza', University of Roma, Italy*

1. Introduction

The skeletal muscle is a very heterogeneous tissue, that is in charge of a broad range of functions such as movement, stability, heat production and cold tolerance. It represents approximately 50% of total body protein, and plays a central role in whole body metabolism (Bassel-Duby & Olson, 2006). In the last two decades, the skeletal muscle, previously considered as a mere protein reservoir, has been shown to release cytokines and other humoral factors (Pedersen & Febbraio, 2008). This tissue plays a pivotal role in the overall energy balance. Indeed, it regulates lipid flux, takes up and stores most of plasma glucose, and modulates insulin sensitivity. In this regard, the skeletal muscle likely plays a crucial role in pathological states characterized by peripheral insulin resistance such as obesity, as also suggested by recent evidence showing the occurrence of a cross-talk between muscle and the adipose tissue (reviewed in Clarke & Henry, 2010).

The human body comprises about six hundred different muscles, composed of multinucleated cells organized to form muscle fibers. The myofiber contains many parallel myofibrils, characterized by alternating light (I) and dark (A) bands. The latters are bisected by a dark region (H zone), while I bands comprise a dark Z line (Z disk). The interspace between two Z disks is termed sarcomere, the functional unit of the myofibril. Myofibril number defines the cross-sectional area (CSA) of the myofiber, and determines the force-generating capacity. Myofibrillar contractile proteins myosin and actin form thick and thin filaments, respectively. Muscle myosin consists of two heavy chains (MyHC), endowed with ATPase activity, and two pairs of light chains (MyLC). Seven different genes coding for embryonic, neonatal and adult MHC isoforms have been described in humans. Myosin is organized in units assembled in a mobile side by side complex, where the head of myosin is at the distal tip of the filament and the tail at the center, rendering the thick filaments bipolar. Myosin heads interact with titin, connecting thick filaments to the Z disk. In the thin filaments, globular actin monomers are arranged in a double helical conformation, associated with tropomyosin, troponin, and nebulin, that regulate the interactions between actin and myosin. Troponin binds to Ca^{2+} released from intacellular stores, releasing tropomyosin and allowing the interaction between actin and myosin heads. Two actin-capping proteins, CapZ and tropomodulin, localized at the positive and negative end of the filament, respectively, prevent actin depolymerization. Finally, the Z disk is anchored to the sarcolemma by the intermediate filament protein desmin (reviewed in Morel & Pinset-Härström).

Contractile force is generated when myosin associates with actin and then rotates to pull the filaments across each other. Sliding of thick and thin filaments results in sarcomere shortening, due to the ATP-dependent cyclic interactions between the so called cross-bridges (protrusions from the myosin molecule, endowed with ATPase activity), and specific binding sites on actin. Contractile force is function of both the isometric length-tension curve and the speed of contraction, that depends on myosin ATPase activity. Two different types of skeletal muscles can be defined: those characterized by oxidative metabolism and by a long ATPase cycle time, termed slow-twitch, and those that produce energy mainly by glycolysis, with a rapid ATPase cycle, the fast-twich type. Different MyHC isoforms characterize the skeletal muscle types: the type I (β, or β-cardiac) isoform is predominantly expressed in slow fibers, while types IIa, and IIx (also known as IID) are expressed in human fast muscles (reviewed in Oldfor, 2007). Such compartmentalization, however, is not absolute, since several muscles in the body have a mixed composition.

Signals from the motor cortex reach the neuromuscular junction, generating an action potential that is transmitted to muscle fibers, leading to contraction. The performance of the motor unit (motor nerve and muscle fibers) is characterized by mechanical (speed of shortening) and metabolic properties (resistance to fatigue). While the former reflects MyHC isoform composition (see above), the latter depends on the maintenance of the energy balance. Speed of shortening and resistance to fatigue are strictly correlated. Indeed, slow muscles, characterized by small CSA, few muscle fibers per motor unit, long ATPase cycle time and high oxidative metabolism, display low speed of shortening, but have the possibility to replenish ATP during contraction. As a consequence, they are recruited during tasks that require low force or power but highly precise and long lasting movements. By contrast, fast muscles, that show large CSA, rapid ATPase cycle associated with glycolytic metabolism, are characterized by higher shortening speed then slow ones, but cannot rapidly replace ATP, and are recruited under circumstances when high power output is needed for short time intervals.

The number of muscle fibers in mammals does not change significantly during life, in the absence of injury or disease. Adult skeletal muscle, composed by post-mitotic cells, is endowed with a marked regenerative potential. Indeed, acute muscle injury is rapidly repaired by newly formed myotubes, since myogenic precursors (satellite cells) are conserved also in the postnatal period (Schultz & McCormick, 1994). In response to regenerative stimuli, satellite cells proliferate to form myoblasts, which divide a limited number of times before terminally differentiating and fusing into multinucleated myotubes (cf. Morgan and Partridge, 2003). Satellite cells are considered an adult stem cell population for the skeletal muscle, that can be renewed without losing the ability to generate differentiated myotubes. However, also other skeletal muscle progenitors have been proposed to contribute to myogenesis. As an example, intramuscular injection of muscle-resident side-population cells has been shown to contribute to myofiber formation. Similarly, CD45/Sca1 positive interstitial muscle-resident cells have been shown to acquire myogenic activity when cocultured with primary myoblasts or in response to muscle injury or Wnt signaling. The existence of distinct subsets of myogenic cells likely suggests that multiple mechanisms may support regeneration in the adult skeletal muscle, although the contribution of these cells to the maintenance or repair of skeletal muscle under physiologic conditions is uncertain, and their therapeutic potential has not been clearly established.

2. Physiological regulation of skeletal muscle mass

In principle, changes of myofiber number and/or dimension, this latter better defined by CSA, result in modulations of the skeletal muscle mass. However, modifications of myofiber number are rarely seen, while variations of CSA may occur frequently. In particular, CSA increases during normal growth or hypertrophy induced, for example, by exercise, and decreases in conditions of inactivity, injury, disease, or aging.

While muscle hypertrophy reflects an accumulation of contractile proteins, the opposite occurs in skeletal muscle atrophy, where both CSA and content of contractile proteins are reduced. On this lines, protein content is the main factor regulating skeletal muscle mass. The amount of proteins in a cell is strictly regulated by the balance between synthesis and degradation rates. In healthy state, protein synthesis and breakdown do not exceed each other, allowing normal protein turnover without modifying the skeletal muscle mass. Although modulations of both sides of turnover eventually converge to produce a new steady-state, physiologic muscle hypertrophy mainly result from increased rates of protein synthesis, which responds earlier than degradation to the inducing stimuli. By contrast, increased breakdown rates are in charge of protein depletion in many situations characterized by muscle atrophy.

A complex interplay among humoral mediators such as insulin and IGFs, and amino acids is involved in regulating the rates of intracellular protein synthesis. In this regard, signaling through the insulin/IGF-1 receptor, as well as increased amino acid levels, have been shown to simultaneously stimulate synthesis and inhibit protein catabolism. Protein synthesis induction by classical anabolic signals such as insulin or IGF-1 relies on the activation of a transduction pathway involving phosphoinositide-3-kinase (PI3K), Akt/PKB, mTOR (mammalian Target Of Rapamycin), and $p70^{S6K}$ (p70 ribosomal S6 kinase). As an example, this pathway has been shown to account for the generation of muscle hypertrophy induced by resistance exercise (reviewed in Adamo & Farrar, 2006). The demonstration that the PI3K/Akt/mTOR pathway is crucial to skeletal muscle increase in size has come from studies reporting that the expression of a constitutively active form of Akt in skeletal muscle cells, or its conditional activation in the skeletal muscle of adult rats, result in the appearance of a hypertrophic phenotype (Rommel et al., 2001; Lai et al., 2004). Similar patterns can be reproduced by administering a mixture of the 3 branched-chain amino acids (BCAAs: leucine, isoleucine, and valine), or even leucine alone. In addition to provide substrate for the assembly of new proteins, amino acids interfere with different transduction pathways involved in the regulation of mRNA translation (Kadowaki & Kanazawa, 2003). In particular, increased leucine intracellular concentrations have been shown to enhance the rate of translation by activating $p70^{S6K}$ and eIF-4F (eukaryotic Initiation Factor 4F), independently from Akt (Lang and Frost, 2005). The body protein-sparing effect of leucine has been suggested by the observation that nitrogen balance in fasting volunteers treated with leucine alone or with BCAA keto acid analogues is improved (Choudry et al., 2006).

The regulation of protein synthesis exerted by amino acids mainly relies on mTOR, a serine/threonine kinase crucially involved in cell growth. mTOR stimulates protein synthesis through three key regulatory proteins: $p70^{S6K}$, 4E-BP1 (eukaryotic initiation factor 4E-binding protein 1) and eIF-4G (eukaryotic Initiation Factor 4G). Reduced mTOR-mediated signaling has been reported in the skeletal muscle of fasted rats compared with the fed state. As expected, also the levels of phosphorylated 4E-BP1 are decreased in fasted animal; this would result in eIF-4E sequestration, inhibiting the assembly of the initiation complex eIF-4F. By contrast, 4E-BP1 is markedly hyperphosphorylated in the skeletal muscle of rats fed a high protein diet, promoting the formation of the eIF-4F complex. Moreover, leucine also promotes phosphorylation of $p70^{S6K}$ (Anthony et al., 2000). Akt activation also induces GSK-3β phosphorylation, thus resulting in its inactivation. GSK-3 β negatively regulates molecules involved in several anabolic processes and most of its effects are mediated by the PI3K/Akt pathway, in which GSK-3β acts both as a downstream target and as a negative regulator (Hanada et al., 2004). Consistently, non-competitive inhibition of GSK-3β, by means of transfection with a dominant-negative cDNA, or by pharmacological compounds, activates the PI3K/Akt pathway, resulting in myotube hypertrophy (Rommel et al., 2001; Van der Velden et al., 2006). In addition, the increased proteolysis observed in muscles isolated from burned rats can be prevented by addition of GSK-3β inhibitors to the incubation medium (Fang et al., 2005). Other studies have shown that GSK-3 β is involved in the pathogenesis of Alzheimer disease, prionic diseases, Hungtington corea, and gp-120 HIV-related neurotoxicity (Jope, 2003). All these considerations suggest that specific GSK-3 β inhibitors could be useful from a clinical point of view in order to correct muscle hypotrophy in wasting diseases such as cancer, HIV, cardiac cachexia, diabetes and to interfere with the pathogenic mechanisms of above cited neurologic diseases.

As for protein breakdown, it is also highly relevant to muscle homeostasis. Indeed, this process not only accounts for the degradation of damaged proteins as well as of regulatory molecules such as cyclins and their inhibitors, but also plays a crucial role in maintaining the right cellular size (Waterlow, 1984). The mobilization of muscle protein may have a physiological significance when aimed at providing substrates for both gluconeogenesis and the synthesis of the acute phase reactants. However, up-regulations of protein degradation rates exceeding protein synthesis may result in skeletal muscle wasting (see below). From this point of view, the intuitive means to counteract the loss of muscle mass resulting from protein hypercatabolism, is to increase protein synthesis. However, in terms of rate equations, protein synthesis is a zero-order process, while degradation of the bulk of cell proteins is a first-order process described by a fractional rate constant. Consequently, under a given set of regulations, the size of the protein pool does not affect the fraction of proteins degraded. This means that, if the breakdown rate constant is higher than physiologic levels, protein loss will occur irrespectively of the protein synthesis rate (cf. Costelli and Baccino, 2003).

3. Mechanisms involved in muscle mass changes

Muscle protein mass is regulated by both anabolic and catabolic signals. In particular, alterations in the balance between the two result in modulations of the skeletal muscle size, towards accretion or depletion when anabolic or catabolic pathways are prevailing, respectively. In this regard, pathological muscle depletion is characterized by a negative nitrogen balance which results from disruption of the equilibrium between anabolism and catabolism, due to reduced synthesis, increased breakdown, or both.

Generally speaking, muscle hypertrophy, either compensatory or due to working overload, is associated with up-regulation of protein synthesis. As reported above, particularly relevant in this regard are the activation of the PI3K/Akt/mTOR pathway induced by engagement of both insulin and IGF-1 receptors, as well as amino acid availability, BCAA in particular (see above). In addition, an important role in skeletal muscle enlargement is played by the protein phosphatase calcineurin (Musarò et al., 1999). As an example, rat myoblasts exposed to IGF-1 show a marked hypertrophic response that involves the enhancement of calcineurin expression, and that can be inhibited by the immunosuppressant agent cyclosporin A (Musarò et al., 1999). Similar observations have been performed also in the whole animal, where muscle hypertrophy induced by functional overload can be prevented by pharmacological inhibition of calcineurin with cyclosporin A or FK506 (Dunn et al., 1999).

While the regulation of protein synthesis is substantially well defined, the mechanisms underlying the activation of cell protein degradation to supraphysiological levels have not been completely elucidated. Intracellular proteolysis in the skeletal muscle is operated by several systems. The lysosomal and the proteasomal ones, are able to degrade proteins into amino acids or small peptides. By contrast, both the Ca^{2+}-dependent and the caspase pathways, characterized by a restricted catabolic specificity, only lead to a limited proteolysis of their substrates.

The ATP-ubiquitin-dependent proteasome system is mainly involved in the degradation of damaged or rapid-turnover proteins. Degradation of its substrates mostly requires the covalent attachment of at least four ubiquitin molecules; the presence of such polyubiquitin chain targets the substrate to the 26S proteasome, a large cytosolic proteolytic complex. Both proteasomal activity and substrate ubiquitylation are ATP-dependent processes. About twenty years ago, the ubiquitin-proteasome system has been shown to contribute significantly also to bulk protein degradation. This has become clear when increased expression of molecules pertaining to this proteolytic system have been reported in experimental conditions characterized by muscle wasting, the more so when two muscle-specific ubiquitin ligases, namely MAFbx/atrogin-1 and MuRF1 have been identified (reviewed in Costelli and Baccino, 2003). The former, in particular, is a component of the SCF complex, involved in targeting proteins for proteasomal degradation; such complex is formed by two molecules, SKP-1 (S) and Cullin-1 (C), that may be associated with a large series of F-box subunits (F), responsible for substrate specificity (Kipreos and Pagano, 2000). The results reported in the literature show that muscle wasting in several conditions such as sepsis, denervation, AIDS, diabetes, and cancer is associated with increased gene expression of both atrogin-1 and MuRF1 (Lecker et al., 2004). While the mechanisms regulating these ubiquitin ligases are not yet completely elucidated, hyperexpression of atrogin-1 has been proposed to depend on reduced signaling through the insulin/IGF-1 anabolic pathway (Sandri et al., 2004; Stitt et al., 2004), while activation of the NF-κB transcription factor, likely cytokine-dependent, seems to drive the increase of MuRF1 mRNA levels (Cai et al., 2004).

The autophagic-lysosomal degradative pathway, relatively non selective, is mostly responsible for the degradation of long-lived proteins as well as for the disposal of damaged organelles (reviewed in Scott & Klionsky, 1998). Autophagy relies on the sequestration of portions of cytoplasm into double-membrane vesicles (autophagosomes). These fuse with lysosomes, where the autophagic body is lysed, its content broken down, and the resulting degradation products made available for recycling (see Scott and Klionsky, 1998). Autophagy has been described in mammalian cells since the 1960s, however the underlying molecular mechanisms have been elucidated in the last years, with the identification of a set of genes named ATG (autophagy-related; Klionsky et al., 2003). Autophagy occurs at a basal rate in normal growth conditions, however, it can be markedly enhanced by specific environmental stresses. A crucial role in the regulation of autophagic rate is played by mTOR (see above). Under nutrient-rich conditions mTOR is active and autophagy is inhibited. By contrast, mTor is inactivated by nutrient starvation, and autophagic degradation is enhanced (Codogno & Mejier, 2005). The contribution of autophagy to skeletal muscle protein breakdown has been recognized only in the last years, although an altered lysosomal function has previously been reported in several myopathies (Bechet et al., 2005). In this regard, the skeletal muscle has been shown to respond to a classical autophagic stimulus such as starvation by increasing the levels of the autophagic marker LC3B-II (Mizushima et al., 2004). These results are consistent and further substantiate previous reports showing that autophagy is the main proteolytic pathway involved in the amino acid-dependent regulation of proteolysis in cultured myotubes (Bechet et al., 2005). On this line, increased gene expression of cathepsins L or B has been reported in the skeletal muscle of septic or tumor-bearing animals (Deval et al., 2001) as well as in muscle biopsies from lung cancer patients (Jagoe et al., 2002). In addition, skeletal muscle wasting in tumor-bearing rats has been shown to be associated with enhanced activity of lysosomal proteases (Greenbaum and Sutherland, 1983; Tessitore et al., 1993). Consistently, administration of leupeptin, an inhibitor of cysteine proteases, counteracts the loss of muscle mass that occurs in sepsis and in experimental cancer cachexia (Ruff and Secrist, 1984; Tessitore et al., 1994). More recently, ATGs have been shown to be induced in muscle by denervation or fasting, through a FoxO3-dependent mechanism (Zhao et al., 2007). In this regard, FoxO3 has been proposed to regulate both autophagy and proteasome-dependent proteolysis (Zhao et al., 2007). However, a sort of hierarchy appears to exist between these two processes, since a parallel study shows that autophagic degradation induced by starvation or FoxO3 overexpression is sufficient to determine muscle depletion even if the ubiquitin-proteasome degradation is blocked using pharmacological or genetic approaches (Mammucari et al., 2007).

Quite intriguing is the role of the Ca^{2+}-dependent proteolytic system in the pathogenesis of muscle protein hypercatabolism. Cysteine proteases called calpains, and a physiological inhibitor named calpastatin, are the components of the Ca^{2+}-dependent proteolytic system. Calpains have been involved in processes such as cell proliferation, differentiation, migration, apoptotic death, and gene expression (Suzuki et al., 2004). A number of proteins, among which protein kinase C, Cdk5, Ca^{2+}/calmodulin-dependent protein kinase IV, calcineurin, titin and nebulin have been proposed as *in vivo* calpain substrates (reviewed in Suzuki et al., 2004). Due to restricted specificity, however, calpain action is limited, and generally leads to irreversible modifications of the substrates, resulting in activity modulations or in increased susceptibility to the action of other degradative pathways (cf. Saido et al., 1994; Williams et al., 1999). Although thiol proteinase inhibitors have been proposed to be ineffective in counteracting muscle protein degradation in experimental cachexia (Temparis et al., 1994; Baracos et al., 1995), other reports have shown that administration of leupeptin is able to protect rats bearing the Yoshida ascites hepatoma AH-130 from muscle wasting, and that Ca^{2+}-dependent proteolysis is activated in muscles and heart of the AH-130 hosts (Costelli et al., 2001, unpublished observations). Similar observations suggesting the involvement of calpains in the pathogenesis of muscle depletion have been reported in septic rats administered dantrolene, an inhibitor of intracellular Ca^{2+} release; such treatment results in prevention of muscle wasting as well as of hyperexpression of calpains and of molecules pertaining to the ubiquitin–proteasome system (Williams et al., 1999; Wray et al., 2002). These reports are particularly intriguing since they propose that Ca^{2+}-dependent proteolysis may be a necessary step to allow the release of myofibrillar proteins from the sarcomere, rendering them susceptible to degradation by the ubiquitin–proteasome system. Finally, a report has demonstrated that hyperexpression of calpastatin partially protects mice from unloaded-induced muscle atrophy (Tidball & Spencer, 2002).

Similarly to calpains, also the caspase system, can only operate a partial proteolysis of its substrates. Caspases are a family of cysteine proteases mostly known for their role in the initiation and execution of the apoptotic process. Few years ago, some studies proposed that caspase 3 could also share with calpains the role as triggers of the initial proteolytic step needed to render myofibrillar proteins available for degradation by the proteasome. In this regard, recombinant caspase-3 has been shown to cleave actomyosin complexes, and caspase-3 inhibitors can prevent the accumulation of actin fragments in the skeletal muscles of diabetic or uremic rats (Du et al., 2004). Consistently with these observation, caspase-3 knock-out mice have been shown to be resistant to denervation-induced muscle atrophy (Plant et al., 2009). In addition, myofibrillar proteins damaged by oxidation appear more susceptible to degradation by caspase-3 (Smuder et al., 2010), while a recent study reports that caspase 3 cleaves specific proteasome subunits in myotube cultures, leading to enhanced proteasome enzymatic activity (Wang et al., 2010). Finally, muscle atrophy that occurs in Duchenne muscular dystrophy or in heart failure has been associated with reduced myonuclei number, suggesting that caspases may contribute to muscle depletion also by inducing apoptotic events (Sandri, 2002).

While several evidence support the concept that hypercatabolism is the major cause of muscle protein depletion, the trigger(s) of such enhanced metabolism remain still elusive. in this regard, humoral mediators are now widely accepted to play a crucial role. Indeed, altered production/release of classical hormones and cytokines generates a complex network that results in inhibition of anabolic and/or anticatabolic signals, favoring the degradative side of protein turnover. Consistently, the muscle wasting pattern observed in experimental and human cachexia or in aging-associated sarcopenia has been shown to be prevented by insulin administration or by local overexpression of IGF-1 (Tessitore et al., 1994; Musarò et al., 2001; Lundholm et al., 2007). On the other side, circulating glucocorticoids are frequently elevated at supraphysiological levels in several chronic pathologies, and have been shown to exert a clear catabolic effect (see Schakman et al., 2009). At least the proteasome and the lysosomal-dependent proteolytic systems are susceptible of regulation by the hormonal milieu. Indeed, insulin is one of the most powerful autophagy inhibitors (Pfeifer, 1977), is able to reduce the expression of both ubiquitin and 14-kDa E2 mRNA, and to down-regulate proteasome activities (Wang et al., 2006). By contrast, glucocorticoid treatment increases the expression of ubiquitin, 14-kDa E2 and 20S proteasome subunit in rat skeletal muscle (see Schakman et al., 2009). Muscle wasting and modulations of ubiquitin expression and proteasome activities have also been reported in experimental animals treated with the cytokines TNFα or IL-1 (Tisdale, 2008). The relevance of cytokines to the onset of muscle wasting at least in cancer cachexia have been demonstrated by studies showing that loss of muscle mass, protein hypercatabolism and ubiquitin hyperexpression can be prevented by administration of antibodies against TNFα, IFNγ or IL-6 (reviewed in Costelli and Baccino, 2003). Consistently with these observations, perturbation in cytokine homeostasis have been reported also in cancer patients, where a positive correlation with both disease progression and mortality rate takes place (Attard-Montalto et al., 1998; Nakashima et al., 1998). In addition, proinflammatory cytokines have been shown to contribute to muscle depletion also in non-neoplastic chronic diseases. Indeed, sepsis is characterized by increased circulating levels of TNFα, IL-1 and IL-6, that appear correlated with severity and lethality. Similarly, a shift towards the proinflammatory side of the cytokine balance has been reported in patients affected by AIDS (Kedzierska & Crowe, 2001), likely accounting for muscle protein hypercatabolism frequently occurring in such patients before the adoption of combined anti-retroviral therapy (HAART; Mangili et al., 2006). Finally, also the sarcopenia and the loss of muscle quality that characterize aging are associated with enhanced levels of proinflammatory mediators (Lee et al., 2007).

In addition to altered protein turnover rates, modulations of the myogenic process have been proposed to contribute to the pathogenesis of muscle wasting. In this regard, one key mediator of muscle depletion, TNFα, has been reported to regulate myogenesis with opposite outcomes. Local increase of TNFα in cardiotoxin-injured muscle has been shown to promote regeneration (Chen et al. 2005), while systemic increase of TNFα in vivo and elevated concentrations of the cytokine in vitro inhibit skeletal myogenesis (Guttridge et al. 2000; Coletti et al. 2002; 2005). In particular, exposure of C2C12 myotube cultures to TNFα leads to down-regulation of both MyoD and myogenin (Guttridge et al., 2000). MyoD appears down-regulated also in a TNFα-dependent experimental model of cancer cachexia (Costelli et al., 2005). A different study has shown that TNFα induces MyoD degradation through an unusual mechanism involving NFκB activation (Guttridge et al. 2000), while recently MyoD hjas been demonstrated to be a substrate of the ubiquitin ligase atrogin-1 (Tintignac et al., 2005). Down-regulation of myogenesis may also depend on impaired stem cell recruitment. In this regard, deregulation of stem cell number or activation has been shown to result in decreased muscle mass (Nicolas et al., 2005). Moreover, TNFα has been proposed to abrogate stem cell function, resulting in delayed or impaired muscle regeneration in mice after injury (Moresi et al., 2008). A compromised regenerative capacity has also been reported in tumor-bearing mice (Coletti et al., 2005; Penna et al., 2010a); such a pattern is associated with the appearance of hematopoietic stem cell infiltration the skeletal muscle, quantitatively more important in the tumor hosts than in controls (Coletti et al., 2005). Muscle atrophy induced in mice by aging or hindlimb suspension has also been associated with loss of muscle precursor cells, that results in reduced regenerative potential (Mitchell and Pavlath 2004).

4. Protein kinases in the pathogenesis of skeletal muscle wasting

Few kinase systems have been involved in the pathogenesis of muscle atrophy, the one regulated by growth factors such as insulin or IGF-1, the Mitogen Activated Protein Kinases (MAPKs), and the energy sensor AMP-activated protein kinase (AMPK).

Insulin/IGF-1 receptors are endowed with an intrinsic tyrosine kinase activity, that is stimulated by interaction with the specific ligands. After engagement, receptor autophosphorylation allows the recruitment of IRS (insulin receptor substrate) factors. Tyrosine-phosphorylated IRS activates PI3K, producing phosphoinositide-3,4,5-triphosphate (PIP3). PIP3 acts on phosphoinositide-dependent kinase 1 (PDK1), which in turn phosphorylates and activates Akt. This kinase is well known for mediating anabolic signals (see above) through the indirect activation of mTOR, that requires the inhibition of TSC (tuberous sclerosis complex). Once phosphorylated, mTOR may participate to two different protein complexes, the Raptor-containing TORC1, sensitive to inhibition by rapamycin, and the Rictor-containing TORC2, which cannot be blocked by rapamycin (reviewed in Schiaffino & Mammucari, 2011). While the latter is required for Akt activation, mTORC1 phosphorylates $p70^{S6K}$, stimulating protein synthesis. In addition to TORC1, protein synthesis induction also relies on Akt-dependent GSK3β inhibition, that consequently removes the blockade impinging on the elongation factor eIF-2B. Active Akt also down-regulates protein breakdown by inactivating FoxO factors, thus inhibiting the transcription of the so called 'atrogenes', among which the muscle-specific ubiquitin ligases atrogin-1 and MuRF1 (Sandri et al., 2004; Stitt et al., 2004). FoxO3, in particular, has been proposed to contribute also to the regulation of LC3, an essential actor in the hyperactivation of the autophagic-lysosomal proteolysis (Zhao et al., 2007). Akt activation is influenced by several regulative mechanisms. Indeed, it is inhibited by $p70^{S6K}$, through IRS inactivation by phosphorylation of serine residues, while it is induced by mTORC2 (see above). The PI3K/Akt pathway plays a pivotal role in modulating the skeletal muscle mass; indeed, it is upregulated in conditions characterized by muscle hypertrophy, while its disruption results in muscle atrophy (Glass, 2010). Not only, a hypertrophic phenotype occurs when Akt is hyperexpressed in skeletal muscle cells or is conditionally activated in the muscle of adult rats (Rommel et al., 2001; Lai et al., 2004). In addition, a protection against denervation-induced atrophy has been shown in transgenic mice overexpressing Akt (Bodine et al., 2001). Perturbations of the IGF-1 signaling pathway have been reported in both *in vitro* and *in vivo* models of muscle atrophy (reviewed in Glass, 2010). Indeed, the levels of active Akt are significantly reduced in C2C12 myotubes exposed to glucocorticoids or nutrient deprivation (Sandri et al., 2004). Decreased activity of the PI3K/Akt pathway has also been shown to occur in muscle wasting induced by denervation (Hornberger et al., 2001), disuse (Sugiura et al., 2005), aging (Clavel et al., 2006) or glucocorticoid treatment (Schakman et al., 2008). By contrast, levels of phosphorylated Akt in the skeletal muscle of tumor-bearing animals are comparable to controls, or even increased (Penna et al., 2010b), although a down-regulation of Akt activation has been reported in patients affected by pancreatic cancer (Schmitt et al., 2007). The maintenance of p-Akt levels in experimental cancer cachexia is particularly intriguing, since circulating IGF-1 and insulin levels are markedly reduced in the tumor-bearing animals (Costelli et al., 2006), and muscle wasting can be prevented by administration of insulin, though not of IGF-1 (Costelli et al., 2006; Tessitore et al., 1994). Akt phosphorylation mainly relies on the balance between the activity of PI3K and the phosphatases PTEN and PP2A. In particular, reduced PTEN activation has been observed in the skeletal muscle of fasted animals, likely to counteract Akt down-regulation, in the attempt to preserve muscle proteins (Hu et al., 2007). However, both phosphatases are comparably expressed in the skeletal muscle of control and tumor-bearing animals. In addition to Akt, also other molecules involved in the regulation of protein synthesis, such as eIF2α, eIF-4B, p70^{S6K}, are in an active state in the skeletal muscle of tumor-bearing animals; however, previous results show that the rates of protein synthesis are not increased, but just maintained at control levels (Costelli et al., 2005, Tessitore et al., 1994). Whether this results from the lack of specific aminoacids or from activation/inactivation of other unknown mechanisms is not clear. In this regard, an inhibition of protein synthesis could result from the atrogin-1-dependent degradation of eIF-3F, a scaffold protein that coordinates both mTOR- and p70^{S6K}-mediated translation. On the same line, protein synthesis has been proposed to be regulated by MuRF-1, independently from the PI3K/Akt pathway (Clarke et al., 2007; Koyama et al., 2008).

Four main MAPKs have been identified in mammals: JNK (1-3) and p38 (α-δ), activated by stress conditions, and the extracellular signal related ERK 1/2 (hereafter referred to as ERK) and ERK5, or big MAPK (Raman et al., 2007). MAPKs are activated by phosphorylation of both threonine and tyrosine residues by MAPK-kinases (MKKs) and inactivated by specific phosphatases such as the MAPK-phosphatase 1 (MKP-1; Raman et al., 2007). MAPKs are recognized as being of crucial importance in the process of myogenesis, although their role in the different steps of new fiber formation and specification still needs to be clarified. As an example, Ras-dependent ERK activation has been shown to lead to MHC-I expression, resulting in slow-fiber type differentiation (Murgia et al., 2000). These observations however, are in contrast with different studies reporting that ERK activation inhibits myotube formation (Miyake et al., 2009), while recent reports show that ERK activation is higher in fast- than in slow-twitch muscles (Shi et al., 2007) and that inhibition of MAPK signaling leads to a shift of fast fibers towards the slow-twitch phenotype (Shi et al., 2008). The activation of p38 appears required to phosphorylate substrates involved in myogenesis, as well as to induce MHC-IIx expression in myoblasts (Meissner et al., 2007). Indeed, p38 modulates the expression of myogenic regulatory factors (MRFs), such as Myf5, and the activities of transcription factors belonging to the MEF2 and MyoD families. A reciprocal regulation has been proposed to exist between p38 and ERK. While the former inhibits ERK, withdrawing myocytes from the cell cycle and enhancing muscle differentiation, ERK inhibition results in marked activation of p38 (Keren et al., 2006). In this regard, the interaction between these two kinases, likely leading to a defective activation of p38, has been proposed to play a role in the development of rhabdomyosarcoma (Puri et al., 2000). In addition to the reciprocal regulation with ERK, a cross-talk between p38 and JNK also takes place. Initially described in cardiomyocytes, it has now been demonstrated also in the skeletal muscle. In particular, p38 has been shown to antagonize the proliferative signal driven in myoblasts by JNK-dependent cyclin D1 transcription, shifting cells towards differentiation (Perdiguero et al., 2007). Consistently, p38 deficient myoblasts are characterized by a prominent JNK phosphorylation, that appears to depend, partially at least, on reduced expression of MKP-1 (Perdiguero et al., 2007). Finally, JNK has been also involved in the activation of caspases in atrophying skeletal muscles (Supinski et al., 2009).

Several situations characterized by muscle wasting, among which aging, type II diabetes, COPD, and inflammatory myopathies are associated with increased MAPK phosphorylation, p38 in particular (reviewed in Glass, 2010). Activation of p38 stimulates atrophy by enhancing the expression of atrogin-1 and MuRF1(Li et al., 2005; Romanello et al., 2010) This is also evident from *in vitro* experiments, showing that the increased expression of atrogin-1 and MuRF-1 induced by TNF-α in C2C12 myotubes as well as the induction of the ubiquitin-specific protease-19 by cigarette smoke in L6 cultures are prevented by p38 inhibitors (Li et al., 2005; Liu et al., 2011). Similarly, atrogin-1 upregulation and muscle mass depletion induced by lipopolysaccharide (LPS) in mice depend on p38 activation; indeed, such effects are inhibited by curcumin administration, that leaves intact LPS ability to modulate both NF-κB and Akt activity (Jin & Li, 2007). LPS exerts its bioactivity through Toll-like receptors (TLR), in particular TLR4, expressed on both macrophages and muscle cells. Signaling through this receptor may significantly impinge on muscle protein degradation for multiple reasones: TLR4 engagement leads to p38 and NF-κB activation; this could result in the upregulation of atrogin-1 and MuRF1 in the muscle, directly, or indirectly, through the release of proinflammatory mediators by macrophages. In addition, TLR4 has recently been involved in the activation of autophagy, increasing autophagosome formation by a p38-dependent mechanism (Doyle et al., 2011). Activation of p38 has also been shown to occur in response to mechanical or electrical stimulation, and functional overload of the skeletal muscle (Boppart et al., 2001; Huey, 2006; Sakamoto et al., 2003;), suggesting that this kinase plays a role in both anabolic and catabolic responses. Among the targets of p38 is MAPK-kinase 2 (MKK2), that appears involved in mediating p38 nuclear export (Gorog et al., 2009). MKK2 is phosphorylated by p38 at two threonine residues, both necessary for the activation (Engel et al., 1995); phosphorylation at T317 allows MKK2 export in a complex containing p38 itself (Ben-Levy et al., 1998; Meng et al., 2002). Heat shock protein 27 (HSP27, Stokoe et al., 1992), involved in the regulation of actin filament dynamics, is a substrate of MKK2 (Guay et al., 1997). Phosphorylation of HSP27 is increased in skeletal muscle hypertrophy and decreased during atrophy (Huey, 2006; Kawano et al., 2007), while HSP27 hyperexpression is able to reduce skeletal muscle depletion due to disuse (Dodd et al., 2009). Finally, MKK2 expression is reduced also in denervation-induced atrophy (Norrby and Tagerud, 2010) . The occurrence of a cross-talk between MKK2/p38 and PI3K/Akt/mTor pathways has been proposed. In this regard, the MKK2/p38 complex exported from the nucleus appears to interact with a cytoplasmic HSP27/Akt complex (Wu et al., 2007). Similarly to Akt, also MKK2 can phosphorylate TSC2 and FoxO1, thus impinging on both protein synthesis and catabolism (reviewed in Rosner et al., 2008).

The involvement of ERK in the pathogenesis of skeletal muscle atrophy is quite controversial. ERK inactivation has been shown to result in muscle atrophy in the rat, irrespective of the fiber type (Shi et al., 2009), and to inhibit the hypertrophic response induced in fast muscles by treatment of the animals with β_2-adrenergic agonists or IGF-1 (Haddad & Adams, 2004; Shi et al., 2007). In addition, reduced levels of phosphorylated ERK have been demonstrated in age-induced sarcopenia (Carlson et al., 2009). In C2C12 myotubes, ERK inhibition appeares required to stimulate ubiquitin ligase expression (Shi et al., 2008). Consistently, ubiquitin hyperexpression induced in L6 myotubes by glucocorticoids has been shown to depend on the activity of both MEK, the kinase upstream of ERK, and the Sp1 transcription factor (Marinovic et al., 2002). Constrasting observations have been reported, however. As an example, ERK activation in C2C12 cultures has been shown to result in reduced myotube size (Rommel et al., 1999), while its inhibition leads to a hypertrophic phenotype similar to that elicited by IGF-1 (Rommel et al., 1999). On the same line, the protection exerted against oxidative stress-induced damage in both C2C12 and L6 myocytes by treatment with IGF-1 has been proposed to involve ERK activity (Yang et al.,

2010). Finally, muscle atrophy due to immobilization by hind-limb suspension has been associated with increased levels of phosphorylated ERK (Kato et al., 2002). In addition to contribute in modulating adult skeletal mass, ERK has also been involved in the myogenic process. Indeed, FGF-induced activation of ERK has been shown to enhance the regenerative capacity of human satellite cells isolated from both young and old subjects, while proliferating fusion-competent myoblasts cannot be observed when ERK is inhibited (Carlson et al., 2009). Several humoral factors, such as IGF-1, proinflammatory cytokines and myostatin can contribute to activate ERK in the skeletal muscle. In particular, recent observations from our laboratory have shown that TNFα-induced myotube reduction in size in C2C12 cultures is associated with ERK activation and increased myostatin expression (Lenk et al., 2009). Similar observation have been reported also in the skeletal muscle of tumor-bearing mice (Penna et al., 2010a). In this regard, myostatin has been previously proposed to activate ERK and to repress differentiation of C2C12 myocytes (Yang et al., 2006), pointing to a causal relationship between myostatin and ERK biological activities.

Increased levels of phosphorylated JNK in the skeletal muscle are characteristically observed in conditions of insulin resistance, such as obesity or type II diabetes (Masharani et al., 2011). JNK activation mediates insulin resistance by enhancing IRS phosphorylation at serine residues, thus inhibiting the transduction of IGF-1/insulin-dependent signals (Masharani et al., 2011). Oxidative stress consequent to lipotoxicity, as well as proinflammatory mediators (cytokines and others, such as homocysteine), derived or not from the adipose tissue, likely participate in activating JNK. This latter, together with ERK and p38 MAPKs, is also activated in the skeletal muscle after exercise; such activation depends on exercise-induced oxidative stress, being prevented by treatment of healthy volunteers with the antioxidant N-acetylcysteine (Petersen et al., 2011). Both JNK and its preferential substrate c-Jun are activated in the muscle of patients with chronic kidney failure (Verzola et al., 2011). Recent reports have shown that the activation of JNK that characterizes denervation-induced muscle atrophy can be prevented by targeted ablation of the adapter protein TRAF6 (Paul et al., 2010). By contrast, no changes in the levels of phosphorylated JNK have been observed in the skeletal muscle of animals bearing experimental tumors (Penna et al., 2010a). The signal transduction pathway dependent on JNK plays a role in the apoptotic response in several cell systems (Dhanasekaran & Reddy, 2008). In this regard, muscle injury induced by cardiotoxin injection has been shown to be initially associated with JNK activation and perturbations in the Bax/Bcl-2 system, and subsequently with classical signs of apoptotic death such as cytochrome c release from mitochondria, caspase activation, PARP cleavage (Sinha-Hikim et al., 2007). The mechanisms underlying JNK-dependent apoptosis in cardiotoxin muscle injury is still unclear, however, increased NO production through iNOS induction might be involved (Sinha-Hikim et al., 2007). Consistently with these observation, diaphragm weakness induced by endotoxin treatment has been associated with JNK phosphorylation and caspase 8 activation (Supinski et al., 2009).

Finally, quite recent evidence support a role for AMPK in the pathogenesis of skeletal muscle wasting. This kinase mainly works as a sensor of intracellular energetic balance, but is also involved in the regulation of protein turnover. AMPK is switched on when the energy state of the cell is low; in the skeletal muscle, also fiber contraction, which is an energy dissipating process, leads to AMPK activation (Mihaylova & Shaw, 2011). The interference exerted by AMPK on protein synthesis is mainly related to its ability to inhibit mTOR signaling (Mihaylova & Shaw, 2011). On the other side, AMPK has been shown to modulate protein degradation rates. Indeed, administration of AICAR, an AMPK agonist, to mice results in increased levels of phosphorylated AMPK, associated with enhanced atrogin-1 expression through a FoxO-dependent mechanism; such a pattern is inhibited by treating the animals with Compound C, an AMPK inhibitor (Nakashima et al., 2007; Romanello et al., 2010). AMPK activation has also been reported in the skeletal muscle of tumor-bearing animals (Penna et al., 2010a ; White et al., 2011), where it is associated with marked alterations of mitochondrial morphology (Penna et al., unpublished observations). The AMPK-dependent pathway links the alterations in the mitochondrial system, including reduced ATP production, with the onset of muscle atrophy. In this regard, energy deficiency could result in AMPK-dependent FoxO activation. Taking into account that FoxO transcription factors have been also involved in the regulation of autophagy, and that this latter process is in charge of sequestration and degradation of damaged organelles, FoxO activation may contribute to mitochondrial loss, further enhancing energy imbalance.

5. Protein kinase inhibitors to prevent skeletal muscle wasting

Muscle wasting is now well accepted to derive from metabolic alterations due to the combined action of several factors that act in a complex network involving different signal transduction pathways. The result of such networking is clearly reflected on muscle protein turnover, ultimately leading to the onset of a protein hypercatabolic state. Muscle wasting in patients affected by chronic diseases, but also in 'healthy' elderly people (sarcopenia of aging), is a highly debilitating condition, that markedly impairs quality of life, recovery from illnesses, and tolerance to therapies. The result is a significant complications in the management of these persons, also with important consequences at the social care level. In this regard, therapeutic approaches aimed at interfering pharmacologically with the onset of tissue wasting need to be pursued. On the bases of results obtained in experimental models, a number of drugs have been proposed to counteract the development of muscle wasting. Among these are protein kinase inhibitors; the rationale for their use stands up from the observations reported by several studies demonstrating that protein kinases are of crucial relevance to the activation/inactivation of mechanisms involved in the depletion/preservation of skeletal muscle mass.

About 25 years ago the first natural kinase inhibitor, namely staurosporine, able to block protein kinase C but also many other kinases, has been discovered. Subsequently, the specific inhibitor of p38, SB203580, has become available, opening the research of heterocyclic "drug-like" structures able to distinguish between different kinases. As an example, SB203580 binds to p38, but not to the closely related JNK (Dar and Shokat, 2011). Up to now, nine kinase inhibitors have currently been approved by the FDA (imatinib, gefitininb, sorafenib, erlotininb, sunitinib, dasatinib, nilotinib, lapatininb, pazopanib, PLX4032), however, there are many other small molecules endowed with similar properties. Most of these inhibitors bind in the ATP site, thus preventing kinase activation. The most relevant use for the inhibitors approved by FDA, mainly working as tyrosine-kinase blockers, is in the antineoplastic therapy. In this regard, imatinib, whose main target is the BCR-Abl kinase, has been the first to be used in the treatment of chronic myelogenous leukemia, while sorafenib is currently administered to patients affected by renal or hepatocellular carcinoma (Dar and Shokat, 2011).

Only recently protein kinase inhibitors have been proposed as a means to counteract the onset of skeletal muscle wasting. In this regard, we have recently demonstrated that treatment of mice bearing the C26 tumor with the ERK inhibitor PD98059 partially but significantly protects tumor hosts from the onset of body weight loss and muscle mass depletion (Penna et al., 2010a). ERK inhibition also results in normalization of atrogin-1 hyperexpression, independently from the state of activation of Akt. Among the targets of ERK is the AP-1 transcription factor, which is activated in tumor-bearing animals (Costelli et al., 2005) and may contribute to muscle atrophy, since this latter is improved inhibiting AP-1 by a c-jun dominant negative (TAM67; Moore-Carrasco et al., 2006) . AP-1 regulated genes may contribute to muscle depletion; as an example, cyclin D1 expression (Moore-Carrasco et al., 2006) could induce satellite cell proliferation, that not necessarily is followed by differentiation, resulting in impaired myogenesis. The differential expression of specific factors defines the phenotype of satellite cells. In particular, while MyoD can be detected more or less throughout the myogenic process, high levels of Pax7 associated with low myogenin expression characterize proliferating satellite cells; an opposite pattern can be observed in differentiating cells (Halevy et al., 2004). Indeed, previous reports have shown that Pax7 hyperexpression results in inhibition of myogenesis (Olguin & Olwin, 2004). Low rates of myogenesis (satellite cell activation and differentiation) participate to the maintenance of physiological skeletal muscle mass (Nicolas et al., 2005). This is confirmed by observations showing that aging- or hindlimb suspension-induced muscle atrophy is associated with a reduced regenerative potential (Mitchell & Pavlath, 2004). Consistently, Pax7 expression is significantly increased in the muscle of C26 hosts with respect to controls, while myogenin levels are reduced. The pattern of Pax7 and myogenin expression in the C26-bearing mice is compatible with an impaired regenerative process and suggests the possibility that activated satellite cells accumulate in tumor host muscle because of either enhanced proliferation or impaired differentiation or both. Altered expression of myogenic factors has previously been reported in AH-130 hepatoma-bearing rats (Costelli et al., 2005), in cancer patients (Ramamoorthy et al., 2009), and in an experimental model of chronic kidney disease (Zhang et al., 2010). In the latter report, downregulation of IGF-1 signaling appears responsible for impaired regeneration (Zhang et al., 2010). The results obtained in our laboratory suggest an alternative mechanism based on ERK activation: when the C26 hosts are treated with PD98059, and ERK is thus inhibited, Pax7 and myogenin expression is restored to control values. These observations suggest that ERK activation likely contributes to maintain satellite cells in an undifferentiated state (Penna et al., 2010a).

Despite several reports have shown that p38 is involved in the induction of atrogenes as well as in the hyperactivation of protein degradation in different model systems, there are no studies demonstrating that its pharmacological inhibition in experimental animals may protect the skeletal muscle from wasting. In this regard, we have tested the effectiveness of the p38 inhibitor SB203580 in preventing skeletal muscle depletion in animals implanted with the C26 tumor. After a subcutaneous inoculum of about 10^6 C26 cells, death of the animals occurs in about 15 days; tumor growth is associated with progressive loss of body weight, depletion of both the skeletal muscle and the adipose tissue mass, as well as with markedly increased circulating levels of IL-6 (Penna et al., 2010a). Control and tumor-bearing mice have been treated daily with subcutaneous injection of 3 or 30 mg/kg initial body weight of SB203580, dissolved in DMSO and then diluted in saline, starting from the day after tumor implantation. No significant differences could be observed as for both muscle force (evaluated by grasping test; Fig. 1A), food intake (Fig. 1B) and tumor mass (C26: 231±93 mg, C26+SB203580 30 mg/kg: 249±31 mg). By contrast, an increase of body weight occurred in the group of tumor bearers that received SB203580 at the dose of 30 mg/kg when compared to both treated or untreated controls (Fig.1C). Body weight accretion, however, does not reflect an effect on the skeletal muscle mass, that remains close to the values of untreated tumor hosts, while the weight of both liver and spleen is significantly higher in treated than in untreated C26 bearers, possibly reflecting drug toxicity (Fig. 1D).

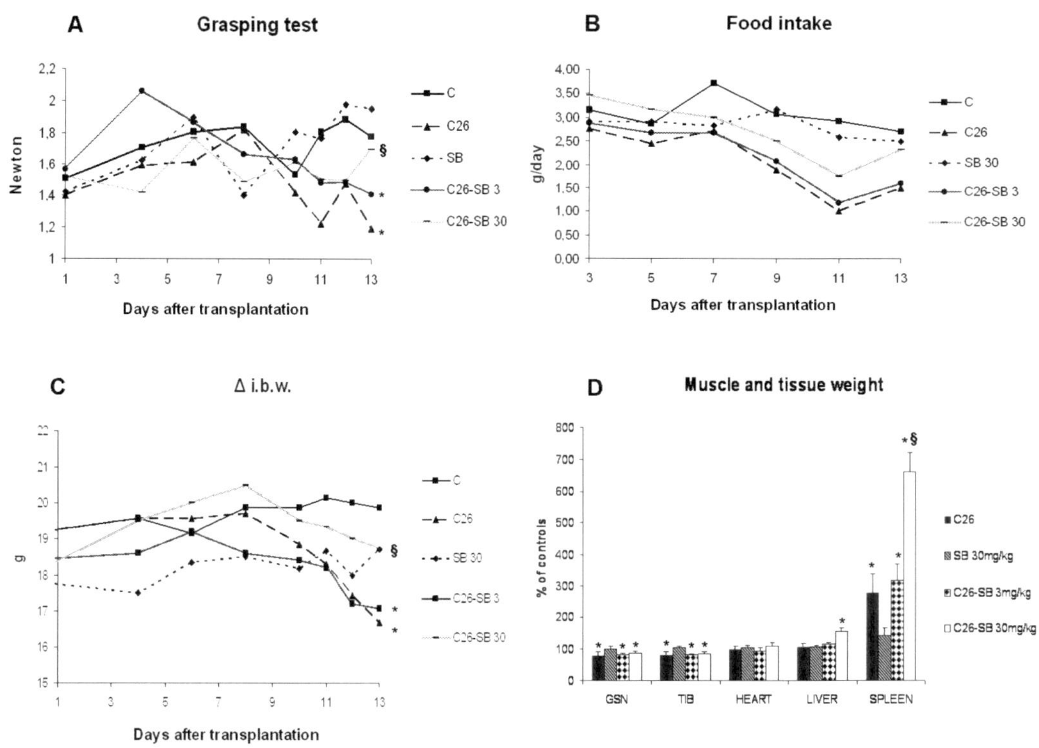

(A): voluntary muscle strength, evaluated by dynamometer, expressed in Newton; (B) food intake, expressed as the means amount of food (g) consumed be the animals every two days; (C) body weight (g), inclusive of the tumor; (D) muscle and tissue weight, expressed as percentages of controls. Date are represented as means ± SD (where not indicated SD is within 10% of the means), n = 8 for each experimental group. Significance of the differences: *p<0.05 vs. untreated controls, § p<0.05 vs. untreated C26 hosts.

Fig. 1. Effect of treatment with SB203580 on cachexia in mice bearing the C26 colon carcinoma.

The results obtained by treating the C26 hosts with SB203580 could sound unexpected, since p38 has been shown to be involved in different mechanisms contributing to skeletal muscle depletion (see above). However, when analyzing the state of activation of MAPKs in two different experimental models, namely rats bearing the AH-130 hepatoma and the C26 hosts, just ERK appears phosphorylated (active; Penna et al., 2010a), suggesting that the usual pattern of p38 involvement in muscle wasting likely does not apply to cancer cachexia. The other way round, p38 inhibition could have lead to a differential modulation of the mechanisms impinging on cancer-associated muscle wasting, mainly protein hypercatabolism and impaired myogenesis, worsening the latter and improving the former, or vice versa, resulting in lack of changes of muscle mass. Further work is warranted to clarify this point.

Finally, imatinib mesylate (IM) administration has been shown to improve muscle pathology in mdx mice, an accepted model for the Duchenne-type muscle dystrophy. In particular, IM-treated mice show a less degree of muscle necrosis, inflammation and fibrosis than control animals. Such effects appear to depend on the inhibition exerted by the drug on the activation of both c-Abl and PDGFR on both peritoneal macrophages and muscle-resident fibroblasts (Huang et al., 2009). Similar observations have been independently reported by another group (Bizario et al., 2009). Both reports suggest that treatment with IM exerts a marked anti-inflammatory effect, lowering the levels of proinflammatory cytokines. Keeping this in mind, we decided to test the effectiveness of IM in preventing muscle wasting in cancer cachexia, where the role played by the inflammatory state is widely accepted. The above cited C26 model has been used. IM has been administered with daily subcutaneous injection at the concentration of 400 mg/kg initial body weight, dissolved in water. Food intake have been recorded daily. At day 15 the animals have been sacrificed to evaluate the effects exerted by the treatment on body and tissue weight. The data reported in Figure 2 show that IM does not induce any detectable modifications when administered to healthy mice, demonstrating that the drug itself does not exert toxic effects on the animals. The treatment, however, is not able to correct the wasting pattern caused by the growth of the C26 tumor. Indeed, loss of body weight (Fig. 2A), cumulative food intake reduction (Fig. 2B), and mass depletion of gastrocnemius, tibialis anterior and heart (Fig. 2C) are comparable between treated and untreated tumor hosts. By contrast, spleen hypertrophy, a constant finding in the C26 hosts (Penna et al., unpublished observations; Fig. 1D, 2C), is completely prevented by treatment with IM (Fig. 2C), confirming the anti-inflammatory effect of this drug. Finally, no significant differences could be observed as for tumor mass between mice administered IM or vehicle (C26: 325±86 mg, C26+IM: 253±79 mg, not statistically significant).

The results show that while able to improve muscle phenotype in mdx mice (Bizario et al., 2009; Huang et al., 2009), IM is ineffective in preventing muscle wasting in tumor-bearing animals, although it likely exert an anti-inflammatory action, as shown by the protection against spleen hypertrophy. These observations may suggest that the tyrosine kinases blocked by IM might not be involved in the pathogenesis of muscle wasting in cancer cachexia. However, the lack of effect could also depend on the different inflammatory situation occurring in the muscle of mdx mice and of tumor-bearing animals. Indeed, while the former is characterized by a marked inflammatory infiltrate, associated with an important fibrotic response, these alterations are quite lacking in the latter.

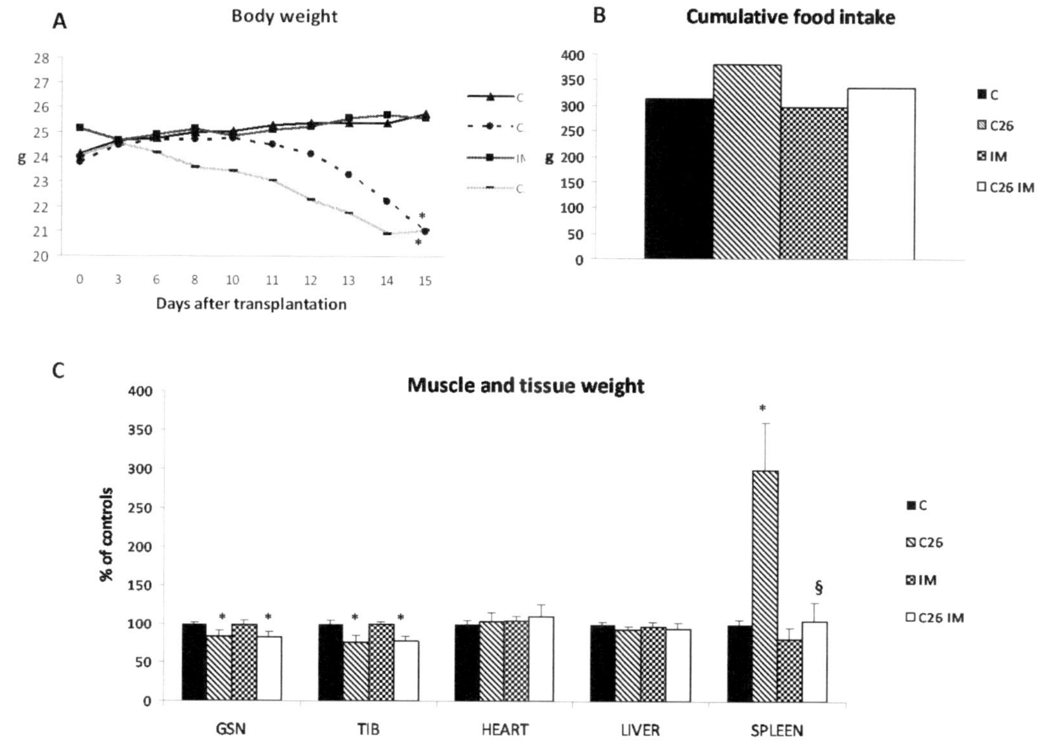

(A): body weight (g), inclusive of the tumor; (B) cumulative food intake (g) over the whole experimental period; (C) muscle and tissue weight, expressed as percentages of controls. Date are represented as means ± SD (where not indicated SD is within 10% of the means), n = 8 for each experimental group. Significance of the differences: *p<0.05 vs. untreated controls, § p<0.05 vs. untreated C26 hosts.

Fig. 2. Effect of treatment with imatinib mesylate (IM) on cachexia in mice bearing the C26 colon carcinoma.

6. Conclusions

Evidence coming from different experimental models have demonstrated the possibility to interfere with the onset of muscle protein hypercatabolism by several means, such as exercise, nutritional, and pharmacological interventions, better if combined. The growing amount of knowledge about the mechanisms underlying the alterations of muscle protein metabolism are highly relevant in this regard. Particular attention deserves the observation that several experimental results point to kinases as crucially involved in the activation/enhancement of the mechanisms leading to skeletal muscle depletion, either in physiological or pathological states. On this line, the availability of specific kinase inhibitors has opened the way to a direct evaluation of their possibility to be used as therapeutic tools to treat conditions characterized by skeletal muscle wasting. At present the results available in the literature are very few, and at least some of them appear encouraging. However, a note of care should be introduced, since inhibiting protein kinases would also impinge on transduction pathways physiologically relevant, rendering unavoidable an accurate estimate of the risk/benefit ratio.

7. References

Adamo, M.L. & Farrar, R.P. (2006). Resistance training, and IGF involvement in the maintenance of muscle mass during the aging process. *Ageing Res Rev*, Vol. 5, No. 3 (Aug), pp. 310-31, 1568-1637

Anthony, J.C., Yoshizawa, F., Anthony, T.G., Vary, T.C., Jefferson, L.S. & Kimball, S.R. (2000). Leucine stimulates translation initiation in skeletal muscle of postabsorptive rats via a rapamycin-sensitive pathway. *J Nutr*, Vol. 130, No. 10 (Oct), pp. 2413-2419, 0022-3166

Attard-Montaldo, S.P., Camacho-Hubner, C., Cotteril, A.M., D'Souza-Li, L., Bartlett, K., Halliday, D. & Eden, O.B. (1998). Changes in protein turnover, IGF-I and IGF binding proteins in children with cancer. *Acta Pediatr*, Vol. 87, No. 1, pp. 54-60, 0803-5253

Baracos, V.E., DeVivo, C., Hoyle, D.H. & Goldberg, A.L. (1995). Activation of the ATP–ubiquitin-proteasome pathway in skeletal muscle of cachectic rats bearing a hepatoma. *Am J Physiol Endocrinol Metab*, Vol. 268, No. 5, pp. E996–E1006, 0193-1849

Bassel-Duby, R. & Olson, E.N. (2006). Signaling pathways in skeletal muscle remodeling. *Annu Rev Biochem*, Vol. 75, pp.19-37, 0066-4154.

Bechet, D., Tassa, A., Taillandier, D., Combaret, L. & Attaix, D. (2005). Lysosomal proteolysis in skeletal muscle. *Int J Biochem Cell Biol*, Vol. 37, No. 10 (Oct), pp. 2098-2114, 1357-2725

Ben-Levy, R., Hooper, S., Wilson, R., Paterson, H.F. & Marshall, C.J. (1998). Nuclear export of the stress-activated protein kinase p38 mediated by its substrate MAPKAP kinase-2. *Curr Biol*, Vol. 8, No. 19 (Sep 24), pp. 1049-1057, 0960-9822.

Bizario, J.C., Cerri, D.G., Rodrigues, L.C., Oliveira, G.L., Nomizo, A., de Araujo, D.D., Fukuhara, P.S., Ribeiro, J.C., de Castro, F.A. & Costa, M.C. (2009). Imatinib mesylate ameliorates the dystrophic phenotype in exercised mdx mice. *JNeuroimmunol*, Vol. 212, No. 1-2 (Jul), pp. 93-101, 0165-5728

Bodine, S.C., Stitt, T.N., Gonzalez, M., Kline, W.O., Stover, G.L., Bauerlein, R., Zlotchenko, E., Scrimgeour, A., Lawrence, J.C., Glass, D.J. & Yancopoulos, G.D. (2001). Akt/mTOR pathway is a crucial regulator of skeletal muscle hypertrophy and can prevent muscle atrophy in vivo. *Nat Cell Biol*, Vol. 3, No. 11 (Nov), pp. 1014-1019,1465-7392

Boppart, M.D., Hirshman, M.F., Sakamoto, K., Fielding, R.A. & Goodyear, L.J. (2001). Static stretch increases c-Jun NH2-terminal kinase activity and p38 phosphorylation in rat skeletal muscle. *Am J Physiol Cell Physiol*, Vol. 280, No. 2 (Feb), pp. C352-C358, 0363-6143

Cai, D., Frantz, J.D., Tawa, N.E., Melendez, P.A., Oh, B.-C., Lidov, H.G.W., Hasselgren, P.-O., Frontera, W.R., Lee, J., Glass, D.J. & Shoelson, S.E. (2004). IKKβ/NF-κB activation causes severe muscle wasting in mice. *Cell*, Vol. 119, No. 2 (Oct), pp. 285-298, 0092-8674

Carlson, M.E., Suetta, C., Conboy, M.J., Aagaard, P., Mackey, A., Kjaer, M. & Conboy, I. (2009). Molecular aging and rejuvenation of human muscle stem cells. *EMBO Mol Med*, Vol. 1, No. 8-9 (Nov), pp. 381-391, 1757-4676

Chen, S.E., Gerken, E., Zhang, Y., Zhan, M., Mohan, R.K., Li, A.S., Reid, M.B. & Li, Y.P. (2005). Role of TNF-{alpha} signaling in regeneration of cardiotoxin-injured muscle. *Am J Physiol Cell Physiol*, Vol. 289, No. 5 (Nov), pp. C1179-1187, 0363-6143

Choudry, H.A., Pan, M., Karinch, A.M. & Souba, W,W. (2006). Branched-chain amino acid-enriched nutritional support in surgical and cancer patients. *J Nutr*, Vol. 136, No. 1 Suppl (Jan), pp. 314S-318S, 0022-3166

Clarke, B.A., Drujan, D., Willis, M.S., Murphy, L.O., Corpina, R.A., Burova, E., Rakhilin, S.V., Stitt, T.N., Patterson, C., Latres, E. & Glass, D.J. (2007). The E3 Ligase MuRF1 degrades myosin heavy chain protein in dexamethasone-treated skeletal muscle. *Cell Metab*, Vol. 6, No. 5 (Nov), pp. 376-385, 1550-4131

Clarke, I.J. & Henry, B.A. (2010). Targeting energy expenditure in muscle as a means of combating obesity. *Clin Exp Pharmacol Physiol*, Vol. 37, No. 1 (Jan), pp. 121-124, 0305-1870

Clavel, S., Coldefy, A.S., Kurkdjian, E., Salles, J., Margaritis, I. & Derijard, B. (2006). Atrophy-related ubiquitin ligases, atrogin-1 and MuRF1 are up-regulated in aged rat Tibialis Anterior muscle. *Mech Ageing Dev*, Vol.127, No. 10 (Oct), pp. 794-801, 0047-6374

Codogno, P. & Meijer, A.J. (2005). Autophagy and signaling: their role in cell survival and cell death. *Cell Death Differ*, Vol. 12, No. 2 Suppl (Nov), pp. 1509- 1518, 1350-9047

Coletti, D., Yang, E., Marazzi, G., & Sassoon, D. (2002). TNFalpha inhibits skeletal myogenesis through a PW1-dependent pathway by recruitment of caspase pathways. *EMBO J*, Vol. 21, No. 4 (Feb), pp. 631-642, 0261-4189

Coletti, D., Moresi, V., Adamo, S., Molinaro, M. & Sassoon, D. (2005). Tumor necrosis factor-alpha gene transfer induces cachexia and inhibits muscle regeneration. *Genesis*, Vol. 43, No. 3 (Nov), pp. 120-128, 1526-954X

Costelli, P., De Tullio, R., Baccino, F.M. & Melloni, E. (2001). Activation of Ca(2+)-dependent proteolysis in skeletal muscle and heart in cancer cachexia. *Br J Cancer*, Vol. 84, No. 7 (Apr), pp. 946–950, 0007-0920

Costelli, P. & Baccino, F.M. (2003). Mechanisms of skeletal muscle depletion in wasting syndromes: role of the ATP-ubiquitin-dependent proteolysis. *Curr Op Clin Nutr Metab Care*, Vol. 6, No. 4 (Jul), pp. 407-412, 1363-1950

Costelli, P., Muscaritoli, M., Bossola, M., Moore-Carrasco, R., Crepaldi, S., Grieco, G., Autelli, R., Bonelli, G., Pacelli, F., Lopez-Soriano, F.J., Argilés, J.M., Doglietto, G.B., Baccino, F.M. & Rossi Fanelli, F. (2005). Skeletal muscle wasting in tumor-bearing rats is associated with MyoD down-regulation. *Int J Oncol*, Vol. 26, No. 6 (Jun), pp. 663-1668, 1019-6439

Costelli, P., Muscaritoli, M., Bossola, M., Penna, F., Reffo, P., Bonetto, A., Busquets, S., Bonelli, G., Lopez-Soriano, F.J., Doglietto, G.B., Argilés, J.M., Baccino, F.M. & Rossi Fanelli, F. (2006). IGF-1 is downregulated in experimental cancer cachexia. *Am J Physiol Regul Integr Comp Physiol*, Vol. 291, No. 3 (Sep), pp. R674-R683, 0363-6119

Dar, A.C. & Shokat, K.M. (2011). The evolution of protein kinase inhibitors from antagonists to agonists of cellular signaling. *Annu Rev Biochem*, Vol. 80, No. 7 (Jun), pp. 769-795, 0066-4154

Deval, C., Mordier, S., Obled, C., Bechet, D., Combaret, L., Attaix, D. & Ferrara, M. (2001). Identification of cathepsin L as a differentially expressed message associated with skeletal muscle wasting. *Biochem J*, Vol. 360, No. 15 (Nov), pp. 143-150, 0264-6021

Dhanasekaran, D.N. & Reddy, E.P. (2008). JNK signaling in apoptosis. *Oncogene*, Vol. 27, No. 48 (Oct), pp. 6245-6251, 0950-9232

Dodd, S.L., Hain, B., Senf, S.M. & Judge, A.R. (2009). Hsp27 inhibits IKKbeta-induced NF-kappaB activity and skeletal muscle atrophy. *FASEB J*, Vol. 23, No. 10 (Oct), pp. 3415-3423, 0892- 6638

Doyle, A., Zhang, G., Abdel Fattah, E.A., Eissa, N.T. & Li, Y.P. (2011). Toll-like receptor 4 mediates lipopolysaccharide-induced muscle catabolism via coordinate activation of ubiquitin- proteasome and autophagy-lysosome pathways. *FASEB J*, Vol. 25, No. 1 (Jan), pp. 99-110, 0892-6638

Du, J., Wang, X., Miereles, C., Bailey, J.L., Debigare, R., Zheng, B., Price, S.R. & Mitch, W.E. (2004). Activation of caspase-3 is an initial step triggering accelerated muscle proteolysis in catabolic conditions. *J Clin Invest*, Vol. 113, No. 1 (Jan), pp. 115-123, 0021-9738

Dunn, S.E., Burns, J.L. & Michel, R.N. (1999). Calcineurin is required for skeletal muscle hypertrophy. *J Biol Chem*, Vol. 274, No. 31 (Jul), pp. 21908-21912, 0021-9258

Fang, C.H., Li, B.G., James, J.H., King, J.K., Evenson, A.R., Ward, G.D. & Hasselgren, P.O. (2005). Protein breakdown in muscle from burned rats is blocked by insulin-like growth factor I and glycogen synthase kinase-3 beta inhibitors. *Endocrinology*, Vol. 146, No. 7 (Jul), pp. 3141-3149, 0013-7227

Engel, K., Schultz, H., Martin, F., Kotlyarov, A., Plath, K., Hahn, M., Heinemann, U. & Gaestel, M. (1995). Constitutive activation of mitogen-activated protein kinase-activated protein kinase 2 by mutation of phosphorylation sites and an A-helix motif. *J Biol Chem*, Vol. 270, No. 45 (Nov), pp. 27213-27221, 0021-9258

Glass, D.J. (2010). Signaling pathways perturbing muscle mass. *Curr Opin Clin Nutr MetabCare*, Vol. 13, No. 3 (May), pp. 225-229, 1363-1950

Gorog, D.A., Jabr, R.I., Tanno, M., Sarafraz, N., Clark, J.E., Fisher, S.G., Cao, X.B., Bellahcene, M., Dighe, K., Kabir, A.M., Quinlan, R.A., Kato, K., Gaestel, M., Marber, M.S. & Heads, R.J. (2009). MAPKAPK-2 modulates p38-MAPK localization and small heat shock protein phosphorylation but does not mediate the injury associated with p38-MAPK activation during myocardial ischemia. *Cell Stress Chaperones*, Vol. 14, No. 5 (Sep), pp. 477-489, 1355- 8145

Greenbaum, L.M. & Sutherland, J.H. (1983). Host cathepsin D response to tumor in the normal and pepstatin-treated mouse. *Cancer Res*, Vol. 43, no. 6 (Jun), pp. 2584-2587, 0008-5472

Guay, J., Lambert, H., Gingras-Breton, G., Lavoie, J.N., Huot, J. & Landry, J. (1997). Regulation of actin filament dynamics by p38 map kinase-mediated phosphorylation of heat shock protein 27. *J Cell Sci*, Vol. 110, No. Pt3 (Feb), pp. 357-368, 0021-9533

Guttridge, D.C., Mayo, M.W., Madrid, L.V., Wang, C.Y. & Baldwin, A.S. Jr. (2000). NFkappaB- induced loss of MyoD messenger RNA: possible role in muscle decay and cachexia. *Science*, Vol. 289, No. 5488 (Sep), pp. 2363-2366, 0036-8075

Haddad, F. & Adams, G.R. (2004). Inhibition of MAP/ERK kinase prevents IGF-I-induced hypertrophy in rat muscles. *J Appl Physiol*, Vol. 96, No. 1 (Jan), pp. 203-210, 8750-7587

Halevy, O., Piestun, Y., Allouh, M.Z., Rosser, B.W., Rinkevich, Y., Reshef, R., Rozenboim, I., Wleklinski-Lee, M. & Yablonka-Reuveni, Z. (2004). Pattern of Pax7 expression during myogenesis in the posthatch chicken establishes a model for satellite cell differentiation and renewal. *Dev Dyn*, Vol. 231, No. 3 (Nov), pp. 489-502, 1058-8388

Hanada, M., Feng, J. & Hemmings, B.A. (2004). Structure, regulation and function of PKB/AKT - a major therapeutic target. *Biochim. Byophis. Acta*, Vol. 1697, No. 1-2 (Mar), pp. 3-16, 0304- 4165

Hornberger, T.A., Hunter, R.B., Kandarian, S.C. & Esser, K.A. (2001). Regulation of translation factors during hindlimb unloading and denervation of skeletal muscle in rats. *Am J Physiol Cell Physiol*, Vol. 281, No. 1 (Jul), pp. C179-C187, 0363-6143

Hu, Z., Lee, I.H., Wang, X., Sheng, H., Zhang, L., Du, J. & Mitch, W.E. (2007). PTEN expression contributes to the regulation of muscle protein degradation in diabetes. *Diabetes*, Vol. 56, No. 10 (Oct), pp. 2449-2456, 0012-1797.

Huang, P., Zhao, X.S., Fields, M., Ransohoff, R.M. & Zhou, L. (2009). Imatinib attenuates skeletal muscle dystrophy in mdx mice. *FASEB J*, Vol. 23, No. 8 (Aug), pp. 2539-2548, 0892-6638

Huey, K.A. (2006). Regulation of HSP25 expression and phosphorylation in functionally overloaded rat plantaris and soleus muscles. *J Appl Physiol*, Vol. 100, No. 2 (Feb), pp. 451- 456, 8750-7587

Kadowaki, M. & Kanazawa, T. (2003). Amino acids as regulators of proteolysis. *J Nutr*, Vol. 133 No. 6 Suppl (Jun), pp. 2052S-2056S, 0022-3166

Kato, K., Ito, H., Kamei, K., Iwamoto, I. & Inaguma, Y. (2002). Innervation-dependent phosphorylation and accumulation of alphaB-crystallin and Hsp27 as insoluble complexes in disused muscle. *FASEB J*, Vol. 16, No. 11 (Sep), pp. 1432-1434, 0892-6638

Kawano, F., Matsuoka, Y., Oke, Y., Higo, Y., Terada, M., Wang, X.D., Nakai, N., Fukuda, H., Imajoh-Ohmi, S. & Ohira, Y. (2007). Role(s) of nucleoli and phosphorylation of ribosomal protein S6 and/or HSP27 in the regulation of muscle mass. *Am J Physiol Cell Physiol*, Vol. 293, No. 1 (Jul), pp. C35-C44, 0363-6143

Kedzierska, K. & Crowe, S.M. (2001). Cytokines and HIV-1: interactions and clinical implications. *Antivir Chem Chemother*, Vol. 12, No. 3 (May), pp. 133-135, 0956-3202

Keren, A., Tamir, Y. & Bengal, E. (2006). The p38 MAPK signaling pathway: a major regulator of skeletal muscle development. *Mol Cell Endocrinol*, Vol. 252, No. 1-2 (Jun), pp. 224-230, 0303- 7207

Kipreos, E.T. & Pagano, M. (2000). The F-box protein family. *Genome Biol*, Vol.1, No. 5, REVIEWS3002, 1474-7596

Klionsky, D.J., Cregg, J.M., Dunn, W.A.Jr., Emr, S.D., Sakai, Y., Sandoval, I.V., Sibirny, A., Subramani, S., Thumm, M., Veenhuis, M. & Ohsumi, Y. (2003). A unified nomenclature for yeast autophagy-related genes. *Dev Cell*, Vol. 5, No. 4 (Oct), pp. 539-545, 1534-5807

Koyama, S., Hata, S., Witt, C.C., Ono, Y., Lerche, S., Ojima, K., Chiba, T., Doi, N., Kitamura, F., Tanaka, K., Abe, K., Witt, S.H., Rybin, V., Gasch, A., Franz, T., Labeit, S. & Sorimachi, H. (2008). Muscle RING-finger protein-1 (MuRF1) as a connector of muscle energy metabolism and protein synthesis. *J Mol Biol*, Vol. 376, No. 5 (Mar), pp. 1224-1236, 0022- 2836

Jagoe, R.T., Redfern, C.P., Roberts, R.G., Gibson, G.J. & Goodship, T.H. (2002). Skeletal muscle mRNA levels for cathepsin B, but not components of the ubiquitin-proteasome pathway, are increased in patients with lung cancer referred for thoracotomy. *Clin Sci (Lond)*, Vol. 102, No. 3 (Mar), pp. 353-361, 0143-5221

Jin, B. & Li, Y.P. (2007). Curcumin prevents lipopolysaccharide-induced atrogin-1/MAFbx upregulation and muscle mass loss. *J Cell Biochem,* Vol. 100, No. 4 (Mar), pp. 960-969, 0730- 2312

Jope, R.S. (2003). Lithium and GSK-3: one inhibitor, two inhibitory actions, multiple outcomes. *Trends Pharmacol Sci,* Vol . 24, No. 9 (Sep), pp. 441-443, 0165-6147

Lai, K.-M., Gonzalez, M., Poueymirou, W.T., Kline, W.O., Na, E., Zlotchenko, E., Stitt, T.N., Economides, A.N., Yancopoulos, G.D. & Glass, D.J. (2004). Conditional activation of akt in adult skeletal muscle induces rapid hypertrophy. *Mol Cell Biol,* Vol. 24, No. 21 (Nov), pp. 9295-9304, 0270-7306

Frost, R.A. & Lang, C.H. (2005). Skeletal muscle cytokines: regulation by pathogen-associated molecules and catabolic hormones. *Curr Opin Clin Nutr Metab Care,* Vol. 8, No. 3 (May), pp. 255-263, 1363-1950

Lecker, S.H., Jagoe, R.T., Gilbert, A., Gomes, M., Baracos, V., Bailey, J., Price, S.R., Mitch, W.E. & Goldberg, A.L. (2004). Multiple types of skeletal muscle atrophy involve a common program of changes in gene expression. *FASEB J,* Vol. 18, No. 1 (Jan), pp. 39-51, 0892-6638

Lee, C.E., McArdle, A. & Griffiths, R.D. (2007). The role of hormones, cytokines and heat shock proteins during age-related muscle loss. *Clin Nutr,* Vol. 26, No. 5 (Oct), pp. 524-534, 0261- 5614

Lenk, K., Schur, R., Linke, A., Erbs, S., Matsumoto, Y., Adams, V. & Schuler, G. (2009). Impact of exercise training on myostatin expression in the myocardium and skeletal muscle in a chronic heart failure model. *Eur J Heart Fail,* Vol. 11, No. 4 (Apr), pp. 342-348, 1388-9842

Li, Y.P., Chen, Y., John, J., Moylan, J., Jin, B., Mann, D.L. & Reid, M.B. (2005). TNF-alpha acts via p38 MAPK to stimulate expression of the ubiquitin ligase atrogin1/MAFbx in skeletal muscle. *FASEB J,* Vol. 19, No. 3 (Mar), pp. 362-370, 0892-6638

Liu, Q., Xu, W.G., Luo, Y., Han, F.F., Yao, X.H., Yang, T.Y., Zhang, Y., Pi, W.F. & Guo, X.J. (2011). Cigarette smoke-induced skeletal muscle atrophy is associated with up-regulation of USP-19 via p38 and ERK MAPKs. *J Cell Biochem,* Vol. 112, No. 9 (Sep), pp. 2307-2316, 0730-2312

Lundholm, K., Körner, U., Gunnebo, L., Sixt-Ammilon, P., Fouladiun, M., Daneryd, P. & Bosaeus, I. (2007). Insulin treatment in cancer cachexia: effects on survival, metabolism, and physical functioning. *Clin Cancer Res,* Vol. 13, No. 9 (May), pp. 2699-2706, 1078-0432

Mammucari, C., Milan, G., Romanello, V., Masiero, E., Rudolf, R., Del Piccolo, P., Burden, S.J., Di Lisi, R., Sandri, C., Zhao, J., Goldberg, A.L., Schiaffino, S. & Sandri, M. (2007). FoxO3 controls autophagy in skeletal muscle in vivo. *Cell Metab,* Vol. 6, No. 6 (Dec), pp. 458-471, 1550-4131

Mangili, A., Murman, D.H., Zampini, A.M. & Wanke, C.A. (2006). Nutrition and HIV infection: review of weight loss and wasting in the era of highly active antiretroviral therapy from the nutrition for healthy living cohort. *Clin Infect Dis,* Vol. 42, No. 6 (Mar), pp. 836-842, 1058-4838

Marinovic, A.C., Zheng, B., Mitch, W.E. & Price, S.R. (2002). Ubiquitin (UbC) expression in muscle cells is increased by glucocorticoids through a mechanism involving Sp1 and MEK1. *J Biol Chem,*Vol. 277, No. 19 (May), pp. 16673-16681, 0021-9258

Masharani, U.B., Maddux, B.A., Li, X., Sakkas, G.K., Mulligan, K., Schambelan, M., Goldfine, I.D. & Youngren, J.F. (2011). Insulin resistance in non-obese subjects is associated with activation of the JNK pathway and impaired insulin signaling in skeletal muscle. *PLoS One*, Vol. 6, No. 5 (May), e19878, 1932-6203

Meissner, J.D., Chang, K.C., Kubis, H.P., Nebreda, A.R., Gros, G. & Scheibe, R.J. (2007). The p38alpha/beta mitogen-activated protein kinases mediate recruitment of CREB-binding protein to preserve fast myosin heavy chain IId/x gene activity in myotubes. *J Biol Chem*, Vol. 282, No. 10 (Mar), pp. 7265-7275, 0021-9258

Meng, W., Swenson, L.L., Fitzgibbon, M.J., Hayakawa, K., Ter Haar, E., Behrens, A.E., Fulghum, J.R. & Lippke, J.A. (2002). Structure of mitogen-activated protein kinase-activated protein (MAPKAP) kinase 2 suggests a bifunctional switch that couples kinase activation with nuclear export. *J Biol Chem*, Vol. 277, No. 40 (Oct), pp. 37401-37405, 0021-9258

Mihaylova, M.M. & Shaw, R.J. (2011). The AMPK signalling pathway coordinates cell growth, autophagy and metabolism. *Nat Cell Biol*, Vol. 13, No. 9 (Sep), pp. 1016-1023, 1465-7392

Miyake, T., Alli, N.S., Aziz, A., Knudson, J., Fernando, P., Megeney, L.A. & McDermott, J.C. (2009). Cardiotrophin-1 maintains the undifferentiated state in skeletal myoblasts. *J Biol Chem*, Vol. 284, No. 29 (Jul), pp. 19679-19693, 0021-9258

Mitchell, P.O. & Pavlath, G.K. (2004). Skeletal muscle atrophy leads to loss and dysfunction of muscle precursor cells. *Am J Physiol Cell Physiol*,Vol. 287, No. 6 (Dec), pp. C1753-C1762, 0363-6143

Mizushima, N., Yamamoto, A., Matsui, M., Yoshimori, T. & Ohsumi, Y. (2004). In vivo analysis of autophagy in response to nutrient starvation using transgenic mice expressing a fluorescent autophagosome marker. *Mol Biol Cell*, Vol. 15, No. 3 (Mar), pp. 1101-1111, 1059- 1524

Moore-Carrasco, R., Garcia-Martinez, C., Busquets, S., Ametller, E., Barreiro, E., López-Soriano, F.J. & Argilés, J.M. (2006). The AP-1/CJUN signaling cascade is involved in muscle differentiation: implications in muscle wasting during cancer cachexia. *FEBS Lett*, Vol. 580, No. 2 (Jan), pp. 691-696, 0014-5793

Morel, J,E, & Pinset-Härström, I. (1975). Ultrastructure of the contractile system of striated skeletal muscle and the processes of muscular contraction. I. Ultrastructure of the myofibril and source of energy. *Biomedicine*, Vol. 22, No. 2 (Mar), pp. 88-96, 0970-2067

Moresi, V., Pristera, A., Scicchitano, B.M., Molinaro, M., Teodori, L., Sassoon, D, Adamo, S. & Coletti, D. (2008). Tumor necrosis factor-alpha inhibition of skeletal muscle regeneration is mediated by a caspase-dependent stem cell response. *Stem Cells*, Vol. 26, No. 4 (Apr), pp. 997-1008, 1066-5099

Morgan, J.E. & Partridge, T.A. (2003). Muscle satellite cells. *Int J Biochem Cell Biol*, Vol. 35, No. 8 (Aug), pp. 1151-1156, 1357-2725

Murgia, M., Serrano, A.L., Calabria, E., Pallafacchina, G., Lomo, T. & Schiaffino, S. (2000). Ras is involved in nerve-activity-dependent regulation of muscle genes. *Nat Cell Biol*, Vol. 2, No. 3 (Mar), pp. 142-147, 1465-7392

Musarò, A., McCullagh, K.J., Naya, F.J., Olson, E.N. & Rosenthal, N. (1999). IGF-1 induces skeletal myocyte hypertrophy through calcineurin in association with GATA-2 and NF-ATc1. Na*ture*, Vol. 400, No. 6744 (Ago), pp. 581-585, 0028-0836

Musarò, A., McCullagh, K., Paul, A., Houghton, L., Dobrowolny, G., Molinaro, M., Barton, E.R., Sweeney, H.L. & Rosenthal, N. (2001). Localized Igf-1 transgene expression sustains hypertrophy and regeneration in senescent skeletal muscle. *Nat Genet*, Vol. 27, No. 2 (Feb), pp. 195-200, 1061-4036

Nakashima, J., Tachibana, M., Ueno, M., Miyajima, A., Baba, S. & Murai, M. (1998). Association between tumor necrosis factor in serum and cachexia in patients with prostate cancer. *Clin Cancer Res*, Vol. 4, No. 7 (Jul), pp. 1743-1748, 1078-0432

Nicolas, N., Marazzi, G., Kelley, K. & Sassoon, D. (2005). Embryonic deregulation of muscle stress signaling pathways leads to altered postnatal stem cell behavior and a failure in postnatal muscle growth. *Dev Biol*, Vol. 281, No. 2 (May), pp. 171-183, 0012-1606

Norrby, M. & Tågerud, S. (2010). Mitogen-activated protein kinase-activated protein kinase 2 (MK2) in skeletal muscle atrophy and hypertrophy. *J Cell Physiol*,Vol. 223, No. 1 (Apr), pp.194-201, 0021-9541

Oldfors, A. (2007). Hereditary myosin myopathies. *Neuromuscul Disord*, Vol. 17, No. 5 (May), pp. 355-367, 0960-8966

Olguin, H.C. & Olwin, B.B. (2004). Pax-7 up-regulation inhibits myogenesis and cell cycle progression in satellite cells: a potential mechanism for self-renewal. *Dev Biol*, Vol. 275, No. 2 (Nov), pp. 375-388, 0012-1606

Paul, P.K., Gupta, S.K., Bhatnagar, S., Panguluri, S.K., Darnay, B.G., Choi, Y. & Kumar, A. (2010). Targeted ablation of TRAF6 inhibits skeletal muscle wasting in mice. *J Cell Biol*, Vol. 191, No. 7 (Dec), pp. 1395-1411, 0021-9525

Pedersen, B.K. & Febbraio, M.A. (2008). Muscle as an endocrine organ: focus on muscle-derived interleukin-6. *Physiol Rev*, Vol. 88, No. 4 (Oct), pp. 1379-1406, 0031-9333

Perdiguero, E., Ruiz-Bonilla, V., Serrano, A.L &, Muñoz-Cánoves,, P. (2007). Genetic deficiency of p38alpha reveals its critical role in myoblast cell cycle exit: the p38alpha-JNK connection. *Cell Cycle*, Vol. 6, No. 11 (Jun), pp. 1298-1303, 1538-4101

Penna, F., Costamagna, D., Fanzani, A., Bonelli, G., Baccino, F.M. & Costelli, P. (2010a). Muscle wasting and impaired myogenesis in tumor bearing mice are prevented by ERK inhibition. *PLoS One*, Vol. 5, No. 10 (Oct), e13604, 1932-6203

Penna. F., Bonetto, A., Muscaritoli, M., Costamagna, D., Minero. V.G., Bonelli, G., Rossi Fanelli, F., Baccino, F.M. & Costelli, P. (2010b). Muscle atrophy in experimental cancer cachexia: Is the IGF-1 signaling pathway involved? *Int J Cancer*, Vol. 127, No. 7 (Oct), 1796-1717, 0020-7136

Petersen, A.C., McKenna, M.J., Medved, I., Murphy, K.T., Brown, M.J., Della Gatta, P. & Cameron- Smith, D. (2011). Infusion with the antioxidant N-acetylcysteine attenuates early adaptive responses to exercise in human skeletal muscle. *Acta Physiol (Oxf)*, doi: 10.1111/j.1748- 1716.2011.02344.x, 1748-1708

Pfeifer, U. (1977). Inhibition by insulin of the physiological autophagic breakdown of cell organelles. *Acta Biol Med Ger*, Vol. 36, No. 1-2, pp. 1691-1694

Plant, P.J., Bain, J.R., Correa, J.E., Woo, M. & Batt, J. (2009). Absence of caspase-3 protects against denervation-induced skeletal muscle atrophy. *J Appl Physiol*, Vol. 107, No. 1 (Jul), pp. 224- 234, 8750-7587

Puri, P.L., Wu, Z., Zhang, P., Wood, L.D., Bhakta, K.S., Han, J., Feramisco, J.R., Karin, M. & Wang, J.Y. (2000). Induction of terminal differentiation by constitutive activation of p38 MAP kinase in human rhabdomyosarcoma cells. *Genes Dev*, Vol. 14, No. 5 (Mar), pp. 574-584, 0890-9369

Raman, M., Chen, W. & Cobb, M.H. (2007). Differential regulation and properties of MAPKs. *Oncogene*,Vol. 26, No. 22 (May), pp. 3100-3112, 0950-9232

Ramamoorthy, S., Donohue, M. & Buck, M. (2009). Decreased Jun-D and myogenin expression in muscle wasting of human cachexia. *Am J Physiol Endocrinol Metab*,Vol. 297, No. 2 (Aug), pp. E392-E401, 0193-1849

Romanello, V., Guadagnin, E., Gomes, L., Roder, I., Sandri, C., Petersen, Y., Milan, G., Masiero, E., Del Piccolo, P., Foretz ,M., Scorrano, L., Rudolf, R. & Sandri M. (2010). Mitochondrial fission and remodelling contributes to muscle atrophy. *EMBO J*, Vol. 29, No. 10 (May), pp.1774-1785, 0261-4189

Rommel, C., Clarke, B.A., Zimmermann, S., Nunez, L., Rossman, R., Reid, K., Moelling, K., Yancopoulos, G.D. & Glass, D.J. (1999). Differentiation stage-specific inhibition of the Raf- MEK-ERK pathway by Akt. *Science*,Vol. 286, No. 5445 (Nov), pp. 1738-1741, 0036-8075

Rommel, C., Bodine, S.C., Clarke, B.A., Rossman, R., Nunez, L., Stitt, T.N., Yancopoulos, G.D. & Glass, D.J. (2001). Mediation of IGF-1-induced skeletal myotube hypertrophy by PI(3)K/Akt/mTOR and PI(3)K/Akt/GSK3 pathways. *Nat Cell. Biol*, Vol. 3, No. 11 (Nov), pp. 1009-1013, 1465-7392

Rosner, M., Hanneder, M., Siegel, N., Valli, A. & Hengstschläger, M. (2008). The tuberous sclerosis gene products hamartin and tuberin are multifunctional proteins with a wide spectrum of interacting partners. *Mutat Res*, Vol. 658, No. 3 (Mar-Apr), pp. 234-246, 1383-5742

Ruff, R.L. & Secrist, D. (1984). Inhibitors of prostaglandin synthesis or cathepsin B prevent muscle wasting due to sepsis in the rat. *J Clin Invest*, Vol. 73, No. 5 (May), pp. 1483-1486, 0021-9738

Saido, T.C., Sorimachi, H. & Suzuki, K. (1994). Calpain: new perspectives in molecular diversity and physiological-pathological involvement. *FASEB J*, Vol. 8, No. 11 (Aug), pp. 814-822, 0892-6638

Sandri, M. (2002). Apoptotic signaling in skeletal muscle fibers during atrophy. *Curr OpinClin Nutr Metab Care*, Vol. 5, No. 3, (May), pp. 249-253, 1363-1950

Sandri, M., Sandri, C., Gilbert, A., Skurk, C., Calabria, E., Picard, A., Walsh, K., Schiaffino, S., Lecker, S.H. & Goldberg, A.L. (2004). Foxo transcription factors induce the atrophy-related ubiquitin ligase atrogin-1 and cause skeletal muscle atrophy. *Cell*, Vol.117, No. 3 (Apr), pp.399-412, 0092-8754

Sakamoto. K., Aschenbach. W.G., Hirshman. M.F. & Goodyear, L.J. (2003). Akt signaling in skeletal muscle: regulation by exercise and passive stretch. *Am J Physiol Endocrinol Metab*, Vol. 285, No. 5 (Nov), pp. E1081-E1088, 0193-1849

Schakman, O., Kalista, S., Bertrand, L., Lause, P., Verniers, J., Ketelslegers, J.M. & Thissen, J.P. (2008). Role of Akt/GSK-3beta/beta-catenin transduction pathway in the muscle anti- atrophy action of insulin-like growth factor-I in glucocorticoid-treated rats. *Endocrinology*,Vol. 149, No. 8 (Aug), pp. 3900-3908, 0013-7227

Schakman, O., Gilson, H., Kalista, S. & Thissen, J.P. (2009). Mechanisms of muscle atrophy induced by glucocorticoids. *Horm Res*, Vol. 72, No. 1 Suppl (Nov), pp. 36-41, 0301-0163

Schiaffino, S. & Mammucari, C. (2011). Regulation of skeletal muscle growth by the IGF1-Akt/PKB pathway: insights from genetic models. *Skelet Muscle*, Vol. 1, No. 1 (Jan), 4, 2044-5040

Schmitt, T.L., Martignoni, M.E., Bachmann, J., Fechtner, K., Friess, H., Kinscherf, R. & Hildebrandt, W. (2007). Activity of the Akt-dependent anabolic and catabolic pathways in muscle and liver samples in cancer-related cachexia. *J Mol Med*, Vol. 85, No. 6 (Jun), pp. 647-654, 0946- 2716

Schultz, E. & McCormick, K.M. (1994). Skeletal muscle satellite cells. *Rev Physiol Biochem Pharmacol*, Vol. 23, pp. 213-257, 0303-4240

Scott, S.V. & Klionsky, D.J. (1998). Delivery of proteins and organelles to the vacuole from the cytoplasm. *Curr Opin Cell Biol*, Vol. 10, No. 4 (Aug), pp. 523-529, 0955-0674

Shi, H., Zeng, C., Ricome, A., Hannon, K.M., Grant, A.L. & Gerrard, D.E. (2007). Extracellular signal-regulated kinase pathway is differentially involved in beta-agonist-induced hypertrophy in slow and fast muscles. *Am J Physiol Cell Physiol*,Vol. 292, No. 5 (May), pp. C1681-C1689, 0363-6143

Shi, H., Scheffler, J.M., Pleitner, J.M., Zeng, C., Park, S., Hannon, K.M., Grant, A.L. & Gerrard, D.E. (2008). Modulation of skeletal muscle fiber type by mitogen-activated protein kinase signaling. *FASEB J*, Vol. 22, No. 8 (Aug), pp. 2990-3000, 0892-6638

Shi, H., Scheffler, J.M., Zeng, C., Pleitner, J.M., Hannon, K.M., Grant, A.L. & Gerrard, D.E. (2009). Mitogen-activated protein kinase signaling is necessary for the maintenance of skeletal muscle mass. *Am J Physiol Cell Physiol*, Vol. 296, No. 5 (May), pp. C1040-C1048, 0363-6143

Sinha-Hikim, I., Braga, M., Shen, R. & Sinha Hikim, A.P. (2007). Involvement of c-Jun NH2-terminal kinase and nitric oxide-mediated mitochondria-dependent intrinsic pathway signaling in cardiotoxin-induced muscle cell death: role of testosterone. *Apoptosis*, Vol. 12, No. 11 (Nov), pp. 1965-1978, 1360-8185

Smuder, A.J., Kavazis, A.N., Hudson, M.B., Nelson, W.B. & Powers, S.K. (2010). Oxidation enhances myofibrillar protein degradation via calpain and caspase-3. *Free Radic Biol Med*, Vol. 49, No. 7 (Oct), pp. 1152-1160, 0891-5849

Stitt, T.N., Drujan, D., Clarke, B.A., Panaro, F., Timofeyva, Y., Kline, W.O., Gonzalez, M., Yancopoulos, G.D. & Glass, D.J. (2004). The IGF-1/PI3K/Akt pathway prevents expression of muscle atrophy-induced ubiquitin ligases by inhibiting FOXO transcription factors. *Mol Cell*, Vol.14, No. 3 (May), pp. 395-403, 1097-2765

Stokoe, D., Engel, K., Campbell, D.G., Cohen, P. & Gaestel, M. (1992). Identification of MAPKAP kinase 2 as a major enzyme responsible for the phosphorylation of the small mammalian heat shock proteins. *FEBS Lett*, Vol. 313, No. 3 (Nov), pp. 307-313, 0014-5793

Sugiura, T., Abe, N., Nagano, M., Goto, K., Sakuma, K., Naito, H., Yoshioka, T. & Powers, S.K. (2005). Changes in PKB/Akt and calcineurin signaling during recovery in atrophied soleus muscle induced by unloading. *Am J Physiol Regul Integr Comp Physiol*, Vol. 288, No. 5 (May), pp. R1273-R1278, 0363-6119

Supinski, G.S., Ji, X. & Callahan, L.A. (2009). The JNK MAP kinase pathway contributes to the development of endotoxin-induced diaphragm caspase activation. *Am J Physiol Regul Integr Comp Physiol*, Vol. 297, No. 3 (Sep), pp. R825-R834, 0363-6119.

Suzuki, K., Hata, S., Kawabata, Y. & Sorimachi, H. (2004). Structure, activation, and biology of calpain. *Diabetes*, Vol. 53, No. 1 Suppl (Feb), pp. S12-S8, 0012-1797

Temparis, S., Asensi, M., Taillandier, D., Aurousseau, E., Larbaud, D., Obled, A., Béchet, D., Ferrara, M., Estrela, J.M. & Attaix, D. (1994). Increased ATP–ubiquitin dependent proteolysis in skeletal muscles of tumor-bearing rats. *Cancer Res*, Vol. 54, No. 21 (Nov), pp. 5568–5573, 0008-5472

Tessitore, L., Costelli, P., Bonetti, G. & Baccino, F.M. (1993). Cancer cachexia, malnutrition, and tissue protein turnover in experimental animals. *Arch Biochem Biophys*, Vol 306, No. 1 (Oct), pp. 52-58, 0003-9861

Tessitore, L., Costelli, P. & Baccino, F.M. (1994). Pharmacological interference with tissue protein hypercatabolism in tumor-bearing rats. *Biochem J*, Vol. 299, No. Pt 1 (Apr), pp. 71-78, 0264- 6021.

Tidball, J.G. & Spencer, M.J. (2002). Expression of a calpastatin transgene slows muscle wasting and obviates changes in myosin isoform expression during murine muscle disuse. *J Physiol*, Vol. 545, No. Pt 3 (Dec), pp. 819–828, 0022-3751

Tintignac, L.A., Lagirand, J., Batonnet, S., Sirri,V., Leibovitch, M.P. & Leibovitch, S.A. (2005). Degradation of MyoD mediated by the SCF (MAFbx) ubiquitin ligase. *J Biol Chem*, Vol. 280, No. 4 (Jan), pp. 2847-2856, 0021-9258

Tisdale, M.J. (2008). Catabolic mediators of cancer cachexia. *Curr Opin Support Palliat Care*, Vol.2, No. 4 (Dec), pp. 256-261, 1751-4258

Van der Velden, J.L., Langen, R.C., Kelders, M.C., Wouters, E.F., Janssen-Heininger, Y. M. & Schols, A.M. (2006). Inhibition of glycogen synthase kinase-3beta activity is sufficient to stimulate myogenic differentiation. *Am. J. Physiol. Cell. Physiol*, Vol. 290, No. 2 (Feb), pp. C453-C462, 0363-6143

Verzola, D., Procopio, V., Sofia, A., Villaggio, B., Tarroni, A., Bonanni, A., Mannucci, I., De Cian, F., Gianetta, E., Saffioti, S. & Garibotto, G. (2011). Apoptosis and myostatin mRNA are upregulated in the skeletal muscle of patients with chronic kidney disease. *Kidney Int*, Vol. 79, No. 7 (Apr), pp. 773-782, 0085-2538

Wang, X., Hu, Z., Hu, J., Du, J. & Mitch, W.E. (2006). Insulin resistance accelerates muscle protein degradation: Activation of the ubiquitin-proteasome pathway by defects in muscle cell signaling. *Endocrinology*, Vol. 147, No. 9 (Sep), pp. 4160-4168, 0013-7227

Wang, X.H., Zhang, L., Mitch, W.E., LeDoux, J.M., Hu, J. & Du, J. (2010). Caspase-3 cleaves specific 19 S proteasome subunits in skeletal muscle stimulating proteasome activity. *J Biol Chem*, Vol. 285, No. 28 (Jul), pp. 21249-21257, 0021-9258

Waterlow,, J.C. (1984). Protein turnover with special reference to man. *Q J Exp Physiol*, Vol. 69, No. 3 (Jul), pp. 409-438, 0958-0670

White, J.P., Baynes, J.W., Welle, S.L., Kostek, M.C., Matesic, L.E., Sato, S. & Carson, J.A. (2011). The Regulation of Skeletal Muscle Protein Turnover during the Progression of Cancer Cachexia in the Apc Mouse. *PLoS One*, Vol. 6, No. 9, e24650, 1932-6203

Williams, A.B., Decourten-Myers, G.M., Fischer, J.E., Luo, G., Sun, X. & Hasselgren, P.O. (1999). Sepsis stimulates release of myofilaments in skeletal muscle by a calcium-dependent mechanism. *FASEB J*, Vol. 13, No. 11 (Aug), pp. 1435-1443, 0892-6638

Wray, C.J., Sun, X., Gang, G.I. & Hasselgren, P.O. (2002). Dantrolene downregulates the gene expression and activity of the ubiquitin–proteasome proteolytic pathway in septic skeletal muscle. *J Surg Res*, Vol. 104, No. 2 (May), pp. 82–87, 0022-4804

Wu, R., Kausar, H., Johnson, P., Montoya-Durango, D.E., Merchant, M. & Rane, M.J. (2007). Hsp27 regulates Akt activation and polymorphonuclear leukocyte apoptosis by scaffolding MK2 to Akt signal complex. *J Biol Chem*, Vol. 282, No. 30 (Jul), pp. 21598-21608, 0021-9258

Yang, W., Chen, Y., Zhang, Y., Wang, X., Yang, N. & Zhu, D. (2006). Extracellular signal-regulated kinase 1/2 mitogen-activated protein kinase pathway is involved in myostatin-regulated differentiation repression. *Cancer Res*, Vol. 66, No. 3 (Feb), pp. 1320-1326, 0008-5472

Yang, S.Y., Hoy, M., Fuller, B., Sales, K.M., Seifalian, A.M. & Winslet, M.C. (2010). Pretreatment with insulin-like growth factor I protects skeletal muscle cells against oxidative damage via PI3K/Akt and ERK1/2 MAPK pathways. *Lab Invest*, Vol. 90, No. 3 (Mar), pp. 391-401, 0023-6837

Zhang, L., Wang, X.H., Wang, H., Du, J. & Mitch, W.E (2010). Satellite cell dysfunction and impaired IGF-1 signaling cause CKD-induced muscle atrophy. *J Am Soc Nephrol*, Vol. 21, No. 3 (Mar), pp. 419-427, 1046-6673

Zhao, J., Brault, J.J., Schild, A., Cao, P., Sandri, M., Schiaffino, S., Lecker, S.H. & Goldberg, A.L. (2007). FoxO3 coordinately activates protein degradation by the autophagic/lysosomal and proteasomal pathways in atrophying muscle cells. *Cell Metab*,Vol. 6, No. 6 (Dec), pp. 472-483, 1550-4131

11

Mathematical Modeling of Syk Activation in Allergen-Stimulated Mast Cells and Basophils

Ambarish Nag[1,*], Michael I. Monine[1,†], Byron Goldstein[1], James R. Faeder[2] and Michael L. Blinov[3]
[1]*Los Alamos National Laboratory*
[2]*University of Pittsburgh School of Medicine*
[3]*University of Connecticut School of Medicine*
U.S.A.

1. Introduction

Mast cells and basophils play major roles in allergic responses of the immediate type (Rivera & Gilfillan, 2006). Allergic individuals produce IgE that is specific for the multivalent foreign molecules (allergens) that trigger their allergic responses. IgE binds with high affinity through its constant region to a monovalent receptor, FcϵRI, that is expressed on the surface of basophils and mast cells and that mediates much of mast cell activation in allergic reactions. Upon exposure to the allergen, the bridging of multiple IgEs by the allergen results in the aggregation of FcϵRI receptors on the cell surface, which in turn triggers a cascade of biochemical reactions that results in the release of preformed mediators from granules together with the synthesis and release of lipid mediators and cytokines. The cytosolic protein tyrosine kinase Syk plays a crucial role in FcϵRI-mediated signaling cascade in mast cells. A clone of the *syk* gene was first isolated on the basis of partial sequenced information of a 40 kDa kinase from porcine spleen (de Castro, 2011; Taniguchi et al., 1991). The product of this gene was identified as a nonreceptor-type protein tyrosine kinase of 72 kDa (Yang et al., 1994) and was named Spleen Tyrosine Kinase (Syk). Syk has two Src-homology 2 domains (SH2) which are separated by a linker region from the kinase domain (Siraganian et al., 2002). FcϵRI is a tetrameric complex composed of an α-chain that binds IgE to form a long-lived complex, and three subunits, a β-chain and two disulfide-linked γ-chains (Blank et al., 1989) that contain immunoreceptor tyrosine-based activation motifs (ITAMs). The FcϵRI receptor lacks intrinsic enzymatic acitivity and requires non-receptor protein tyrosine kinase activity in propagating cellular signals in allergen-stimulated mast cells and basophils. The unphosphorylated β-chain of the FcϵRI receptor associates weakly with the Src family kinase Lyn, which is anchored to the inner layer of the plasma membrane (Vonakis et al., 1997; 2001). Upon receptor aggregation, Lyn transphosphorylates tyrosines in the β and γ ITAMs. Syk from the cytosol binds with high affinity through its two SH2 domains to the doubly phosphorylated γ ITAM (Benhamou et al., 1993; Hutchcroft et al., 1992). Recent experimental

---

*Currently at National Renewable Energy Laboratory, U.S.A.
†Currently at Novartis Institutes for Biomedical Research, Inc., U.S.A.

findings (de Castro et al., 2010; Siraganian et al., 2010) indicate that the binding of Syk to phosphorylated γ ITAM causes a conformational change of the Syk molecule and enhances its enzymatic activity. The conformation change of the activated Syk molecule exposes its COOH-terminal region, thereby enabling the phosphorylation of two conserved tyrosines (Tyr-624 and Tyr-625 in rat Syk). These two phosphotyrosines maintain the γ ITAM bound Syk in an open conformation, which allows the phosphorylation on multiple tyrosines by both Lyn and Syk (Siraganian et al., 2002), with a second Syk molecule playing the major role in transphosphorylating two tyrosines, Tyr-519 and Tyr-520, in the Syk activation loop (Zhang et al., 2000; 1998). This yields a fully activated Syk that phosphorylates the adaptor proteins LAT (Linker for Activation of T cells) and NTAL (Non-T-cell Activation Linker) that function as scaffolds, organizing other signaling proteins that are responsible for signaling events further downstream (Rivera, 2005). The phosphorylation of the two tyrosine residues in the activation loop of Syk is critical to the propagation of FcϵRI signaling, since the substitution of these tyrosines by phenylalanine abrogates signaling and degranulation in mast cells (Zhang et al., 1998).

In order to predict the strength of the allergen-induced signaling in mast cells and basophils, it is important to understand the subtle kinetic effects underlying Syk activation, and how these kinetic effects influence the level of Syk activation, since this event couples the initial signaling events in the cascade with the downstream signaling events. Although many of the biochemical reactions leading from Syk activation to histamine release have been investigated in great detail, a complete model of the signaling cascade still eludes us. A detailed mathematical model of the early signaling events, up to and including Syk activation, that are triggered when IgE-FcϵRI complexes are exposed to a bivalent or trivalent ligand on rat basophilic leukemia (RBL) cells has been developed by Faeder et al. (2003). In the model, full activation of a Syk molecule requires its transphosphorylation by a second Syk molecule, the two Syk molecules being bound to different receptors in a receptor dimer (Zhang et al., 2000). In this chapter, we review our research (Nag, Faeder & Goldstein, 2010; Nag, Monine, Blinov & Goldstein, 2010) on the exploration of the different factors that influence the kinetics and the level of Syk activation in stimulated mast cells and basophils using the mathematical model of Faeder et al. (2003).

First, we use the model of Faeder et al. (2003) to investigate investigate the effects of the extracellular allergen concentration and the cellular FcϵRI and Syk expression levels on the extent of Syk activation. Using model simulations, we explore how the extent of Syk activation varies as a function of the concentration of the allergen, for given FcϵRI and Syk concentrations, and for a given set of kinetic parameters characterizing binding, unbinding, phosphorylation and dephosphorylation events (Nag, Faeder & Goldstein, 2010).

For bivalent allergens binding to the monovalent IgE-FcϵRI complex, a plot of the equilibrium concentration of receptors in aggregates versus the log of the free allergen concentration, the cross-linking curve, is symmetric and bell-shaped. One might naively expect that the variation of Syk activation with the equilibrium allergen concentration should be qualitatively similar to the variation in the number of cross-linked FcϵRI receptors. We define the Syk activation curve as a plot of the number of fully activated Syk molecules in a cell versus the log of the equilibrium free allergen concentration. Our model predicts that if the mechanism by which Syk is fully activated involves the transphosphorylation of Syk by Syk, then Syk activation curves can be either bell-shaped or double humped, depending on the relative cellular concentrations of Syk and FcϵRI. For many physiological Syk expression levels, the predicted Syk activation curve follows the receptor cross-linking curve. However, for limiting Syk concentration, an excess of cross-linked receptors over Syk molecules over a range of free equilibrium allergen concentrations can occur and result in high-dose inhibition of Syk activation (Hlavacek et al., 2003). At other allergen concentrations, the extent of cross-linking is lower so that there is enough Syk for two Syk molecules to be bound to the same receptor dimer. This predicted differential behavior at different allergen concentrations gives rise to the variable shape (unimodal/multimodal) of the Syk activation curve (Hlavacek et al., 2003). More interestingly, for some reasonable parameter regimes, the bimodal Syk activation curves can be non-symmetric with respect to the ligand concentration. We discuss how, even though it is only receptor aggregates that trigger responses, differences in total ligand concentration, that are not reflected in the extent of receptor aggregation, can lead to subtle kinetic effects that yield qualitative differences in the levels of Syk activation.

Secondly, we use our mathematical model to evaluate the role of serial engagement (Davis, 1995; Valitutti et al., 1995) in Syk activation in mast cells (Nag, Monine, Blinov & Goldstein, 2010). The terms serial triggering and serial engagement were introduced in immunology when Valitutti et al. (Davis, 1995; Valitutti et al., 1995) reported that within the contact area between an antigen presenting cell (APC) and a T cell, a few antigenic peptides bound to major histocompatibility complex molecules (pMHC) mediated the internalization of hundreds of T cell receptors (TCRs). The concept of serial engagement was introduced to describe the ability of a single peptide, bound to a MHC molecule, to sequentially interact with TCRs within the contact region between a T cell and an APC.

The observation that TCRs undergo serial engagement, coupled with the kinetic proofreading model for cell signaling (Goldstein et al., 2008; McKeithan, 1995), led to the hypothesis that for T cell activation there should be an optimal range of half-lives for the pMHC-TCR bond (Lanzavecchia et al., 1999). The basic idea of kinetic proofreading is that for a TCR to become activated it must remain associated with a pMHC long enough for a set of biochemical modifications to occur. In case of dissociation of the pMHC from the TCR before the necessary modifications have been completed, signaling is frustrated and activation is not achieved. In order to obtain a measurable signaling response from a T cell, multiple TCRs must be activated. Therefore, at low pMHC density, a single pMHC needs to trigger multiple TCR before it diffuses out of the contact region. If the pMHC dissociates too rapidly it will encounter many TCRs but activate only a few. On the other hand, if it remains bound too long it will activate those it encounters but the frequency of encounters will be small. The recognition that the pMHC-TCR bond half-life has opposing effects on kinetic proofreading and serial engagement led to the prediction that to achieve an optimal rate of TCR activation there should be an optimal half-life, or equivalently an optimal dissociation rate constant k_{off}, (Coombs & Goldstein, 2005; Lanzavecchia et al., 1999). Although some studies have found an optimal half-life for T cell activation (Carreno et al., 2007; Coombs et al., 2002; Kalergis et al., 2001), there are other results that are not consistent with this model (reviewed in (Aleksic et al., 2010; Stone et al., 2009)).

In addition to surface ligands, soluble multivalent ligands can serially engage cell surface receptors involving repeated binding and dissociating of ligand sites from receptors before all ligand sites become free and the ligand leaves the surface. Although there has been a significant effort directed at unraveling the role of serial engagement of TCRs in activating T-cells, the role of serial engagement by soluble multivalent ligands, of other multichain immune recognition receptors in cell activation has remained largely unexplored.

The immunological term avidity was coined more than fifty years ago to distinguish between the binding properties of an antibody and its monovalent Fab fragment (reviewed in (Karush, 1989)). It was observed that an IgG, at low concentrations, could bind to a surface containing multiple binding sites (epitopes) with an apparent affinity that was orders of magnitude greater than the equilibrium constant for binding of one of its Fab sites to an epitope (Gopalakrishnan & Karush, 1974; Greenbury et al., 1965; Hornick & Karusch, 1972). Avidity arises when the density of surface binding sites is sufficiently high that multivalent ligands are observed to dissociate from the surface more slowly than their monovalent constituent parts, thus exhibiting a high apparent affinity. When a site on a doubly bound IgG dissociates, the epitope it was bound to will diffuse away. If this newly freed site on the IgG binds to another epitope before its second site dissociates, the IgG will remain bound to the surface. Thus, avidity can arise by ligands serially engaging epitopes, traversing from epitope to epitope, along the surface.

Avidity can arise in another way that does not require serial engagement. For example, the cross-linking IgE-FcϵRI complexes on rat basophilic leukemia (RBL) cells by a highly multivalent antigen leads to the rapid formation of large stable aggregates of immobilized receptors (Andrews et al., 2008; Menon et al., 1986). This observation suggests the possibility that a site on a bound ligand can dissociate from, and rebind to the same immobilized receptor many times, effectively increasing its off rate constant, before all its sites on the ligand become free and the ligand dissociates from the cell. Recent multi-color tracking experiments indicate that small aggregates of IgE-FcϵRI complexes, comprising two to four receptors, remain mobile on the RBL surface even at ligand doses that result in degranulation (Andrews et al., 2009).

A major difficulty in assessing the role of serial engagement of receptors in cell signaling is that the phenomena cannot be studied in isolation. To gauge the role of serial engagement in mast cell signaling, we use the detailed mathematical model of the early events triggered when IgE-FcϵRI complexes are exposed to a bivalent or trivalent ligand on RBL cells (Faeder et al., 2003). Simulations using this model has enabled us to delineate how different alterations in the binding properties of the ligand, that result in the same rates of serial engagement of receptors, can alter receptor activation in different ways.

2. Mathematical model

2.1 Components

We briefly review our mathematical model for the early signaling events initiated by ligand-induced aggregation of FcϵRI-IgE complexes on mast cells and basophils. The model is summarized in Fig. 1. The external stimulus in the model is constituted by a symmetric bivalent ligand, such as a monoclonal anti-IgE, that cross-links two IgE molecules, each bound to a FcϵRI receptor. We assume that aggregates of receptors are limited to dimers. The tetrameric FcϵRI is modeled as three subunits, with the disulphide-bonded pair of γ chains treated as one unit. The extracellular region of the α chain binds to the Fc portion of IgE. The binding of IgE to the α subunit of the FcϵRI receptor is not explicitly considered, as the half-life for the dissociation of the IgE-FcϵRI complex is much longer than the processes we consider (Kulczycki & Metzger, 1974). We therefore ignore dissociation and treat the IgE-FcϵRI complex as a single unit which we refer to as the receptor.

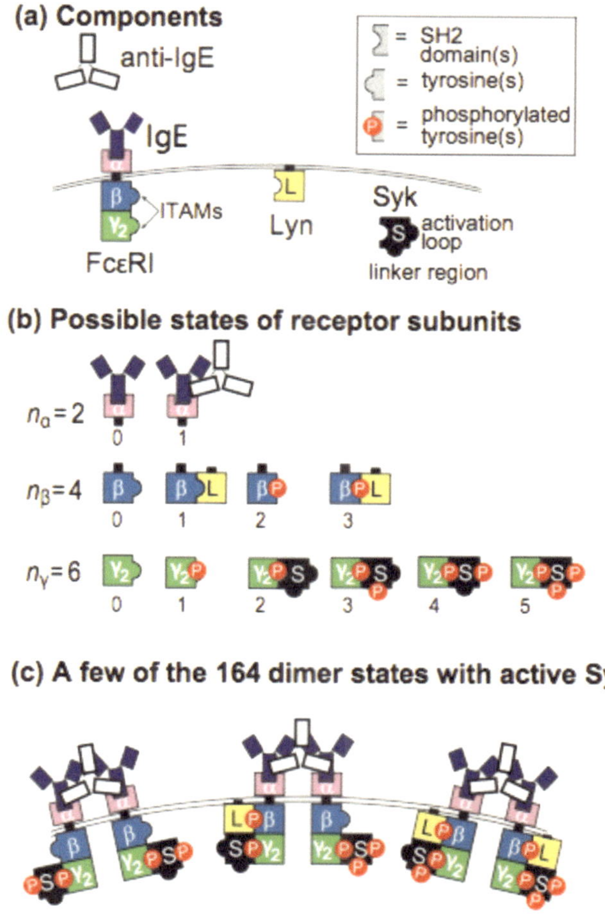

Fig. 1. Components and states of the model, modified from Faeder et al. (2003). The bivalent ligand is a monoclonal anti-IgE that can dimerize receptors.

The β subunit and the γ dimer contain the ITAMs that, upon phosphorylation, become binding sites for the SH2 domains of the two kinases, Lyn and Syk. A simplifying assumption of our model is that multiple tyrosine residues on receptor subunits and Syk are treated as single units of phosphorylation, as indicated in Fig. 1a. For example, the separate tyrosines in the β ITAM are lumped together, as are those in the γ ITAM, so that in the model an ITAM is either phosphorylated or unphosphorylated. Sites of Syk tyrosine phosphorylation are lumped into two units: the activation loop, which is phosphorylated by Syk, and the linker region, which is phosphorylated by Lyn. Fig. 1b shows the states of the β chain and the lumped γ chains included in the model. Each unit can be phosphorylated or unphosphorylated, and can be associated with a kinase in any of several states. In the model, each β chain and γ dimer can bind only a single kinase molecule at a time.

The model permits a large number of receptor states (Figs. 1b and 1c). Since the state of each subunit is independent of the states of the other subunits, the total number of monomer states is $n_\alpha n_\beta n_\gamma = 48$. When a monovalent ligand is present as well as a bivalent ligand, the number of monomeric states increases to 72. In a dimer, each subunit must be engaged with the ligand, so the total number of dimer states is $n_\beta n_\gamma (n_\beta n_\gamma + 1)/2 = 300$. In addition, there are six nonreceptor states, free ligand, free Lyn, and Syk in each of its four possible states of phosphorylation, to yield a total of 354 distinct chemical species in the model. Our model is composed of 354 chemical species and the 3680 chemical reactions that connect them. (For details of the chemical reactions see Fig. 2 in Faeder et al. (2003).)

2.2 Network structure and parameters

The required input for the model are the rate constants associated with each of the reactions, as well as the initial concentrations of each of the components, which are specific to the cell type being modeled. We consider a bivalent allergen that differs from the ligand used by Faeder et al. (2003) in that it binds reversibly to the receptor. Other than a different bivalent ligand, the current reaction network is the same as in Faeder et al. (2003). Lyn can associate with a receptor in a dimer in two possible ways, weakly with the unphosphorylated β chain of the receptor and strongly to the phosphorylated β ITAM. Syk associates with FcϵRI through an interaction between its tandem SH2 domains and the doubly phosphorylated γ ITAM. Lyn, associated with a receptor aggregate, can transphosphorylate the β and γ ITAMs on an adjacent receptor. In the model, all Lyn molecules available to the receptor are in an active form. The available Lyn is considerably less than the total cellular Lyn. Syk is phosphorylated in the model by either Lyn or Syk through transphosphorylation. Lyn phosphorylates Syk tyrosines located in the linker region, while Syk phosphorylates the activation loop tyrosines. Although Lyn may be responsible *in vivo* for a small portion of the Syk activation loop phosphorylation, full Syk activation loop phosphorylation requires kinase-active Syk (Hong et al., 2002).

In the model, phosphorylated units that are not protected through association with an SH2 domain can be dephosphorylated with a common rate constant, d, termed the intrinsic rate constant for dephosphorylation. Dephosphorylation is blocked when an SH2 domain is associated with the phosphorylated site. At long times, the model system goes to a steady state, not an equilibrium. In the model, tyrosines are constantly being phosphorylated and dephosphorylated. As a result, maintaining a constant level of phosphorylation requires a constant input of energy. Since the modifications of the intracellular domains of the receptors do not influence the extracellular binding, the distribution of ligand-receptor aggregates goes to the same equilibrium as would be obtained if no chemical modifications occurred.

2.3 Implementation of rule-based modeling

The reaction rules and their associated rate constants were specified in the syntax of the second-generation version of BioNetGen (Blinov et al., 2004; Faeder et al., 2009), which uses graph theoretic methods to automatically generate the associated network of kinetic balances (ordinary differential equations). The open-source software (available through http://bionetgen.org) uses standard numerical algorithms to solve the generated system of equations and obtain the time courses of all the species until the system attains a steady state. In our simulations, all the kinetic parameters except the ones related to ligand binding and unbinding are taken from Faeder et al. (2003), where a complete discussion of how the parameters were obtained is given.

3. The role of cellular FcϵRI and Syk expression levels in determining the Syk activation response to extracellular stimulation of mast cells and basophils

In the current investigation we model the Syk dose-response curves in human basophils. A broad distribution of Syk expression levels in human basophils, ranging from 5000 to 60,000 molecules per cell, has been reported in a recent survey (MacGlashan Jr, 2007). MacGlashan and co-workers have observed that a low expression level of Syk attenuates the calcium response (MacGlashan Jr & Lavens-Phillips, 2001), an event which is more proximal to Syk activation in the signaling cascade than histamine release. Maximal histamine release from human basophils shows a strong correlation with Syk activation (MacGlashan Jr, 2007). The number of FcϵRI per human basophil varies from 1000 to 1000,000 with a typical value of about 100,000 FcϵRI (MacGlashan Jr, 2007). Experimental observations (MacGlashan Jr, 2007) suggest that the median Lyn expression level in human basophils is about 100,000 molecules per cell. Using simulations of the detailed model, we examine the Syk activation curve for different sets of receptor and Syk expression levels, keeping the available Lyn fixed at 100,000 per cell.

3.1 The shape of the Syk activation curve is predicted to depend on the abundance of Syk relative to the number of FcϵRI

At equilibrium, the receptor cross-linking curve is always symmetric and bell shaped (Dembo & Golstein, 1978). The shape of the Syk activation curve is predicted to depend on the concentrations of Syk and FcϵRI receptors. The number of FcϵRI per human basophils varies from 10^3 to 10^6 with a typical value of about 100,000 FcϵRI (MacGlashan Jr, 2007). The corresponding Syk concentration ranges from 5×10^3 to 6×10^4 per cell (MacGlashan Jr, 2007). At the lower boundary of the range (5×10^3 Syk per cell), a receptor population of 7×10^4 per cell results in a predicted bimodal Syk activation curve (Fig. 2a). At the upper boundary (6×10^4 Syk per cell), the same level of receptors results in a bell-shaped Syk activation curve (Fig. 2b).

In the model, full activation of a Syk molecule requires its transphosphorylation by a second Syk, which occurs when two Syk molecules are bound to different receptors cross-linked by a bivalent ligand. We refer to a receptor dimer with two bound Syk molecules, one on each receptor, as a Syk dimer. Formation of Syk dimers is essential for Syk activation. Double humped Syk activation curves are predicted to occur (Fig. 2a) if the concentration of receptors dimers becomes sufficiently high that the number of receptor dimers with two Syks bound decreases with increasing receptor dimer concentration.

Studies of histamine release dose response curves (log-log plots of the fraction of a cell's total histamine released in a given time versus the concentration of ligand the cell is exposed to) lead to the definition of two types of inhibition of histamine release. In Type I inhibition, commonly known as antigen excess inhibition, histamine release is inhibited when receptor crosslinks are decreased. In the more interesting Type II inhibition, histamine release is inhibited when receptor crosslinks are increased. This type of inhibition was first observed by Becker et al. (Becker et al., 1973) who exposed human basophils to anti-IgE and found that inhibition of histamine release was correlated with increased aggregation of IgE. Applying the same terminology for Syk activation, Type I inhibition occurs in our model for parameter regimes in which the number of Syk molecules is comparable to, or exceeds, the number of receptors in dimers, so that the dose-response curve (Syk activation curve) follows the

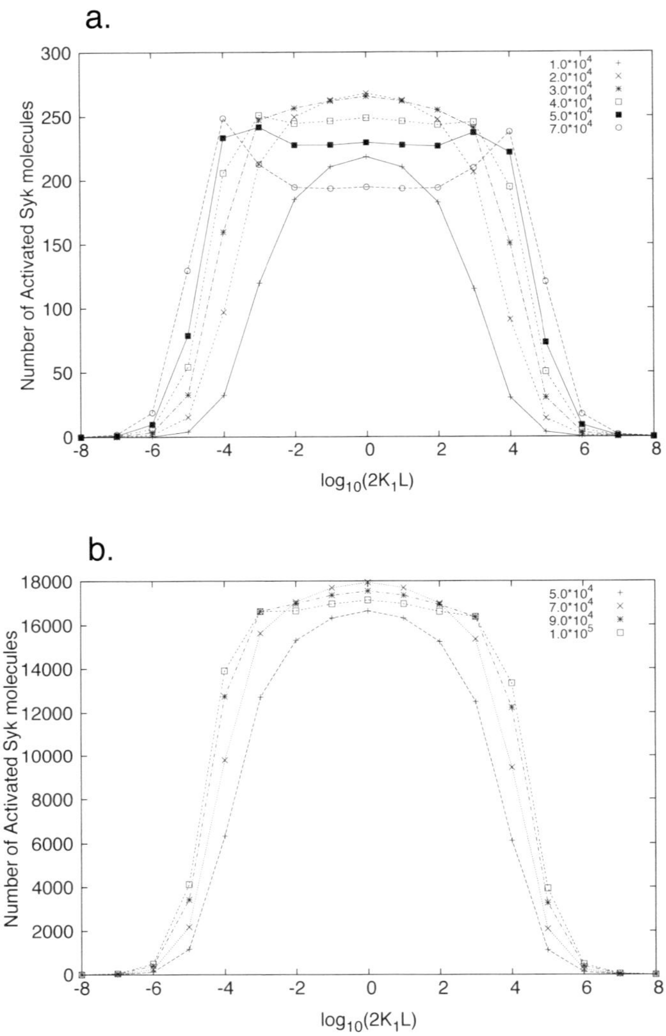

Fig. 2. Syk Activation curves for $k_{+1} = 2.5 \times 10^6$ M^{-1}s^{-1}, $k_{+2} = 8 \times 10^{-9}$ cm^{2}s^{-1}, $k_{-1} = k_{-2} = 0.01$ s^{-1}, and $[\text{Lyn}]_{tot} = 10^5$ per cell. Each curve corresponds to a different number of receptors per cell given in the legend. The number of activated Syk molecules on the y axis are given in numbers per cell. (a) $[\text{Syk}]_{tot} = 5 \times 10^3$ per cell; (b) $[\text{Syk}]_{tot} = 6 \times 10^4$ per cell. The number of activated Syk molecules on the y axis are the numbers per cell. In (a) the ratio of total receptors to total Syk varies from 2-14 while in (b) the same ratio varies from 0.83-1.67.

receptor cross-linking curve at all the ligand concentrations (Fig. 3a). This type of inhibition can be traced to the binding sites on IgE being blocked by the ligand at sufficiently high concentrations, so that the number of receptor cross-links is reduced. Type II inhibition in our model system results from an excess of FcϵRI-bound IgE dimers compared to the number of available Syk molecules, and is characterized by decreasing signaling response

(Syk activation) at ligand concentrations where receptor cross-linking is increasing. The reduction in Syk activation, with increasing ligand concentration, near the center of the Syk activation curve in Fig 3b is a Type II inhibition. Although we used a bivalent ligand in our simulations to illustrate the two types of inhibition, the valence of the ligand is not critical as long as it is multivalent.

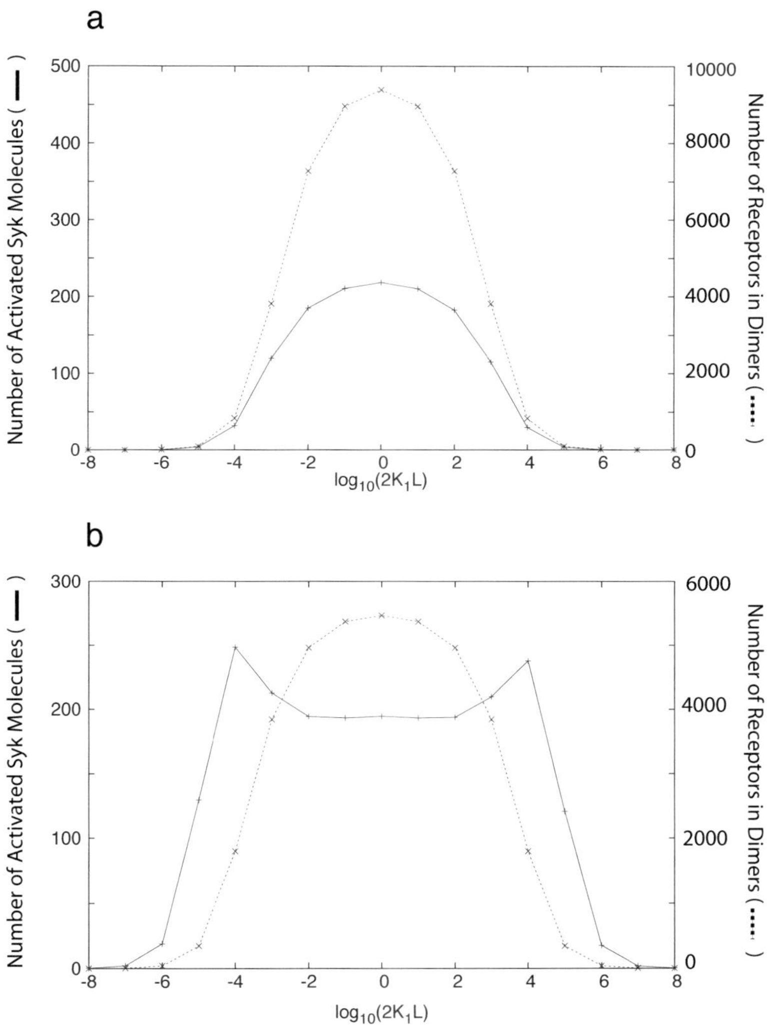

Fig. 3. Comparison of receptor cross-linking and Syk activation curves to determine the type of antigen excess inhibition. $k_{+1} = 2.5 \times 10^6 \mathrm{M}^{-1}\mathrm{s}^{-1}$, $k_{+2} = 8 \times 10^{-9}\ \mathrm{cm}^2\mathrm{s}^{-1}$, $k_{-1} = k_{-2} = 0.01\ \mathrm{s}^{-1}$; $[\mathrm{Syk}]_{tot} = 5 \times 10^3$ per cell; $[\mathrm{Lyn}]_{tot} = 10^5$ per cell. (a) $[\mathrm{Fc}\epsilon\mathrm{RI}]_{tot} = 1 \times 10^4$. The dashed line represents (Number of receptors in dimers)/20.0 and the solid line represents the number of activated Syk molecules. (b) $[\mathrm{Fc}\epsilon\mathrm{RI}]_{tot} = 7 \times 10^4$. The dashed line represents (Number of receptors in dimers)/250.0 and the solid line represents the number of activated Syk molecules.

3.2 Effect of monovalent ligands on Syk activation

A way to distinguish between Type I and Type II inhibition is by comparing responses to anti-IgE in the presence and absence of Fab fragments (Magro & Alexander, 1974; Magro & Bennich, 1977). Adding monovalent ligand that competes for the same binding site as a multivalent ligand will always reduce the number of cross-links. Thus, if the response increases in the presence of monovalent ligand, the inhibition is Type II, while if it decreases it is of Type I. To demonstrate, we have incorporated monovalent ligand into the model, with the monovalent ligand having the same rate constants for binding and dissociation as a single anti-IgE Fab site. The reference case we use for studying the effects of adding monovalent ligands on the signaling response, measured by Syk activation in our model, is characterized by the parameter values $k_{-1} = k_{-2} = 0.01 s^{-1}$, $k_{+1} = 2.5 \times 10^6$ M^{-1} s^{-1}, $k_{+2} = 8 \times 10^{-9}$ molecule^{-1} cm^2 s^{-1} and total Syk, Lyn and FcϵRI numbers per cell of 5×10^3, 1×10^5 and 1×10^5 respectively. We use this parameter set because it yields a bimodal Syk activation curve in the absence of the monovalent ligand. We add monovalent ligand to a maximum concentration of 6.6 nM which corresponds to 4×10^6 molecules per cell. We consider the effects of adding monovalent ligands on Syk activation at two bivalent ligand concentrations, 2×10^{-5} nM and 2×10^{-3} nM, corresponding to $2K_1L$ values of 10^{-5} and 10^{-3} respectively. Fig. 4 shows that adding monovalent IgE ligand in increasing concentrations leads to a gradual lowering in the Syk activation level at $2K_1L = 10^{-5}$ (Type I inhibition). In contrast, adding monovalent ligand in increasing concentrations at $2K_1L = 10^{-3}$ results in a gradual increase of the Syk activation level (Type II inhibition).

On plotting the Syk activation level as a function of the added monovalent ligand concentration for two values 10^{-5} and 10^{-3} of $2K_1L$, we obtain Fig. 4b which is qualitatively similar to the plot of histamine release as a function of the amount of added monomer anti-IgE (Fab) in Fig. 1 of Ref (Magro & Alexander, 1974) as shown in Fig. 4c. The similarity between the predicted variation of Syk activation and the experimentally observed variation of histamine release (Goldstein & Wofsy, 1994; Magro & Alexander, 1974) with added monomer concentration suggests one possible mechanism for Type II inhibition.

3.3 A symmetric cross-linking curve can lead to an asymmetric Syk activation curve

For the ligand-receptor binding parameters in Table 1, and FcϵRI and Syk concentrations of 1×10^5 and 7×10^3 per cell respectively, we observe a Syk activation curve with two prominent maxima at $2K_1L = 10^{-4}$ and $2K_1L = 10^4$ (Fig. 5a). This parameter set is used because it accentuates the asymmetry between the two maxima. The extent of receptor cross-linking at these two ligand concentrations are equal. The mean lifetime of the receptor dimer is independent of the ligand concentration, which implies that kinetic proofreading effects at these two ligand concentrations should be identical. Still, in the steady state, the model predicts that there is greater Syk activation at the lower ligand concentration ($2K_1L = 10^{-4}$), than at the higher ($2K_1L = 10^4$).

Even though all signaling events are initiated by receptor aggregation, it appears that the concentrations of free and bound receptors which are not in aggregates influence signaling. To investigate the origin of the asymmetry, we vary both the the ligand-IgE association (k_{+1}) and dissociation (k_{-1}) constants, such that the corresponding equilibrium constant K_1, and therefore the equilibrium cross-linking curve, remains unchanged. If the ligand-IgE dissociation constant is gradually reduced from 0.3 ($10^{-0.5}$) to 0.0003 ($10^{-3.5}$)s^{-1}, the Syk

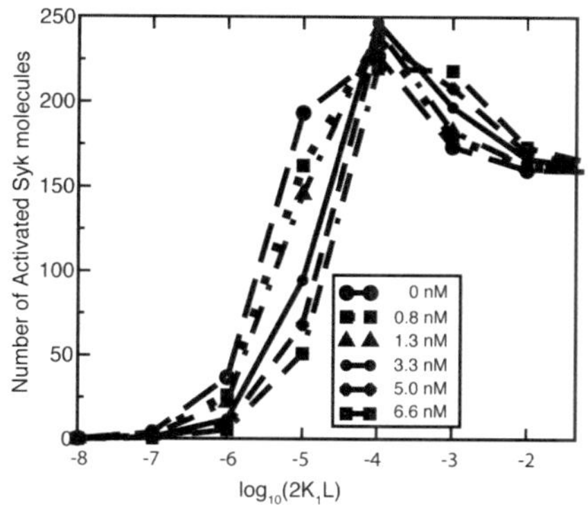

(a)

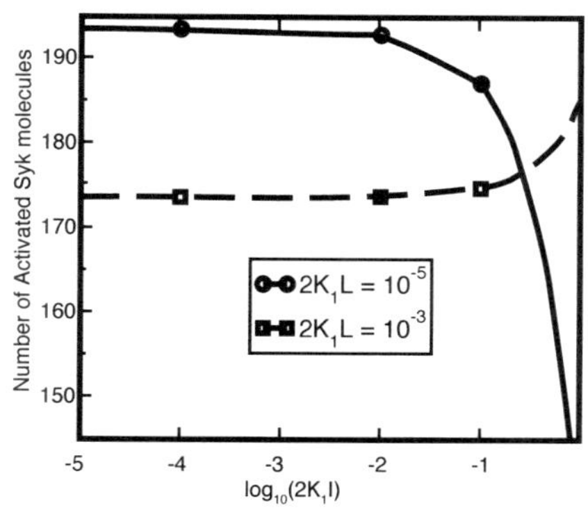

(b)

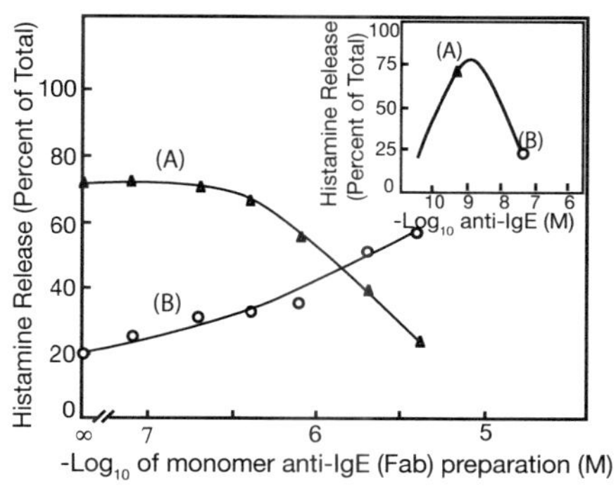

(c)

Fig. 4. (a)The effect of adding different concentrations of a monovalent IgE ligand on the extent of Syk activation at $2K_1L = 10^{-5}$ (Syk activation increases with $2K_1L$) and at $2K_1L = 10^{-3}$ (Syk activation decreases with $2K_1L$). The concentration of monovalent inhibitor (Fab fragments when the ligand is anti-IgE) present in the simulation is given in the figure. (b) The variation of the extent of Syk activation as a function of the concentration per cell of a monovalent inhibitor, I, at two ligand concentrations given by $2K_1L = 10^{-5}$ and $2K_1L = 10^{-3}$. The ligand-receptor binding parameters used are $k_{+1} = 2.5 \times 10^6$ M^{-1}s^{-1}, $k_{+2} = 8 \times 10^{-9}$ cm^{2}s^{-1}, $k_{-1} = k_{-2} = 0.01$s^{-1}. $[\text{Syk}]_{tot} = 5 \times 10^3$ per cell; $[\text{Lyn}]_{tot} = 10^5$ per cell; $[\text{Fc}\epsilon\text{RI}]_{tot} = 10^5$ per cell. (c) Modified Figure 1 from Magro and Alexander (Magro & Alexander, 1974). Plot of histamine release due to 5×10^{-10} M anti-IgE (A) and 5×10^{-8} M (B) incubated with increasing concentrations of monomer anti-IgE (Fab). The insert shows the positioning on the dose response curve of the two anti-IgE concentrations.

activation at $2K_1L = 10^{-4}$ remains practically unaltered, whereas the Syk activation at $2K_1L = 10^4$ increases and approaches the extent of Syk activation at $2K_1L = 10^{-4}$ (Fig. 5b). On the other hand, if the ligand-IgE dissociation constant is increased from 0.3 to 10^3 s^{-1}, the Syk activation at $2K_1L = 10^4$ remains unchanged and that at $2K_1L = 10^{-4}$ decreases and approaches the Syk activation level at $2K_1L = 10^4$ for $k_{-1} = 0.3$ s^{-1} (Fig. 5c). In order to determine the source of asymmetry in the Syk activation curve, we consider a pair of reduced models, which are discussed in the following section.

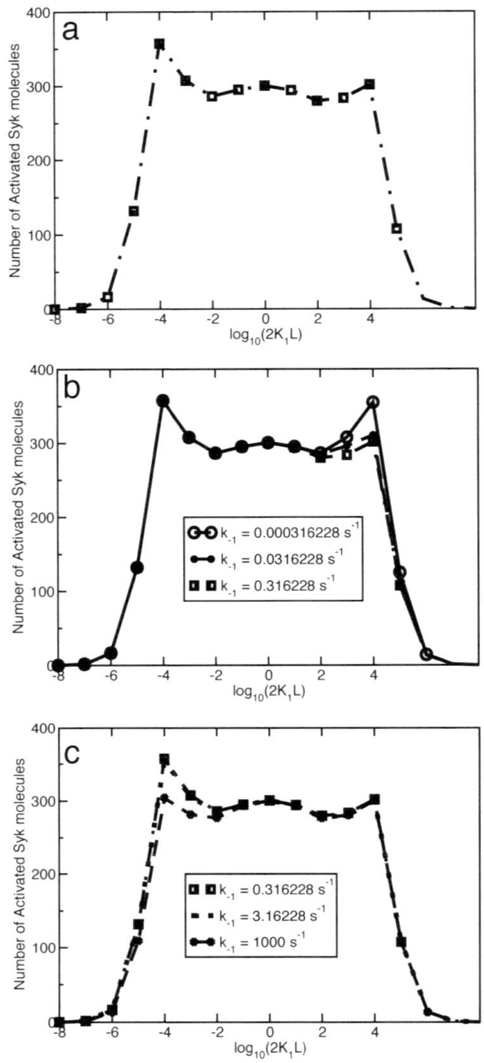

Fig. 5. (a) Syk Activation curve for $k_{+1} = 2.5 \times 10^7\ \mathrm{M}^{-1}\ \mathrm{s}^{-1}$; $k_{+2} = 8 \times 10^{-9}\mathrm{cm}^2\mathrm{s}^{-1}$; $k_{-1} = 0.316(10^{-0.5})\ \mathrm{s}^{-1}$; $k_{-2} = 0.0316(10^{-1.5})\mathrm{s}^{-1}$; (b)-(c) Syk Activation curves for $k_{+2} = 8 \times 10^{-9}\mathrm{cm}^2\mathrm{s}^{-1}$; $k_{-2} = 0.0316(10^{-1.5})\mathrm{s}^{-1}$; (b)The k_{+1} and k_{-1} values are varied such that $K_1 = k_{+1}/k_{-1}$ is fixed at 7.9 $\times 10^7\ \mathrm{M}^{-1}$. The legends show the k_{-1} values used and the k_{+1} values are accordingly adjusted so that K_1 remains fixed. For parts (a), (b) and (c), $[\mathrm{Syk}]_{tot} = 7 \times 10^3$ per cell; $[\mathrm{Lyn}]_{tot} = 10^5$ per cell; $[\mathrm{Fc\epsilon RI}]_{tot} = 10^5$ per cell.

3.4 Reduced models of Syk binding to cross-linked receptors

We consider two reduced models, 1 and 2, where the kinetic effects involving Lyn and the β subunit of the receptor have been eliminated. Both models correspond to the limit where the available Lyn is in large excess. In model 1, cross-linked receptors transphosphorylate each other on the γ subunit with a transphosphorylation rate that is the same as that of

Parameter	Description	Value(s)
k_{+1}	ligand-receptor binding forward rate constant	$2.5 \times 10^7 M^{-1} s^{-1}$
k_{-1}	ligand-receptor binding reverse rate constant	$0.316(10^{-0.5}) s^{-1}$
k_{+2}	ligand-receptor cross-linking forward rate constant	8.0×10^{-9} cm^{2}mole^{-1}s^{-1}
k_{-2}	ligand-receptor cross-linking reverse rate constant	$0.0316(10^{-1.5}) s^{-1}$

Table 1. Ligand-receptor binding parameters which accentuate the asymmetry in the Syk activation curves. Other rate constants used in our calculations are taken from Faeder et al. (2003). The bimolecular rate constants k_{+1} and k_{+2} are divided by the inverse cell density (1.0×10^{-6} ml) and the cell surface area (8.0×10^{-6} cm^2) to convert them to unimolecular rate constants.

Lyn bound through it SH2 domain to the phosphorylated β subunit of the receptor. As in the full model, bound Syk protects the γ subunit from being dephosphorylated. The rate of dephosphorylation of unprotected phosphorylated γ subunits is the same as in the full model. As seen in Fig. 6a, model 1 exhibits an asymmetry in a plot of Syk dimer formation as a function of the log of the ligand concentration.

In model 2, in addition to assuming Lyn is in excess, all receptor γ subunits are always phosphorylated, thereby eliminating the kinetic effects of receptor phosphorylation. This corresponds to the additional limit that the rate of receptor dephosphorylation goes to zero. For example, if in the full model, the available Lyn is increased from 10^5 to 10^8 per cell and the rate constant for dephosphorylation reduced from 20 s^{-1} to 10^{-4} s^{-1}, then a plot of Syk dimer formation for the full model and model 2 are identical (Fig. 6b). When this second assumption is made, the asymmetry in the Syk dimer curve is eliminated, indicating that the kinetics of phosphorylation and dephosphorylation of the γ subunit plays a role in creating the asymmetry.

3.5 The asymmetry in Syk activation originates from asymmetry in Syk dimer formation.

Since model 1 captures the asymmetry in Syk activation predicted by the full model, we use it to analyze the source of asymmetry in Syk dimer formation. As shown in Fig. 7b for model 1, one can form a Syk dimer (d_0) in a number ways. We have obtained expressions for the mean time required by any particular Syk molecule to end up in a Syk dimer, starting as either cytosolic Syk or as part of a complex with other molecule(s).

3.5.1 Mean time for a Syk molecule to enter into a Syk dimer

We calculate for reduced model 1, the mean times taken by a particular (tagged) Syk molecule to end up in a Syk dimer (Syk-FcϵRI-Ligand-FcϵRI-Syk) starting as either free Syk or as part of a Syk containing complex other than the Syk dimer. In Fig. 7a, we assign notations to different Syk containing species in the reduced model. The mean time taken to form a Syk dimer from species i is given by τ_i. The kinetic parameters used are described in Table 2. We define the mean time of incorporation of the tagged Syk molecule in species d_0, starting from species i, in the following fashion. This mean time is divided into two parts, the first part corresponding to the mean time of transition from species i to all other species containing the tagged Syk molecule. This is the inverse of the sum of rates of all processes that transform species i to the

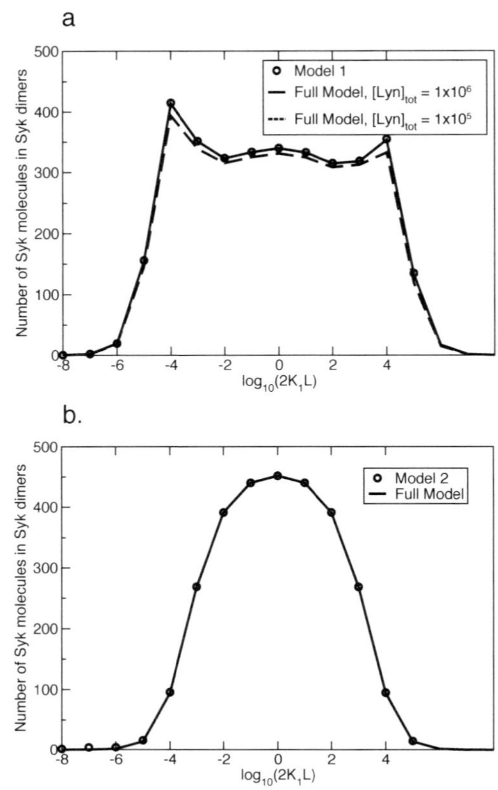

Fig. 6. Comparison of number of Syk dimers per cell obtained using reduced model 1 and 2, with the corresponding numbers obtained using the full model. $[\mathrm{Syk}]_{tot} = 7 \times 10^3$ per cell; $[\mathrm{Fc}\epsilon\mathrm{RI}]_{tot} = 10^5$ per cell. We use the same ligand-receptor and Syk-pγ binding and unbinding parameters as in Table 2 for both the reduced models and the full model. (a) We consider two different total cellular concentrations of Lyn in the full model, $[\mathrm{Lyn}]_{tot} = 10^5$ and $[\mathrm{Lyn}]_{tot} = 10^6$ per cell. (b) Comparison of reduced model 2 to the full model when Lyn is in large excess and the rate of dephosphorylation is negligible ($[\mathrm{Lyn}]_{tot} = 10^8$ per cell, $k_{-p} = 10^{-4}\mathrm{s}^{-1}$).

other species. The second part represents the mean remaining time to incorporate the tagged Syk in the Syk dimer species d_0. In terms of the species from Fig. 7a and the kinetic parameters from Table 2, the mean times of transition from species a_0, b_0, c_0, c_1 and s_0 respectively.

$$m_{a0} = \frac{1}{k_{+2}(b_0 + b_1 + b_2) + k_{-1} + k_{-S}} \tag{1}$$

$$m_{b0} = \frac{1}{k_{+2}(a_0 + a_1 + a_2) + 2k_{+1}L + k_{-S}} \tag{2}$$

$$m_{c0} = \frac{1}{2k_{-2} + k_{-S} + k_{-p} + k_{+S}s_0} \tag{3}$$

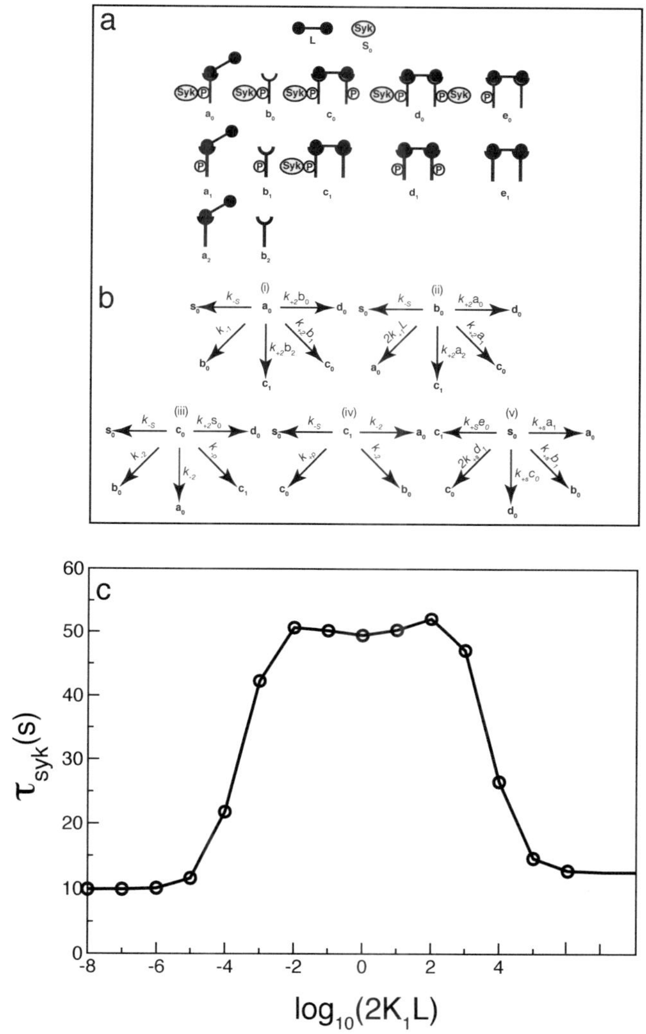

Fig. 7. (a) Schematic representation of the different Syk containing species in the reduced model 1. We tag a particular Syk molecule and calculate the mean time required by this Syk molecule to be incorporated into a Syk dimer (d_0). (b) (i-v) shows the different routes by which a tagged Syk molecule can be transferred from one species to another. (c) Mean time τ_{Syk} taken by any particular Syk molecule to be incorporated into a Syk dimer, starting as cytosolic Syk (s_0) or as other specie(s) (a_0, b_0, c_0, c_1) in (a). The rate parameters are taken from Table. 2. $[\text{Syk}]_{tot} = 7 \times 10^3$ per cell; $[\text{Fc}\epsilon\text{RI}]_{tot} = 10^5$ per cell.

$$m_{c1} = \frac{1}{2k_{-2} + k_{-S} + k_{+p}} \tag{4}$$

$$m_{s0} = \frac{1}{k_{+S}(a_1 + b_1 + 2d_1 + c_0 + e_0)} \tag{5}$$

Parameter	Description	Value(s)
k_{+1}	ligand-receptor binding forward rate constant	$2.5 \times 10^7 M^{-1}s^{-1}$
k_{-1}	ligand-receptor binding reverse rate constant	$0.316(10^{-0.5})s^{-1}$
k_{+2}	ligand-receptor cross-linking forward rate constant	8.0×10^{-9} $cm^2 mole^{-1}s^{-1}$
k_{-2}	ligand-receptor cross-linking reverse rate constant	$0.0316(10^{-1.5})s^{-1}$
k_{+gP}	receptor γ phosphorylation forward rate constant	$3.0\ s^{-1}$
k_{-gP}	receptor γ dephosphorylation rate constant	$20s^{-1}$
k_{+S}	Syk-phosphorylated receptor γ binding forward rate constant	6.0×10^{-5} $molecule^{-1}s^{-1}$
k_{-S}	Syk-phosphorylated receptor γ reverse rate constant	$0.13s^{-1}$

Table 2. Kinetic parameters used in reduced model with receptor γ phosphorylation-dephosphorylation. The bimolecular rate constants are divided by the inverse cell density (1.0×10^{-6} ml) and the cell surface area (8.0×10^{-6} cm^2) to convert them to unimolecular rate constants.

In terms of the above quantitites, τ_i can be defined as follows.

$$\tau_{a0} = m_{a0}[1 + (k_{+2}b_1\tau_{c0} + k_{+2}b_2\tau_{c1} + k_{-1}\tau_{b0} + k_{-S}\tau_{s0})] \quad (6)$$

$$\tau_{b0} = m_{b0}[1 + (k_{+2}a_1\tau_{c0} + k_{+2}a_2\tau_{c1} + 2k_{+1}L\tau_{a0} + k_{-S}\tau_{s0})] \quad (7)$$

$$\tau_{c0} = m_{c0}[1 + (k_{-2}\tau_{a0} + k_{-2}\tau_{b0} + k_{-S}\tau_{s0} + k_{-p}\tau_{c1})] \quad (8)$$

$$\tau_{c1} = m_{c1}[1 + (k_{-2}\tau_{a0} + k_{-2}\tau_{b0} + k_{-S}\tau_{s0} + k_{+p}\tau_{c0})] \quad (9)$$

$$\tau_{s0} = m_{s0}[1 + (k_{+S}a_1\tau_{a0} + k_{+S}b_1\tau_{b0} + 2k_{+S}d_1\tau_{c0} + k_{+S}e_0\tau_{c1})] \quad (10)$$

The above equations are solved for τ_{a0}, τ_{b0}, τ_{c0}, τ_{c1} and τ_{s0} using parameters from Table 2 for different L values corresponding to $\log_{10}(2K_1L)$ values from -8 to 8. The expression levels of Syk and FcϵRI receptor per cell are taken as 7×10^3 and 10^5 respectively. The mean time for any tagged Syk molecule to be incorporated into a Syk dimer (d_0) is given by

$$\tau_{Syk} = p_{a0}\tau_{a0} + p_{b0}\tau_{b0} + p_{c0}\tau_{c0} + p_{c1}\tau_{c1} + p_{s0}\tau_{s0} \quad (11)$$

where

$$\begin{aligned} p_{a0} &= a_0/[\text{Syk}]_{tot};\ p_{b0} = b_0/[\text{Syk}]_{tot};\ p_{c0} = c_0/[\text{Syk}]_{tot}; \\ p_{c1} &= c_1/[\text{Syk}]_{tot};\ \ p_{s0} = s_0/[\text{Syk}]_{tot} \end{aligned} \quad (12)$$

and $[\text{Syk}]_{tot} = a_0 + b_0 + c_0 + c_1 + 2d_0 + s_0$. The quantity τ_{Syk} is plotted as a function of $\log_{10}2K_1L$ in Fig. 7c. This mean time τ_{Syk} is not symmetric with respect to reflection about $2K_1L = 1$. The mean time at $2K_1L = 10^{-4}$ is less than its counterpart at $2K_1L = 10^4$, as shown in Fig 7c.

3.6 The asymmetry in Syk activation depends on the receptor-ligand dissociation constant

In Fig. 8a, we vary k_{-1} and k_{+1} keeping their ratio K_1 fixed, and calculate the average time τ_{Syk} required by a Syk to end up in a Syk dimer. At $k_{-1} = 10^{-3.5}$ s^{-1}, the τ_{Syk} values at $2K_1L = 10^{-4}$ and $2K_1L = 10^4$ are close to each other, so that there is little difference in the extent of Syk activation at these two ligand concentrations as was seen in Fig. 5b. As k_{-1} is increased from $10^{-3.5}$ s^{-1}, τ_{Syk} at $2K_1L = 10^{-4}$ remains unchanged while τ_{Syk} at $2K_1L = 10^4$ increases. This is consistent with our results from the full model, that Syk activation at

$2K_1L = 10^{-4}$ is essentially unchanged, while Syk activation at $2K_1L = 10^4$ decreases as k_{-1} is increased and K_1 is kept fixed (Fig. 5b). The difference between the τ_{Syk} values at the two ligand concentrations increases, as does the asymmetry in Syk activation obtained using the full model. When k_{-1} is increased beyond 0.316 ($10^{-0.5}$) s^{-1}, τ_{Syk} at $2K_1L = 10^{-4}$ increased while τ_{Syk} at $2K_1L = 10^4$ reaches a plateau. As a result, the difference in the τ_{Syk} values at these two ligand concentrations decreases with increasing k_{-1}. As k_{-1} is increased to 10^3 s^{-1}, τ_{Syk} at both $2K_1L = 10^{-4}$ and $2K_1L = 10^4$ approaches the same value. This is the same trend exhibited by Syk activation in the full model (Fig. 5c). The strong correlation between τ_{Syk} from reduced model 1 and Syk activation in the full model supports the idea that the asymmetry in the Syk activation profile originates from the asymmetry in τ_{Syk}.

The question remains, how can the kinetics of ligand binding, which occurs on the outside of the cell, give rise to an asymmetry in the Syk activation curve, when the receptor cross-linking curve is symmetric? To answer the question we considered two receptors, one free and one bound. These receptors might, for example, have come from a receptor dimer when it dissociated and their cytoplasmic domains may be associated with Syk or Lyn or both. We show how to calculate two quantities at equilibrium, τ_R and τ_B, the mean times respectively of an unbound receptor to enter into an aggregate and for a bound receptor that is not cross-linked to a second receptor to enter into an aggregate (Fig. 9a).

3.6.1 Mean time for a receptor to enter into a receptor dimer

We consider the case the cross-linking of monovalent receptors by a bivalent ligand. There are two possible transitions an unbound receptor can undergo. It can interact with a bound receptor to form a dimer or it can bind to a ligand in solution to become a bound receptor. Similarly two transitions are possible for a bound receptor that is not in an aggregate. It can associate with a free receptor to form a dimer or it can detach from its ligand and become a free receptor. These transitions are illustrated in Fig. 9a where the rates of transition are given.

The mean times τ_R and τ_B will be found by solving two linear algebraic equations that are functions of the equilibrium concentrations of the free receptors, R, and the bound receptors not in dimers, B. At equilibrium, these concentrations are calculated from the single site equilibrium constants for binding and crosslinking, K_1 and K_2 respectively, and the total receptor and ligand concentrations, R_T and L_T. However, for our purposes, it is more useful to express the results in terms of the free ligand concentration.

If there are R_T total receptors on the cell surface and there are no processes that remove or add new receptors the the total number of receptors is conserved and

$$\begin{aligned} R_T &= R + B + 2D = R + 2K_1LR + 2(2K_1L)(K_2/2)R^2 \\ &= (1 + 2K_1L)R + 2K_1K_2LR^2 \end{aligned} \tag{13}$$

where D is the concentration of receptor dimers. Defining $r = R/R_T$ as the fraction of free receptors Eq. (13) becomes

$$1 = (1 + 2K_1L)r + 2K_1K_2R_TLr^2 \tag{14}$$

It can be shown that w, the fraction of receptors not in dimers is given by

$$w = (1 + 2K_1L)r = \frac{-1 + \sqrt{1 + 4\delta}}{2\delta} \tag{15}$$

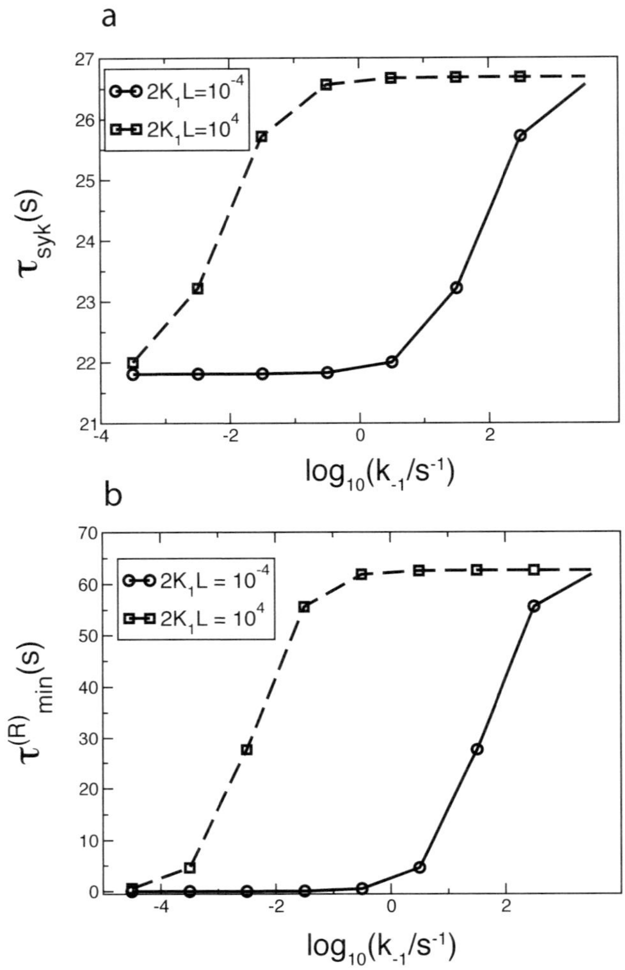

Fig. 8. Variation with k_{-1} of (a) average time (τ_{Syk}) required by any particular Syk molecule to end up in a Syk dimer and (b) minimum time ($\tau^{(R)}_{min}$) required by a receptor to be incorporated into a receptor dimer, at ligand concentrations given by $2K_1L = 10^{-4}$ and $2K_1L = 10^4$. The k_{-1} and k_{+1} are varied in such a way that $K_1 = k_{+1}/k_{-1}$ remains fixed at 7.9×10^7 M^{-1}. Other rate constants are taken from Table 2. $[\text{Syk}]_{tot} = 7 \times 10^3$ per cell; $[\text{Fc}\epsilon\text{RI}]_{tot} = 10^5$ per cell.

where

$$\delta = \frac{2K_1K_2R_TL}{(1+2K_1L)^2} \tag{16}$$

It follows that

$$R = wR_T\frac{1}{1+2K_1L} \tag{17}$$

$$B = wR_T\frac{2K_1L}{1+2K_1L} \tag{18}$$

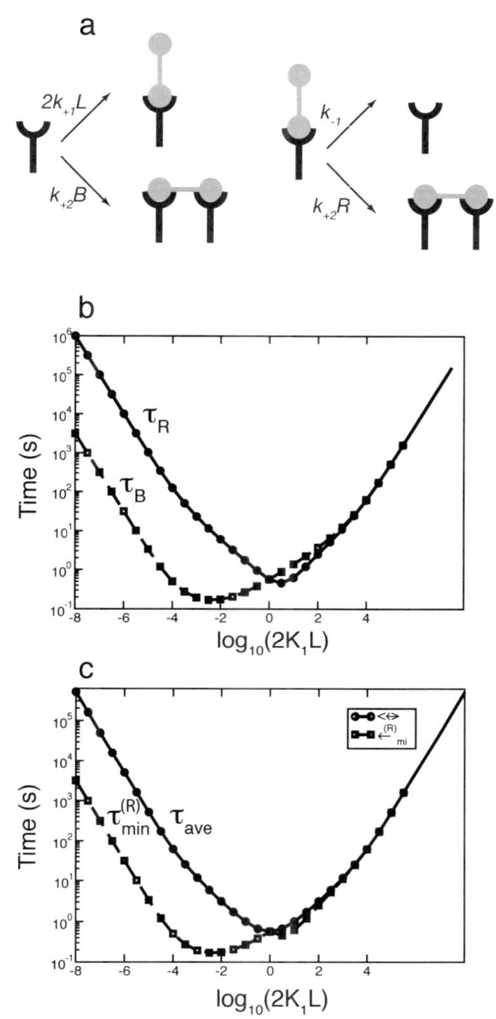

Fig. 9. (a) Rate constants that characterize the transitions that a free receptor and a bound receptor not in an aggregate can undergo. (b) The mean times for a free receptor (τ_R) and a bound receptor (τ_B) to find a partner and form a dimer as functions of the bivalent ligand concentration. (c) The mean τ_{ave} of τ_R and τ_B and the minimum $\tau_{min}^{(R)}$ of τ_R and τ_B as a function of the ligand concentration. In parts (b) and (c), the ligand receptor binding and cross-linking parameters are taken from Table 2 and $[\text{Fc}\epsilon\text{RI}]_{tot} = 10^5$ per cell.

We next write down the equations for the mean times.

$$\tau_B = \frac{1}{k_{+2}R + k_{-1}} + \frac{k_{-1}}{k_{+2}R + k_{-1}}\tau_R \tag{19}$$

$$\tau_R = \frac{1}{k_{+2}B + 2k_{+1}L} + \frac{2k_{+1}L}{k_{+2}B + 2k_{+1}L}\tau_B \tag{20}$$

The first term in each equation is the mean time to make a transition, while the second term is the mean remaining time after the first transition until the receptor is incorporated into a dimer. We define the following quantities:

$$t_1 = \frac{1}{k_{+2}R + k_{-1}} \qquad t_2 = \frac{1}{k_{+2}B + 2k_{+1}L} \tag{21}$$

$$p_1 = \frac{k_{-1}}{k_{+2}R + k_{-1}} \qquad p_2 = \frac{2k_{+1}L}{k_{+2}B + 2k_{+1}L} \tag{22}$$

The ts are mean times to make a transition while the ps are probabilities of making specific transitions.

Eqs. (19) and (20) hold whether or not the system is in equilibrium. At equilibrium, there are simplifications. For example, at equilibrium, $p_1 = p_2$, i.e., the probability of a ligand dissociating from a bound receptor is equal to the probability of a free receptor binding a ligand. This must be so if equilibrium is to be maintained. Solving Eqs. (19) and (20), we have

$$\tau_B = \frac{t_1 + p_1 t_2}{1 - p_1 p_2} \tag{23}$$

$$\tau_R = \frac{t_2 + p_2 t_1}{1 - p_1 p_2} \tag{24}$$

3.6.2 Asymmetry in Syk activation profile results from differential minimum time for a receptor to reform a dimer at different ligand concentrations

In Fig. 9b we show the mean times τ_R and τ_B as functions of the bivalent ligand concentration. Neither are symmetric, nor is their minimum, although their sum is symmetric (Fig.9c). Monovalent binding is responsible for the asymmetry. Consider the mean times at two free ligand concentrations, L_+ and L_-, where $\log(2K_1L_+) = -\log(2K_1L_-)$. At L_-, the bound receptor gets incorporated into a dimer more rapidly than the unbound receptor because it is present in the lowest concentration, and so it more rapidly finds a partner to crosslink. At L_+ the opposite is true. The asymmetry that arises in the minimum time for a receptor to reform a dimer, $\tau_{min}^{(R)}$, results from an asymmetric effect of the free ligand concentration on the lifetime of the minority receptor (the receptor at lowest concentration). At high ligand concentrations, minority receptors (unbound) are rapidly converted to bound receptors giving them less time to enter into a dimer. Thus, they may have to undergo interconversion between unbound and bound forms many times before forming a dimer. At low ligand concentrations, the initial conversion of the minority receptors (bound) to the unbound form is independent of the ligand concentration, depending only on the rate of dissociation.

In Fig. 8b we keep the equilibrium constant $K_1 = k_{+1}/k_{-1}$ fixed and, for the same set of parameters as in Fig. 8a, vary k_{-1}. We observe that the dependence of $\tau_{min}^{(R)}$ on $\log_{10}(k_{-1}/s^{-1})$ at the two ligand concentrations given by $2K_1L = 10^{-4}$ and $2K_1L = 10^4$ is qualitatively similar to that of τ_{Syk}. Since receptor dimerization is upstream from, and necessary for Syk activation, we infer that the dependence of τ_{Syk} on k_{-1} results from the corresponding dependence of $\tau_{min}^{(R)}$ on k_{-1}.

4. Serial engagement of IgE-FcϵRI complexes can enhance Syk activation in mast cells

As we have mentioned earlier in this chapter, mast cell responses mediated by FcϵRI are initiated when a multivalent ligand, an allergen for example, aggregates the receptors. In order to maintain the cellular response, it is not required that new aggregates constantly form. Kent et al. (Kent et al., 1994) exposed RBL cells to covalently cross-linked oligomers of IgE for a short duration and then added excess monomeric IgE to prevent further aggregation of receptors. Under these conditions, aggregated receptors continued to propagate the cellular signal. Phosphorylation of the β and γ immunoreceptor tyrosine-based activating motifs (ITAMs), phosphorylation of the protein tyrosine kinase Syk, and the release of histamine containing granules were sustained for substantial lengths of time after new aggregate formation was blocked. Although serial engagement of IgE-FcϵRI complexes is not necessary for mast cell signaling, the question still remains whether serial engagement can enhance mast cell response at low allergen concentrations. After deriving an expression for the rate at which a ligand of valence N serially engages receptors on a surface, we have refined this question and used the model of Faeder et al. (2003) to answer it.

4.1 Extension of mathematical model to simulate Syk activation in RBL cells stimulated by trivalent ligands

Earlier in this chapter, we have discussed how the model of Faeder et al. (2003) has been used to simulate the early response of RBL cells to the addition of a reversible bivalent ligand that binds to, and dimerizes, IgE-Fc complexes on RBL cell surfaces. The model consists of a network of 354 distinct chemical species connected by 3680 chemical reactions, 21 rate constants and three concentrations, the surface concentrations of FcϵRI and available Lyn, and the total concentration of Syk. With the exception of the rate constants that describe the interaction of the bivalent ligand with IgE, k_{+1}, k_{-1}, k_{+2} and k_{-2}, the parameters used in the simulations are the same as those given in Table I of Faeder et al. (2003). In the simulations for the current investigation, we take $k_{\mathrm{off}} = k_{-1} = k_{-2}$ and the total concentrations of Lyn, Syk and FcϵRI are specific for RBL cells.

We now generalize our model to simulate the early signaling events triggered by the addition of a symmetric trivalent ligand with three identical binding sites interacting with IgE bound to FcϵRI on RBL cells. There is now an additional surface crosslinking reaction involving a trivalent ligand with two of its sites bond crosslinking a third IgE-FcϵRI, described by rate constants k_{-3} and k_{+3}. In the simulations, we take $k_{\mathrm{off}} = k_{-1} = k_{-2} = k_{-3}$ and $k_{+2} = k_{+3}$. It is assumed that a Lyn bound to a receptor can randomly transphosphorylate any other receptor in the aggregate with the same rates, as it was assumed for the dimeric aggregate. For the trivalent ligand, the model expands to 2954 distinct chemical species connected by 49948 reactions.

4.2 Theory

First, we obtain an expression for the rate of serial engagement of a multivalent ligand of valence N binding to free mobile receptors on a cell surface that has a concentration of R free receptors. This expression allows us to choose parameter values that define ligands with different rates of serial engagement. We then use these parameter values to simulate the early response of mast cells to ligands that serial engage receptors at different rates. We use as our measure of signaling response, receptors that have fully activated Syk associated with their γ ITAMs. Syk becomes partially active upon binding through its two SH2 domains to the doubly phosphorylated γ ITAM (Shiue et al., 1995). In our model, Syk is fully active when it has been transphosphorylated on its activation loop tyrosines by a second Syk (reviewed in Faeder et al. (2003)). Finally, we use the results of the simulations to evaluate the role of serial engagement in early responses of mast cell signaling.

4.2.1 Serial engagement of a N-valent ligand

The rate of serial engagement equals the number of receptors the ligand encounters from the time t_a it first attaches to a cell surface receptor until the time t_b when its last bond breaks, divided by the time this process lasted, $t_b - t_a$. Figure 10 illustrates the reactions a five valent ligand can undergo once it is initially bound to a receptor on a surface. The mean time an N-valent ligand, that initially has one site bound to a receptor, remains on a surface when the concentration of free receptors is R is (Hlavacek et al., 2002; 1999)

$$t_N = \frac{1}{k_{\text{off}}} \left[\frac{(1 + K_2 R)^N - 1}{N K_2 R} \right] \tag{25}$$

where $K_2 = k_{+2}/k_{\text{off}}$ is the equilibrium cross-linking constant.

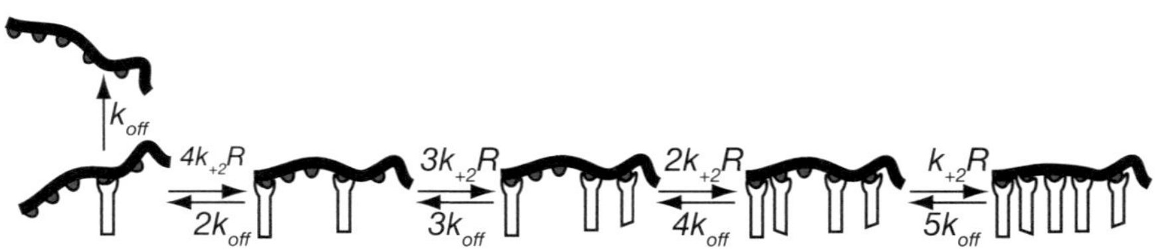

Fig. 10. The multiple reactions a ligand of valence five ($N = 5$) can undergo from the time it first binds a cell surface receptor until the last ligand-receptor bond breaks and the ligand leaves the surface. R is the concentration of free receptors. In these reactions, k_{+2} is the single-site rate constant for addition of a receptor to a ligand-receptor complex and k_{off} is the corresponding reverse rate constant. By taking all the forward and reverse rate constants to be the same, we have assumed that these rate constants are independent of the bound state of the ligand.

In the following section, we derive the following expression for n_N, the mean number of receptors an N-valent ligand that initially has one site bound to a receptor, serial engages before it dissociates from the cell surface.

4.2.2 Average number of receptors serially engaged by a multivalent ligand

An expression for the average number of receptors a multivalent ligand serially engages, n_N, is obtained by solving a set of ordinary differential equations (ODEs). We consider a cell with monovalent receptors on its surface that is in equilibrium with a homogeneous concentration of N-valent ligands. At equilibrium, there is a distribution of bound ligands, some with one site bound, some with two sites bound, up to some with N sites bound. We consider the ensemble of N-valent ligands that, at $t = 0$, have one and only one site bound to a receptor. We let $B_i(t)$ be the fraction of these ligands that have i sites bound at time t or equivalently, the probability that a ligand has i sites bound at time t when initially it had one site bound. We keep track of the number of receptors a ligand serial engages by keeping track of how many receptors dissociate from a ligand until the ligand leaves the cell surface. Since the rate at which a receptor dissociates from a ligand with i sites bound is ik_{off}, the averaged number of receptors a ligand will serial engage from the time it binds to the cell until the time it leaves the cell is:

$$n_N = k_{\text{off}} \int_0^\infty \Big(2B_2(t') + 3B_3(t') \cdots + NB_N(t')\Big)dt' \tag{26}$$

Before solving the general problem, we illustrate our approach by finding the number of receptors a bivalent ligand serial engages during its time on the cell surface.

The equilibrium concentration of free receptors we call R. For a bivalent ligand, we have the following set of ODEs:

$$\frac{dB_1}{dt} = -k_{\text{off}}B_1 - k_{+2}RB_1 + 2k_{\text{off}}B_2 \tag{27}$$

$$\frac{dB_2}{dt} = k_{+2}RB_1 - 2k_{\text{off}}B_2 \tag{28}$$

The initial conditions are that $B_1 = 1$ and $B_2 = 0$.

The average number of receptors a bivalent ligand engages during its time on the cell surface is

$$n_2 = 2k_{\text{off}} \int_0^\infty B_2(t')dt' \tag{29}$$

As is standard, we use Laplace transforms ($\mathcal{L}$), to transform the problem into solving a set of algebraic equations, taking s to be the transform variable and $\bar{B}_i(s) = \mathcal{L}[B_i(t)]$. Note that

$$n_N = \lim_{s\to 0} k_{\text{off}}\Big(2\bar{B}_2(s) + 3\bar{B}_3(s) \cdots + N\bar{B}_N(s)\Big) \tag{30}$$

This follows because the Laplace transform of $\int_0^t f(u)du$ is $\bar{f}(s)/s$ and

$$\lim_{t\to\infty} g(t) = \lim_{s\to 0} s\bar{g}(s) \tag{31}$$

Taking Laplace transforms of both sides of Eqs. (27) and (28), we have

$$\begin{aligned} s\bar{B}_1 - B_1(0) &= -k_{\text{off}}\bar{B}_1 - k_{+2}R\bar{B}_1 + 2k_{\text{off}}\bar{B}_2 \\ sB_2 &= k_{+2}R\bar{B}_1 - 2k_{\text{off}}\bar{B}_2 \end{aligned}$$

We can take the limit before we solve the equations. Setting $s = 0$, $k = k_{+2}R/k_{\text{off}} = K_2R$ and $B_1(0) = 1$,

$$\begin{aligned} \frac{1}{k_{\text{off}}} &= (1+k)\bar{B}_1 - 2\bar{B}_2 \\ 0 &= k\bar{B}_1 - 2\bar{B}_2 \end{aligned}$$

Solving, we have that $B_2 = k/(2k_{\text{off}})$ and

$$n_2 = \lim_{s\to 0} k_{\text{off}} 2B_2(s) = k = K_2R \tag{32}$$

For a ligand with valence N,

$$\frac{dB_i}{dt} = (i+1)k_{\text{off}}B_{i+1} + (N-i+1)k_{+2}RB_{i-1} - ik_{\text{off}}B_i - (N-i)k_{+2}RB_i \tag{33}$$

B_i can appear through dissociation of a receptor from any of $i+1$ receptor-occupied sites of B_{i+1}, which corresponds to a term $(i+1)k_{\text{off}}B_{i+1}$, and through engaging of a receptor by any of $(N-(i-1))$ free sites of B_{i-1}, which corresponds to a term $(N-i+1)k_{+2}RB_{i-1}$. B_i can disappear by losing a receptor from any of its i receptor-bound sites and by adding a receptor to any of $(N-i)$ free sites, which corresponds to substraction of terms $ik_{\text{off}}B_i$ and $(N-i)k_{+2}RB_i$ respectively.

Putting $k = k_{+2}R/k_{\text{off}}$, $\tau = k_{\text{off}}t$, $B_0 = B_{N+1} = 0$, and collecting terms,

$$\frac{dB_i}{d\tau} = (i+1)B_{i+1} - (i+k(N-i))B_i + k(N-i+1)B_{i-1},\ i = 1,\ldots,N \tag{34}$$

(Note that $B_0 = B_{N+1} = 0$ implies that the corresponding Laplace transforms $\bar{B}_0 = \bar{B}_{N+1} = 0$.)

As before, taking Laplace transforms and setting the transform variable equal to zero, we obtain

$$\begin{aligned}
-1 &= 2\bar{B}_2 - (1+k(N-1))\bar{B}_1 \\
0 &= 3\bar{B}_3 - (2+k(N-2))\bar{B}_2 + (N-1)k\bar{B}_1 \\
0 &= 4\bar{B}_4 - (3+k(N-3))\bar{B}_3 + (N-2)k\bar{B}_2 \\
&\vdots \\
0 &= -N\bar{B}_N + k\bar{B}_{N-1}
\end{aligned}$$

First, we solve this system of equations for $\bar{B}_1$ by summing all the equations:

$$\begin{aligned}
-1 &= \sum_{i=1}^{N}\left((i+1)\bar{B}_{i+1} - (i+k(N-i))\bar{B}_i + k(N-i+1)\bar{B}_{i-1}\right) \\
&= \sum_{i=2}^{N} i\bar{B}_i - \sum_{i=1}^{N}(i+k(N-i))\bar{B}_i + \sum_{i=1}^{N-1} k(N-i)\bar{B}_i \\
&= \sum_{i=1}^{N}\left(i - (i+k(N-i)) + k(N-i)\right)\bar{B}_i - \bar{B}_1
\end{aligned}$$

Each term in the sum on the right hand side equals zero, so that $\bar{B}_1 = 1$. To find $\bar{B}_2$, we sum up all equations except the first one:

$$\begin{aligned}
0 &= \sum_{i=2}^{N}\left((i+1)\bar{B}_{i+1} - (i+k(N-i))\bar{B}_i + k(N-i+1)\bar{B}_{i-1}\right) \\
&= \sum_{i=3}^{N} i\bar{B}_i - \sum_{i=2}^{N}(i+k(N-i))\bar{B}_i + \sum_{i=1}^{N-1} k(N-i)\bar{B}_i \\
&= \sum_{i=2}^{N}\left(i - (i+k(N-i)) + k(N-i)\right)\bar{B}_i - 2\bar{B}_2 + k(N-1)\bar{B}_1
\end{aligned}$$

	$K_2R = 0.1$			$K_2R = 1.0$			$K_2R = 10.0$		
N	$k_{\text{off}}t_N$	n_N	$r_Nk_{\text{off}}^{-1}$	$k_{\text{off}}t_N$	n_N	$r_Nk_{\text{off}}^{-1}$	$k_{\text{off}}t_N$	n_N	$r_Nk_{\text{off}}^{-1}$
2	1.05	0.10	0.095	1.50	1.00	0.67	6.00	10.00	1.67
3	1.10	0.21	0.190	2.33	3.00	1.29	44.33	120.00	2.71
4	1.16	0.33	0.285	3.75	7.00	1.87	366.00	1330.00	3.63
5	1.22	0.46	0.380	6.20	15.00	2.42	3221.00	14640.00	4.55

Table 3. The influence of the parameter K_2R (the product of the equilibrium cross-linking constant and the free receptor concentration) on the mean time a N-valent ligand remains bound to the surface (t_N), the mean number of receptors the ligand engages while it remains bound to the surface (n_N), and the rate the ligand serial engages receptors (r_N). Increasing K_2R increases the rate at which a ligand cross-links a receptor compared to the rate at which a receptor dissociates from the ligand.

Again, the first sum on the right hand side equals zero so that $\bar{B}_2 = k(N-1)\bar{B}_1/2 = k(N-1)/2$. Similarly, $\bar{B}_3 = k(N-2)\bar{B}_2/3 = k^2(N-1)(N-2)/(1\cdot 2\cdot 3)$. Following this approach, it is straightforward to show that

$$\bar{B}_j = k^{j-1}\frac{(N-j+1)(N-j+2)\dots(N-1)}{j!} = k^{j-1}\frac{(N-1)!}{j!(N-j)!} \tag{35}$$

and therefore, that

$$n_N = \sum_{j=2}^{N} j\bar{B}_j = \sum_{j=2}^{N} k^{j-1}\frac{(N-1)!}{(j-1)!(N-j)!} \tag{36}$$

4.2.3 Average rate of serial engagement

The average rate, r_N, an N-valent ligand serially engages receptors is

$$r_N = n_N/t_N \tag{37}$$

In Table 3 we evaluate these expressions for $N = 1-5$ and $K_2R = 0.10, 1.0$ and 10.0. The parameter K_2R is the product of the surface equilibrium cross-linking constant and the surface concentration of free receptors. Larger values of K_2R favor aggregate formation. Note that for large values of K_2R, as the valence is increased, the number of serial engagements and the mean time the ligand remains bound to the cell rises rapidly but their ratio, the rate of serial engagement, increases only modestly. From Eqs (25)-(37) it can be shown that the maximal rate of serial engagement achievable by an N-valent ligand when k_{off} is fixed and $k_{+2}R$ is increased is Nk_{off}. When $k_{+2}R$ is fixed and k_{off} is increased, the maximal rate is $k_{+2}R$.

4.2.4 The rates of serial engagement of receptors for bivalent and trivalent ligands

We now focus on bivalent and trivalent ligands interacting with monovalent receptors. From Eqs. (25) - (37) we have that the rate of serial engagement of a bivalent ligand

$$r_2 = 2k_{\text{off}}\frac{K_2R}{2+K_2R} \tag{38}$$

A second way this result can be obtained is to note that one receptor is bound and released by a bivalent ligand in one cycle time t_c so that $r_2 = 1/t_c$. The cycle time is the sum of the mean times for a site on a doubly bound ligand to open and for the singly bound ligand to bind to another receptor, i.e., $t_c = 1/(2k_{\rm off}) + 1/(k_{+2}R)$. (When the ligand is a pMHC rather than a bivalent ligand the two in Eq. (38) becomes a one (Wofsy et al., 2001).) This simple approach for calculating r_2 cannot easily be generalized to ligands with valence greater than two because for these ligands, there is more than one recycling time.

From Eqs. (25) - (37), we have the rate of serial engagement of a trivalent ligand

$$r_3 = 3k_{\rm off}\frac{2K_2R + (K_2R)^2}{3 + 3K_2R + (K_2R)^2} \tag{39}$$

It is easy to see from these expressions that for large values of K_2R, the rate of serial engagement approaches its maximal value of $Nk_{\rm off}$. We use Eqs. (38) and (39) to pick the parameter values that characterize the ligands for our simulations.

There are three separate ways to vary the ligand properties to enhance serial engagement, increase $k_{\rm off}$, increase k_{+2} or increase the valence of the ligand:

1. Increasing $k_{\rm off}$ enhances the rate at which the ligand serial engages receptors but this has the deleterious effect of enhancing kinetic proofreading by reducing the time a receptor remains in an aggregate. Once a receptor leaves an aggregate, unless it can rapidly enter into a new aggregate, all modifications it has undergone will be reversed and it will return to its basal state (Mao & Metzger, 1997; Peirce & Metzger, 2000).

2. Increasing k_{+2} reduces the time it takes a free site on a ligand in a ligand-receptor complex to bind to a new receptor. This enhances the effects of serial engagement by making it more likely that a ligand-receptor complex will form a new dimer before the ligand dissociates and the receptor returns to its basal state.

3. Increasing the valence of the ligand has a similar effect to increasing the forward rate constant. For an N-valent ligand bound through a single site to a receptor, the forward rate of binding a second receptor and forming a dimer is proportional to $(N-1)k_{+2}$. However, increasing the valence introduces other effects as well. A higher valent ligand can form larger aggregates and large aggregates are more effective at signaling than smaller aggregates, even when the ligands are oligomers of IgE that do not undergo serial engagement (Fewtrell & Metzger, 1980). In RBL cells, Lyn, the kinase that is responsible for ITAM phosphorylation, is limiting (Torigoe et al., 1997; Wofsy et al., 1997). Therefore, larger aggregates have a higher probability than receptor dimers of containing a receptor associated with Lyn, and a Lyn in an aggregate can serial phosphorylate all its neighboring receptors in the aggregate.

4.3 Simulations of mast cell responses to bivalent and trivalent ligands for different rates of serial engagement

We use our model to assess the role of serial engagement in mast cell signaling. The predicted outputs of the model that we use in our study are the concentrations (number per cell) of the phosphorylated β and γ ITAMs of FcϵRI, and the concentration of Syk bound to a phosphorylated γ ITAM and transphosphorylated on its activation loop tyrosines by a Syk bound to an adjacent receptor. We refer to Syk bound to the γ ITAM and phosphorylated on its activation loop tyrosines as fully activated Syk or phosphorylated Syk. Using antibodies specific for phosphorylated tyrosines in the activation loop of Syk, it has been shown that upon FcϵRI aggregation, these tyrosines became phosphorylated and this phosphorylation depended on the kinase activity of Syk (Zhang et al., 2000). The presence of these activation loop tyrosines are necessary for Syk-mediated propagation of FcϵRI signaling (Zhang et al., 1998).

4.3.1 Increasing serial engagement by increasing k_{off} reduces Syk activation

In Figure 11 we compare the predicted levels of β and γ ITAM phosphorylation and full Syk activation induced by four different hypothetical ligands, two bivalent (circles) and two trivalent (squares), that have the same single site forward rate constants $k_{+2} = 2.0 \times 10^{-11}$ cm^2/s. (In the simulations, we take the surface area of the cell $A = 8 \times 10^{-6}$ cm^2 and the total number of receptors per cell $N_T = 4 \times 10^5$ so that the receptor density $R_T = 5 \times 10^{10}$ cm^{-2}. Therefore, at low ligand concentrations, when most of the receptors are free so that $R \approx R_T$, the rate at which a bound ligand with one site free cross-links a receptor is $k_{+2}R = 1.0$ s^{-1}.) Each pair of bivalent and trivalent ligands have dissociation rate constants $k_{\text{off}} = 0.01$ s^{-1} (solid lines and symbols) and 10 s^{-1} (dashed lines and open symbols). For the dimer, increasing k_{off} from 0.01 s^{-1} to 10 s^{-1} corresponds to increasing the rate of serial engagement from $r_2 = 0.017$ s^{-1} to 0.10 s^{-1} (Eq. 38), while for the trimer, the increase is from $r_3 = 0.027$ s^{-1} to 0.20 s^{-1} (Eq. 39). In addition to increasing the rate of serial engagement, increasing k_{off} reduces the lifetime of the bond between a site on the ligand and a receptor site and thus, increases the effects of kinetic proofreading (McKeithan, 1995). As previously discussed (Faeder et al., 2003), to equalize the comparison, the levels of phosphorylation are plotted as a function of the number of receptors in aggregates. Figure 11a shows the ligand concentrations needed to achieve the same level of receptor aggregation for the different ligands. As k_{off} is increased while keeping $k_{+2}R_T$ fixed, for both bivalent and trivalent ligands, a higher ligand concentration is required to yield the same number of aggregated receptors, since the rate at which the receptor cross-link is broken is enhanced, whereas the rate at which the cross-links form remains unaltered. In Figure 11d we see that as the lifetime of a receptor in an aggregate is decreased, i.e., k_{off} is increased, full Syk activation is dramatically reduced for both bivalent and trivalent ligands. Kinetic proofreading dominates any positive effects serial engagement may have on Syk activation. This result is similar to that seen in experiments using fast and slowly dissociating multivalent ligands that aggregate IgE on RBL cells (Torigoe et al., 1998).

Figure 11b and Figure 11c show that the proximal signaling events, β and γ ITAM phosphorylation, compared with Syk phosphorylation, are effected less by the changes in the ligand-receptor lifetime as expected from a kinetic proofreading model (McKeithan, 1995). The reduction in Syk phosphorylation is pronounced because it is much further down the signaling cascade than phosphorylation of the receptor γ-chain. A number of reversible steps must occur to go from γ-chain phosphorylation to Syk phosphorylation. A Syk molecule has to bind to a phosphorylated γ-chain before the γ-chain is dephosphorylated. For transphosphorylation of the γ-bound Syk molecule to occur, a second Syk molecule must bind, which requires a second γ-chain to be phosphorylated in the same aggregate. Finally, transphoshorylation among the Syk molecules must occur before either of the Syk molecules dissociates.

4.3.2 Increasing serial engagement by increasing k_{+2} increases Syk activation

In Figure 12 we compare the levels of β and γ ITAM phosphorylation and full Syk activation induced by four different hypothetical ligands, two bivalent (circles) and two trivalent

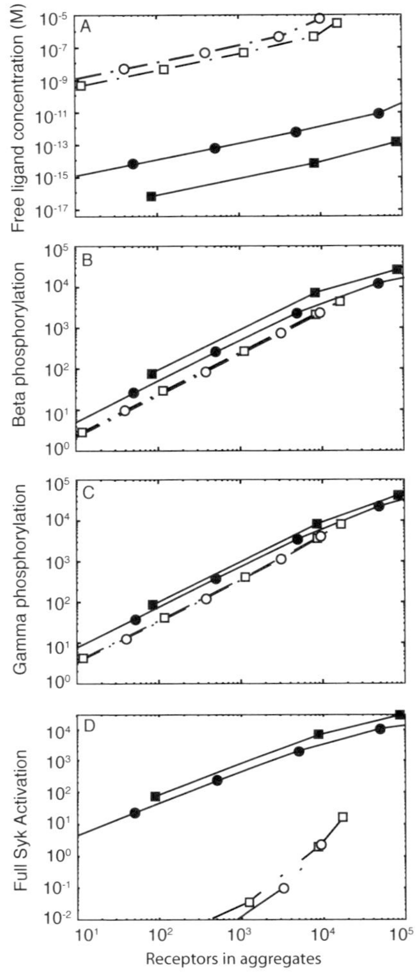

Fig. 11. Simulation of the effects of increasing the rate of serial engagement of bivalent and trivalent ligand by increasing the ligand-receptor dissociation constant, k_{off}. The four curves correspond to $k_{+2}R_T = 1.0\ s^{-1}$ and bivalent (circles) and trivalent (squares) ligands with $k_{off} = 0.01\ s^{-1}$ (solid lines and symbols) and $10\ s^{-1}$ (dashed lines and open symbols). The additional parameters used in the simulations are given in Table I of Faeder et al. (2003). The x-axis indicates levels of FcϵRI aggregation, given as the number of receptors per cell in aggregates. One percent of the cell's receptors in aggregates corresponds to 4×10^3 receptors. (A) Free ligand concentrations corresponding to specified levels of receptor aggregation. (B), (C), and (D) are respectively the predicted number of receptors per cell with the β ITAM phosphorylated, the γ ITAM phosphorylated, and Syk bound to the phosphorylated γ ITAM and transphosphorylated on its activation loop by another Syk.

(squares), that have the same dissociation rate constant $k_{off} = 0.1\ s^{-1}$ and different surface cross-linking constants. Each pair of bivalent and trivalent ligands have surface cross-linking constants such that $k_{+2}R_T = 0.1\ s^{-1}$ (solid lines and symbols) and $100\ s^{-1}$ (dashed lines and open symbols). As in Figure 11, we compare ligands at different concentrations that result in the same number of receptors being in aggregates at equilibrium. Since k_{off} is fixed, the lifetime of a receptor in an aggregate is the same for all four ligands. As $k_{+2}R_T$ is increased from 0.1 s^{-1} to 100 s^{-1}, the rate of serial engagement for the bivalent ligand increases from 0.067 s^{-1} to 0.20 s^{-1} (Eq.(38)), and for the trivalent ligand from 0.13 s^{-1} to 0.30 s^{-1} (Eq.(39)). (These rates of serial engagement were calculated for low ligand concentrations, when $R \approx R_T$ and $k_{+2}R \approx k_{+2}R_T$.) We see that the simulations predict that increasing the rate of serial engagement while holding k_{off} fixed has a minor effect on receptor phosphorylation (12b and c), but a more pronounced effect on the phosphorylation of Syk by Syk, i.e, on full Syk activation (12d).

4.3.3 Increasing the valence of a ligand increases Syk activation, partly as a result of increased serial engagement

From Eqs. (38) and (39) we can calculate the increase in the rate of serial engagement as a result of increasing the valence of a ligand from two to three while keeping its rate constants unchanged. The factor by which the rate of serial engagement increases, r_3/r_2, depends only on the parameter $k_{+2}R$, and monotonically decreases from 2 to 3/2 as $k_{+2}R$ increases from zero to infinity. In Figure 13, as in 12, we compare the predicted full Syk activation from simulations where the stimulating ligand is either bivalent (circles) or trivalent (squares), the rate of serial engagement is varied by varying k_{+2}, and for both ligands, the lifetime of a receptor bound to a ligand is the same, 10 s ($k_{off} = 0.10$ s^{-1}). The range over which the rate of serial engagement varies in Figure 13 corresponds to $k_{+2}R = 0.01 - 100$ s^{-1}. In Figure 13a, the two top curves and two bottom curves correspond to 1000 and a 100 receptors per cell in aggregates. For the same rate of serial engagement and the same number of receptors in aggregates, the model predicts that the trivalent ligand is more effective than the bivalent ligand at activating Syk.

In Figure 13b, we re-plot the simulations for the upper two curves in Figure 13a. To illustrate how we can asses the contribution of serial engagement in enhancing Syk phosphorylation of Syk when the valence is increased, we consider the point labeled 1 on the Syk activation curve for the bivalent ligand in Figure 13b. This point corresponds to the serial engagement rate for a bivalent ligand with $k_{+2}R = 0.01$ s^{-1} and $k_{off} = 0.10$ s^{-1} . When the valence is increased from two to three with the rate constants unchanged, the rate of serial engagement for the trimer increases above that of the dimer to the point on the trivalent ligand curve labeled 2. We see that there would be an increase in Syk phosphorylation of Syk for a trivalent ligand even if there was no increase in the rate of serial engagement. This is indicated by the difference B-A on the y-axis and is a result of effects other than serial engagement. The increase in Syk activation that is attributable to the increase in serial engagement is indicated by the difference C-B on the y-axis, which can be substantial.

Lyn is strongly regulated by Cbp and Csk (Kawabuchi et al., 2000; Ohtake et al., 2002) on RBL cells and as a result, the amount of Lyn available to the receptor is in short supply (Torigoe et al., 1997; Wofsy et al., 1997). Although Lyn regulation is not in the model, the Lyn concentration available to the receptor is chosen so that Lyn is limiting. In the model, in the absence of the ligand, less than five percent of the receptors are associated with Lyn which is consistent with the observations of Yamashita et al. (Yamashita et al., 1994). Upon aggregation, large receptor aggregates are more likely to contain a Lyn than small receptor aggregates, and once a Lyn is in an aggregate it can transphosphorylate all receptors in its proximity. It is

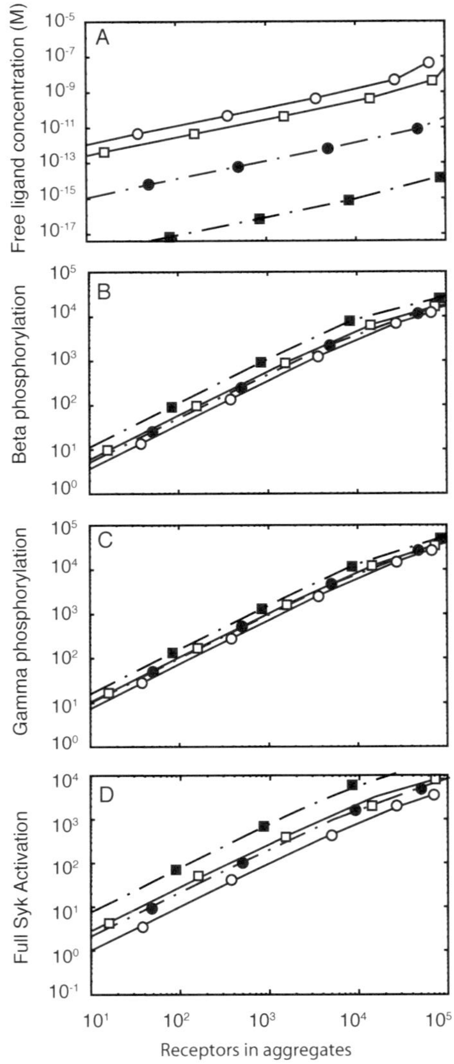

Fig. 12. Simulation of the effects of increasing the rate of serial engagement of bivalent and trivalent ligand by increasing the rate of crosslinking, $k_{+2}R$. The four curves correspond to $k_{\text{off}} = 0.1\ \text{s}^{-1}$ and bivalent (circles) and trivalent (squares) ligands with $k_{+2}R = 0.1\ \text{s}^{-1}$ (solid lines and symbols) and $100\ \text{s}^{-1}$ (dashed lines and open symbols). All other parameters are the same as in Figure 11. The x-axis indicates levels of FcϵRI aggregation, given as the number of receptors per cell in aggregates. (A) Free ligand concentrations corresponding to specified levels of receptor aggregation. (B), (C), and (D) are respecitively the predicted number of receptors per cell with the β ITAM phosphorylated, the γ ITAM phosphorylated, and Syk bound to the γ ITAM and transphosphorylated by Syk on its activation loop.

these effects that the model captures when the valence of the ligand is increased from two to three and that account for the increase in Syk phosphorylation that is not a result of serial

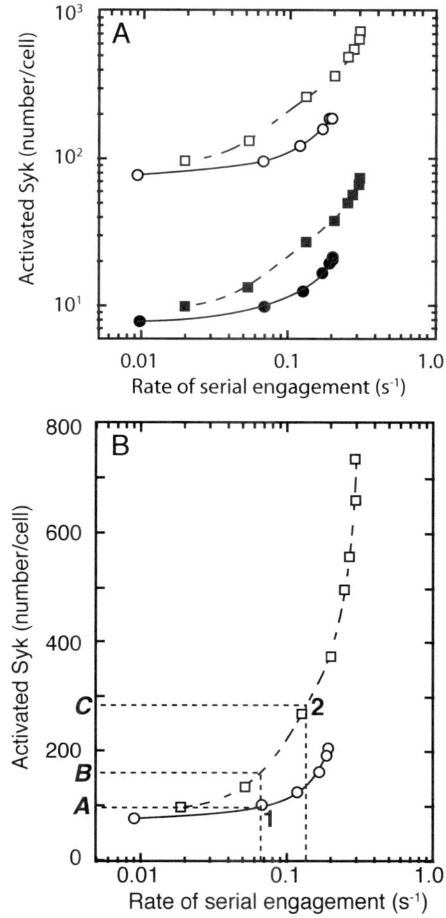

Fig. 13. Predicted number of activated Syk (Syk transphosphorylated by Syk) as a function of the rate of serial engagement. (A) The circles refer to bivalent ligand simulations and the squares to trivalent ligand simulations. Open and solid symbols refer to 1000 and 100 receptors in aggregates respectively. (B) Replot of the bivalent and trivalent curves in (A) for 1000 receptors in aggregates. The symbol 1 corresponds to a dimer with $k_{+2}R = 0.01\ s^{-1}$ and $k_{off} = 0.10\ s^{-1}$. The symbol 2 corresponds to a trimer with the same rate constants. $A = 100$, $B = 161$ and $C = 290$ activated Syk per cell. The net increase in activated Syk, $(C - A)$ is 190 Syk per cell, of which $C - B = 129$ Syk per cell can be attributed to serial engagement.

engagement. We investigate this further by considering ligands that cannot serial engage receptors.

4.3.4 Serial engagement is not necessary for Syk activation

Although the aggregation of FcεRI is an absolute requirement for mast cell signaling mediated by FcεRI, serial engagement is not, as has been demonstrated using oligomers of IgE (Kent et al., 1994). As a test of our model, we have carried out simulations with bivalent and trivalent oligomers of IgE considered as ligands, taking the dissociation rate constant $k_{off} = 0$, which ensures that there is no serial engagement. (Since the half-life for dissociation of an IgE-Fcϵ complex is close to a day (Kulczycki & Metzger, 1974), this is an excellent approximation in simulating experiments that last a few hours or less.) The predicted results for γ-phosphorylation and Syk activation for IgE dimers and trimers at one hour are shown in Figures 14a and 14b. These results are compared to histamine release dose response curves for RBL-2H3 cells exposed to IgE oligomers for one hour. Because of the slow forward rate constant for IgE binding to FcϵRI (Wofsy et al., 1995), equilibrium is achieved at one hour only for IgE concentrations greater than about 5 μg/ml. The simulations indicate that to achieve measureable histamine release from RBL cells requires the activation of substantial numbers of Syk molecules. For example, the simulations predict that 10% histamine release with trivalent ligands, about 25% of maximal release, occurs when approximately 10,000 Syk are fully activated.

5. Discussion

Mutivalent ligands (allergens) bind to, and aggregate, IgE complexed with FcϵRI receptors expressed on the surface of mast cells and basophils. We have used a generalized version of the model of Faeder et al. (2003) of the early cell signaling events (seconds to minutes) mediated by the cross-linking of IgE-FcϵRI complexes to study the dependence of the activation of Syk on the cellular concentrations of Syk and IgE-FcϵRI complexes and to investigate the role of serial engagement in Syk activation. The model was originally developed for covalently cross-linked dimers of IgE binding to FcϵRI on the surface of RBL cells (Faeder et al., 2003). For bivalent ligands binding reversibly to IgE-FcϵRI complexes, the model applies when IgE acts as a monovalent receptor. We restrict the use of the model to concentrations of bivalent ligands for which the binding of two ligands to one IgE is negligible. In the model, full activation of Syk can only occur when two Syk molecules are bound to a receptor dimer formed by a bivalent ligand. A Syk molecule becomes fully activated when it is *transphosphorylated* on its activation loop tyrosines (Tyr^{518} and Tyr^{519} in murine Syk) by a Syk bound to the adjacent receptor in the dimer. Using antibodies specific for phosphorylated tyrosines in the activation loop of Syk, Zhang et al. (Zhang et al., 2000) showed that upon FcϵRI aggregation, these tyrosines became phosphorylated and the extent of phosphorylation depended on the kinase activity of Syk. Further, the presence of these activation loop tyrosines was necessary for Syk-mediated propagation of FcϵRI signaling (Zhang et al., 1998). Syk is capable of phosphorylating its activation loop tyrosines both *in vitro* (Tsang et al., 2008; Zhang et al., 2000) and *in vivo* Zhang et al. (2000), but it is still unresolved whether transphosphorylation (phosphorylation of one Syk molecule by another in the same complex) or autophosphorylation (phosphorylation of a Syk molecule by the kinase domain of the same molecule) is the mechanism by which the Syk activation loop tyrosines are phosphorylated in mast cells and basophils.

If the mechanism of Syk activation is Syk transphosphorylation of Syk, as postulated by Faeder et al. (2003), the Syk activation curve can be either bell-shaped or double-humped, depending on the cellular concentrations of Syk and IgE-FcϵRI complexes (see Figure 2). On the other hand, if Syk activation is achieved via autophosphorylation, the model predicts that only a bell-shaped Syk activation curve is possible. In the model, it is assumed that because of steric hindrance only one Syk molecule can bind per receptor, even though there are two disulfide-linked γ-chains that, when fully phosphorylated, present two binding sites for Syk.

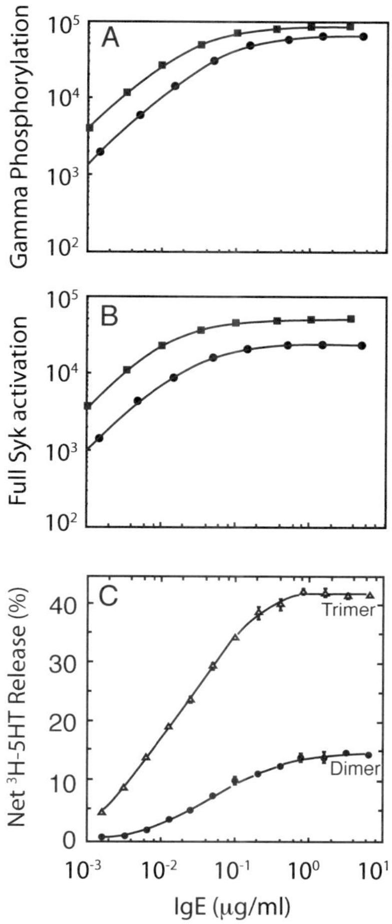

Fig. 14. Predicted γ and Syk phosphorylation, and measured RBL-2H3 cell degranulation, for dimeric and trimer oligomers of IgE. In (A) and (B) the circles refer to IgE dimer simulations and the squares to IgE trimer simulations. The rate constants characterizing the oligomers are taken to be $k_{\text{off}} = 0\ \text{s}^{-1}$, $k_{+1} = 8 \times 10^4\ \text{M}^{-1}\ \text{s}^{-1}$ and $k_{+2}R_T = 100\ \text{s}^{-1}$. The simulations correspond to exposure of the ligands to RBL-2H3 cells for one hour. (C) Experimentally determined degranulation of RBL-2H3 cells after one hour of exposure to either dimers or trimers of IgE (Fewtrell & Metzger, 1980).

If two Syk can be bound simultaneously to the same receptor on two separate γ-chains, and if they can transphosphorylate each other to achieve full Syk activation, the prediction is still that Syk activation curves can be either bell-shaped or double humped depending on the ratio of the receptor to Syk concentrations.

Unfortunately, at present there are no experimentally-determined Syk activation curves to help us distinguish among possible mechanisms for the phosphorylation of the Syk activation loop tyrosines. If we look further downstream than Syk activation at histamine release from basophils, which requires kinase active Syk, the data are consistent with the Syk transphosphorylation mechanism. For example, although most histamine release curves show only a single maximum, double-humped histamine release curves have been observed (Delisi & Siraganian, 1979; Weyer et al., 1982). (Histamine release curves are plots of the fraction of a cell's total histamine that is released in a given time versus the logarithm of the concentration of ligand the cell is exposed to.) In addition, for ligand concentrations where receptor aggregate formation is high and histamine release low, the addition of monovalent hapten, which reduces the number of receptors in dimers, increases histamine release (Magro & Alexander, 1974; Magro & Bennich, 1977). In model simulations, the same manipulation increases Syk activation (Figure 4).

To study the properties of Syk activation, we performed simulations over a range of ligand concentrations for different Syk and receptor concentrations. From the simulations we found that even though the receptor cross-linking curve was symmetric with respect to the log of the ligand concentration in the steady state (Figure 15), the Syk activation curve often was not. This was easiest to see when the Syk activation curve was doubled-humped (Figure 5). The asymmetry in the steady state occured when, at two different ligand concentrations yielding identical total numbers of receptors in dimers and identical lifetimes of a receptor in a dimer, the levels of Syk activation differed. This was puzzling since all that differed on the cell surface at the two ligand concentrations was the number of receptors bound monovalently. Why should this matter since basophils and mast cells don't respond to monovalently bound ligand? The answer is subtle and concerns the competition between kinetic proofreading, that activation requires the receptor to undergo multiple binding and chemical modification events while the receptor is in an aggregate, and serial engagement, that ligands can bind multiple receptors before dissociating from the cell surface.

Once a receptor dimer breaks up, the two receptors, one with a ligand attached and the other free, will diffuse away from each other. If Syk or Lyn is bound to the receptors, these kinases will dissociate with half-lives of about ten seconds, leaving behind unprotected ITAM phosphotyrosines to be rapidly dephosphorylated. Thus, the receptors would return to their basal state. If, however, a receptor can re-form a dimer before it has been transformed back to its basal state, it can partially compensate for the effects of kinetic proofreading. The asymmetry in the Syk activation curve arises because the minimum time required by a receptor to form a dimer is dependent on the free ligand concentration (Figure 9). At low ligand concentration, and therefore low numbers of bound receptors, a singly bound receptor will collide with many free receptors and have ample opportunity to rapidly form a dimer. At high ligand concentration, and therefore low numbers of free receptors, a free receptor will have many bound receptors to collide with, but, because the free ligand concentration is high, the free receptor may bind a ligand from solution before it has time to form a dimer. Thus, at low ligand concentration, re-formation of dimers is more effective than at high ligand concentration at reducing the effects of kinetic proofreading. In our simulations, the double-humped Syk activation curves (Figure 5) show that Syk activation can be higher at the first maximum (at the lower ligand concentration) than at the second maximum (at the higher ligand concentration) where the extent of receptor cross-linking is the same at both maxima.

To further probe the underlying mechanism of the asymmetry in the Syk activation curve, we considered our model in the limit when Lyn is in large excess and γ-ITAM phosphorylation follows immediately upon dimer formation. In this limit, simulations again showed an asymmetry in the double-humped Syk activation curve (Fig.6a). Since for full Syk activation to occur, a Syk dimer (Syk-FcϵRI-ligand-FcϵRI-Syk) must form, we calculate the mean time

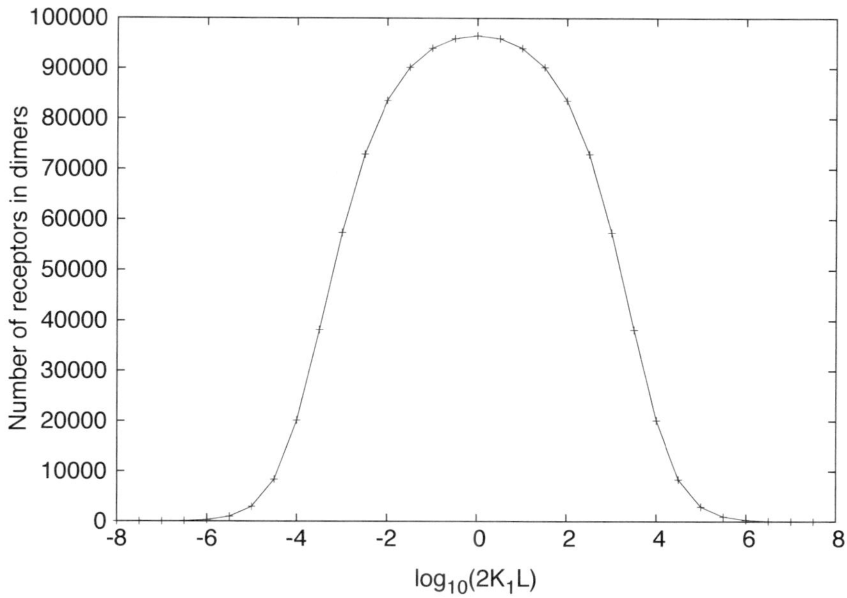

Fig. 15. Receptor cross-linking (dimerization) curve for $k_{+1} = 2.5 \times 10^7$ M^{-1} s^{-1}; $k_{+2} = 8 \times 10^{-9}$cm^{2}s^{-1}; $k_{-1} = 0.316(10^{-0.5})$ s^{-1}, $k_{-2} = 0.0316(10^{-1.5})$ s^{-1}. The receptor aggregation is maximum at $L = 1/2K_1$, i.e. at $\log_{10}(2K_1 L) = 0$, where $K_1 = k_{+1}/k_{-1}$ is the equilibrium constant for the extracellular bivalent ligand, present at concentration L, for binding to the Fab part of the IgE.

for a Syk molecule that is not in a Syk dimer to be incorporated into a Syk dimer. The time for this to occur showed a similar asymmetry as that of the minimum time for a receptor to enter a dimer. It appears that the asymmetry in the Syk activation curve is a manifestation of the underlying competition that is constantly occurring between kinetic proofreading and serial engagement in signaling cascades triggered by multivalent ligands aggregating cell surface receptors.

The asymmetry we observed in simulations of Syk activation curves was always small. Although the explanation of the effect is interesting, whether it is ever important in cell signaling is an open question. Since amplifications of cell signals can be highly nonlinear, it is possible that small differences in the concentration of activated kinases early in a signaling cascade can have profound effects on later cell responses.

A major function of activated Syk in mast cells and basophils is to phosphorylate a set of tyrosines on the transmembrane adapter protein LAT (linker for the activation of T cells). LAT's three distal tyrosines, when phosphorylated, all bind the adapter protein Grb2, and Grb2 mediates the aggregation of LAT (Houtman et al., 2006). The size of the LAT aggregates that form depends strongly on whether or not all three of the Grb2 binding site tyrosines are phosphorylated (Nag et al., 2009). It remains to be seen whether small differences in the amount of activated Syk can lead to large differences in the distribution of LAT aggregates that form downstream of Syk.

The model prediction of differential Syk activation at low and high extracellular ligand concentrations, which arises from differential compensation by serial engagement for the effects of kinetics proofreading at the two ligand concentrations, has led us to explore in greater detail, the role of serial engagement in enhancing Syk activation in mast cells and basophils. When surface densities of receptors are sufficiently high, multivalent ligands can bind to cell surfaces and serial engage numerous receptors before dissociating, i.e., before all of the ligand's binding sites are simultaneously free. Even with a valence as low as two, a ligand can engage multiple receptors from the time it first attaches to the surface until the time it dissociates (see Table 3). A single binding site on a ligand may bind a receptor, dissociate from it, and repeat the cycle with a new receptor multiple times before the ligand leaves the surface.

Using our detailed mathematical model, we have investigated the role of serial engagement of FcϵRI in mast cell signaling. We use the model for low concentrations of the multivalent ligand so that the binding of two ligands to one IgE is negligible. For the purpose of exploring the role of serial engagement in enhancing Syk activation, we have included in our model, cross-linking reactions between a trivalent ligand and an IgE-FcϵRI complex. The concentration of receptors is chosen to be high enough that there is a significant population of free receptors so that serial engagement is favored. A basic assumption of the model is that when a ligand is bound to more than one receptor and a bond opens, the freed receptor diffuses away before the site on the ligand can rebind to it. This is a reasonable assumption as long as receptors remain mobile on the cell surface. Receptors in small aggregates on the RBL cell surface remain mobile (Andrews et al., 2009), but not the receptors in large aggregates (Andrews et al., 2008; Menon et al., 1986). We expect serial engagement to be significantly reduced and exert little effect on cell signaling when the surface receptors are immobile. We have only considered the case when receptors remain mobile and our results do not apply to large aggregates where receptors are immobile. The expression we have derived for the rate of serial engagement for a ligand of valence N, Eq. (37), holds only when receptors are mobile.

We considered three ways in which serial engagement could be enhanced: by increasing the rate at which a ligand binding site dissociates from a receptor; by increasing the rate at which a free site on a ligand that is bound to a receptor can cross-link another receptor; and by increasing the valence of the ligand. Model simulations showed that increasing the rate of serial engagement by increasing the rate of dissociation reduced the concentration of activated Syk (see Figure 11d). Increasing the rate of dissociation of a ligand-receptor bond decreases the lifetime of a receptor in an aggregate and accentuates the effect of kinetic proofreading. The further down the signaling pathway an event occurs, the more pronounced are the effects of kinetic proofreading on the event, as can be seen by comparing the predicted γ-chain phosphorylation (Figure 11c) with the predicted Syk phosphorylation (Figure 11d). The positive effects serial engagement has on signaling are outweighed by kinetic proofreading when the lifetime of the ligand-receptor bond is decreased(McKeithan, 1995; Torigoe et al., 1998).

Simulations predict that increasing the rate at which cross-linking occurs, while keeping the lifetime of the bond between a ligand site and receptor constant, so as not to enhance kinetic proofreading, increases Syk activation. In Figure 13a, a single curve corresponds to a series of bivalent or trivalent ligands with increasing rates of serial engagement achieved by increasing cross-linking constants. The ligands have the same dissociation rate constants and

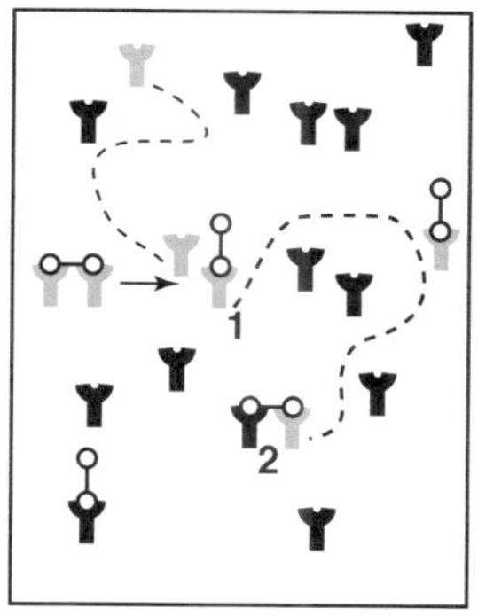

Fig. 16. At low ligand concentration, when most receptors are unbound, a receptor-ligand complex that dissociates from an aggregate (1), can rapidly enter a new aggregate (2), provided that the rate of cross-linking is greater than the rate of dissociation of its ligand-receptor bond. If this is not the case the ligand-receptor complex will be rapidly converted to a free receptor, reducing it chance of quickly finding a binding partner.

their concentrations have been chosen so that they aggregate the same number of receptors on the cell surface, yet the faster the ligands are able to serial engage receptors, the more Syk they are able to activate. This model prediction supports the hypothesis used to explain the previously discussed asymmetric Syk activation profile, that serial engagement can partially nullify the effects of kinetic proofreading by allowing receptors that have dissociated from an aggregate to enter new aggregates before they fully return to their basal state. Previously we estimated that when a receptor leaves an aggregate its unprotected phosphotyrosines are dephosphorylated in less than a second, and any Syk or Lyn that are bound to phosphorylated ITAMS dissociate with half-lives of the order of 10-15 seconds (Faeder et al., 2003). Even within an aggregate, phosphorylation and dephosphorylation is constantly occurring (Kent et al., 1994). If a receptor can enter into a new aggregate before all the modifications it has undergone have been reversed, full activation of Syk will be more efficient.

As depicted in Figure 16 for a bivalent ligand at low concentration, a receptor that remains bound to a ligand when an aggregate breaks up is much more likely to be rapidly incorporated into a new aggregate than a receptor with its binding site free, provided the rate for cross-linking a new receptor is equal to, or faster than the rate of dissociation of the ligand-receptor bond, i.e., $k_{+2}R \geq k_{\text{off}}$, or equivalently $K_2R \geq 1$. Even when this condition on the rate constants is met, serial engagement can only partially compensate for the effects of kinetic proofreading since free receptors that leave an aggregate will most likely return to the basal state before they form new aggregates.

The model also predicts that increasing the valence of a ligand from two to three, while keeping the rate constants the same, increases Syk activation (Figure 13). We have derived expressions for the rates of serial engagement of dimers and trimers, Eqs. (38) and (39), which show that increasing the valence while keeping all other quantities fixed raises the rate of serial engagement. However, the predicted increase in Syk activation is not solely due to the increase in serial engagement. Both experiment (Fewtrell & Metzger, 1980) and the model predictions (Figure 14) indicate that trimeric oligomers of IgE are more effective than dimers at activating RBL cells, even though these ligands cannot serial engage receptors. This is because

Lyn is limiting in RBL cells, with most receptors not associated with a Lyn (Torigoe et al., 1997; Wofsy et al., 1997; Yamashita et al., 1994). Large aggregates are more likely than small aggregates to have a Lyn bound to a receptor in the aggregate and initiate signaling. Further, the larger the aggregate, the higher is the number of receptors that can be phosphorylated by a single Lyn. Serial engagement can further accentuate this signaling. We have estimated the additional contribution to Syk activation from the increase in serial engagement when a trivalent ligand is substituted for a bivalent ligand and shown that it can be substantial (Figure 13b).

Ligand-induced receptor aggregation initiates a chemical cascade that involves chemical reactions that build and use transient molecular scaffolds. Upon aggregation of FcϵRI on mast cells, the cytoplasmic domains of the receptor become sites for the coalescence of the kinases Lyn and Syk, with Syk undergoing rapid activation by phosphorylation. The activated Syk phosphorylates conserved tyrosines on adaptor molecules such as LAT to form scaffolding molecules. However, the structures formed around the cytoplasmic domains of the receptor, as well as other scaffolding proteins, are ephemeral with components going on and off rapidly. How successful this construction will be depends on the lifetime of a receptor in an aggregate. If the lifetime is too short, most of the chemical cascades that are initiated will not go to completion, and signaling will be dampened or completely prevented. This is the idea behind kinetic proofreading (reviewed in (Goldstein et al., 2008)), introduced in the context of cell signaling by McKeithan (McKeithan, 1995), and kinetic proofreading has highlighted the role of the pMHC-TCR bond dissociation rate constant in T-cell signaling. We have used a mathematical model of the initial steps in the chemical cascade triggered by FcϵRI aggregation on mast cells to investigate the role of serial engagement in signaling, and shown that for a series of ligands with the same dissociation rate constant, increasing the forward rate constant for cross-linking, or increasing the ligand valence increases serial engagement and Syk activation. Serial engagement is able to partially reverse the effects of kinetic proofreading and enhance mast cell signaling.

6. Conclusion

Faeder et al. have developed detailed mathematical model of the early signaling events, up to and including Syk activation, that are triggered when IgE-FcϵRI complexes are exposed to a bivalent or trivalent ligand on rat basophilic leukemia (RBL) cells. We have used this model to investigate the effect of cellular Syk and FcϵRI concentrations on the level of Syk activation. Using model simulations, we show that for a subset of physiological FcϵRI and Syk concentrations, differences in total extracellular ligand concentration can lead to subtle kinetic effects that yield qualitative differences in the levels of Syk activation. Since these kinetic effects arise from differential serial engagement rates at different ligand concentrations, we have assessed the contribution of serial engagement to Syk activation in mast cells and basophils, and shown that for a series of ligands with the same dissociation rate constant, increasing the forward rate constant for cross-linking, or increasing the ligand valence increases serial engagement and Syk activation. Serial engagement is able to partially reverse the effects of kinetic proofreading and enhance mast cell signaling. To summarize, we review in this chapter, our research directed at determining the possible factors that influence the levels of Syk activation in mast cells and basophils. Syk activation is a crucial step in the signal transduction cascade in mast cells and basophils that gives rise to the allergic response in individuals. An in-depth understanding of the strength of the signal transduced via this biochemical network, and its underlying kinetics, will be of immense assistance in a network level approach for design of allergy drugs that targets multiple molecules in the signal transduction cascade.

7. Acknowledgements

This work was supported by NIH Grant R37-GM035556 and by the Department of Energy through contract W-7405-ENG-36. MLB received partial supported from NIH Grants R01 GM076570 and NIH U54 RR022232 . The authors have no financial conflict of interest.

8. References

Aleksic, M., Dushek, O., Zhang, H., Shenderov, E., Chen, J.-L.and Cerundolo, V., Coombs, D. & van der Merwe, P. A. (2010). Dependence of T cell antigen recognition on TCR/pMHC confinement time, *Immunity* 32: 163–174.

Andrews, N. L., Lidke, K. A., Pfeiffer, J. R., Burns, A. R., Wilson, B. S., Oliver, J. M. & Lidke, D. S. (2008). Actin restricts FcϵRI diffusion and facilitates antigen-induced receptor immobilization, *Nature Cell Biol.* 10: 955–962.

Andrews, N. L., Pfeiffer, J. R., Martinez, A. M., Haaland, D. M., Davis, R. W., Kawakami, T., Oliver, J. M., Wilson, B. S. & Lidke, D. S. (2009). Small, mobile FcϵRI aggregates are signaling competent., *Immunity* 31: 469–479.

Becker, K., Ishizaka, T., Metzger, H., Ishizaka, K. & Grimley, P. (1973). Surface IgE on human basophils during histamine release, *J. Exp. Med.* 138: 394–409.

Benhamou, M., Ryba, N. J. P., Kihara, H., Nishikata, H. & Siraganian, R. P. (1993). Protein-tyrosine kinase $p72^{Syk}$ in high-affinity IgE receptor signaling: identification of a component of pp72 and association with the receptor γ chain after receptor agregation, *J. Biol. Chem.* 268: 23318–23324.

Blank, U., Ra, C., Miller, L., White, K., Metzger, H. & Kinet, J. (1989). Complete structure and expression in transfected cells of high affinity IgE receptor, *Nature* 337: 187–189.

Blinov, M. L., Faeder, J. R., Goldstein, B. & Hlavacek, W. S. (2004). Bionetgen: Software for rule-based modeling of signal transduction based on the interactions of molecular domains, *Bioinformatics* 20: 3289–3292.

Carreno, L. J., Bueno, S. M., P., B., Nathenson, S. G. & Kalergis, A. M. (2007). The half-life of the T-cell receptor/peptide-major histocompatibility complex interaction can modulate T-cell activation in response to bacterial challenge., *Immunol.* 121: 227–237.

Coombs, D. & Goldstein, B. (2005). T cell activation: Kinetic proofreading, serial engagement and cell adhesion., *J. Comput. Appl. Math.* 184: 121–139.

Coombs, D., Kalergis, A. M., Nathenson, S. G., Wofsy, C. & Goldstein, B. (2002). Activated TCR remain marked for internalization after dissociation from peptide-MHC, *Nat. Immunol.* 3: 926–931.

Davis, M. M. (1995). Serial engagement proposed, *Nature* 375: 104.

de Castro, R. (2011). Regulation and function of syk tyrosine kinase in mast cell signaling and beyond, *J. Signal Transduct.* 2011(507291): 1–9.

de Castro, R., Zhang, J., Jamur, M., Oliver, C. & Siraganian, R. (2010). Tyrosines in the Carboxyl Terminus Regulate Syk Kinase Activity and Function, *J. Biol. Chem.* 285: 26674–26684.

Delisi, C. & Siraganian, R. (1979). Receptor cross-linking and histamine release. ii. interpretation and analysis of anomalous dose response patterns, *J. Immunol.* 122: 2293–2299.

Dembo, M. & Golstein, B. (1978). Theory of Equilibrium Binding of Symmetric Bivalent Haptens to Cell Surface Antibody: Application to Histamine Release from Basophils, *J. Immunol.* 121: 345–353.

Faeder, J. R., Blinov, M. L. & Hlavacek, W. S. (2009). Rule-based modeling of biochemical systems with BioNetGen., *Methods Mol. Biol.* 500: 113–167.

Faeder, J. R., Hlavacek, W. S., Reischl, I., Blinov, M. L., Metzger, H., Redondo, A., Wofsy, C. & Goldstein, B. (2003). Investigation of early events in FcϵRI-mediated signaling using a detailed mathematical model., *J. Immunol.* 170: 3769–3781.

Fewtrell, C. & Metzger, H. (1980). Larger oligomers of IgE are more effective than dimers in stimulating rat basophilic leukemia cells, *J. Immunol.* 125: 701–710.

Goldstein, B., Coombs, D., Faeder, J. R. & Hlavacek, W. S. (2008). Kinetic proofreading model, *Adv. Exp. Med. Biol.* 640: 82–94.

Goldstein, B. & Wofsy, C. (1994). Aggregation of cell surface receptors, *Lectures on Mathematics in the Life Sciences* 24: 109–135.

Gopalakrishnan, P. V. & Karush, F. (1974). Antibody affinity VII. Multivalent interaction of anti-lactoside antibody, *J. Immunol.* 113: 769–778.

Greenbury, C. L., Moore, D. H. & Nunn, A. C. (1965). The reaction with red cells of 7S antibody. Its subunits and their recombinants., *Immunology* 8: 420–431.

Hlavacek, W., Faeder, J., Blinov, M., Perelson, A. & Goldstein, B. (2003). The Complexity of Complexes in Signal Transduction, *Biotechnolgy and Bioengineering* 84: 783–794.

Hlavacek, W. S., Percus, J. K., Percus, O. E., Perelson, A. S. & Wofsy, C. (2002). Retention of antigen on follicular dendritic cells and B lymphocytes through complement-mediated multivalent ligand-receptor interactions: theory and application to HIV treatment, *Math. Biosci.* 176: 185–202.

Hlavacek, W. S., Wofsy, C. & Perelson, A. S. (1999). Dissociation of HIV-1 from follicular dendritic cells during HAART: Mathematical analysis, *Proc. Natl. Acad. Sci. USA* 96: 14681–14686.

Hong, J. J., Yankee, T. M., Harrison, M. L. & Geahlen, R. L. (2002). Regulation of signaling in B cells through the phosphorylation of Syk on linker-region tyrosines. A mechanism for negative signaling by the Lyn tyrosine kinase, *J. Biol. Chem.* 277: 31703–31714.

Hornick, C. L. & Karusch, F. (1972). Antibody affinity III. The role of multivalence., *Immunochem.* 9: 325–340.

Houtman, J. C., Yamaguchi, H., Barda-Saad, M., Braiman, A., Bowden, B., Appella, E., Schuck, P. & Samelson, L. E. (2006). Oligomerization of signaling complexes by the multipoint binding of GRB2 to both LAT and SOS1, *Nat. Struct. Mol. Biol.* 13: 798–805.

Hutchcroft, J. E., Geahlen, R. H., G., D. G. & Oliver, J. M. (1992). Fcϵri-mediated tyrosine phosphorylation and activation of the 72-kDa protein-tyrosine kinase, PTK72, in RBL-2H3 rat-tumor mast cells, *Proc. Natl. Acad. Sci. USA* 89: 9107–9111.

Kalergis, A. M., Boucheron, N., Doucey, M. A., Palmieri, E., Goyarts, E. C., Vegh, Z., Luescher, I. F. & Nathenson, S. G. (2001). Efficient cell activation requires an optimal dwell-time of interaction between the TCR and the pMHC complex, *Nat. Immunol.* 2: 229–234.

Karush, F. (1989). The affinity of antibody: Range, variability, and the role of multivalence, *in* G. W. Litman & R. A. Good (eds), *Comprehensive Immunology 5: Immunoglobulins*, Plenum Press, New York, NY, pp. 85–116.

Kawabuchi, M., Satomi, Y., Takao, T., Shimonishi, Y., Nada, S., Nagai, K., Tarakhovsky, A. & Okada, M. (2000). Transmembrane phosphoprotein cbp regulates the activities of Src-family tyrosine kinases, *Nature* 404: 999–1003.

Kent, U. M., Mao, S.-Y., Wofsy, C., Goldstein, B., Ross, S. & Metzger, H. (1994). Dynamics of signal transduction after aggregation of cell-surface receptors: Studies on the type I receptor for igE, *Proc. Natl. Acad. Sci. USA* 91: 3087–3091.

Kulczycki, A. J. & Metzger, H. (1974). The interaction of IgE with rat basophilic leukemia cells. II. quantitative aspects of the binding reaction., *J. Exp. Med.* 140: 1676–1695.

Lanzavecchia, A., Lezzi, G. & Viola, A. (1999). From TCR engagement to T cell activation: A kinetic view of T cell behavior, *Cell* 96: 1–4.

MacGlashan Jr, D. (2007). Relationship between spleen tyrosine kinase and phosphatidylinositol 5′ phosphatase expression and secretion from human basophils in the general population, *J. Allergy Clin. Immunology* 119: 626–633.

MacGlashan Jr, D. & Lavens-Phillips, S. (2001). Characteristics of the free cytosolic calcium timelag following igE-mediated stimulation of human basophils: significance for the non-releasing basophil phenotype, *J. Leukoc. Biol.* 69: 224–232.

Magro, A. & Alexander, A. (1974). Histamine Release: In vitro studies of the inhibitory region of the dose-response curve, *J. Immunol.* 112: 1762–1765.

Magro, A. & Bennich, H. (1977). Concanavalin A induced histamine release from human basophils in vitro, *Immunology* 33: 51–58.

Mao, S.-Y. & Metzger, H. (1997). Characterization of protein-tyrosine phosphatases that dephosphorylate the high affinity receptor for IgE, *J. Biol. Chem* 272: 14067–14073.

McKeithan, K. (1995). Kinetic proofreading in T-cell receptor signal transduction., *Proc. Natl. Acad. Sci. USA* 92: 5042–5046.

Menon, A. K., Holowka, D., Webb, W. W. & Baird, B. (1986). Cross-linking of receptor-bound IgE to aggregates larger than dimers leads to rapid immobilization, *J. Cell Biol.* 102: 541–550.

Nag, A., Faeder, J. & Goldstein, B. (2010). Shaping the response: the role of fcepsilonri and syk expression levels in mast cell signalling, *IET Syst. Biol.* 4: 334–347.

Nag, A., Monine, M., Blinov, M. & Goldstein, B. (2010). A detailed mathematical model predicts that serial engagement of ige-fc epsilon ri complexes can enhance syk activation in mast cells, *J. Immunol.* 185: 3268–3276.

Nag, A., Monine, M., Faeder, J. R. & Goldstein, B. (2009). Aggergation of membrane proteins by cytosloic cross-linkers:theory and simulation of the Lat-grb2-Sos1 system, *Biophys. J.* 96: 2604–2623.

Ohtake, H., Ichikawa, N., Okada, M. & Yamashita, T. (2002). Cutting Edge:transmembrane Phosphoprotein Csk-Binding Protein/Phosphoprotein Associated With Glycosphingolipid-Enriched Microdomains as a Negative Feedback Regulator of Mast Cell Signaling Through the FcϵRI, *J. Immunol.* 168: 2087–2090.

Peirce, M. & Metzger, H. (2000). Detergent resistant microdomains offer no refuge for proteins phosphorylated by the IgE receptor, *J. Biol. Chem* 275: 34976–34982.

Rivera, J. (2005). NTAL/LAB and LAT: a balancing act in mast-cell activation and function, *Trends Immunol.* 117: 1214–1225.

Rivera, J. & Gilfillan, A. M. (2006). Molecular regulation of mast cell activation, *J. Allergy Clin. Immunol.* 117: 1214–1225.

Shiue, L., Zoller, M. J. & Brugge, J. S. (1995). Syk is activated by phosphotyrosine-containing peptides representing the tyrosine-based activation motifs of the high affinity receptor for IgE, *J. Biol. Chem.* 270: 10498–10502.

Siraganian, R., de Castro, R., Barbu, E. & Zhang, J. (2010). Mast cell signaling: The role of protein tyrosine kinase Syk, its activation and screening methods for new pathway participants, *FEBS Letters* 584: 4933–4940.

Siraganian, R., Zhang, J., Suzuki, K. & Sada, K. (2002). Protein tyrosine kinase Syk in mast cell signaling, *Mol. Immunol.* 38: 1229–1233.

Stone, J. D., Chervin, A. S. & Kranz, D. M. (2009). T-cell receptor binding affinities and kinetics: impact on T-cell activity and specificity, *Immunol.* 126: 165–176.

Taniguchi, T., Kobayashi, T., Kondo, J., Takahashi, K. et al. (1991). Molecular Cloning of a Porcine Gene *syk* That Encodes a 72-kDa Protein-Tyrosine Kinase Showing High Susceptibility to Proteolysis, *J. Biol. Chem.* 266: 15790–15796.

Torigoe, C., Goldstein, B., Wofsy, C. & Metzger, H. (1997). Shuttling of initiating kinase between discrete aggregates of the high affinity receptor for IgE regulates the cellular response, *Proc. Natl. Acad. Sci. USA* 94: 1372–1377.

Torigoe, C., Inman, J. K. & Metzger, H. (1998). An unusual mechanism for ligand antagonism, *Science* 281: 568–572.

Tsang, E., Giannetti, A., Shaw, D., Dinh, M., Tse, J., Gandhi, S. et al. (2008). Molecular mechanism of the syk activation switch, *J. Biol. Chem.* 283: 32650–32659.

Valitutti, S., Müller, S., Cella, M., Padovan, E. & Lanzavecchia, A. (1995). Serial triggering on many T-cell receptors by a few peptide-MHC complexes, *Nature* 375: 148–151.

Vonakis, B. M., Chen, H., Haleem-Smith, H. & Metzger, H. (1997). The Unique Domain as the Site on Lyn Kinase for Its Constitutive Association with the High Affinity Receptor for IgE, *J. Biol. Chem.* 272: 24072–24080.

Vonakis, B. M., Haleem-Smith, H., Benjamin, P. & Metzger, H. (2001). Interaction between the unphosphorylated receptor with high affinity for IgE and Lyn kinase, *J. Biol. Chem.* 276: 1041–1050.

Weyer, A., Dandeu, J., Marhand, F. & David, B. (1982). In vitro histamine release from human basohils triggered by a purified allergen from Dermatophagoides Farinac: Bimodel aspect of the dose response curve, *Ann. Immunol. (Paris)* 133D(1): 87–94.

Wofsy, C., Coombs, D. & Goldstein, B. (2001). Calculations show substantial serial engagement of T cell receptors, *Biophys. J.* 80: 606–612.

Wofsy, C., Kent, U. M., Mao, S.-Y., Metzger, H. & Goldstein, B. (1995). Kinetics of tyrosine phosphorylation when IgE dimers bind to Fcϵ receptors on rat basophilic leukemia cells, *J. Biol. Chem.* 270: 20264–20272.

Wofsy, C., Torigoe, C., Kent, U. M., Metzger, H. & Goldstein, B. (1997). Exploiting the Difference Between Intrinsic and Extrinsic Kinases: Implications for Regulation of Signaling by Immunoreceptors, *J. Immunol.* 159: 5984–5992.

Yamashita, T., Mao, S. & Metzger, H. (1994). Aggregation of the high-affinity IgE receptor and activation of p53/p56^{lyn} protein-tyrosine kinase, *Proc. Natl. Acad. Sci.* 91: 11251–11255.

Yang, C., Yanagi, S., Wang, X., Sakai, K., Taniguchi, T. & Yamamura, H. (1994). Purification and characterization of a protein-tyrosine kinase p72syk from porcine spleen, *Eur. J. Biochem.* 221: 973–978.

Zhang, J., Billingsley, M. L., Kincaid, R. L. & Siraganian, R. P. (2000). Phosphorylation of Syk activation loop tyrosine is essential for Syk function, *J. Biol. Chem.* 275: 35442–35447.

Zhang, J., Kimura, T. & Siraganian, R. (1998). Mutations in the activation loop tyrosines of protein tyrosine kinase Syk abrogate intracellular signaling but not kinase activity, *J. Immunol.* 161: 4366–4374.

12

Roles of Kinases in Osteoblast Function

Tetsuya Matsuguchi
Kagoshima University,
Japan

1. Introduction

Osteoblasts, as well as osteocytes that are a terminally differentiated form of osteoblasts, are responsible for bone formation by producing bone matrix proteins, which subsequently induce tissue mineralization. Osteoblasts differentiate and mature from their progenitors in response to various regulatory factors including bone morphogenetic proteins (BMPs), Insulin-like growth factor 1 (IGF-1), fibroblast growth factor 2 (FGF-2), parathyroid hormone (PTH), tumor necrosis factor-α (TNF-α), Wnts, and extracellular matrix (ECM) signals. Matrix mineral deposition occurs at the terminal stage of osteoblast differentiation and is associated with maximal expression of osteocalcin (Ocn).

Although several transcription factors, such as runt-related transcription factor 2 (Runx2), Osterix (Osx), and activating transcription factor (ATF) 4, have been demonstrated to be crucial to bone development, the molecular details of intracellular signals controlling stage-specific osteoblast differentiation remain enigmatic. It is well recognized that the mitogen activated protein kinase (MAPK) superfamily, including p44/p42 extracellular signal-regulated kinases (ERKs), p38 kinases, and p54/p46 c-Jun N-terminal kinases (JNKs), integrates signals from a diverse range of extracellular stimuli, and plays important roles in cellular functions such as proliferation, differentiation, and cell death in a variety of cell types. Despite intensive investigation, the physiological roles of MAPKs in osteogenic differentiation have not been clearly demonstrated. AMP-activated protein kinase (AMPK) is a serine-threonine kinase widely known as an essential regulator of energy homeostasis of cells. Interestingly, however, non-metabolic functions of AMPK, including its roles in cell differentiation have recently been demonstrated. Several recent studies including ours have indicated that the differentiation of osteoblasts is functionally associated with the AMPK activity.

Being located in the skeletons, osteoblasts are constantly exposed to various forms of mechanical stresses. Many previous studies have indicated that mechanical stresses on osteoblasts affect their cell differentiation process, which may explain why moderate exercises contribute to healthy bone development. However, mechanical stresses often induce tissue inflammation in bones, especially when they are excessive. We and others have recently revealed that osteoblasts respond to mechanical stresses by expressing various inflammatory cytokines and chemokines, which is dependent on the activation of kinases, such as MAPKs.

In this chapter, I will first briefly review recent and basic knowledge about osteoblast functions. I will then try to review the contribution of various kinase signals, including MAPKs and AMPK, to the osteoblast differentiation process. Finally, I will review the known mechanisms of kinase activation by mechanical stresses in osteoblasts, as well as how the mechanical stress-activated kinases affect the physiological functions of osteoblasts, including cell proliferation, differentiation, survival and the expression of chemokines and cytokines.

2. The basic picture of osteoblast functions

In vertebrates, bone tissue constitutes the skeleton and provides mechanical support, motility, and protection of internal organs. Bone tissue also acts as a reservoir of biologically essential minerals including calcium, phosphate, and magnesium. Osteoblasts, as well as osteocytes are responsible for bone formation by producing bone matrix proteins, which subsequently induce tissue mineralization (Raisz and Kream 1983). Osteoblasts are derived from bone marrow mesenchymal stem cells (MSCs), and terminally differentiate into osteocytes in bone tissue. In contrast, osteoclasts, which are derived from bone marrow hematopoietic stem cells, are in charge of bone resorption. The growth and maintenance of bone mass are thus regulated by the coordinated actions of osteoblasts and osteoclasts. Bone remodelling is a dynamic and constant process even in adults, replacing approximately 15% of the trabecular and 3% of the cortical bones every year (Manolagas and Jilka 1995). Disturbance of this delicate balance in bone remodelling leads to various bone disorders. The most well-known is probably osteoporosis which affects over 200 million people world-wide (Cooper, Campion, and Melton 1992) and is considered as a serious public health concern. Osteoporosis is caused by the relative increase of bone resorption resulting in decreased bone density and increased risk of fractures. In contrast, relative increase of bone formation causes disorders such as osteopetrosis and osteosclerosis.

The coordinated balance of bone remodelling is established by multiple coupling mechanisms between osteoblasts and osteoclasts. The most noticeable mechanism is through RANK ligand (RANKL) expressed on the surface of osteoblasts. RANKL signals through RANK, a cell surface receptor expressed on osteoclasts (Yasuda et al. 1998; Lacey et al. 1998; Boyle, Simonet, and Lacey 2003). In a complex interplay with macrophage colony-stimulating factor (M-CSF), which is also produced by osteoblasts, RANKL induces the maturation and activation of osteoclasts. Osteoblasts also secrete osteoprotegerin (Opg) that acts as an inhibitory decoy receptor for RANKL, and the expression ratio of RANKL to Opg by osteoblasts dictates osteoclastogenesis (Manolagas and Jilka 1995; Boyle, Simonet, and Lacey 2003). Thus osteoblasts are the central players of bone metabolism controlling both bone formation and resorption.

2.1 Molecular mechanisms of osteoblast differentiation

Osteoblasts are derived from MSCs, which have the potential to differentiate into myoblasts, fibroblasts, and chondrocytes, in addition to osteoblasts. Osteoblasts differentiate and mature from their progenitors in response to various regulatory factors including BMPs, Insulin-like growth factor 1 (IGF-1), fibroblast growth factor 2 (FGF-2), parathyroid hormone (PTH), tumor necrosis factor-α (TNF-α), Wnts, and ECM signals (Deng et al. 2008). Osteoblasts produce various bone matrix proteins during differentiation (Fig. 1). Matrix mineral deposition occurs at the terminal stage of osteoblast differentiation and is associated with maximal expression of Ocn (Hauschka et al. 1989).

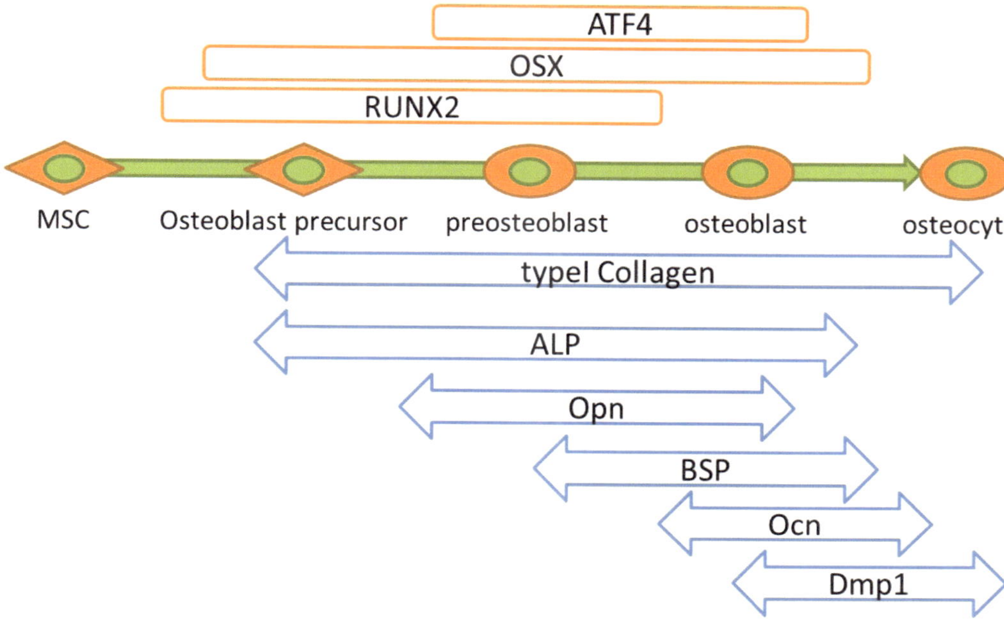

Fig. 1. Schematic presentation of osteoblast differentiation. Osteoblast differentiation is controlled by the expression of several transcription factors including Runx2, Osx, and ATF4. A series of proteins termed "osteogenic markers" contributing to bone matrix formation are produced during osteoblast differentiation in stage-specific manners. These proteins include ColI, Opn, and Ocn.

Differentiation of MSCs into different lineages is governed by the expression of lineage-specific transcription factors. Specifically, osteoblast differentiation is controlled by transcription factors including Runx2 and Osx. Gene deficiency of either Runx2 or Osx causes defective bone formation and lack of mature osteoblasts in mice, suggesting that both of these transcription factors are necessary for osteoblast differentiation in vivo (Ducy et al. 1997; Komori et al. 1997; Nakashima et al. 2002). The Runx regulatory element is found in the gene promoter regions of major bone matrix proteins including collagen type I (ColI), osteopontin (Opn), bone sialoprotein (Bsp), and Ocn (Marie 2008), indicating that the expression of these proteins is directly regulated by Runx2. Furthermore, recent reports have indicated that Runx2 interacts with some other transcription factors, which either activates or repress the transcriptional activity of Runx2 (Karsenty 2008). Thus Runx2 functions as a transducer through which various signals can modulate osteoblast differentiation. A typical example is Runx2-Smad interaction. Two Smad members, Smad1 and Smad5 which are activated by BMPs, interact with Runx2 and enhance its osteoblastogenetic activity.

The essential role of Osx, a zinc-finger protein with homology to the Sp1/Kruppel transcription factor family, in osteogenesis relies on its ability to regulate the expression of a number of osteogenic markers such as Opn, Dickkoph-related protein 1 (Dkk1), and ColI. Furthermore, transcriptional regulators such as Nuclear factor of activated T-cells, cytoplasmic (NFATc) or Nucleolar protein 66 (NO66), a jumonji family histone demethylase, have been shown to interact with Osx and regulate its transcriptional responses (Koga et al. 2005; Sinha et al. 2010). Osx has been generally believed as a downstream transcription factor of Runx2, as Osx-deficient osteoblasts still express Runx2 (Nakashima et al. 2002) and the transcription of Osx is positively regulated by Runx2 (Celil, Hollinger, and Campbell 2005). Interestingly, although expression of Osx in vivo requires Runx2, BMP-2 is still able to stimulate Osx expression in Runx2-deficient cells (Lee et al. 2003). Moreover, recent data have indicated that BMP-2 activates expression of Osx through Runx2-dependent as well as -independent mechanisms involving Dlx5 and Msx2 (Ulsamer et al. 2008; Matsubara et al. 2008).

Other than Runx2 and Osx, various transcription factors have been identified as important regulators of osteoblast differentiation. They include ATF4, activator protein 1 (AP1), homeobox proteins (Msx1, Msx2, Dlx5, Dlx6), and CCAAT/enhancer-binding proteins (C/EBPs) (Marie 2008). ATF4, also known as cAMP-response element-binding protein 2 (CREB2), is a basic leucine-zipper transcriptional factor that belongs to ATF/CREB protein family (Karpinski et al. 1992). Similar to Runx2 and Osx, an essential role of ATF4 in bone development in vivo was revealed by a study using ATF4-deficient mice, which showed a marked reduction in bone mineralization and trabecular development throughout life (Yang et al. 2004). The promoting effects of ATF4 on bone development is not through the modulation of Runx2 or Osx expression level, as ATF4-deficient mice showed normal expression levels of these two transcriptional factors (Yang et al. 2004).

2.2 Osteoblasts as immune-regulatory cells

Bone tissue often becomes the site of inflammation caused by infection, auto-immune responses, and mechanical stresses. Recent lines of evidence have indicated that osteoblasts are important immune-regulatory cells controlling inflammatory responses in bone tissue. We and others have previously shown that osteoblasts express several members of Toll-like receptors (TLRs) which sense microbe infection by binding pathogen-associated molecular patterns (PAMPs) such as lipopolysaccharide (LPS) (Kikuchi et al. 2001; Kikuchi et al. 2003; Amcheslavsky et al. 2005). Stimulation of osteoblasts through TLRs induce the production of RANKL and TNF-α, both of which activate osteoclast-mediated bone resorption (Kikuchi et al. 2001; Kikuchi et al. 2003; Hayashi et al. 2003).

It has been reported that in inflammatory bone tissues, osteoblasts also express chemokines including monocyte chemoattractant protein (MCP)-1, macrophage-inflammatory protein (MIP)-1, RANTES (Lisignoli et al. 2002) and interleukin (IL)-8 (Bendre et al. 2003). These chemokines attract monocytes, macrophages and T cells through CC chemokine receptor 1 (CCR1), CCR2, CCR3, CCR5, and CXC chemokine receptor1 (CXCR1) (Yoshie, Imai, and Nomiyama 2001). Interestingly, IL-8 has recently been shown to increase the motility of osteoclasts and directly induce their differentiation through both RANKL-dependent and independent pathways (Bendre et al. 2003).

2.3 Effects of mechanical stresses on osteoblast functions

Mechanical loading to the skeleton is important for the development and maintenance of strong and healthy bones. Significant bone loss is often encountered due to long-term immobilization, which has become an increasing social problem. Being located in the skeletons, osteoblasts are constantly exposed to various forms of mechanical stresses. Many previous studies have indicated that mechanical stresses on osteoblasts affect their cell proliferation, survival, and differentiation processes, which may explain why moderate exercises contribute to healthy bone development. However, mechanical stresses often induce tissue inflammation in bones, especially when they are excessive. We and others have recently revealed that osteoblasts respond to mechanical stresses by expressing various inflammatory cytokines and chemokines (Bandow et al. 2007; Maeda et al. 2007).

3. Kinases involved in osteoblast differentiation

As briefly discussed in 2.1, several transcription factors including Runx2, Osx, and ATF4 have been demonstrated to be crucial for bone development. However, the molecular details of intracellular kinases controlling stage-specific osteoblast differentiation remain enigmatic. The biggest reason, I presume, is that the expression of most kinases remains relatively constant during osteoblast differentiation, unlike some transcription factors whose expression is clearly induced in differentiation stage-specific manners. Here I list some of the reported kinases whose activities are functionally associated with osteoblast differentiation.

3.1 MAPKs

It is well recognized that the MAPK superfamily, including p44/p42 ERKs, p38 kinases, and p54/p46 JNKs, integrates signals from a diverse range of extracellular stimuli, and plays important roles in cellular functions such as proliferation, differentiation, and cell death in a variety of cell types (Widmann et al. 1999). Activation of MAPKs requires dual phosphorylation of Tyr and Thr residues in the activation loop of the molecules, which is catalized by a family of dual specificity kinases termed MAPK kinases (MKKs). Conversely, their dephosphorylation is catalyzed by dual specificity phosphatases termed MAPK phosphatases (MKPs). Among MKP family members, some show highly selective substrate specificity, whereas others efficiently inactivate all three classes of MAPKs. We previously cloned an MKP termed MKP-M (also referred to as DUSP16 or MKP-7) (Matsuguchi et al. 2001; Masuda et al. 2001; Tanoue et al. 2001), which preferentially inactivates JNKs (Matsuguchi et al. 2001; Masuda et al. 2001).

The role of ERKs in osteoblast differentiation has been studied with inconsistent results. In recent reports, matrix mineralization of MC3T3-E1 cells (a mouse osteoblastic cell line) and preosteocytic MLO-A5 cells was increased by an inhibitor of MEK1/2 that are upstream activators of ERKs (Kono et al. 2007). In contrast, selective expression of a constitutively active form of ERK1 by a transgenic approach accelerated in vitro differentiation of mouse calvaria cells, as well as in vivo bone development in mice (Ge et al. 2007). We additionally observed that matrix mineralization by MC3T3-E1 cells was enhanced by an MEK1/2 inhibitor, whereas the same inhibitor clearly inhibited matrix mineralization of primary osteoblasts from mouse calvaria, indicating the regulatory role of ERKs in osteogenic differentiation may vary depending on the cell types or differentiation stages of osteoblasts (Matsuguchi et al. 2009).

Being similar to ERKs, the role of p38 kinase in osteoblast differentiation is also disputable. BMP2 is known to activate p38 kinase in osteoblasts. Kinase activity of p38 kinase was reported to be stimulatory (Gallea et al. 2001) or inhibitory (Vinals et al. 2002) in BMP-2-induced osteoblast differentiation. Noticeably, it has recently been reported that p38 kinase phosphorylates Osx at Ser-73/77 (Ortuno et al. 2010). Phosphorylatied Osx showed increased ability to recruit coactivators including p300 and SWI/SNF leading to the formation of transcriptionally active complexes, indicating the possibility that p38 exerts promotion of osteoblast differentiation by phosphorylating Osx.

JNK was originally identified by its ability to specifically phosphorylate the transcription factor c-Jun on its N-terminal transactivation domain (Hibi et al. 1993). JNK consists of three isoforms (JNK1, 2 and 3) deriving from distinct genes. Among them, JNK1 and JNK2 are ubiquitously expressed while JNK3 is mainly expressed in brain, testis, and heart (Kyriakis and Avruch 2001). Alternative splicing at the C-terminus yields 46 kDa and 54 kDa polypeptides for each JNK isoform. The 46kDa form is predominant for JNK1, whereas the 54kDa form is predominant for JNK2 and JNK3. Although JNK1 and JNK2 possess structural similarities and many overlapping biological functions, recent evidences have revealed some functional differences between the two kinases (Sabapathy et al. 2004; Liu, Minemoto, and Lin 2004). We have recently found that the terminal osteoblastic differentiation was significantly inhibited by the treatment with a specific JNK inhibitor as well as the over-expression of MKP-M, a JNK-specific MKP (Matsuguchi et al. 2009). Conversely, enhanced matrix mineralization was observed by inducible overexpression of $p54^{JNK2}$ in an isoform-specific manner. More specifically, JNK inhibition significantly suppressed the late stage molecular events of osteoblastic differentiation, such as gene expression of Ocn and Bsp. In contrast, earlier differentiation events including alkaline phosphatase (ALP) activation and Opn expression were not inhibited by JNK inactivation. As for transcriptional factors, induction of ATF4 expression during osteoblastic differentiation was significantly inhibited, whereas the expression levels of Runx2 and Osx, were not significantly affected by JNK inactivation. Thus JNK activity is essential for the late-stage differentiation of osteoblasts.

3.2 Phosphoinositide3-Kinase (PI3K)-akt signaling pathway

Various cytokines and growth factors activate PI3K to produce phosphatidylinositol 3,4,5-phosphate (PIP_3), which then recruits and activates Akt. Akt has been shown to be activated by BMPs in osteoblasts (Ghosh-Choudhury et al. 2002). Deletion of Akt1, the major form of Akt in osteoblasts, in mice results in osteopenia caused by the inhibition of differentiation and the increase of apoptosis of osteoblasts (Kawamura et al. 2007). Conversely, osteoblast-specific deletion of Pten, an inhibitory phosphatase of PI3K, by Ocn-Cre system in mice induced progressive increases in bone volume and density throughout life (Liu et al. 2007). The isolated Pten-defective osteoblasts showed accelerated differentiation capacity and decreased susceptibility to apoptosis. These results indicated that the PI3K-Akt signal is promotive of osteoblasts differentiation and inhibitory to osteoblast apoptosis. Consistently, it has been shown that Akt is essential for Runx2 indcution in osteoblasts by IGF-I, a known osteogenic growth factor.

3.3 Glycogen synthase kinase 3 (GSK3) in Wnt signals

Wnt family proteins play important roles in both embryogenesis and postnatal development. Wnts bind their cell surface receptor, Frizzled (Frz) in combination with the co-receptors, low density lipoprotein receptor-related protein (LRP) 5/6. The activated receptors induce the inactivation of a serine/threonine kinase, GSK3. In the steady-state cells, GSK3 form a complex with β-Catenin and phosphorylates it leading to its ubiquitin-mediated degradation (Cadigan and Liu 2006). When Wnt is signal is on, GSK3-mediated β-Catenin degradation is off, leading to the accumulation of β-Catenin in the nucleus. Nuclear β-Catenin forms a complex with a transcription factor, TCF/LEF, activating the transcription of a series of Wnt-target genes including Runx2 in osteoblasts. This GSK3-mediated cannonical pathway is responsible for the osteogenic activity of Wnts, as inhibitors of GSK3 stimulate osteoblastgenesis in vitro and in vivo (Kulkarni et al. 2006).

Involvemennt of GSK3 in the synergestic effects of Wnt and BMP-4 on osteoblastic differentiation has recently been reported (Fukuda et al. 2010). Wnt3a, a canonical Wnt, synergistically stimulated ColI expression and ALP activity in the presence of BMP-4. Interestingly, overexpression of β-Catenin did not affect BMP-4-induced ALP activity, whereas inhibition or stimulation of GSK3β activity resulted in either stimulation or suppression of ALP activity, respectively, in the presence of BMP-4. This finding indicated that GSK3β mediates synergy of BMP-4 and Wnt3a to promote osteoblastic differentiation through a β-Catenin-independent mechanism.

3.4 AMP-Activated Protein Kinase (AMPK)

AMP-activated protein kinase (AMPK) is a serine-threonine kinase widely known as a regulator of cell metabolism (Hardie 2007). AMPK is a hetero-trimeric protein consisting of the catalytic α subunit and the regulatory β and γ subunits in a 1:1:1 stoichiometric ratio. As several isoforms have been identified for each subunit (α1-2; β1-2; γ1-3), 12 combinations are possible, which show relative tissue specificity (Cheung et al. 2000). AMPK is activated by cellular stresses with increased AMP:ATP ratio as in calorie restriction. Subsequently, ATP-generating pathways are activated, while ATP-consuming mechanisms are inhibited, thereby restoring the cellular AMP:ATP ratio. AMPK is thus considered as an essential regulator of energy homeostasis, and termed as a metabolic "energy sensor" of the biological system. The activation of AMPK is through phosphorylation of the catalytic α subunit by an upstream kinase (Hawley et al. 1996). The most characterized upstream AMPK kinase (AMPKK) is LKB1, which is a reported target of metformin, a type 2 diabetes drug (Shaw et al. 2005).

Intriguingly, recent studies have revealed non-metabolic functions of AMPK, such as the regulation of cell polarity and mitosis (Lee et al. 2007), and mammal longevity (McCarty 2004). Several lines of evidence have also indicated the involvement of AMPK in the regulation of cellular differentiation. Activation of AMPK was suggested to be inhibitory to the differentiation of adipocytes (Dagon, Avraham, and Berry 2006; Giri et al. 2006; Habinowski and Witters 2001; Hwang et al. 2005; Tong et al. 2008), and myoblasts (Fulco et al. 2008), and promotive to the differentiation of endothelial progenitor cells (Li et al. 2008).

We have recently explored the role of AMPK in osteoblast differentiation using in vitro differentiation models (Kasai et al. 2009). We found that the phosphorylation of AMPKα was progressively decreased during osteoblastic differentiation. Conversely, osteoblast differentiation was significantly inhibited by glucose restriction and treatment with metformin, both of which are known activators of AMPK, as well as the forced expression of a constitutively active form of AMPKα. Metformin stimulation of osteoblasts significantly inhibited gene expression of Runx2 along with osteogenic markers including Ocn, Bsp, and Opn. Our result is consistent with a recent report showing the inhibitory effects of AMPK activators on Wnt/β-Catenin signaling in human osteoblastic cells (Takatani et al. 2011). In contrast to our result, treatment by metformin was previously reported to promote osteoblast differentiation (Cortizo et al. 2006; Kanazawa et al. 2008). We presume that the inconsistent results were due to the different dose of metformin used in experiment. The previous reports used metformin at much lower concentrations than we did, which might not be sufficient to induce sustained AMPKα phosphorylation (Kasai et al. 2009). Thus the effects of AMPK activity on osteoblast differentiation may vary in a dose-dependent fashion.

4. Kinases involved in mechanotransduction of osteoblasts

It has been well established that the load-induced mechanical stresses significantly affect cellular functions of osteoblasts. Mechanical stresses have been reported to be associated with increased proliferation, accelerated differentiation, and inhibition of apoptosis of osteoblasts (Aguirre et al. 2006; Robling and Turner 2009; Turner et al. 2009). Recent lines of evidence have also revealed that mechanical stresses induce the expression of RANKL, inflammatory cytokines, and chemokines by osteoblasts (Bandow et al. 2007; Maeda et al. 2007). However, the reception and intracellular signals transducing these stresses (termed as mechanotransduction) of osteoblasts are not well understood. What makes things more complicated is that there are various types of mechanical stresses. They include compression (static or cyclic), tension (static and cyclic), shear stress (including fluid shear stress), ultrasound, and others. It should be considered that the mechanotrasduction mechanisms for different types of stresses may significantly vary. Moreover, intensity and frequency for each type of mechanical stress may also affect the induced molecular events in cells. Here, I will review some of the kinases previously identified as mechanotransducers of osteoblasts.

4.1 MAPKs

4.1.1 ERKs

It has been reported that in osteoblasts ERKs are activated by mechanical stimuli including fluid shear stress (You et al. 2001; Kapur, Baylink, and Lau 2003; Weyts et al. 2002), compression by centrifugation (Hatton et al. 2003), cyclic stretch (Jessop et al. 2002), magnaetic drag force (Pommerenke et al. 2002), and ultrasound (Bandow et al. 2007) (Fig. 2). At least in our hands, it is the MAPK most evidently activated by cyclic stretch (unpublished result) and ultrasound (Bandow et al. 2007). Osteoblast proliferation induced by cyclic stretch requires the activation of ERKs as the treatment with a specific inhibitor of the ERK activating pathway abrogated the proliferation-promoting activity of cyclic stretch (Boutahar et al. 2004). ERK activation has also been reported to be essential for the fluid shear stress to stimulate osteoblast differentiation (Kido et al. 2009). Furthermore, a previous report has indicated that the stretch-induced anti-apoptotic effect on osteoblasts requires the activation of ERKs but not PI3K or p38 kinase (Plotkin et al. 2005). This anti-apoptotic effect involves the nuclear translocation of ERKs and new gene transcription (Plotkin et al. 2005).

The upstream mechanisms mediating mechanical activation of ERKs have been proposed for some experimental systems (Fig. 2). It has previously been shown that integrins are essential

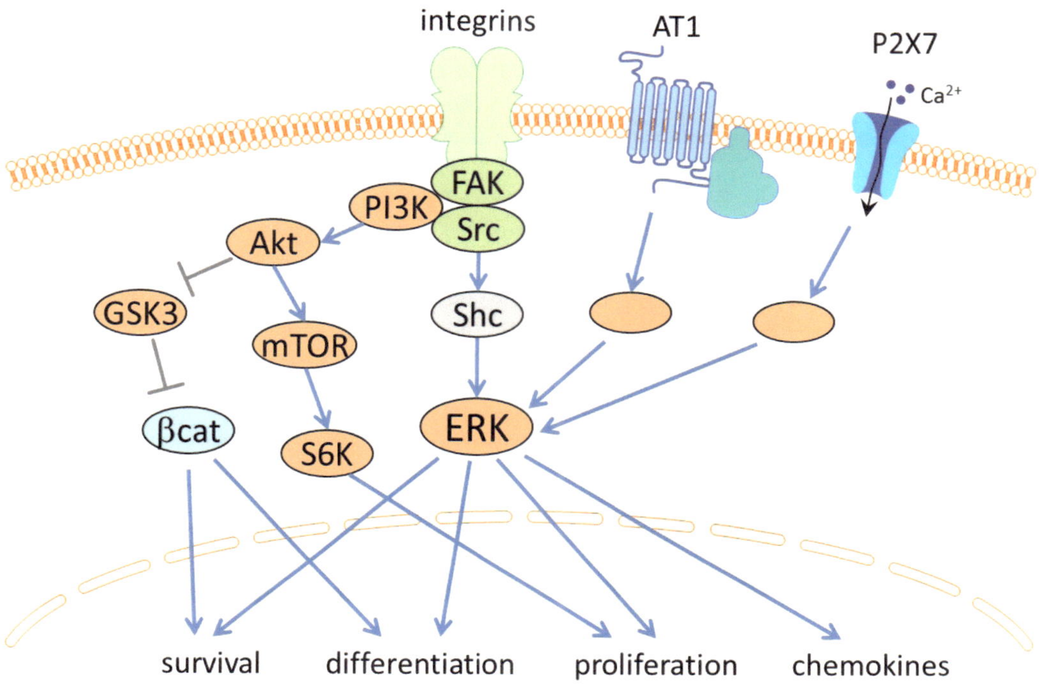

Fig. 2. Working model: Mechanical stress-sensing mechanisms of osteoblasts. ERKs are activated by various types of mechanical stress and initiate multiple cellular functions.

for the activation of ERKs in osteoblasts by cyclic stretch (Schmidt et al. 1998; Zhang et al. 2003). We have recently reported that the treatment of osteoblasts by low-intensity pulsed ultrasound (LIPUS) induced ERK activation and the expression of chemokine and RANKL (Bandow et al. 2007). These LIPUS-induced cellular events were abrogated by candesaltan, a specific inhibitor of angiotensin II type 1 receptor (AT1). AT1 is a known mechanoreceptor in cardiomyocytes (Zou et al. 2004). Osteoblasts express AT1 mRNA and protein and the level of expression increased significantly during osteoblast differentiation (Bandow et al. 2007). These findings indicate that AT1 functions as a mechanoreceptor of in osteoblasts as well as in cardiomyocytes. Another example is P2X7, a two-transmembrane ionotropic receptor. P2X7 has been reported as a mechanotransducer of osteoblasts, which is essential for the ERK activation by fluid shear stress (Liu et al. 2008), indicating that ERK activation by fluid shear stress is dependent on intracellular Ca^{2+}.

In addition, multiple autocrine and paracrine factors have been implicated to transduce ERK activation by mechanical stresses. For example, ERK activation in osteoblasts by fluid shear stress has been reported to be dependent on nitric oxide (NO) synthesis (Rangaswami et al. 2009; Kapur, Baylink, and Lau 2003). According to another report, however, synthesis of NO and prostacyclin is essential for ERK activation in osteoblasts induced by mechanical stretch, but not by fluid shear stress (Jessop et al. 2002).

4.1.2 JNKs and p38 Kinase

Transient activation of JNKs in osteoblasts has been reported to be induced by cyclic stretch (Danciu et al. 2003), compression (Maeda et al. 2007), and fluid shear stress (Lee et al. 2008; Wang et al. 2011). It has been reported that JNK activity is important for compression-induced IL-8 expression (Maeda et al. 2007). On the other hand, transient activation of p38 kinase in osteoblasts has been reported to be induced by cyclic stretch (Kusumi et al. 2005), compression (Maeda et al. 2007), fluid shear stress (Lee et al. 2008; Yeh et al. 2010), and magnetic force (Yuge et al. 2003). Recent reports indicated that p38 activation is essential for Opg expression induced by cyclic stretch (Kusumi et al. 2005), and compression-induce IL-8 expression (Maeda et al. 2007). Intriguingly, acceleration of osteoblast differentiation induced by magnetic force is mainly attributed to p38 kinase activation, while the activity of ERKs or JNKs is dispensable (Yuge et al. 2003). Thus p38 kinase may be specifically involved in the mechanotransduction of magnetic force in osteoblasts.

Activation of JNKs and p38 kinase by cyclic stretch in osteoblasts, however, is not as universally observed as that of ERKs and sometimes controversial. For example, a previous report showed that the application of cyclic stretch to human osteoblasts induced phosphorylation of ERKs, but not JNKs or p38 kinase (Zhu et al. 2009). Additionally, we found that cyclic stretch did not induce JNK or p38 kinase phosphorylation in mouse osteoblast cell lines or mouse primary osteoblasts, whereas it induced significant ERK phosphorylation (unpublished results). Thus, activation of JNKs and p38 kinase by cyclic stretch may be cell line-specific or dependent on stretch conditions.

4.2 Integrin-associated kinases

Cells are surrounded by ECM, which profoundly affects cellular functions. Most effects from ECM are mediated through integrins, cell surface receptors responsible for cell attachment to and reception of signals from ECM. As integrins lack intracellular enzymatic domains, signals from ECM are transduced by various intracellular signaling molecules, the most crucial of which is focal adhesion kinase (FAK) (Fig. 2). Integrin engagement induces activation and the subsequent autophosphorylation of FAK, which then interacts with another tyrosine kinase, Src. These molecular events eventually lead to the activation of ERKs (Bellido 2010). Integrins have been shown as important mechanotransducers in various cell types including osteoblasts.

4.2.1 FAK

As FAK-null mice are embryonic lethal (Furuta et al. 1995), mice with conditional knock-out of FAK in osteoblasts have been generated (Kim et al. 2007). Although these mice were normal in basal osteoblast differentiation, they were defective in bone healing due to decreased matrix deposition, which may be caused by defective mechanotransduction of osteoblasts. The role of FAK in osteoblasts has also been explored in vitro. Fluid shear stress-induced osteoblast differentiation as well as activation of ERK and JNK was blocked by the expression of a dominant negative FAK mutant, indicating the essential role of FAK in mechanotransduction of fluid shear stress (Wang et al. 2011). It has also been reported that $FAK^{-/-}$ osteoblasts are unresponsive to fluid shear stress in ERK activation and Opn expression (Young et al. 2011).

4.2.2 Src

A well-known tyrosine kinase, Src, is involved in integrin signals downstream of FAK. The role of Src in mechanotransduction of osteoblasts has recently been examined. Fluid shear stress induced increased phosphorylation at Tyr-416 of Src in osteoblasts (Morgan et al. 2011). Inhibition of Src, however, had no effects on fluid shear stress-induced Opn expression or ERK activation. These findings indicate that Src is activated by fluid shear stress but not necessary for the proper mechanotransduction of osteoblasts. Another recent report indicated that protein kinase G (PKG), which is activated by NO-mediated production of cGMP, induced the activation of Src in mechanically stimulated osteoblasts (Rangaswami et al. 2010). Activation of Src by PKG is necessary for the interaction of Src with β3 integrin, indicating that Src functions as a converging point of integrin and NO-PKG signals in osteoblasts.

4.3 PI3K-Akt activating pathway

Sustained activation of PI3K in osteoblasts by fluid shear stress has been reported (Lee et al. 2010) (Fig. 2). The resultant activation of Akt, mTOR, and p70S6 kinase leads to cell proliferation. The fluid shear stress-induced activation of PI3K appears to be induced by a complex formation of PI3K p85 subunit with $\alpha_V\beta_3/\beta_1$ integrins, Shc, and FAK. It has also been reported that PI3K-Akt pathway activated by fluid shear stress in osteoblasts and osteocytes, induces increased nuclear translocation of β-Catenin (Kitase et al. 2010; Xia et al. 2010), as Akt phosphorylates and inactivates GSK3 (Wang, Brown, and Martin 2011). The increased β-Catenin signal by fluid shear stress is essential for anti-apoptotic effects against glucocorticoid, indicating a possible mechanism how mechanical loading sustains healthy bone structure.

4.4 LIM Kinase 2 (LIMK2)

When osteoblasts are loaded with mechanical stress, the structure of actin cytoskeleton is reorganized, which results in changes of cell stiffness (Jaasma et al. 2007). LIM kinase (LIMK) is a unique protein kinase that has two LIM domain repeats at the N-terminus, followed by a PDZ domain and a catalytic kinase domain. LIMK protein family includes two members, LIMK1 and LIMK2. LIMK1 is highly expressed in neural tissues (Takahashi, Funakoshi, and Nakamura 2003), whereas LIMK2 is widely expressed in various tissues (Acevedo et al. 2006). LIMK2 phosphorylates and inactivates cofilin, an actin-depolymerizing factor, thereby inducing actin cytoskeleton reorganization (Takahashi, Funakoshi, and Nakamura 2003). A recent report explored the functional role of LIMK2 in osteoblasts treated with fluid shear stress (Fu et al. 2008). Gene silencing of LIMK2 in osteoblasts by siRNA clearly decreased reorganization of actin cytoskeleton induced by fluid shear stress. This finding indicates an important role of LIMK2 in osteoblasts with mechanical loading.

5. Conclusion

Bone is a dynamic tissue where constant remodelling is going on. Osteoblasts are unique cells with multiple aspects. They (1) differentiate into osteocytes, the main cell type in bone tissue, (2) produce proteins for the formation of bone matrix, (3) recruit osteoclasts and control their function, thus playing a regulatory role in bone remodelling, (4) sensing the degree and nature of mechanical loading, which are necessary for the proper maintenance of bone structure suitable for the environment, (5) sense infection and recruit inflammatory cells for host defence. As I have discussed in this review, many of the multiple functions of osteoblasts are controlled by a list of kinases. It should be noted that the same kinases, such as MAPKs and PI3Ks, are involved in many functions of osteoblasts. Thus, in osteoblasts, a lot of crosstalk exists among signal transduction pathways through kinases. If this complicated network is broken due to kinase malfunction, it certainly leads to bone metabolic disorders. In concluding, this review may provide insights into a new list of therapeutic target molecules for diseases that have become threatening social problems, such as osteoporosis.

6. Acknowledgement

The author thanks Norika Chiba, Ph. D. (Cedars-Sinai Medical Center and David Geffen School of Medicine, University of California at Los Angeles) for her useful suggestions. Our research projects are supported by grants from Ministry of Education, Science and Culture of the Japanese Government.

7. References

Acevedo, K., N. Moussi, R. Li, P. Soo, and O. Bernard. 2006. "LIM kinase 2 is widely expressed in all tissues." *J Histochem Cytochem* no. 54 (5):487-501. doi: 10.1369/jhc.5C6813.2006.

Aguirre, J. I., L. I. Plotkin, S. A. Stewart, R. S. Weinstein, A. M. Parfitt, S. C. Manolagas, and T. Bellido. 2006. "Osteocyte apoptosis is induced by weightlessness in mice and precedes osteoclast recruitment and bone loss." *J Bone Miner Res* no. 21 (4):605-15. doi: 10.1359/jbmr.060107.

Amcheslavsky, A., H. Hemmi, S. Akira, and Z. Bar-Shavit. 2005. "Differential contribution of osteoclast- and osteoblast-lineage cells to CpG-oligodeoxynucleotide (CpG-ODN) modulation of osteoclastogenesis." *J Bone Miner Res* no. 20 (9):1692-9. doi: 10.1359/JBMR.050515.

Bandow, K., Y. Nishikawa, T. Ohnishi, K. Kakimoto, K. Soejima, S. Iwabuchi, K. Kuroe, and T. Matsuguchi. 2007. "Low-intensity pulsed ultrasound (LIPUS) induces RANKL, MCP-1, and MIP-1beta expression in osteoblasts through the angiotensin II type 1 receptor." *J Cell Physiol* no. 211 (2):392-8. doi: 10.1002/jcp.20944.

Bellido, T. 2010. "Antagonistic interplay between mechanical forces and glucocorticoids in bone: a tale of kinases." *J Cell Biochem* no. 111 (1):1-6. doi: 10.1002/jcb.22660.

Bendre, M. S., D. C. Montague, T. Peery, N. S. Akel, D. Gaddy, and L. J. Suva. 2003. "Interleukin-8 stimulation of osteoclastogenesis and bone resorption is a mechanism for the increased osteolysis of metastatic bone disease." *Bone* no. 33 (1):28-37.

Boutahar, N., A. Guignandon, L. Vico, and M. H. Lafage-Proust. 2004. "Mechanical strain on osteoblasts activates autophosphorylation of focal adhesion kinase and proline-rich tyrosine kinase 2 tyrosine sites involved in ERK activation." *The Journal of biological chemistry* no. 279 (29):30588-99. doi: 10.1074/jbc.M313244200.

Boyle, W. J., W. S. Simonet, and D. L. Lacey. 2003. "Osteoclast differentiation and activation." *Nature* no. 423 (6937):337-42. doi: 10.1038/nature01658.

Cadigan, K. M., and Y. I. Liu. 2006. "Wnt signaling: complexity at the surface." *J Cell Sci* no. 119 (Pt 3):395-402. doi: 10.1242/jcs.02826.

Celil, A. B., J. O. Hollinger, and P. G. Campbell. 2005. "Osx transcriptional regulation is mediated by additional pathways to BMP2/Smad signaling." *J Cell Biochem* no. 95 (3):518-28. doi: 10.1002/jcb.20429.

Cheung, P. C., I. P. Salt, S. P. Davies, D. G. Hardie, and D. Carling. 2000. "Characterization of AMP-activated protein kinase gamma-subunit isoforms and their role in AMP binding." *Biochem J* no. 346 Pt 3:659-69.

Cooper, C., G. Campion, and L. J. Melton, 3rd. 1992. "Hip fractures in the elderly: a world-wide projection." *Osteoporosis international : a journal established as result of cooperation between the European Foundation for Osteoporosis and the National Osteoporosis Foundation of the USA* no. 2 (6):285-9.

Cortizo, A. M., C. Sedlinsky, A. D. McCarthy, A. Blanco, and L. Schurman. 2006. "Osteogenic actions of the anti-diabetic drug metformin on osteoblasts in culture." *Eur J Pharmacol* no. 536 (1-2):38-46. doi: S0014-2999(06)00196-8 [pii] 10.1016/j.ejphar.2006.02.030.

Dagon, Y., Y. Avraham, and E. M. Berry. 2006. "AMPK activation regulates apoptosis, adipogenesis, and lipolysis by eIF2alpha in adipocytes." *Biochem Biophys Res Commun* no. 340 (1):43-7. doi: S0006-291X(05)02711-7 [pii] 10.1016/j.bbrc.2005.11.159.

Danciu, T. E., R. M. Adam, K. Naruse, M. R. Freeman, and P. V. Hauschka. 2003. "Calcium regulates the PI3K-Akt pathway in stretched osteoblasts." *FEBS Lett* no. 536 (1-3):193-7.

Deng, Z. L., K. A. Sharff, N. Tang, W. X. Song, J. Luo, X. Luo, J. Chen, E. Bennett, R. Reid, D. Manning, A. Xue, A. G. Montag, H. H. Luu, R. C. Haydon, and T. C. He. 2008. "Regulation of osteogenic differentiation during skeletal development." *Front Biosci* no. 13:2001-21. doi: 2819 [pii].

Ducy, P., R. Zhang, V. Geoffroy, A. L. Ridall, and G. Karsenty. 1997. "Osf2/Cbfa1: a transcriptional activator of osteoblast differentiation." *Cell* no. 89 (5):747-54. doi: S0092-8674(00)80257-3 [pii].

Fu, Q., C. Wu, Y. Shen, S. Zheng, and R. Chen. 2008. "Effect of LIMK2 RNAi on reorganization of the actin cytoskeleton in osteoblasts induced by fluid shear stress." *J Biomech* no. 41 (15):3225-8. doi: 10.1016/j.jbiomech.2008.08.010.

Fukuda, T., S. Kokabu, S. Ohte, H. Sasanuma, K. Kanomata, K. Yoneyama, H. Kato, M. Akita, H. Oda, and T. Katagiri. 2010. "Canonical Wnts and BMPs cooperatively induce osteoblastic differentiation through a GSK3beta-dependent and beta-catenin-independent mechanism." *Differentiation* no. 80 (1):46-52. doi: 10.1016/j.diff.2010.05.002.

Fulco, M., Y. Cen, P. Zhao, E. P. Hoffman, M. W. McBurney, A. A. Sauve, and V. Sartorelli. 2008. "Glucose restriction inhibits skeletal myoblast differentiation by activating SIRT1 through AMPK-mediated regulation of Nampt." *Dev Cell* no. 14 (5):661-73. doi: S1534-5807(08)00074-9 [pii]
10.1016/j.devcel.2008.02.004.

Furuta, Y., D. Ilic, S. Kanazawa, N. Takeda, T. Yamamoto, and S. Aizawa. 1995. "Mesodermal defect in late phase of gastrulation by a targeted mutation of focal adhesion kinase, FAK." *Oncogene* no. 11 (10):1989-95.

Gallea, S., F. Lallemand, A. Atfi, G. Rawadi, V. Ramez, S. Spinella-Jaegle, S. Kawai, C. Faucheu, L. Huet, R. Baron, and S. Roman-Roman. 2001. "Activation of mitogen-activated protein kinase cascades is involved in regulation of bone morphogenetic protein-2-induced osteoblast differentiation in pluripotent C2C12 cells." *Bone* no. 28 (5):491-8. doi: S875632820100415X [pii].

Ge, C., G. Xiao, D. Jiang, and R. T. Franceschi. 2007. "Critical role of the extracellular signal-regulated kinase-MAPK pathway in osteoblast differentiation and skeletal development." *J Cell Biol* no. 176 (5):709-18. doi: jcb.200610046 [pii] 10.1083/jcb.200610046.

Ghosh-Choudhury, N., S. L. Abboud, R. Nishimura, A. Celeste, L. Mahimainathan, and G. G. Choudhury. 2002. "Requirement of BMP-2-induced phosphatidylinositol 3-kinase and Akt serine/threonine kinase in osteoblast differentiation and Smad-dependent BMP-2 gene transcription." *The Journal of biological chemistry* no. 277 (36):33361-8. doi: 10.1074/jbc.M205053200.

Giri, S., R. Rattan, E. Haq, M. Khan, R. Yasmin, J. S. Won, L. Key, A. K. Singh, and I. Singh. 2006. "AICAR inhibits adipocyte differentiation in 3T3L1 and restores metabolic alterations in diet-induced obesity mice model." *Nutr Metab (Lond)* no. 3:31. doi: 1743-7075-3-31 [pii]
10.1186/1743-7075-3-31.

Habinowski, S. A., and L. A. Witters. 2001. "The effects of AICAR on adipocyte differentiation of 3T3-L1 cells." *Biochem Biophys Res Commun* no. 286 (5):852-6. doi: 10.1006/bbrc.2001.5484
S0006-291X(01)95484-1 [pii].

Hardie, D. G. 2007. "AMP-activated/SNF1 protein kinases: conserved guardians of cellular energy." *Nat Rev Mol Cell Biol* no. 8 (10):774-85. doi: 10.1038/nrm2249.

Hatton, J. P., M. Pooran, C. F. Li, C. Luzzio, and M. Hughes-Fulford. 2003. "A short pulse of mechanical force induces gene expression and growth in MC3T3-E1 osteoblasts via an ERK 1/2 pathway." *J Bone Miner Res* no. 18 (1):58-66.
doi: 10.1359/jbmr.2003.18.1.58.

Hauschka, P. V., J. B. Lian, D. E. Cole, and C. M. Gundberg. 1989. "Osteocalcin and matrix Gla protein: vitamin K-dependent proteins in bone." *Physiol Rev* no. 69 (3):990-1047.

Hawley, S. A., M. Davison, A. Woods, S. P. Davies, R. K. Beri, D. Carling, and D. G. Hardie. 1996. "Characterization of the AMP-activated protein kinase kinase from rat liver and identification of threonine 172 as the major site at which it phosphorylates AMP-activated protein kinase." *The Journal of biological chemistry* no. 271 (44):27879-87.

Hayashi, S., T. Yamada, M. Tsuneto, T. Yamane, M. Takahashi, L. D. Shultz, and H. Yamazaki. 2003. "Distinct osteoclast precursors in the bone marrow and extramedullary organs characterized by responsiveness to Toll-like receptor ligands and TNF-alpha." *Journal of immunology* no. 171 (10):5130-9.

Hibi, M., A. Lin, T. Smeal, A. Minden, and M. Karin. 1993. "Identification of an oncoprotein- and UV-responsive protein kinase that binds and potentiates the c-Jun activation domain." *Genes Dev* no. 7 (11):2135-48.

Hwang, J. T., I. J. Park, J. I. Shin, Y. K. Lee, S. K. Lee, H. W. Baik, J. Ha, and O. J. Park. 2005. "Genistein, EGCG, and capsaicin inhibit adipocyte differentiation process via activating AMP-activated protein kinase." *Biochem Biophys Res Commun* no. 338 (2):694-9. doi: S0006-291X(05)02159-5 [pii] 10.1016/j.bbrc.2005.09.195.

Jaasma, M. J., W. M. Jackson, R. Y. Tang, and T. M. Keaveny. 2007. "Adaptation of cellular mechanical behavior to mechanical loading for osteoblastic cells." *J Biomech* no. 40 (9):1938-45. doi: 10.1016/j.jbiomech.2006.09.010.

Jessop, H. L., S. C. Rawlinson, A. A. Pitsillides, and L. E. Lanyon. 2002. "Mechanical strain and fluid movement both activate extracellular regulated kinase (ERK) in osteoblast-like cells but via different signaling pathways." *Bone* no. 31 (1):186-94.

Kanazawa, I., T. Yamaguchi, S. Yano, M. Yamauchi, and T. Sugimoto. 2008. "Metformin enhances the differentiation and mineralization of osteoblastic MC3T3-E1 cells via AMP kinase activation as well as eNOS and BMP-2 expression." *Biochem Biophys Res Commun* no. 375 (3):414-9. doi: S0006-291X(08)01546-5 [pii] 10.1016/j.bbrc.2008.08.034.

Kapur, S., D. J. Baylink, and K. H. Lau. 2003. "Fluid flow shear stress stimulates human osteoblast proliferation and differentiation through multiple interacting and competing signal transduction pathways." *Bone* no. 32 (3):241-51.

Karpinski, B. A., G. D. Morle, J. Huggenvik, M. D. Uhler, and J. M. Leiden. 1992. "Molecular cloning of human CREB-2: an ATF/CREB transcription factor that can negatively regulate transcription from the cAMP response element." *Proc Natl Acad Sci U S A* no. 89 (11):4820-4.

Karsenty, G. 2008. "Transcriptional control of skeletogenesis." *Annual review of genomics and human genetics* no. 9:183-96. doi: 10.1146/annurev.genom.9.081307.164437.

Kasai, T., K. Bandow, H. Suzuki, N. Chiba, K. Kakimoto, T. Ohnishi, S. Kawamoto, E. Nagaoka, and T. Matsuguchi. 2009. "Osteoblast differentiation is functionally associated with decreased AMP kinase activity." *J Cell Physiol* no. 221 (3):740-9. doi: 10.1002/jcp.21917.

Kawamura, N., F. Kugimiya, Y. Oshima, S. Ohba, T. Ikeda, T. Saito, Y. Shinoda, Y. Kawasaki, N. Ogata, K. Hoshi, T. Akiyama, W. S. Chen, N. Hay, K. Tobe, T. Kadowaki, Y. Azuma, S. Tanaka, K. Nakamura, U. I. Chung, and H. Kawaguchi. 2007. "Akt1 in osteoblasts and osteoclasts controls bone remodeling." *PLoS One* no. 2 (10):e1058. doi: 10.1371/journal.pone.0001058.

Kido, S., R. Kuriwaka-Kido, T. Imamura, Y. Ito, D. Inoue, and T. Matsumoto. 2009. "Mechanical stress induces Interleukin-11 expression to stimulate osteoblast differentiation." *Bone* no. 45 (6):1125-32. doi: 10.1016/j.bone.2009.07.087.

Kikuchi, T., T. Matsuguchi, N. Tsuboi, A. Mitani, S. Tanaka, M. Matsuoka, G. Yamamoto, T. Hishikawa, T. Noguchi, and Y. Yoshikai. 2001. "Gene expression of osteoclast differentiation factor is induced by lipopolysaccharide in mouse osteoblasts via Toll-like receptors." *Journal of immunology* no. 166 (5):3574-9.

Kikuchi, T., Y. Yoshikai, J. Miyoshi, M. Katsuki, T. Musikacharoen, A. Mitani, S. Tanaka, T. Noguchi, and T. Matsuguchi. 2003. "Cot/Tpl2 is essential for RANKL induction by lipid A in osteoblasts." *J Dent Res* no. 82 (7):546-50.

Kim, J. B., P. Leucht, C. A. Luppen, Y. J. Park, H. E. Beggs, C. H. Damsky, and J. A. Helms. 2007. "Reconciling the roles of FAK in osteoblast differentiation, osteoclast remodeling, and bone regeneration." *Bone* no. 41 (1):39-51. doi: 10.1016/j.bone.2007.01.024.

Kitase, Y., L. Barragan, H. Qing, S. Kondoh, J. X. Jiang, M. L. Johnson, and L. F. Bonewald. 2010. "Mechanical induction of PGE2 in osteocytes blocks glucocorticoid-induced apoptosis through both the beta-catenin and PKA pathways." *J Bone Miner Res* no. 25 (12):2657-68. doi: 10.1002/jbmr.168.

Koga, T., Y. Matsui, M. Asagiri, T. Kodama, B. de Crombrugghe, K. Nakashima, and H. Takayanagi. 2005. "NFAT and Osterix cooperatively regulate bone formation." *Nat Med* no. 11 (8):880-5. doi: 10.1038/nm1270.

Komori, T., H. Yagi, S. Nomura, A. Yamaguchi, K. Sasaki, K. Deguchi, Y. Shimizu, R. T. Bronson, Y. H. Gao, M. Inada, M. Sato, R. Okamoto, Y. Kitamura, S. Yoshiki, and T. Kishimoto. 1997. "Targeted disruption of Cbfa1 results in a complete lack of bone formation owing to maturational arrest of osteoblasts." *Cell* no. 89 (5):755-64. doi: S0092-8674(00)80258-5 [pii].

Kono, S. J., Y. Oshima, K. Hoshi, L. F. Bonewald, H. Oda, K. Nakamura, H. Kawaguchi, and S. Tanaka. 2007. "Erk pathways negatively regulate matrix mineralization." *Bone* no. 40 (1):68-74. doi: S8756-3282(06)00629-6 [pii]
10.1016/j.bone.2006.07.024.

Kulkarni, N. H., J. E. Onyia, Q. Zeng, X. Tian, M. Liu, D. L. Halladay, C. A. Frolik, T. Engler, T. Wei, A. Kriauciunas, T. J. Martin, M. Sato, H. U. Bryant, and Y. L. Ma. 2006. "Orally bioavailable GSK-3alpha/beta dual inhibitor increases markers of cellular differentiation in vitro and bone mass in vivo." *J Bone Miner Res* no. 21 (6):910-20. doi: 10.1359/jbmr.060316.

Kusumi, A., H. Sakaki, T. Kusumi, M. Oda, K. Narita, H. Nakagawa, K. Kubota, H. Satoh, and H. Kimura. 2005. "Regulation of synthesis of osteoprotegerin and soluble receptor activator of nuclear factor-kappaB ligand in normal human osteoblasts via the p38 mitogen-activated protein kinase pathway by the application of cyclic tensile strain." *J Bone Miner Metab* no. 23 (5):373-81. doi: 10.1007/s00774-005-0615-6.

Kyriakis, J. M., and J. Avruch. 2001. "Mammalian mitogen-activated protein kinase signal transduction pathways activated by stress and inflammation." *Physiological reviews* no. 81 (2):807-69.

Lacey, D. L., E. Timms, H. L. Tan, M. J. Kelley, C. R. Dunstan, T. Burgess, R. Elliott, A. Colombero, G. Elliott, S. Scully, H. Hsu, J. Sullivan, N. Hawkins, E. Davy, C. Capparelli, A. Eli, Y. X. Qian, S. Kaufman, I. Sarosi, V. Shalhoub, G. Senaldi, J. Guo, J. Delaney, and W. J. Boyle. 1998. "Osteoprotegerin ligand is a cytokine that regulates osteoclast differentiation and activation." *Cell* no. 93 (2):165-76.

Lee, D. Y., Y. S. Li, S. F. Chang, J. Zhou, H. M. Ho, J. J. Chiu, and S. Chien. 2010. "Oscillatory flow-induced proliferation of osteoblast-like cells is mediated by alphavbeta3 and beta1 integrins through synergistic interactions of focal adhesion kinase and Shc with phosphatidylinositol 3-kinase and the Akt/mTOR/p70S6K pathway." *The Journal of biological chemistry* no. 285 (1):30-42. doi: 10.1074/jbc.M109.010512.

Lee, D. Y., C. R. Yeh, S. F. Chang, P. L. Lee, S. Chien, C. K. Cheng, and J. J. Chiu. 2008. "Integrin-mediated expression of bone formation-related genes in osteoblast-like cells in response to fluid shear stress: roles of extracellular matrix, Shc, and mitogen-activated protein kinase." *J Bone Miner Res* no. 23 (7):1140-9. doi: 10.1359/jbmr.080302.

Lee, J. H., H. Koh, M. Kim, Y. Kim, S. Y. Lee, R. E. Karess, S. H. Lee, M. Shong, J. M. Kim, J. Kim, and J. Chung. 2007. "Energy-dependent regulation of cell structure by AMP-activated protein kinase." *Nature* no. 447 (7147):1017-20. doi: 10.1038/nature05828.

Lee, M. H., T. G. Kwon, H. S. Park, J. M. Wozney, and H. M. Ryoo. 2003. "BMP-2-induced Osterix expression is mediated by Dlx5 but is independent of Runx2." *Biochemical and biophysical research communications* no. 309 (3):689-94.

Li, X., Y. Han, W. Pang, C. Li, X. Xie, J. Y. Shyy, and Y. Zhu. 2008. "AMP-activated protein kinase promotes the differentiation of endothelial progenitor cells." *Arterioscler Thromb Vasc Biol* no. 28 (10):1789-95. doi: ATVBAHA.108.172452 [pii] 10.1161/ATVBAHA.108.172452.

Lisignoli, G., S. Toneguzzi, F. Grassi, A. Piacentini, M. Tschon, S. Cristino, G. Gualtieri, and A. Facchini. 2002. "Different chemokines are expressed in human arthritic bone biopsies: IFN-gamma and IL-6 differently modulate IL-8, MCP-1 and rantes production by arthritic osteoblasts." *Cytokine* no. 20 (5):231-8.

Liu, D., D. C. Genetos, Y. Shao, D. J. Geist, J. Li, H. Z. Ke, C. H. Turner, and R. L. Duncan. 2008. "Activation of extracellular-signal regulated kinase (ERK1/2) by fluid shear is Ca(2+)- and ATP-dependent in MC3T3-E1 osteoblasts." *Bone* no. 42 (4):644-52. doi: 10.1016/j.bone.2007.09.058.

Liu, J., Y. Minemoto, and A. Lin. 2004. "c-Jun N-terminal protein kinase 1 (JNK1), but not JNK2, is essential for tumor necrosis factor alpha-induced c-Jun kinase activation and apoptosis." *Mol Cell Biol* no. 24 (24):10844-56. doi: 24/24/10844 [pii] 10.1128/MCB.24.24.10844-10856.2004.

Liu, X., K. J. Bruxvoort, C. R. Zylstra, J. Liu, R. Cichowski, M. C. Faugere, M. L. Bouxsein, C. Wan, B. O. Williams, and T. L. Clemens. 2007. "Lifelong accumulation of bone in mice lacking Pten in osteoblasts." *Proceedings of the National Academy of Sciences of the United States of America* no. 104 (7):2259-64. doi: 10.1073/pnas.0604153104.

Maeda, A., K. Soejima, K. Bandow, K. Kuroe, K. Kakimoto, S. Miyawaki, A. Okamoto, and T. Matsuguchi. 2007. "Force-induced IL-8 from periodontal ligament cells requires IL-1beta." *J Dent Res* no. 86 (7):629-34.

Manolagas, S. C., and R. L. Jilka. 1995. "Bone marrow, cytokines, and bone remodeling. Emerging insights into the pathophysiology of osteoporosis." *The New England journal of medicine* no. 332 (5):305-11. doi: 10.1056/NEJM199502023320506.

Marie, P. J. 2008. "Transcription factors controlling osteoblastogenesis." *Archives of biochemistry and biophysics* no. 473 (2):98-105. doi: 10.1016/j.abb.2008.02.030.

Masuda, K., H. Shima, M. Watanabe, and K. Kikuchi. 2001. "MKP-7, a novel mitogen-activated protein kinase phosphatase, functions as a shuttle protein." *J Biol Chem* no. 276 (42):39002-11.

Matsubara, T., K. Kida, A. Yamaguchi, K. Hata, F. Ichida, H. Meguro, H. Aburatani, R. Nishimura, and T. Yoneda. 2008. "BMP2 regulates Osterix through Msx2 and Runx2 during osteoblast differentiation." *The Journal of biological chemistry* no. 283 (43):29119-25. doi: 10.1074/jbc.M801774200.

Matsuguchi, T., N. Chiba, K. Bandow, K. Kakimoto, A. Masuda, and T. Ohnishi. 2009. "JNK activity is essential for Atf4 expression and late-stage osteoblast differentiation." *J Bone Miner Res* no. 24 (3):398-410. doi: 10.1359/jbmr.081107.

Matsuguchi, T., T. Musikacharoen, T. R. Johnson, A. S. Kraft, and Y. Yoshikai. 2001. "A novel mitogen-activated protein kinase phosphatase is an important negative regulator of lipopolysaccharide-mediated c-Jun N-terminal kinase activation in mouse macrophage cell lines." *Mol Cell Biol* no. 21 (20):6999-7009.

McCarty, M. F. 2004. "Chronic activation of AMP-activated kinase as a strategy for slowing aging." *Med Hypotheses* no. 63 (2):334-9. doi: 10.1016/j.mehy.2004.01.043.

Morgan, J. M., A. Wong, D. C. Genetos, and C. E. Yellowley. 2011. "Src is sufficient, but not necessary, for osteopontin induction in osteoblasts." *Biorheology* no. 48 (1):65-74. doi: 10.3233/BIR-2011-0582.

Nakashima, K., X. Zhou, G. Kunkel, Z. Zhang, J. M. Deng, R. R. Behringer, and B. de Crombrugghe. 2002. "The novel zinc finger-containing transcription factor osterix is required for osteoblast differentiation and bone formation." *Cell* no. 108 (1):17-29. doi: S0092867401006225 [pii].

Ortuno, M. J., S. Ruiz-Gaspa, E. Rodriguez-Carballo, A. R. Susperregui, R. Bartrons, J. L. Rosa, and F. Ventura. 2010. "p38 regulates expression of osteoblast-specific genes by phosphorylation of osterix." *The Journal of biological chemistry* no. 285 (42):31985-94. doi: 10.1074/jbc.M110.123612.

Plotkin, L. I., I. Mathov, J. I. Aguirre, A. M. Parfitt, S. C. Manolagas, and T. Bellido. 2005. "Mechanical stimulation prevents osteocyte apoptosis: requirement of integrins, Src kinases, and ERKs." *Am J Physiol Cell Physiol* no. 289 (3):C633-43. doi: 10.1152/ajpcell.00278.2004.

Pommerenke, H., C. Schmidt, F. Durr, B. Nebe, F. Luthen, P. Muller, and J. Rychly. 2002. "The mode of mechanical integrin stressing controls intracellular signaling in osteoblasts." *J Bone Miner Res* no. 17 (4):603-11. doi: 10.1359/jbmr.2002.17.4.603.

Raisz, L. G., and B. E. Kream. 1983. "Regulation of bone formation." *N Engl J Med* no. 309 (1):29-35.

Rangaswami, H., N. Marathe, S. Zhuang, Y. Chen, J. C. Yeh, J. A. Frangos, G. R. Boss, and R. B. Pilz. 2009. "Type II cGMP-dependent protein kinase mediates osteoblast mechanotransduction." *The Journal of biological chemistry* no. 284 (22):14796-808. doi: 10.1074/jbc.M806486200.

Rangaswami, H., R. Schwappacher, N. Marathe, S. Zhuang, D. E. Casteel, B. Haas, Y. Chen, A. Pfeifer, H. Kato, S. Shattil, G. R. Boss, and R. B. Pilz. 2010. "Cyclic GMP and protein kinase G control a Src-containing mechanosome in osteoblasts." *Sci Signal* no. 3 (153):ra91. doi: 10.1126/scisignal.2001423.

Robling, A. G., and C. H. Turner. 2009. "Mechanical signaling for bone modeling and remodeling." *Crit Rev Eukaryot Gene Expr* no. 19 (4):319-38.

Sabapathy, K., K. Hochedlinger, S. Y. Nam, A. Bauer, M. Karin, and E. F. Wagner. 2004. "Distinct roles for JNK1 and JNK2 in regulating JNK activity and c-Jun-dependent cell proliferation." *Mol Cell* no. 15 (5):713-25. doi: 10.1016/j.molcel.2004.08.028 S1097276504005143 [pii].

Schmidt, C., H. Pommerenke, F. Durr, B. Nebe, and J. Rychly. 1998. "Mechanical stressing of integrin receptors induces enhanced tyrosine phosphorylation of cytoskeletally anchored proteins." *The Journal of biological chemistry* no. 273 (9):5081-5.

Shaw, R. J., K. A. Lamia, D. Vasquez, S. H. Koo, N. Bardeesy, R. A. Depinho, M. Montminy, and L. C. Cantley. 2005. "The kinase LKB1 mediates glucose homeostasis in liver and therapeutic effects of metformin." *Science* no. 310 (5754):1642-6. doi: 10.1126/science.1120781.

Sinha, K. M., H. Yasuda, M. M. Coombes, S. Y. Dent, and B. de Crombrugghe. 2010. "Regulation of the osteoblast-specific transcription factor Osterix by NO66, a Jumonji family histone demethylase." *EMBO J* no. 29 (1):68-79. doi: 10.1038/emboj.2009.332.

Takahashi, H., H. Funakoshi, and T. Nakamura. 2003. "LIM-kinase as a regulator of actin dynamics in spermatogenesis." *Cytogenet Genome Res* no. 103 (3-4):290-8. doi: 10.1159/000076815.

Takatani, T., M. Minagawa, R. Takatani, K. Kinoshita, and Y. Kohno. 2011. "AMP-activated protein kinase attenuates Wnt/beta-catenin signaling in human osteoblastic Saos-2 cells." *Mol Cell Endocrinol* no. 339 (1-2):114-9. doi: 10.1016/j.mce.2011.04.003.

Tanoue, T., T. Yamamoto, R. Maeda, and E. Nishida. 2001. "A Novel MAPK phosphatase MKP-7 acts preferentially on JNK/SAPK and p38 alpha and beta MAPKs." *J Biol Chem* no. 276 (28):26629-39.

Tong, J., M. J. Zhu, K. R. Underwood, B. W. Hess, S. P. Ford, and M. Du. 2008. "AMP-activated protein kinase and adipogenesis in sheep fetal skeletal muscle and 3T3-L1 cells." *J Anim Sci* no. 86 (6):1296-305. doi: jas.2007-0794 [pii] 10.2527/jas.2007-0794.

Turner, C. H., S. J. Warden, T. Bellido, L. I. Plotkin, N. Kumar, I. Jasiuk, J. Danzig, and A. G. Robling. 2009. "Mechanobiology of the skeleton." *Sci Signal* no. 2 (68):pt3. doi: 10.1126/scisignal.268pt3.

Ulsamer, A., M. J. Ortuno, S. Ruiz, A. R. Susperregui, N. Osses, J. L. Rosa, and F. Ventura. 2008. "BMP-2 induces Osterix expression through up-regulation of Dlx5 and its phosphorylation by p38." *The Journal of biological chemistry* no. 283 (7):3816-26. doi: 10.1074/jbc.M704724200.

Vinals, F., T. Lopez-Rovira, J. L. Rosa, and F. Ventura. 2002. "Inhibition of PI3K/p70 S6K and p38 MAPK cascades increases osteoblastic differentiation induced by BMP-2." *FEBS Lett* no. 510 (1-2):99-104. doi: S0014579301032367 [pii].

Wang, B., T. Du, Y. Wang, C. Yang, S. Zhang, and X. Cao. 2011. "Focal adhesion kinase signaling pathway is involved in mechanotransduction in MG-63 cells." *Biochemical and biophysical research communications* no. 410 (3):671-6. doi: 10.1016/j.bbrc.2011.06.054.

Wang, H., J. Brown, and M. Martin. 2011. "Glycogen synthase kinase 3: a point of convergence for the host inflammatory response." *Cytokine* no. 53 (2):130-40. doi: 10.1016/j.cyto.2010.10.009.

Weyts, F. A., Y. S. Li, J. van Leeuwen, H. Weinans, and S. Chien. 2002. "ERK activation and alpha v beta 3 integrin signaling through Shc recruitment in response to mechanical stimulation in human osteoblasts." *J Cell Biochem* no. 87 (1):85-92. doi: 10.1002/jcb.10278.

Widmann, C., S. Gibson, M. B. Jarpe, and G. L. Johnson. 1999. "Mitogen-activated protein kinase: conservation of a three-kinase module from yeast to human." *Physiol Rev* no. 79 (1):143-80.

Xia, X., N. Batra, Q. Shi, L. F. Bonewald, E. Sprague, and J. X. Jiang. 2010. "Prostaglandin promotion of osteocyte gap junction function through transcriptional regulation of connexin 43 by glycogen synthase kinase 3/beta-catenin signaling." *Mol Cell Biol* no. 30 (1):206-19. doi: 10.1128/MCB.01844-08.

Yang, X., K. Matsuda, P. Bialek, S. Jacquot, H. C. Masuoka, T. Schinke, L. Li, S. Brancorsini, P. Sassone-Corsi, T. M. Townes, A. Hanauer, and G. Karsenty. 2004. "ATF4 is a substrate of RSK2 and an essential regulator of osteoblast biology; implication for Coffin-Lowry Syndrome." *Cell* no. 117 (3):387-98. doi: S0092867404003447 [pii].

Yasuda, H., N. Shima, N. Nakagawa, K. Yamaguchi, M. Kinosaki, S. Mochizuki, A. Tomoyasu, K. Yano, M. Goto, A. Murakami, E. Tsuda, T. Morinaga, K. Higashio, N. Udagawa, N. Takahashi, and T. Suda. 1998. "Osteoclast differentiation factor is a ligand for osteoprotegerin/osteoclastogenesis-inhibitory factor and is identical to TRANCE/RANKL." *Proceedings of the National Academy of Sciences of the United States of America* no. 95 (7):3597-602.

Yeh, C. R., J. J. Chiu, C. I. Lee, P. L. Lee, Y. T. Shih, J. S. Sun, S. Chien, and C. K. Cheng. 2010. "Estrogen augments shear stress-induced signaling and gene expression in osteoblast-like cells via estrogen receptor-mediated expression of beta1-integrin." *J Bone Miner Res* no. 25 (3):627-39. doi: 10.1359/jbmr.091008.

Yoshie, O., T. Imai, and H. Nomiyama. 2001. "Chemokines in immunity." *Adv Immunol* no. 78:57-110.

You, J., G. C. Reilly, X. Zhen, C. E. Yellowley, Q. Chen, H. J. Donahue, and C. R. Jacobs. 2001. "Osteopontin gene regulation by oscillatory fluid flow via intracellular calcium mobilization and activation of mitogen-activated protein kinase in MC3T3-E1 osteoblasts." *The Journal of biological chemistry* no. 276 (16):13365-71. doi: 10.1074/jbc.M009846200.

Young, S. R., J. M. Hum, E. Rodenberg, C. H. Turner, and F. M. Pavalko. 2011. "Non-overlapping functions for Pyk2 and FAK in osteoblasts during fluid shear stress-induced mechanotransduction." *PLoS One* no. 6 (1):e16026. doi: 10.1371/journal.pone.0016026.

Yuge, L., A. Okubo, T. Miyashita, T. Kumagai, T. Nikawa, S. Takeda, M. Kanno, Y. Urabe, M. Sugiyama, and K. Kataoka. 2003. "Physical stress by magnetic force accelerates differentiation of human osteoblasts." *Biochemical and biophysical research communications* no. 311 (1):32-8.

Zhang, J., W. Li, M. A. Sanders, B. E. Sumpio, A. Panja, and M. D. Basson. 2003. "Regulation of the intestinal epithelial response to cyclic strain by extracellular matrix proteins." *FASEB J* no. 17 (8):926-8. doi: 10.1096/fj.02-0663fje.

Zhu, J., X. Zhang, C. Wang, and X. Peng. 2009. "Periprosthetic strain magnitude-dependent upregulation of type I collagen synthesis in human osteoblasts through an ERK1/2 pathway." *Int Orthop* no. 33 (5):1455-60. doi: 10.1007/s00264-009-0735-z.

Zou, Y., H. Akazawa, Y. Qin, M. Sano, H. Takano, T. Minamino, N. Makita, K. Iwanaga, W. Zhu, S. Kudoh, H. Toko, K. Tamura, M. Kihara, T. Nagai, A. Fukamizu, S. Umemura, T. Iiri, T. Fujita, and I. Komuro. 2004. "Mechanical stress activates angiotensin II type 1 receptor without the involvement of angiotensin II." *Nat Cell Biol* no. 6 (6):499-506. doi: 10.1038/ncb1137.

13

The Role of Mitogen-Activated Protein Kinase in Treatment Strategies for Fear and Drug Addiction

Robyn Mary Brown, Andrew J. Lawrence and Jee Hyun Kim*
Florey Neuroscience Institutes, University of Melbourne, Australia

1. Introduction

Fear-avoidance and reward-seeking are necessary motivations to guide survival; however, both can lead to maladaptive behavior when expressed inappropriately, manifesting as anxiety disorders and drug addiction. Both disorders are a major worldwide public health concern with a high co-morbidity (M. S. O'Brien et al., 2005; Wasserman et al., 1997). Specifically, the World Health Organization (WHO) pin-pointed generalized anxiety disorder and substance abuse as the most common mental disorders across the world, ranking them highly as a cause of disease burden (WHO, 2001). Furthermore, in the recent United States National Comorbidity Survey Replication study, it was reported that 18.1% and 8.9% of adults met the 12-month Diagnostic and Statistical Manual of Mental Disorders (DSM-IV; American Psychiatric Association, 2000) criteria for an anxiety or substance abuse disorder, respectively (Kessler et al., 2005). These staggering statistics also highlight the economic costs of anxiety disorders and substance abuse. For example, one estimate puts the yearly economic cost of alcohol abuse to be $148 billion, and the economic cost of harmful drug use and dependence has been estimated to be $98 billion in the United States alone (Harwood et al., 1998). Another study reported the economic cost of anxiety disorders to be $42 billion in the United States (Greenberg et al., 1999).

Given the prevalence and the financial burden of anxiety disorders and substance abuse, a significant amount of research has been dedicated to finding effective treatments for those disorders. At present, the use of pharmacological agents has become a standard approach to attempt to ameliorate aspects of drug addiction and anxiety (Jupp & Lawrence, 2010; Reinblatt & Riddle, 2007). Unfortunately, the prevalence of these disorders still remain extremely high, although pharmacotherapy may be an effective approach to the treatment of drug addiction and anxiety disorders for some individuals. Furthermore, relapse is high in both substance abuse and anxiety disorders, even after pharmacotherapy (C. P. O'Brien, 1997b; Rachman, 1989). Thus, it has been suggested that cognitive-behavioral therapies that rely on learning mechanisms may potentially lead to more effective treatments for both

* Corresponding Author

types of disorders (Myers et al., 2011; Sutton et al., 2003). Therefore, much effort has been placed in finding an effective pharmacotherapy to combine with cognitive-behavioral therapies to effectively treat these mental disorders and prevent their relapse (M. Davis et al., 2006; Hofmann, 2007).

The importance of cognitive-behavioral therapies in treating anxiety disorders and substance abuse has been highlighted following observations that learning and memory processes play a significant role in acquisition and relapse in these disorders both in clinical settings and in the laboratory (Bouton et al., 2001; Mineka & Zinbarg, 1996; Rasmusson & Charney, 1997). For example, a common form of anxiety, posttraumatic stress disorder (PTSD), develops in vulnerable people following exposure to a traumatic event, and it is believed that these symptoms arise in part through a learning process in which cues present at the time of trauma acquire the ability to elicit fear (M. Davis & Whalen, 2001). Those cues can generalize over time to other settings and trigger pervasive fear and anxiety in inappropriate situations (Rothbaum & M. Davis, 2003). Drug addiction also involves learning - cues that are associated with the rewarding properties of a drug can trigger drug use in an individual (Rescorla, 1988; Robinson & Berridge, 2001). For example, drug paraphernalia such as a syringe, can trigger powerful drug craving that contributes to the maintenance of, and relapse to, drug use (A. R. Childress et al., 1986). Discovering that memory processes underlie anxiety disorders and drug addiction has not only led to a more successful modeling of anxiety and drug use using rodents in the laboratory, but also has guided pharmacological research. That is, considerable progress has already been made in delineating the neural bases of memory storage (Kandel, 2001, 2009), and research on anxiety disorders and drug addiction is now focused on characterizing the specific effects of proteins that have already been identified to be involved in general learning and memory. Mitogen-activated protein kinase (MAPK), in particular, refers to the entire superfamily of signaling cascades including the extracellular signal-regulated kinase (ERK) (Sweatt, 2001). The MAPK cascade is distinguished by a unique core series of three kinases: the first is MAP kinase kinase kinase (MAPKKK) that phosphorylates (i.e., activates) the second, a MAP kinase kinase (MEK). MEKs in turn activate a MAP kinase. The two main isoforms of MAPK are ERK1 and ERK2, which are 44 kDa and 42 kDa respectively (L. Lu et al., 2006). MAPK is well-known for its critical role in learning and memory as well as cellular growth, differentiation and survivability (Ahn & Krebs, 1990; Hoshi et al., 1988; Impey et al., 1999; Kornhauser & Greenberg, 1997; Sweatt, 2001). The present chapter will provide an in-depth review of the existing findings on the relationship between MAPK and fear and drug-seeking behavior using animal models. The aim of our chapter is for the audience to gain a better understanding of the significance of MAPK in anxiety disorders and drug addiction in order to facilitate the development of pharmacological adjuncts to assist in treatment of these disorders. The next section will first briefly review the research that revealed the molecular cascade underlying memory storage and formation, and the importance of MAPK signaling in that cascade.

2. The molecular biology of memory formation and storage

Memory formation has long been though to be sub-served by a mechanism called 'synaptic plasticity', which refers to molecular and morphological changes that occur in dendrites and axons that either enhances or reduces the effectiveness of electrical and/or chemical communication between neurons across synapses (Konorski, 1948). Due to the seminal work on Aplysia by the Nobel laureate Kandel and his colleagues, we now know that synaptic plasticity requires various molecular cascades that ultimately result in modifications of synaptic structure and efficacy (Kandel 2001, 2009). The best-studied mechanism that may potentiate these changes is N-methyl-D-aspartate receptor (NMDAR)-dependent long-term potentiation (LTP) (Sweatt, 1999). LTP refers to a lasting enhancement of the strength of synaptic connection as a result of brief repetitive activation of that synaptic pathway (Bliss & Collingridge, 1993). For example, experimentally-induced LTP involves a few trains of high-frequency stimuli to the connection between two neurons (Huang & Kandel, 1994). We know that LTP is important for memory storage because LTP disruption has been shown to impair memory consolidation in a region-specific manner. In the hippocampus, for example, antagonism of NMDARs impairs both LTP induction as well as hippocampal-dependent spatial learning in rats (Morris et al., 1986).

NMDA-dependent LTP is generally divided into three phases: short-term potentiation (or initial LTP), early LTP (E-LTP), and late LTP (L-LTP) (Sweatt, 1999). Short-term potentiation refers to the first stage of LTP that is independent of protein kinase activity for its induction or expression (Roberson et al., 1996). E-LTP starts at around 30 mins upon induction and typically lasts 60-90 mins. E-LTP requires protein kinase activation but is independent of gene transcription and protein synthesis (Frey et al., 1993). The last phase of LTP, L-LTP, lasts many hours and requires gene transcription and protein synthesis for its maintenance, and such processes are critical in the modification of synapses (Frey, et al., 1993). Although it appears that short-term potentiation is necessary for the subsequent phases of LTP, not much is known about the mechanisms underlying it (Roberson et al., 1996), therefore, the present review will focus on the molecular cascade related to E-LTP and L-LTP.

Typically, excitatory synaptic transmission occurs chemically via presynaptic release of glutamate into the synapse. Glutamate then binds to 2-amino-3-(5-methyl-3-oxo-1,2- oxazol-4-yl)propanoic acid receptors (AMPARs) on the postsynaptic membrane, triggering the influx of positively-charged of sodium ions into the postsynaptic neuron (Agranoff & Siegel, 1999). Such influx of ions causes depolarization of the cell, called the excitatory postsynaptic potential (EPSP). Likewise, the induction of E-LTP begins with NMDA receptor activation that leads to an influx of positively-charged calcium ions through ligand- and voltage-gated calcium channels into the postsynaptic neuron, causing depolarization (Ascher & Nowak, 1986). If calcium influx does not reach a threshold, E-LTP fails to occur (Lynch, 2004). The rapid rise in intracellular calcium concentration initiates the short- or long-lasting activation of several proteins that appear to be necessary for E-LTP. For example, blocking phosphorylation (i.e., activation) of calcium/calmodulin-dependent protein kinase II (CaMKII) and protein kinase C (PKC) impairs the induction and maintenance of E-LTP (Sweatt, 1999). Protein kinase A (PKA) and mitogen-activated protein kinase (MAPK) phosphorylation also contribute to the maintenance of E-LTP (Lynch, 2004; Sweatt, 1999). Additionally, protein kinase Mζ (PKMζ), which is calcium independent, becomes activated during the maintenance of E-LTP. Those protein kinases carry out phosphorylation of existing AMPARs to increase their activity, as well as mediating insertion of AMPARs into the postsynaptic membrane (Malenka & Bear, 2004).

As mentioned previously, L-LTP is a natural extension of E-LTP. That is, long-lasting activation of some of the protein kinases mentioned above can induce L-LTP. L-LTP can last many hours, even days, and requires gene transcription and *de novo* protein synthesis in the postsynaptic neuron (Frey et al., 1988; Lynch, 2004; Stanton & Sarvey, 1984). Those characteristics distinguish L-LTP from E-LTP because E-LTP does not require either. For example, administration of the protein synthesis inhibitor anisomycin blocks L-LTP, but not E-LTP, in the hippocampus (Frey et al., 1988). Gene transcription necessary for L-LTP is triggered by the persistent activation of MAPK. MAPK can translocate to the nucleus and activate nuclear molecules such as cyclic adenosine monophosphate (cAMP)-response element binding protein (CREB), which is a gene transcription factor (Lynch, 2004; Sweatt, 1999). ERK-mediated changes in transcription factor activity can ultimately result in the protein synthesis and morphological changes in neurites observed in L-LTP (Figure 1). For example, activity-regulated cytoskeletal element (Arc; also termed Arg3.1), is an immediate early gene that requires MAPK for its transcription (Chotiner et al., 2010). Arc has been widely implicated in experience-dependent synaptic plasticity because Arc is tightly coupled to the activation of signaling pathways implicated in learning, rather than being involved in normal cellular activity (Fletcher et al., 2006). Indeed, Arc expression in the hippocampus is critical for both LTP maintenance and the formation of hippocampus-dependent memory (Guzowski et al., 2000) and MEK inhibition suppresses experience-induced Arc mRNA levels (Chortiner et al., 2010). Further, Arc transcription can lead to protein synthesis via activation of receptors such as Notch (Alberi et al., 2011). Notch receptors and ligands are transmembrane proteins present at synapses that play a major role in changing neurite structure (Sestan et al., 1999). Blocking any one of the steps in the molecular cascade mentioned above impairs memory consolidation and or/learning and LTP (Kandel, 2001). Taken together, it appears that formation of long-term memory is linked to a molecular cascade involving pre-synaptic glutamate release, NMDA-mediated calcium entry, activation of protein kinases, gene expression and transcription, and protein synthesis (Figure 1).

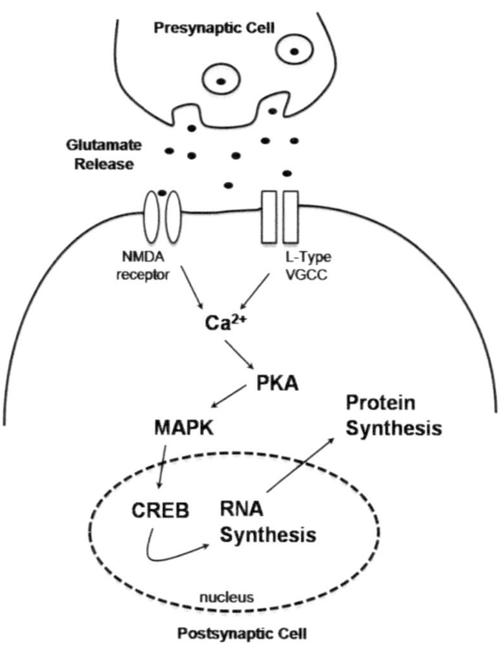

Fig. 1. Schematic diagram illustrating intracellular signalling pathway involved in the neuronal plasticity implicated in the formation of memory.

It is important to note that MAPK is critical for both L-LTP and E-LTP; MEK inhibitors significantly attenuate E-LTP and block L-LTP (S. Davis, Salin et al., 2000; English & Sweatt, 1997; Schafe et al., 2008; Sweatt, 1999; Wu et al., 1999). That is, MAPK signaling appears to be a molecular link between E-LTP and L-LTP. Indeed, many signaling cascades involved in E-LTP, including CaMKII and PKC, can converge on ERK (Kelleher et al., 2004). The strong possibility of MAPK being a link between E-LTP and L-LTP has significant implications for its role in memory. Historically, E-LTP and L-LTP are believed to be mechanisms subserving short-term memory (STM) and long-term memory (LTM), respectively, because both E-LTP and STM endure for a few hours, whereas L-LTP persists for many hours and even days, similar to LTM (Kandel, 2009). Thus, blockade of MAPK may sever any possible link between STM and LTM, and critically affect memory consolidation and storage. Therefore, it is not surprising that much attention has been given to the role of MAPK in animal models of anxiety disorders and drug addiction in the past couple of decades, as it became increasingly apparent that memory processes underlie these disorders.

3. MAPK in conditioned fear

Among the early advocates of the idea that anxiety disorders emerge from traumatic memories were Watson and Rayner (1920) who proposed that humans are born with very few 'emotional reaction patterns', and the way in which stimuli come to evoke emotion throughout one's life is through learning, or conditioning, processes. That is, animals learn to respond in distinct ways to different stimuli according to previous experiences with those stimuli. According to this notion, simple aversive learning and memory processes are critical for triggering anxiety disorders. To support this idea, in the famous 'little Albert' experiment, a baby boy was conditioned to fear a rat by sounding an aversive loud noise each time the baby touched the rat (Watson & Rayner, 1920). Before this experience, the baby was keen to touch and play with the rat. After the experience, however, the baby displayed signs of fear and anxiety (e.g., crying and reaching for mum) when in the presence of the rat. The simple pairing of the rat and the aversive noise induced fear learning in little Albert. This associative learning is referred to as classical or Pavlovian conditioning because Pavlov was the first to demonstrate such associative learning in animals (Konorski, 1948; I. P. Pavlov, 1927). Typically, Pavlovian *fear* conditioning involves pairings of a conditioned stimulus (CS), such as a tone, with an aversive unconditioned stimulus (US), such as a footshock. The CS can be discrete or diffuse, such as a localized light or a context, respectively. Initially, the CS is a neutral stimulus that has little effect on animals. On the other hand, the US is a stimulus that already elicits fear responses via physiological, autonomic and behavioral changes. These fear responses triggered by the US are referred to as unconditioned responses. When the CS and the US are repeatedly paired, the CS starts to elicit such autonomic and behavioral fear responses on its own without the US, and these responses are referred to as conditioned responses (CRs).

Such fear conditioning is readily acquired in animals, and is not easily forgotten (Gale et al., 2004). Because this form of learning can be acquired even in very immature animals (e.g., 12-day-old rats; (Sullivan et al., 2000), Pavlovian conditioning provides a model for investigating the neurobiological bases of anxiety disorders, which often emerge during childhood/early adolescence in humans (Kessler, et al., 2005; Newman et al., 1996). Apart from Pavlovian conditioned fear, however, there are other laboratory models of anxiety disorders. For example, Grillon and colleagues proposed that anxiety disorders are caused by unpredictable aversive events that lead to hyperactive fear responses that are generalized across different situations and environments (Grillon et al., 2004). Specifically, if an aversive US is not signalled reliably by a discrete CS, subjects show more fear to the training context compared to subjects that received the US signalled reliably by a CS. The elevated fear to the context exhibited by the unsignalled group reflects generalized anxiety according to this theory. Another laboratory model for anxiety disorders relies on the concept of 'unlearned' fear, which refers to the innate fear that animals have to various stimuli without any previous learning experiences with these stimuli (Boulis & M. Davis, 1989). Unlearned fear is triggered by a diffuse cue rather than a specific cue; therefore, unlearned fear can serve as a good model for anxiety disorders that may not be triggered by a specific object or a situation. Additionally, unlearned fear may explain some specific phobias that may be an exaggerated, debilitating form of natural fear (e.g. acrophobia).

Nevertheless, Pavlovian fear conditioning is the most utilized neurobiological model for anxiety, as it reflects the most common form of anxiety disorder, namely post-traumatic stress disorder (PTSD). This is partly because this preparation allows a variety of discrete stimuli to be used as the CS or the US, so that the role of each sensory modality and different neural structures in learning can be examined. Also, Pavlovian conditioning is easily acquired in many different animals across a range of different ages, unlike contextual conditioning in Grillon's model for anxiety (Grillon et al., 2004). Further, considering that anxiety disorders may be normal fear gone astray (Rosen & Schulkin, 1998), using Pavlovian conditioning as a model for fear learning provides a tool for understanding how the fear system works. It is no surprise then since Pavlov (1927) researchers have investigated the neurobiology underlying conditioned fear. Although much work has been done with the rat, it should be noted that similar neural processes are involved in learned fear across a range of species (Davis, 2000; LeDoux, 2000; Price, 2003). These findings have pin-pointed the amygdala as a central structure that is essential for the acquisition, storage and expression of conditioned fear.

The amygdala is made up of many sub-nuclei including the central (CeA) and basolateral (BLA) nuclei. The BLA is further divided into basal (BA) and lateral (LA) nuclei. During fear conditioning, perceptual (e.g., olfactory, visual) and somatosensory (e.g., noxious) information regarding the CS (discrete as well as contextual information) and US from hippocampal, cortical and thalamic structures is transmitted to the LA (LeDoux, 1993; LeDoux et al., 1990; McDonald et al., 1996; Price, 2003; C. Shi & Davis, 2001). As the result of CS-US convergence, neuronal activity in the LA is modified (Quirk et al., 1997; Quirk et al., 1995; Repa et al., 2001). Interestingly, once the CS and the US have been associated, the fear memory is stored in the BLA via similar intracellular molecular changes involved in LTP discussed earlier (Bailey et al., 1999; Campeau et al., 1991; Farb et al., 1992; Lee & Kim, 1998; C. H. Lin et al., 2001; Maren et al., 2003). Further, in vitro electrophysiological recordings show that LTP in the BLA is blocked by inhibition of MAPK activity without affecting basal transmission (Huang et al., 2000; Schafe et al., 2000). Hence, fear conditioning results in an enhanced functional and structural connectivity between sensory pathways and the BLA, such that future presentation of the CS alone is sufficient to activate the BLA. Such enhanced functional and structural connectivity requires MAPK signaling, as evidenced by many empirical demonstrations of the relationship between MAPK and conditioned fear.

3.1 The role of MAPK in the acquisition of conditioned fear

The involvement of MAPK in conditioned fear was initially illustrated by correlational studies that showed increases in phosphorylated MAPK (pMAPK) following fear conditioning via Western blotting and immunohistochemistry (Atkins et al., 1998; Gresack et al., 2009; Schafe & LeDoux, 2000; Sindreu et al., 2007). In one of the first studies examining the role of MAPK and conditioned fear, Sweatt and his colleagues first gave context-shock or cue-shock training to rats (Atkins et al., 1998). The memory of conditioned fear was assessed by CS-elicited freezing fear 24 hours later. Freezing is a species-specific defense response in rodents, and is characterized by an immobile, crouching position accompanied by hypervigilance, and is operationally defined as the absence of movement other than that required for respiration (Blanchard & Blanchard, 1972). At test, both context- and cue-shock trained rats exhibited robust freezing to the context or the cue, respectively. In a follow-up experiment, Western blotting revealed significant increases in pMAPK in the hippocampus (a known neural structure important for contextual and spatial learning) of rats that received either context-shock or cue-shock training compared to sham-trained rats that were merely placed in the context. This increase was significant 60 mins following conditioning, but not 1 min or 24 hours following conditioning. Interestingly, pre-conditioning systemic injection of the NMDA antagonist MK801 blocked this conditioning-related increase in pMAPK, as well as impairing both cue- and context-elicited freezing in a separate group of rats tested 24 hrs later. This finding suggests that the increase in MAPK phosphorylation in the hippocampus following fear conditioning is part of the NMDA-dependent molecular cascade that may be responsible for the formation of long-term memory. Additionally, Sindreau et al., (2007) also reported increased pERK in the CA1 area of the hippocampus following contextual fear conditioning in mice. Peak pERK was detected 30 mins following conditioning, which is different from Atkins et al. (1998) but can possibly be attributed to inter-species differences. Importantly, double immuno-labeling using the neuronal marker NeuN, interneuron marker GAD69, or astrocyte marker GFAP revealed that learning-induced ERK activation was exclusively neuronal (CA1 pyramidal cells) and not glial or interneuronal (Sindreu, et al., 2007). Further, activation of MAPK, PKA and CREB kinase was observed in the same neurons, further supporting that the molecular cascade underlying memory is likely to be involved in contextual fear memory consolidation.

The first study to examine phosphorylation of MAPK in the amygdala, a structure central to both contextual and cued fear conditioning, reported that tone-shock training significantly increased pMAPK-labelled neurons in the LA, compared to tone-only and shock-only control groups (Schafe & LeDoux, 2000). Additionally, Western blot revealed increases in phosphorylated ERK (pERK) in the conditioned group. Consistent with Atkins et al. (1998), both immunohistochemical and western blot data showed that pMAPK levels were the strongest 60 mins following conditioning. However, no rats were tested for their long-term memory following conditioning or tone-/shock-only training, making it difficult to attribute the increases in pMAPK/pERK in the LA to associative fear memory formation. Another study has provided such evidence for increases in pMAPK/pERK levels in the amygdala following Pavlovian fear conditioning as well as another type of fear learning. Specifically, Radwanska et al. (2002) trained rats on a two-way active avoidance procedure, in which footshocks were given in a compartment of a two-compartment shuttle box. During each training trial the auditory CS was presented and the rat could 'avoid' the shock by moving into the other compartment within 5 seconds of CS onset. Failure to change the compartment within this time led to the onset of footshock US. Rats were then either sacrificed for immunohistochemistry, or were tested for avoidance memory 24 hours later. At test, rats showed good retention of two-way avoidance learning by showing increased avoidance behavior upon CS-onset at test compared to training. Increased pERK was also observed in the ventral portion of the dorsal LA compared to naive rats. In a separate, but identical experiment, increased activation of the immediate early-gene, Fos, was also observed in the same area. These results were replicated using Pavlovian fear conditioning (Di Benedetto et al., 2009; Radwanska et al., 2002). The authors suggested that the increases in pERK and c-Fos in the ventral-dorsal LA may be indicative of fear memory storage as a prior electrophysiology study observed that this specific area of the amygdala contained neurons that displayed extinction-resistant firing patterns (Repa et al., 2001).

MAPK phosphorylation in the LA as a potential marker for storage of fear memory is supported by a very recent study that examined fear memory retrieval and forgetting in developing rats (Kim et al., 2012). In that study, a memory phenomenon called 'infantile amnesia' was examined, which refers to the more rapid forgetting displayed by young animals compared to older animals (Campbell & Spear, 1972). Typically, no age differences in retention are observed when testing occurs soon after training, indicating that animals of all ages are equally able to acquire and express memory. However, when testing occurs after a delay, retention increases dramatically with maturation. This general finding has since been replicated numerous times, and is of substantial biological significance because it is an ubiquitous phenomenon that occurs in all altricial species, including humans (for a review, see Campbell & Spear, 1972). Interestingly, infantile amnesia reflects a failure to retrieve the fear memory at test because various pretest "reminder" treatments are able to reinstate the memory. For example, a single pre-test shock unconditioned stimulus (US) can reverse infantile forgetting (Campbell & Jaynes, 1966; Kim & Richardson, 2007a, 2007b). Kim et al. (2012) first showed that when postnatal day (P) 16 and 23 rats receive tone CS-shock pairings, both ages exhibited similar levels of conditioned fear when tested immediately after conditioning; when tested after 2 days, however, P16 rats showed poor CS-elicited freezing relative to P23 rats. When rats were assessed for pMAPK after test, both P23 and P16 rats given paired presentations exhibited significant elevation of pMAPK-immunoreactive (ir) neurons in the BLA compared to rats given unpaired presentations, despite P16 rats showing poor CS-elicited freezing (i.e., forgetting). In contrast, only the P23-paired group showed an elevated number of pMAPK-ir neurons in the medial prefrontal cortex (mPFC), indicating that MAPK phosphorylation in the mPFC tracks memory expression. These findings provided initial evidence suggesting that while the mPFC is involved in memory retrieval, MAPK phosphorylation in the BLA may be a persisting neural signature of fear memory storage that is forgotten but can be retrieved with reminder treatments.

The studies described above show whether learning or memory retrieval is followed by MAPK activation, providing correlational evidence for the involvement of MAPK in fear memory consolidation. In a different type of correlational study, Gresack et al. (2009) used sex differences to examine whether stronger memory is associated with greater MAPK activation following learning because it is well-established that there are sex differences in contextual fear conditioning due to differences in hippocampal dendritic and synaptic morphology and cell excitability (Juraska, 1998; Madeira et al., 1991; M. D. Smith et al., 2002). It was first shown that male rats exhibit more long-term retention of contextual fear conditioning than female rats, as tested by context-elicited freezing 24 hours following context-shock training (Gresack, et al., 2009). Western blot then showed that pERK levels in ventral, but not dorsal, hippocampus were higher in males than females, relative to same-sex controls, 60 minutes following fear conditioning. Further, there were no sex differences in context conditioning-induced pERK increases in the amygdala, suggesting that the sex difference in pERK may be restricted to the ventral hippocampus. These findings imply different strengths of memory may be subserved by quantitative differences in MAPK activation.

Taken together, the correlational studies show phosphorylation of MAPK following fear conditioning and expression. However, they do not indicate whether MAPK is necessary for long-term fear memory. The direct upstream activator of MAPK is MEK, and various fear studies using MEK inhibitors support the hypothesis that MAPK regulation is necessary for memory consolidation (Atkins et al., 1998; Izquierdo et al., 2000; Schafe et al., 1999; Selcher et al., 1999). First, it was shown that systemic injection of SL327, a MEK inhibitor that crosses the blood-brain barrier (Favata et al., 1998), dose-dependently decreased pMAPK in the hippocampus in rats (Atkins et al., 1998). More importantly, systemic injection of SL327 one-hour before fear conditioning significantly impaired both cue- and context-elicited freezing when tested 24 hours later, compared vehicle injection. Open-field and shock sensitivity tests indicated that the effects of SL327 were not due to changes in anxiety levels or sensory perception. Importantly, memory impairment was not due to state-dependent mechanisms because injection of SL327 immediately after conditioning also blocked cue- and context-freezing. These findings were replicated in mice (Selcher et al., 1999). Finally, an intracerebroventricular (ICV) infusion study using the selective MEK inhibitor PD098059 (Seger & Krebs, 1995), definitively showed that memory impairments due to inhibition of MAPK activity was central rather than peripheral. Specifically, ICV infusion of PD098059 prior to tone-conditioning significantly reduced both tone- and context-elicited freezing when tested 24 hours, but not 1 hr, after conditioning (Schafe et al., 1999). The failure to see any memory deficit at the 1 hr test strongly suggests that MAPK activation during conditioning is necessary for LTM but not for STM. This is consistent with the idea that MAPK is particularly important for L-LTP, a putative neural mechanism underlying long-term memory. It should also be noted that in all three studies mentioned in this paragraph, there were no discernible effects of the drug during the conditioning session.

Since Schafe et al. (1999), quite a few studies have examined the necessity of MAPK activation in various neural structures that form the neural circuitry underlying conditioned fear, including the BLA, hippocampus, cortical and thalamic structures. Schafe et al., (2000) showed that bilateral intra-BLA infusion of the selective MEK inhibitor U0126 (Seger, et al., 1995) 30 mins before tone-shock conditioning dose-dependently attenuated tone-elicited fear when rats were tested 24 hours later. There was no effect of drug during conditioning and STM was unaffected. Di Benedetto et al. (2009) replicated this finding in mice using fear-potentiated startle (FPS) as the measure of conditioned fear. The startle response is evoked by an intense, unexpected stimulus (startle pulse; e.g., a brief but loud, sharp noise), and in studies with rodents, a whole body jump is usually taken as the magnitude of startle (M. Davis et al., 1993). FPS is the increase in startle response when the fear-associated stimulus (e.g., CS previously associated with shock) is present compared to when it is absent. When U0126 was infused bilaterally into the LA 30 mins before conditioning, FPS was significantly impaired when tested 24 hrs later compared to vehicle infusion (Di Benedetto, et al., 2009). Immunohistochemical double-labeling showed that conditioning enhanced pMAPK was exclusively found in neuronal cells and not glial cells. In both of these studies, reacquisition of conditioned fear after test was not affected, indicating that the infusion of MEK inhibitors did not permanently damage the amygdala, and any drug effects were specific to the initial fear memory consolidation. Lastly, a study showed that for cued-fear conditioning, MAPK phosphorylation is critical not only in the BLA, but also in thalamic inputs into the BLA (Apergis-Schoute et al., 2005). In that study, rats were given intra-thalamic infusions of U0126 either 30 mins before or immediately after cued-fear conditioning. At both infusion intervals, findings revealed that long-term memory to the CS (assessed at 24 h) was impaired, whereas short-term memory (assessed at 1-3 h) was intact. In additional experiments, it was shown that intra-thalamic infusion of U0126 before LTP-inducing stimulation of thalamic inputs to the LA significantly disrupted LTP in the LA. Taken together, it appears that MAPK activation in the BLA and its thalamic inputs is critical for cued-fear memory consolidation.

In contrast to the BLA, intra-hippocampal infusion studies provide mixed results. For example, bilateral infusion of the MEK inhibitors PD098059 and U0126 into the dorsal hippocampus 30 mins before contextual fear conditioning was shown to have no effects on STM (tested at 1 hr) and LTM (tested at 24 hrs) (Ahi et al., 2004). On the other hand, disruption of the biphasic increase in pERK following context conditioning (at 0-1 h and 9-12 h post-conditioning) via bilateral infusion of U0126 into the dorsal hippocampus led to an impairment in freezing to the context when tested 24 hours following conditioning. STM, assessed 1 h after conditioning, was unaffected (Trifilieff et al., 2006). Both of these studies were performed in mice. It may be the case that the critical phase of MAPK activation in the hippocampus following contextual fear conditioning is rather late (i.e., hours after conditioning) compared to MAPK activation in the BLA.

Despite the inconclusive literature on the necessity of hippocampal MAPK activation on contextual fear memory consolidation, findings using the inhibitory step-down avoidance task are much clearer. Inhibitory step-down avoidance is a hippocampus-dependent task (I. Izquierdo & Medina, 1993), and its training typically involves placing the rat onto a platform, and presenting a footshock when the rat steps down. The rat learns to avoid stepping down and exploring the box. The latency to step down is used as a measure for fear memory in subsequent tests. Izquierdo and his colleagues have shown that bilateral infusion of PD098059 into the dorsal hippocampus 180 mins following training significantly impaired avoidance performance when tested 24 hrs later (Walz, Roesler, Barros et al., 1999). This effect was replicated again in the hippocampus (Walz et al., 2000), as well as in other brain structures such as the amygdala, entorhinal and parietal cortices at different time intervals following conditioning (Walz, Roesler, Barros, et al., 1999; Walz, Roesler, Quevedo et al., 1999; Walz, et al., 2000), suggesting differences in time-specificity of critical MAPK activation in different cortical structures. The same researchers also examined hippocampal MAPK involvement in the expression of inhibitory learning shortly after training compared to long-term memory expression. Bilateral infusion of PD098059 into the dorsal hippocampus 10 mins prior to test significantly attenuated retrieval of avoidance memory when the test occurred 31 days after training, whereas STM test 3 hrs following test was unaffected (L. A. Izquierdo et al., 2000). This finding complements Kim et al. (2012) that showed retrieval of conditioned fear involves phosphorylation of MAPK. It appears that this phosphorylation of MAPK is not an indication of memory reconsolidation; rather, it is implicated in retrieval processes. Further studies are necessary to disambiguate this issue.

One study examined whether *increasing* ERK phosphorylation via an experimental manipulation affects fear memory consolidation. Giovannini et al. (2003) first showed that bath application of H2 and H3 histamine receptor agonists increased pERK in hippocampal brain slices. In a follow-up experiment, rats received bilateral hippocampal infusion of H2 or H3 receptor agonists amthamine/RAMH immediately after context-shock training. When the rats were tested 72 hours later, both H2 and H3 agonists improved rats' freezing to the trained context. Infusions of the MEK inhibitor U0126 blocked any affects of H2 and H3 agonists, further supporting the idea that any memory enhancement caused by H2/H3 agonists is due to ERK phosphorylation (Giovannini et al., 2003). Although these results do not show that MAPK phosphorylation in the hippocampus is necessary for hippocampal-dependent learning, they provide some evidence that MAPK phosphorylation in the hippocampus plays a critical role in fear memory consolidation.

Overall, the localized infusion studies described above provide a convincing case for the necessity of MAPK in fear conditioning, measured by various types of experimental paradigms. Those studies typically used MEK inhibitors that specifically inhibit MAP kinases p44MAPK (or ERK1) *and* p42MAPK (or ERK2) without distinguishing between the two. Therefore, research using genetic knockout (KO) or knockdown mouse models provided further clues as to which MAPK subtype is particularly important for fear learning. In the very first study using the genetic KO model, Sweatt and colleagues examined the p44 MAPK/ERK1 KO mice and observed no deficits in their hippocampal LTP or contextual- or cued-fear conditioning (Selcher et al., 2001). The authors concluded that the effects of MEK inhibition on fear learning must be due to the inhibition of ERK2 isoform of MAPK rather than ERK1. The same group of researchers then used mutant mice that displayed decreased ERK 1/2 phosphorylation in the hippocampus (Shalin et al., 2004). The mutant mice showed the same open field, rotarod, shock sensitivity behavior as the wildtypes, indicating that their motor coordination, general anxiety, and sensory perception were not different from controls. Further, within-session conditioning freezing levels did not differ between the mutant and the wildtype. However, the mutant mice displayed significant impairment in context-, but not cue-, elicited freezing when tested 24 hours later. Recently, a separate group reported that full ERK2 KO mice show almost a 100% prenatal mortality rate, thus used a specific ERK2 mutant mouse with partially reduced (20-40%) ERK2 (Satoh et al., 2007). Those ERK2 mutant mice showed the same ERK1 level, anatomy and synaptic density as wildtypes. There were also no differences in open-field, elevated plus maze, Y-maze, wheel-running or circadian activity. Further, no differences were observed in within-session freezing during fear conditioning between the mutant and wildtype, indicating no differences in shock sensitivity. At test (48 h) however, the mutant mice were impaired in both cued- and context-elicited freezing compared to wildtype, although there were no differences when measured 2 h after training (STM test). These results suggest that ERK2, but not ERK1, plays an essential role in long-term memory formation.

Taken together, MAPK cascade, especially the one involving the ERK2 isoform, appears essential in fear conditioning. There is also overwhelming evidence that MAPK activation following learning is critical for memory consolidation and LTM, as STM is not affected by MEK inhibition. The next section will review in detail the role of MAPK in an animal model of treating anxiety disorders, namely fear extinction.

3.2 The role of MAPK in extinction of conditioned fear

At present, the most effective treatment for anxiety disorders is exposure therapies, which are based on the process of extinction (Hofmann, 2007). Extinction refers to the decrease or suppression in responses expressed to a stimulus due to repeated exposure to that stimulus without any reinforcement. As with conditioning, Pavlov was the first person to document extinction in the laboratory (see Pavlov, 1927, for review). He observed that animals show a robust CR after CS-US pairings; however, the CR dramatically decreased if the CS was presented repeatedly without the US. *Fear* extinction hence refers to the decrease in fear conditioned responding due to repeated presentation of the CS without any aversive outcome.

In a typical extinction session, a CS is first paired with a shock, and repeated presentation of the CS in the absence of the shock leads to a reduction of conditioned fear both in magnitude and frequency. Early theoretical models of Pavlovian conditioning suggested that this decreased response to the CS after extinction was due to the 'unlearning' or 'erasure' of the original CS-US association (Rescorla & Wagner, 1972). That is, the association between the CS and US is severed so that the CS becomes neutral again. However, it is now widely believed that the CS-US association remains intact after extinction, and that extinction reflects a new learning (CS-no US) process that inhibits the original CS-elicited behavior. The primary evidence for this view comes from behavioral studies that show performance to an extinguished CS recovers without subsequent re-training of the CS-US association, namely in the cases of *reinstatement*, *renewal* and *spontaneous recovery* (Bouton, 2002). Reinstatement refers to the recovery of extinguished fear responses via a pre-test reminder US, usually a footshock. Renewal refers to the recovery of extinguished fear responses when the subject is tested outside the extinction context. Spontaneous recovery refers to the recovery of extinguished fear responses due to the passage of time following extinction training. These findings suggest that extinction is a new learning that is mediated by internal, physical and temporal contexts. Strong evidence for new learning also comes from pharmacological and molecular studies demonstrating that extinction shares similar neural mechanisms with acquisition of learning. Specifically, formation of a new long-term memory and extinction are both linked to a molecular cascade involving pre-synaptic glutamate release, NMDA-mediated calcium entry, activation of protein kinases, gene expression and transcription, and protein synthesis (Baker & Azorlosa, 1996; Lattal et al., 2006; C. H. Lin, et al., 2001). This molecular cascade has been found to be necessary for the consolidation of fear extinction memory in neural structures important for extinction. In recent years, local infusion studies, lesion studies, in *vivo* electrophysiology studies and immunohistochemical studies in specific neural structures strongly suggest that extinction of conditioned fear involves a circuit including the amygdala and mPFC (Herry & Garcia, 2002; Milad & Quirk, 2002; Morgan et al., 1993; Sotres-Bayon et al., 2009). The hippocampus has been also incorporated into this neural circuitry, to account for the context effects on extinction (Bouton et al., 2006; Corcoran & Maren, 2001; Myers & Davis, 2007; Sotres-Bayon et al., 2006).

Because extinction is an active learning process that involves many of the neurobiological substrates that subserve normal learning and memory, there is now accumulating evidence for the involvement of MAPK in extinction of conditioned fear. Indeed, mice exhibit significant increases in pMAPK-labelled neurons in the BLA 15, 60 and 360 mins following extinction of conditioned freezing responses compared to control groups that were not trained or did not receive any extinction (Herry et al., 2006). Additionally, increases in phosphorylation of ERK1 and ERK2 are also observed in the hippocampus 15 - 60 mins following extinction training compared to untrained mice (Fischer et al., 2007). These findings show that extinction triggers MAPK phosphorylation not unlike fear conditioning.

Interestingly, detecting the activation of MAPK has recently been used to dissociate the neural structures involved in extinction of conditioned fear in developing rats. Kim and colleagues have provided substantial evidence that fear extinction processes are dissociated across development, and have suggested that this is because some neural structures mature later than others (Kim & Richardson, 2010). To summarize, P24 rats display adult-like extinction in that they show renewal, reinstatement and spontaneous recovery of extinguished stimuli. However, pre-weanling aged rats (e.g., P17) do not show any of those behavioral phenomena. Pharmacological studies also show that reducing NMDA, γ-aminobutyric acid (GABA), and opioid neurotransmission impairs extinction in P24 rats, but extinction in P17 rats is only affected by the blocking of opioid neurotransmission. When phosphorylation of MAPK was detected using immunohistochemistry, P24 rats showed significantly increased pMAPK levels in the amygdala and mPFC compared to untrained, reactivated and context-exposure controls 1 hr following extinction (Kim et al., 2009). In contrast, extinction in P17 rats led to increases in pMAPK only in the amygdala, but not in the mPFC. The authors concluded that the differential involvement of the mPFC as shown by MAPK phosphorylation was at least partly responsible for the developmental dissociation of extinction-related phenomena. Kim and colleagues also showed that adolescent (P35) rats show impairments in long-term extinction of conditioned fear compared to pre-adolescent (P24) and adult (P70) rats, a finding that was accompanied by significantly lower levels of pMAPK in the infralimbic cortex (IL) of the mPFC in P35 rats compared to the other ages 1 hr following extinction training (Kim et al., 2011). Importantly, the long-term extinction deficit displayed by adolescent rats was overcome if adolescent rats were given double the amount of extinction training, and as a consequence pMAPK levels were also increased, supporting the notion that MAPK is directly involved in the consolidation of extinction memory.

The first study that provided evidence for the necessity of MAPK in fear extinction used the MEK inhibitor PD98059 for localized infusions (K. T. Lu et al., 2001). In that study, PD98059 was infused bilaterally into the BLA or the hippocampus 10 mins before extinction training. When rats were tested 24 hours later, results showed that BLA, but not hippocampal, infusion significantly impaired maintaining the extinction memory. This finding was replicated using the MEK inhibitor U0126 (C. H. Lin, Yeh, Lu et al., 2003). In those studies FPS was used as a measure for conditioned fear response, which typically does not allow the assessment of within-session extinction fear levels due to the obnoxious level of the startle pulse. Therefore, it is hard to conclude whether the infusion of MEK inhibitors disrupted fear retrieval and/or extinction acquisition. Using freezing as an index for fear, Herry et al. (2006) demonstrated in mice that pre-extinction bilateral infusions of U0126 into the BLA completely blocked within-session extinction acquisition compared to vehicle infusion. Further, a high level of CS-elicited freezing was maintained at test the next day in the drug infusion group. This finding suggests that MAPK is important for the acquisition of extinction, rather than its consolidation, which is contrary to the role of MAPK in conditioned fear. It may be the case that the role of MAPK phosphorylation in the BLA is dissociated across conditioning and extinction.

In fact, studies indicate that the mPFC, rather than the BLA, is critical for the consolidation of fear extinction. For example, immediate, but not late (2 or 4 h), post-extinction infusion of PD098059 bilaterally in the mPFC resulted in a full return of conditioned freezing compared to vehicle infusions when rats were tested for their long-term extinction (Hugues et al., 2004). In a follow-up study, Hugues et al. (2006) first showed that hippocampal inputs to the mPFC display LTP-like changes following fear extinction compared to before extinction training. Further, LTP-like changes were blocked by PD098059 infusion into the mPFC. Additionally, Western blot revealed decreases in pERK in the mPFC following PD098059 infusions. These findings imply that MAPK activation in the mPFC is important for LTP-like changes in the mPFC following extinction, which may be the underlying mechanism for extinction memory consolidation (Hugues et al., 2006).

Hippocampal MAPK involvement in fear extinction is still equivocal. For example, Fischer et al. (2007) reported that immediate, post-extinction infusion of PD098059 or U0126 both impaired long-term extinction of contextual conditioned fear. On the other hand, Izquierdo and colleagues found that bilateral infusions of PD098959 prior to extinction only affected inhibitory avoidance expression during extinction, without impairing any long-term extinction (Szapiro et al., 2003). However, when they infused the p38MAPK inhibitor, SB203580, immediately after extinction training of inhibitory avoidance behavior, rats failed to show any extinction memory the next day (Rossato et al., 2006). This effect was temporally-specific, so that infusion 3 hours following extinction had no effects. The authors concluded that unlike inhibitory avoidance conditioning, extinction of inhibitory avoidance behavior is not dependent on p44MAPK and p42MAPK activation in the hippocampus.

Overall, the studies described above provide further support for the idea that fear extinction involves a new learning that relies on similar molecular cascades as fear learning. This premise led to the discovery that increasing the neurotransmission involved in memory consolidation can enhance and facilitate the consolidation of extinguished fear memory. Specifically, systemic, as well as intra-BLA, injection of the NMDA partial agonist D-cycloserine (DCS) facilitates extinction (Walker et al., 2002). It appears that increasing NMDA neurotransmission during extinction leads to a stronger extinction memory when tested the next day. Follow-up studies clarified that drugs acting on NMDA receptors are most likely to be affecting the consolidation of extinction (Ledgerwood et al., 2003; Santini et al., 2001). Using freezing as an index of fear, Ledgerwood et al. (2003) demonstrated that intra-BLA infusions of DCS had no effect on within-session extinction of conditioned freezing compared to vehicle, whilst facilitating extinction at test. Interestingly, this facilitation effect of DCS on fear extinction appears to be due, at least partly, to increases in MAPK phosphorylation (Matsuda et al., 2010; Yang & Lu, 2005). In Yang and Lu (2005), systemic DCS injection 15 mins prior to extinction training significantly facilitated extinction the next day compared to rats injected with saline. Intra-BLA infusion of MEK inhibitors, PD098059 or U0126, however, abolished any DCS effects, when infused 5 mins following the DCS injection. Western blot revealed that compared to the context-exposure control group, extinction training significantly elevated pMAPK in the BLA. Further, administration of DCS enhanced the effect of extinction training on MAPK phosphorylation, which in turn was blocked by MEK inhibitors. These results strongly suggest that the DCS effect on extinction of conditioned fear is mediated by the MAPK-dependent signalling cascades in the amygdala. This general finding has been replicated in mice (Matsuda, et al., 2010).

Taken together, these studies provide compelling evidence for the critical role of MAPK in conditioned fear, especially cued-conditioning. There is accumulating evidence indicating that fear extinction involves activation and / or plasticity of inhibitory circuits in the amygdala (Chhatwal et al., 2005; Marsicano et al., 2002; Rosenkranz et al., 2003; Royer & Pare, 2002). It may be the case that MAPK activation acts directly on the molecular processes underlying synaptic plasticity and dendritic integration of neurons in the inhibitory circuit. For example, acquisition of extinction and its consolidation are protein synthesis-dependent (Berman & Dudai, 2001; C. H. Lin, Lee et al., 2003; Suzuki et al., 2004). Further, recent studies implicate the MAPK pathway in controlling rapid local translation during the stabilization of newly formed hippocampal memories, which is consistent with a role for MAPK / ERK-mediated control of local protein synthesis during the acquisition of fear extinction (Kelleher et al., 2004; Thomas & Huganir, 2004). Importantly, extinction induces up-regulation of the protein phosphatase calcineurin in the BLA, which depends on *de novo* protein synthesis (C. H. Lin, Yeh, Lu, et al., 2003). In fact, inhibition of calcineurin interferes with extinction (C. H. Lin, Yeh, Leu et al., 2003). Those observations indicate that acquisition of extinction may involve, at least in part, a MAPK-dependent control of mRNA translation in the BLA. Additionally, MAPK has been implicated in other forms of synaptic plasticity including long-term depression (LTD) (G. M. Thomas & Huganir, 2004). Therefore, it may be the case that MAPK is necessary not only for the maintenance of CS-no US memory in the extinction neural circuitry, but also for the reversal (that is, depotentiation) of fear conditioning-induced LTP (H. C. Lin et al., 2010; S. C. Mao et al., 2006; Rosenkranz & Grace, 2002; Rumpel et al., 2005). In either case, pharmacological modulation of MAPK activation could potentially provide novel therapeutic strategies for the treatment for anxiety disorders, assuming a lack of off-target effects following long-term systemic treatment.

4. MAPK and drug addiction

Drug addiction is more than mere drug use, rather it is a chronic, relapsing disorder that consists of a compulsive pattern of drug-seeking and drug-taking behavior that takes place at the expense of other activities. A key question in addiction research is identifying what changes in the brain underpin the transition from casual to compulsive drug use. Over the last 20 years much research has accumulated describing the various long-lasting cellular, molecular and neurochemical adaptations which occur with repeated drug administration (Kalivas, 2009; Kalivas & O'Brien, 2008; Robinson & Berridge, 1993; M. J. Thomas et al., 2008). Furthermore, there is a compelling body of evidence linking various models of addiction with synaptic plasticity in brain areas involved in reward and reinforcement (Kauer & Malenka, 2007; Kelley, 2004). Drug-induced neuroadaptations are thought to be critical in the transition to addiction, though it remains to be fully elucidated how, and to what extent, these changes contribute to prolonged relapse vulnerability after cessation of drug use (Shaham & Hope, 2005).

A central problem facing the treatment of drug addiction is the enduring vulnerability to relapse displayed by users despite months or even years of abstinence (Dackis & O'Brien, 2001; Wagner & Anthony, 2002). Even with successful detoxification, abstinence remains an unobtainable goal for many addicted individuals with up to 90% relapsing within 12 months (DeJong, 1994; Deroche-Gamonet et al., 2004; Gossop et al., 1989). Drug craving, a subjective affective state experienced by humans, which motivates them to seek out drugs, is associated with relapse propensity (Markou et al., 1993; C. P. O'Brien, 1997a). Craving can be induced in addicted individuals by exposure to drug-related paraphernalia, images or environmental contexts (A. Childress et al., 1988; Grant et al., 1996). Repeated exposure to such cues and contexts during the initiation and maintenance of drug use is therefore thought to result in these cues acquiring incentive motivational and conditioned reinforcing value (C. O'Brien et al., 1992). Once formed, these pathological associations may ultimately contribute to the precipitation of craving and relapse upon re-exposure to drug-associated stimuli. In addition to drug-associated cues, relapse can be triggered by other factors such as stress and exposure to the drug itself (S. A. Brown et al., 1995; C. O'Brien, et al., 1992).

Research into potential treatment strategies for addiction, therefore, primarily focuses on relapse prevention. Animal models of relapse have yielded invaluable information regarding the neurobiology of drug-seeking and the persistent neuroadaptations which likely underpin this behavior, revealing a number of key molecules that appear to play a critical role. ERK is one of these molecules and the following sections in this chapter will cover the evidence which supports a role for ERK in mediating relapse to drug-seeking and the drug-induced changes which potentially underlie this behavior.

4.1 Addiction neurocircuitry

Despite their varying pharmacological profiles and properties, drugs of abuse all share the common feature of acutely enhancing neurotransmission in the mesocorticolimbic dopamine system. This interaction can be direct, as in the case of psychostimulants, or indirect, as in the case of opioids, but each ultimately result in increased levels of extracellular dopamine in the terminal fields of these neurons and this increase in synaptic dopamine is thought to be responsible for their reinforcing properties (Di Chiara & Imperato, 1988; Wise, 1996). The mesocorticolimbic system is comprised of dopaminergic neurons, the cell bodies of which originate in the ventral tegmental area (VTA) and project to the nucleus accumbens (NAc) and other limbic regions such as the hippocampus and amygdaloid nuclei. A further subdivision of these dopaminergic neurons innervates cortical areas including the mPFC, anterior cingulate and orbitofrontal cortices (OFC) (Le Moal, 1991). Different terminal regions are thought to be involved in the different aspects of addictive behavior. The NAc is thought to be involved in the mediation of the acute reinforcing effects of drugs (Di Chiara, 2002; Volkow et al., 2003), whereas changes in the PFC, OFC and anterior cingulate are related to the decreases in inhibitory control and hyper-responsiveness to drug-related cues (Kalivas & Volkow, 2005). The amygdaloid nuclei and hippocampus play an important role in conditioning and learning associations between drugs and drug-related stimuli (Fuchs et al., 2007; Fuchs et al., 2005; Rogers & See, 2007; See, 2005).

In addition to dopaminergic input, the striatum receives excitatory glutamatergic input from the BLA, the hippocampus, the mPFC, the OFC, cingulate cortex and midline thalamic nuclei (Alexander et al., 1986). Dopaminergic and glutamatergic inputs converge on the dendritic spines of GABAergic striatal efferent neurons (Sesack & Pickel, 1992; Totterdell & Smith, 1989). This allows dopamine to play a modulatory role in regulating the excitatory input from glutamate to the dendritic spines (A. D. Smith & Bolam, 1990). While dopamine is thought to be responsible for the acute effects of addictive drugs, persistent vulnerability to relapse is believed to be a result of enduring neuroadaptations in corticostriatal circuitry (Kalivas, 2009). The VTA also receives glutamatergic input from structures such as the PFC (Omelchenko & Sesack, 2007). Excitatory inputs from the PFC play an important role in regulating the activity of VTA neurons and therefore the extracellular levels of dopamine within forebrain regions (Carr & Sesack, 2000).

4.2 The role of MAPK in responses to drug-associated cues and contexts

A process of classical conditioning occurs with repeated drug use: the multiple cues and contexts associated with drug use (previously neutral stimuli) become conditioned stimuli (CS) through repeated pairings with the drug (the unconditioned stimulus, US). As this learning about drug-associated stimuli develops, the cues gain increasingly greater control over behavior and drug-associated cues can acquire the capacity to evoke salient and powerful memories of the drug-taking experience that induce craving and ultimately relapse (Jentsch & Taylor, 1999). In animals these cues have been shown to increase operant drug self-administration and precipitate relapse-like behavior (Aguilar et al., 2009; Crombag et al., 2008; Shaham et al., 2003). In the conditioned place preference (CPP) paradigm, a variety of species (even crayfish) will preferentially choose to spend more time in an environment where they have received a non-contingent injection of drug over an environment where they have received vehicle (Nathaniel et al., 2009; Tzschentke, 1998). Animal models of relapse have also demonstrated that the motivational properties of drug-associated stimuli cues can augment or "incubate" over time (Grimm et al., 2001), thereby highlighting the need to acknowledge the role of drug-associated cues when developing therapeutic interventions.

CPP is based on classical conditioning principles whereby animals are conditioned to associate a distinct environment (i.e., CS) with a drug experience (i.e., US) (Tzschentke, 1998). A place preference describes a situation where the subject spends more time in a drug-paired environment when given the choice after a conditioning period. As researchers, we make an inference regarding the rewarding or hedonic properties of a drug based on its ability to induce approach (Bardo & Bevins, 2000). Though conditioned approach towards specific stimuli, especially drugs of abuse, obviously occurs in humans, this phenomenon has never been clinically validated (Bardo, et al., 2000). As with operant self-administration, in the reinstatement version of this paradigm the association between the CS and US can be extinguished and then reinstated either by non-contingent 'priming' injections of the drug (Parker & McDonald, 2000) or stress (L. Lu et al., 2002). In the case of CPP however, there is no reinstatement of an instrumental task, rather the reinstatement of conditioned approach to the CS (the environment in which the appetitive stimulus was experienced).

Activation of ERK has been shown to play a role in the associative learning processes that occur upon exposure to an environment previously paired with a drug experience. Thus, ERK activation has been shown to be required for the acquisition of CPP induced by cocaine (Valjent et al., 2000), MDMA (Salzmann et al., 2003), THC (Valjent et al., 2001), amphetamines (Gerdjikov et al., 2004; Mizoguchi et al., 2004) and morphine (Ozaki et al., 2004), but not ethanol (Groblewski, Ryabinin et al., 2011). It is most likely that the NAc is a site of action, at least for amphetamines, as intra-NAc injections of the MEK inhibitor PD98059 impair both methamphetamine and amphetamine-induced CPP (Gerdjikov, et al., 2004; Mizoguchi, et al., 2004). This is supported by the observation of increased activation of ERK in the NAc and dorsal striatum of mice that have been conditioned to methamphetamine (Mizoguchi, et al., 2004) and the NAc, dorsal striatum and prefrontal cortex of mice that have been conditioned to cocaine (Valjent, Corbille et al., 2006). Another study found that exposing mice to a previously cocaine-paired context increased phosphorylation of ERK in the dorsal hippocampus, but not the NAc. Enhanced CREB and DARPP-32 phosphorylation in the NAc, however, was observed (Tropea et al., 2008). Pre-treatment with the MEK inhibitor SL327 has also been shown to abolish the conditioned locomotor response of mice placed in a context previously paired with cocaine or D-amphetamine (Valjent, Corvol et al., 2006), indicating the role of ERK can be generalised to other contextually-driven behaviors.

It also appears that the NAc mediates consolidation of learned associations between a drug's rewarding effects (US) and the drug-paired context (CS). If rats are injected with a MEK inhibitor either systemically or into the NAc core after the test for the expression of cocaine or morphine CPP, subsequent CPP is abolished (Miller & Marshall, 2005; Valjent, Corbille, et al., 2006). Erasure of CPP expression is accompanied with blockade of ERK phosphorylation in the NAc and dorsal striatum (Valjent, Corbille, et al., 2006). This effect can be mirrored with a protein synthesis inhibitor thus providing evidence that ERK exerts this effect on memory reconsolidation through regulation of protein synthesis, presumably by controlling transcription.

4.3 The involvement of MAPK in the actions of drugs of abuse & drug-induced plasticity

The MAPK signalling cascade has been implicated in responses to most drugs of abuse, supporting the notion that drug-induced plasticity might share common mechanisms with learning and memory processes (Berke & Hyman, 2000; Girault et al., 2007; Hyman, 2005). ERK is expressed abundantly throughout the brain, including mesocorticolimbic regions (Ortiz et al., 1995). Acute and chronic administration of most drugs of abuse such as cocaine, morphine, nicotine and Δ^9-tetrahydrocannabinol (THC) has been shown to increase activation of ERK in reward-related areas such as the VTA, NAc, PFC, central and extended amygdala and hippocampus (reviewed in Girault et al., 2007; Zhai et al., 2008). The exception is alcohol, where data are equivocal. Acute administration has actually been shown to decrease phosphorylation of ERK in the cortex and hippocampus (Chandler & Sutton, 2005) but increase phosphorylation in other areas (Bachtell et al., 2002). Chronic ethanol intake has been shown to attenuate ERK phosphorylation in the striatum, whereas withdrawal for one day potentiates ERK phosphorylation in the same region (Cui et al., 2011).

The activation of ERK in response to repeated administration of drugs of abuse suggests that ERK plays a role in drug-induced plasticity. One measure of drug-induced plasticity is the development and expression of behavioral sensitization (an enhancement of the psychomotor-activating properties of a drug). Indeed, activation of ERK in the brain mesocorticolimbic area has been shown to be involved in sensitization and hence drug-induced plasticity (Pierce et al., 1999; X. Shi & McGinty, 2006; Valjent, et al., 2000; Valjent, Corvol, et al., 2006). Thus, pre-treatment with the MEK inhibitor SL327 prior to the repeated administration of cocaine or amphetamine prevents the subsequent expression of sensitization to their locomotor-activating effects upon receiving a challenge dose of the drug (Valjent, Corvol, et al., 2006). In contrast, the expression of sensitization was not affected when SL327 was administered before the challenge (Valjent, Corvol, et al., 2006). Therefore it appears ERK is more critical for the development of sensitization to psychostimulants. ERK activation in the VTA is implicated in this process as intra-VTA infusions of PD98059 attenuate the development of sensitization to cocaine (Pierce, et al., 1999).

ERK in the NAc also appears to be involved in drug-induced plasticity as cocaine sensitized rats display increased phospho-ERK2 during withdrawal from cocaine which is normalized upon cocaine challenge (Boudreau et al., 2007). ERK2 phosphorylation corresponded with accumbens AMPA receptor surface expression (Boudreau, et al., 2007). The authors suggest the possibility that ERK contributes to both rapid (LTP) and slowly induced (cocaine withdrawal) increases in AMPA receptor surface expression, thus contributing to addiction-related plasticity (Boudreau, et al., 2007). Similarly, withdrawal following development of sensitization to cocaine has been shown to be associated with time-dependent increases in ERK activity which correlate with the increased expression of NMDAR subunits (Schumann & Yaka, 2009). Furthermore, the increase in both GluR1 expression and ERK activity is blocked with intra-NAc infusions of a MEK inhibitor 21 d after withdrawal from cocaine. These data suggest that the development of cocaine psychomotor sensitization triggers a delayed increase in the expression of NMDAR subunits in the NAc, which in turn enhances the activity of ERK (Schumann, et al., 2009).

4.4 Drugs of abuse and mechanisms of ERK signalling

The activation of ERK by drugs of abuse in mesocorticolimbic regions has been shown to require both glutamate and dopamine signalling as it is prevented by both dopamine D1 and NMDA receptor antagonists (Berhow et al., 1996; Jenab et al., 2005; Valjent, et al., 2000; Valjent, et al., 2001; Valjent et al., 2005), and in the case of amphetamine, by group I metabotropic glutamate (mGlu) receptor antagonists (Choe et al., 2002). This is not surprising as ERK is found downstream of all these receptors. Thus, the activation of dopamine D1 receptors by dopamine increases cyclic adenosine monophosphate (cAMP) and thus cAMP-dependent protein kinase A (Bilbao et al. 2008). Activated cAMP-dependent protein kinase (Bilbao, et al. 2008) then activates ERK through the the dopamine- and cAMP-regulated phosphoprotein 32 kDa (DARPP-32) signaling cascade (Greengard et al., 1999; Svenningsson et al., 2004). In striatal neurons, the mGlu5 receptor activates ERK through Ca2+-dependent and -independent mechanisms, the latter involving the scaffolding protein Homer (L. Mao et al., 2005). Activation of NMDA receptors and the resulting Ca^{2+} influx activate Ca^{2+}/calmodulin-dependent protein kinase II (CaMKII). CaMKII then activates Ras-GTP, which initiates a number of kinase cascades, including Raf-1, MAPK/ERK kinases and ERK (G. M. Thomas, et al., 2004; Xia et al., 1996). ERK is also activated by opioid receptors most likely via a mechanism involving Ras and beta-gamma subunits of Gi/o proteins (Belcheva et al., 1998). Once activated, ERK translocates to the cell nucleus where it regulates the activity of the transcription factors CREB and Elk-1 which control the transcription of many genes implicated in addiction, such as c-fos, ΔFosB, zif268, Homer and preprodynorphin (Benavides & Bibb, 2004; Choe & McGinty, 2001; S. Davis, Vanhoutte et al., 2000; Girault, et al., 2007; Hurd & Herkenham, 1993; L. Lu, et al., 2006; McClung & Nestler, 2003; Nestler, 2001; Radwanska et al., 2005). Downstream effects of many drugs of abuse on these genes have been shown to be mediated via ERK (Bachtell, et al., 2002; Salzmann, et al., 2003; Valjent, et al., 2000).

Evidence has accumulated which supports a coincident interaction between D1 and NMDA receptors with respect to cocaine-driven ERK signaling in the striatum. Thus, cocaine-induced ERK activation has been shown to require the coincident stimulation of D1 and NMDA receptors (Valjent, et al., 2000; Valjent, et al., 2005) and this appears to be controlled by Ras-guanine nucleotide-releasing factor 1 (Ras-GRF1), a neuronal specific activator of Ras-ERK signaling (Fasano et al., 2009). The relevance of this Ras-GRF1-mediated ERK signaling to reward and drug-induced plasticity is demonstrated by the observation that CPP and locomotor sensitization are significantly attenuated in Ras-GRF1–deficient mice, and conversely heightened in over-expressing transgenic animals (Fasano, et al., 2009).

More recently, the D1 or cocaine-mediated activation of ERK in striatal neurons was shown to depend on a potentiation of NMDA receptor function independent of cAMP (Pascoli et al., 2011). Furthermore, NMDA receptor potentiation involves tyrosine phosphorylation of the NR2B subunit via Src-family kinase (SFKs). This mechanism, demonstrated in cultured striatal neurons, was also observed *in vivo*, as ERK phosphorylation in response to D1 receptor agonism in mice was blocked by pretreatment with an NMDA receptor antagonist (Pascoli, et al., 2011). The observed potentiation by the D1 receptor potentially explains why NMDA receptor stimulation is required for ERK activation by cocaine and other drugs of abuse. As the authors suggest this mechanism has potentially significant functional consequences, providing a possible mechanism for facilitation of long-term potentiation of glutamatergic synapses in D1 receptor-expressing striatal neurons (Andre et al., 2010; Shen et al., 2008). This was evidenced by the attenuation of cocaine-induced sensitization and CPP upon inhibition of this D1R/Src family kinases/NR2B pathway (Pascoli, et al., 2011).

4.5 MAPK in reward learning and synaptic plasticity

As discussed earlier in this chapter, MAPK has been shown to be intimately involved in synaptic plasticity and hence learning and memory processes (for detailed review see G. M. Thomas, et al., 2004). As with fear-related disorders, drug addiction has been proposed by some to be a form of pathological learning (Berke, et al., 2000). Reinforcement as a concept is ultimately defined by learning processes - it is the strengthening of associations through experience. Reinforcement can also be viewed as a process that enhances memory consolidation for the association (Landauer, 1969). Either way, it is the "stamping in" (and hence learning) of associations that ultimately leads to reinforcement. The associations may manifest in a variety of ways: associations between a stimulus and a response (Thorndike, 1898), associations between a US and CS (I.P. Pavlov, 1927), or associations between a response and an outcome (Skinner, 1938). With repeated exposure drugs cause aberrant "stamping in" of associations between drug-taking, the resultant drug high and associated cues and contexts.

The striatum is a key structure which mediates instrumental (action-outcome) learning and it is glutamate-dopamine signaling which appears particularly important (Horvitz, 2002). Dopamine release in the dorsal striatum is thought to enable coding of action-outcome associations facilitating potentiation of corticostriatal synapses during learning (Di Filippo et al., 2009; Reynolds et al., 2001). The induction of corticostriatal LTP is dependent on the coincident activation of D1 and NMDA receptors (Calabresi et al., 1992; Kerr & Wickens, 2001; Reynolds, et al., 2001). Thus, corticostriatal LTP has been shown to require activation of ERK (Ferguson et al., 2006; Mazzucchelli et al., 2002; Sgambato et al., 1998) and indeed, the requirement for coincident activation of NMDA and D1 receptors for ERK activation has led some to propose that ERK functions to selectively induce plasticity at corticostriatal synapses during highly salient learning situations like reward learning, when coordinated dopamine and NMDA receptor activation is likely to occur (Girault, et al., 2007; Shiflett & Balleine, 2011).

The involvement of ERK in instrumental learning and corticostriatal LTP (Shiflett et al., 2010; G. M. Thomas, et al., 2004) suggests that ERK may play a role in the action-outcome learning which occurs with initial drug use. Indeed, a recent study showed that clinically relevant concentrations of ethanol dose-dependently attenuated the high frequency stimulation-induced LTP and ERK activation in the dorsomedial striatum (Xie et al., 2009). Relatively little is understood regarding synaptic plasticity in habit formation, but it may involve corticostriatal LTD (Shiflett, et al., 2011). The induction of LTD in the striatum was recently shown to require ERK activation (Cui, et al., 2011) which matches with findings regarding LTD in other brain regions such as hippocampus (Thiels et al., 2002). This recent study also showed that chronic ethanol intake by rats attenuated both ERK phosphorylation and LTD induction, whereas withdrawal for one day potentiated ERK phosphorylation and LTD induction (Cui, et al., 2011). By showing that chronic ethanol intake and withdrawal differentially altered synaptic plasticity in the dorsolateral striatum through the ERK signaling pathway, these findings suggest that chronic alcohol use may impact on the habit-formation process, and this neural maladaptation may consequently lead to habitual drug-seeking behavior (Cui, et al., 2011).

Persistent vulnerability to relapse in addiction is thought to be a result of a specific impairment in LTD at cortico-accumbal synapses (Kasanetz et al. 2010; Martin et al., 2006; Moussawi et al., 2009). In addition to LTD, cortico-accumbal LTP is abolished *in vivo* after withdrawal from cocaine self-administration (Knackstedt et al., 2010; Moussawi, et al., 2009) The loss of both LTP and LTD *in vivo* after cocaine self-administration has led to the hypothesis that the inability of the medium spiny neurons in the NAc core to undergo bidirectional prefrontal-induced neuroplasticity may reduce the capacity of drug addicts to adaptively organize a behavioral response that incorporates environmental contingencies that mediate against relapse (Kalivas, 2009; M. J. Thomas, et al., 2008). The following section will describe the literature supporting a role for ERK in relapse to drug-seeking behavior.

4.6 The involvement of MAPK in drug-seeking behavior

The operant self-administration procedure is a widely accepted model to study drug reinforcement and drug-seeking because it has both face and construct validity (Feltenstein & See, 2008). In the operant paradigm the animal is trained to perform an instrumental task to self-administer a drug (generally by pressing a lever or via a nose-poke) for a period of time, thereby developing an action-outcome association between the operant task and the ensuing reward (conditioned response). Rodents will readily self-administer most drugs of abuse, including cocaine (R. M. Brown & Lawrence, 2009; McPherson et al., 2010). Accordingly, the abuse potential of a substance in humans can be predicted from self-administration in rodents (Collins et al., 1984). Moreover, similar to the laboratory model of anxiety disorders, the self-administration paradigm can incorporate a Pavlovian conditioning component by introducing CSs that signal drug delivery to represent cues/contexts that are critically related to relapse of drug-seeking (Adams et al., 2010; Jupp et al., 2011).

4.6.1 The role of MAPK in the reinstatement model of drug-seeking

As with extinction of fear, the reinforced response (e.g. lever press) can be subsequently extinguished whereby no drug is administered despite the animal responding on the drug-paired manipulandum. Responding will typically spike in what has been termed an 'extinction burst' and then gradually decline to a level whereby the behavior is deemed extinguished. Extinction learning is an active process that reduces the value or salience of these conditioned cues and contexts. Longer and repeated cue/contextual re-exposure without the associated fear or drug experience reduces conditioned responding. Extinction learning can be effective for reducing cue- and context-induced symptoms in addiction and anxiety and can improve outcomes in these disorders. Reinstatement of drug-seeking behavior as measured by responding on the drug-paired manipulandum can be precipitated in a number of ways: with non-contingent 'priming' injections of the drug (Stretch et al., 1971), stressors (Carroll, 1985; Shaham & Stewart, 1996), and drug-associated cues (W. M. Davis & Smith, 1976). In humans, drug craving and relapse have shown to be triggered by similar factors (S. A. Brown, et al., 1995; C. O'Brien, et al., 1992), giving this model an element of construct validity. This model also exhibits predictive validity, as drugs such as naltrexone that decrease relapse in abstinent drug users also decrease reinstatement of drug seeking behavior in rats (Comer et al., 2006)

As mentioned previously, drug-associated cues induce ERK phosphorylation in components of the mesocorticolimbic system (L. Lu et al., 2005; Miller, et al., 2005) and ERK has been shown to play a role in drug-induced and synaptic plasticity (G. M. Thomas, et al., 2004; Valjent, Corvol, et al., 2006) thus it is not surprising that evidence exist that support a role for ERK in cue-elicited reinstatement of drug-seeking. Cue-induced reinstatement of alcohol-seeking is associated with enhanced activation of ERK in the BLA and NAc shell (Schroeder et al., 2008). The mGlu5 receptor antagonist 2-Methyl-6-(phenylethynyl)pyridine (MPEP) was shown to block both the cue-induced drug-seeking and the associated increase in phospho-ERK (Schroeder, et al., 2008), indicating that ERK activation is dependent on mGlu5 receptor signalling and plays a key role in the effect of MPEP on drug-seeking.

A recent study demonstrated a critical role for ERK in the ability of BNDF infusions in the dmPFC to attenuate reinstatement of both cue-induced and primed cocaine-seeking, as well as cocaine-seeking following abstinence (Whitfield et al., 2011). Vehicle-infused rats showed significant decreases in phospho-ERK and CREB at the end of the cocaine self-administration period that were reversed by intra-dmPFC BDNF. This ability of BDNF to reverse cocaine-mediated decreases in ERK and CREB phosphorylation was blocked by both the MEK inhibitor U0126 and the tyrosine kinase inhibitor K252a, thus demonstrating the behavioral effects of BDNF were most likely mediated via a TrkB receptor-ERK pathway. This study elucidates a mechanism whereby BDNF/TrkB activates ERK-regulated CREB phosphorylation in the dmPFC to counteract the neuroadaptations induced by cocaine self-administration and subsequent relapse to cocaine-seeking (Whitfield, et al., 2011).

There are no studies as yet which have specifically examined the role of ERK in extinction learning related to operant responding. There is one CPP study that showed heightened ERK activation in the amygdala following extinction of cocaine-induced CPP (Chen & Xu, 2010), which is consistent with fear extinction literature showing a key role for the amygdala (Myers, et al., 2007). In addition, dopamine D3 receptor mutant mice showed delayed CPP extinction compared with wild-type mice and this was associated with sustained ERK activation in the NAc and PFC (Chen, et al., 2010). This suggests that extinction of cocaine conditioning involves dopamine D3 receptor/ERK signaling though this finding is not necessarily readily generalized to other drugs of abuse as the MEK inhibitor SL327 has been shown to have no effect on extinction of an ethanol-induced CPP in mice (Groblewski, Franken et al., 2011).

4.6.2 The role of MAPK in the abstinence model of relapse and incubation of craving

Of the millions of people suffering from addiction, only a small proportion actually seek treatment in a formal setting. For example, in 2009 out of 23.5 million Americans who were considered to have an illicit drug or alcohol use problem, only 2.6 million (11.2%) received treatment at a specialty facility (SAMHSA, 2010). In this sense then, the reinstatement model of relapse is only relevant to those addicts who engage in new learning to overcome their addiction. The bulk of drug dependent individuals will attempt to overcome their addiction by simply attempting to abstain from drug use (also known as going 'cold turkey'). Therefore another relevant model of relapse is one that measures drug-seeking following a period of abstinence.

In 1986 it was suggested that in humans, craving induced by drug cues actually increases over the first several weeks of the withdrawal period and remains elevated over extended periods of abstinence (Gawin & Kleber, 1986). Scientists have since discovered an analogous phenomenon in laboratory animals which has been termed 'incubation of craving'. It has been demonstrated with opiates (Shalev et al., 2001), psychostimulants (Grimm, et al., 2001; Shepard et al., 2004) and ethanol (Bienkowski et al., 2004), that cue-induced drug-seeking increases over time over the first months of withdrawal. This phenomenon has also been observed with the natural reinforcer sucrose (Grimm et al., 2002), though in this case the effect peaks after 1–4 weeks, but decays after 2 months, while for cocaine, the responding peaks after 1 month and persists for up to 3 months in rats (L. Lu et al., 2004). In humans, this incubation of drug craving ultimately results in an increased propensity to relapse as a function of time spent in withdrawal. A more appropriate paradigm for this sort of relapse behavior is a modification of the reinstatement model known as the abstinence model (e.g. Brown et al., 2009; Lu et al., 2009; reviewed in Reichel and Bevins, 2009). After a period of self-administration the animal undergoes withdrawal for a period of time (typically a number of weeks) in the home cage. At the end of this period drug-seeking behavior is measured by re-introducing the subject to the operant chamber and measuring responding under extinction conditions (i.e. in the absence of drug).

Numerous cellular and molecular changes are produced by chronic drug use, which are thought to regulate drug-seeking behavior, yet only a small number thus far have been shown to be associated with time-dependent increases in drug-seeking behavior. One of these neuroadaptations is activation of ERK. Studies by Lu and colleagues have revealed that it is ERK phosphorylation in the CeA, but not BLA, that underpins incubation of cocaine and opiate craving as inhibition of CeA, but not BLA, ERK phosphorylation decreases drug-seeking after 30 days of drug withdrawal but not after 1 day of withdrawal (Li et al., 2008; L. Lu, et al., 2005). After 30 withdrawal days, inhibition of CeA ERK phosphorylation by U0126 or the NMDA receptor antagonist AP-5 decreased cue-induced cocaine seeking.

Subsequently, other studies have shown a similar enhancement of ERK phosphorylation in the NAc core (Edwards et al., 2011) and mPFC (Koya et al., 2008) as a result of exposure to a cocaine-paired context after extended abstinence in animals that had previously self-administered cocaine. In the former study, the activation of ERK was found only in rats that had self-administered cocaine and not their yoked controls, demonstrating the effect was not merely the result of cocaine exposure (Edwards, et al., 2011). Though the precise role of increased ERK activity in the NAc and mPFC in incubation of craving remains to be investigated, a key role for ERK in the amygdala has been found in Pavlovian fear conditioning and memory reconsolidation (Nader et al., 2000). ERK in the amygdala also appears to be critically involved in the long-term neuroadaptations which result in heightened reactivity to drug-associated cues. This is supported by the observation that re-exposure to an ethanol-associated context and discrete cues following abstinence increases activation of ERK in the BLA and ventral lateral amygdala (Radwanska et al., 2008). Collectively, these findings indicate that activation of ERK, particularly in the amygdala appears to be a key substrate for incubation of drug craving.

5. Concluding remarks

The studies highlighted in the present chapter clearly demonstrate the learning and memory principles underlying fear and drug-seeking behavior. Additionally, drug addiction has been emphasized as a brain disorder that involves enduring neuroadaptations that alter signal transduction within neurons as well as electrical and chemical communications between neurons and neural structures. Considering the significance of the MAPK cascade in cellular growth, differentiation and neuronal survivability (Ahn & Krebs, 1990; Hoshi et al., 1988; Impey et al., 1999; Kornhauser & Greenberg, 1997), perhaps it is not surprising that MAPK signalling also plays a seemingly important role in both the acquisition and treatment of anxiety and drug-seeking behavior as pointed out in the present chapter. Given the prevalence and financial burden of anxiety disorders and drug addiction, we propose that further research into the MAPK signalling cascade could be a fruitful strategy in finding effective treatments for anxiety disorders and drug addiction.

6. References

Adams, C. L., Short, J. L., & Lawrence, A. J. (2010). Cue-conditioned alcohol seeking in rats following abstinence: Involvement of metabotropic glutamate 5 receptors. *Br J Pharmacol, 159*, 534-542.

Agranoff, B. W., & Siegel, G. J. (1999). *Basic neurochemistry: Molecular, cellular, and medical aspects.* Philadelphia: Lippincott-Raven.

Aguilar, M. A., Rodriguez-Arias, M., & Minarro, J. (2009). Neurobiological mechanisms of the reinstatement of drug-conditioned place preference. *Brain Res Rev, 59,* 253-277.

Ahi, J., Radulovic, J., & Spiess, J. (2004). The role of hippocampal signaling cascades in consolidation of fear memory. *Behav Brain Res, 149,* 17-31.

Ahn, N. G., & Krebs, E. G. (1990). Evidence for an epidermal growth factor-stimulated protein kinase cascade in swiss 3t3 cells. Activation of serine peptide kinase activity by myelin basic protein kinases in vitro. *J Biol Chem, 265,* 11495-11501.

Alberi, L., Liu, S., Wang, Y., Badie, R., Smith-Hicks, C., Wu, J., et al. (2011). Activity-induced notch signaling in neurons requires arc/arg3.1 and is essential for synaptic plasticity in hippocampal networks. *Neuron, 69,* 437-444.

Alexander, G. E., DeLong, M. R., & Strick, P. L. (1986). Parallel organization of functionally segregated circuits linking basal ganglia and cortex. *Annu Rev Neurosci, 9,* 357-381.

Andre, V. M., Cepeda, C., Cummings, D. M., Jocoy, E. L., Fisher, Y. E., William Yang, X., et al. (2010). Dopamine modulation of excitatory currents in the striatum is dictated by the expression of d1 or d2 receptors and modified by endocannabinoids. *Eur J Neurosci, 31,* 14-28.

Apergis-Schoute, A. M., Debiec, J., Doyere, V., LeDoux, J. E., & Schafe, G. E. (2005). Auditory fear conditioning and long-term potentiation in the lateral amygdala require erk/map kinase signaling in the auditory thalamus: A role for presynaptic plasticity in the fear system. *J Neurosci, 25,* 5730-5739.

Ascher, P., & Nowak, L. (1986). A patch-clamp study of excitatory amino acid activated channels. *Adv Exp Med Biol, 203,* 507-511.

American Psychiatric Association. (2000). *Diagnostic and statistical manual of mental disorders* (4th ed.). Washington DC: Author.

Atkins, C. M., Selcher, J. C., Petraitis, J. J., Trzaskos, J. M., & Sweatt, J. D. (1998). The mapk cascade is required for mammalian associative learning. *Nat Neurosci, 1,* 602-609.

Bachtell, R. K., Tsivkovskaia, N. O., & Ryabinin, A. E. (2002). Alcohol-induced c-fos expression in the edinger-westphal nucleus: Pharmacological and signal transduction mechanisms. *J Pharmacol Exp Ther, 302,* 516-524.

Bailey, D. J., Kim, J. J., Sun, W., Thompson, R. F., & Helmstetter, F. J. (1999). Acquisition of fear conditioning in rats requires the synthesis of mrna in the amygdala. *Behav Neurosci, 113,* 276-282.

Baker, J. D., & Azorlosa, J. L. (1996). The nmda antagonist mk-801 blocks the extinction of pavlovian fear conditioning. *Behav Neurosci, 110,* 618-620.

Bardo, M. T., & Bevins, R. A. (2000). Conditioned place preference: What does it add to our preclinical understanding of drug reward? *Psychopharmacology (Berl), 153,* 31-43.

Belcheva, M. M., Vogel, Z., Ignatova, E., Avidor-Reiss, T., Zippel, R., Levy, R., et al. (1998). Opioid modulation of extracellular signal-regulated protein kinase activity is ras-dependent and involves gbetagamma subunits. *J Neurochem, 70,* 635-645.

Benavides, D. R., & Bibb, J. A. (2004). Role of cdk5 in drug abuse and plasticity. *Ann N Y Acad Sci, 1025,* 335-344.

Berhow, M. T., Hiroi, N., & Nestler, E. J. (1996). Regulation of erk (extracellular signal regulated kinase), part of the neurotrophin signal transduction cascade, in the rat mesolimbic dopamine system by chronic exposure to morphine or cocaine. *J Neurosci, 16*, 4707-4715.

Berke, J. D., & Hyman, S. E. (2000). Addiction, dopamine, and the molecular mechanisms of memory. *Neuron, 25*, 515-532.

Berman, D. E., & Dudai, Y. (2001). Memory extinction, learning anew, and learning the new: Dissociations in the molecular machinery of learning in cortex. *Science, 291*, 2417-2419.

Bienkowski, P., Rogowski, A., Korkosz, A., Mierzejewski, P., Radwanska, K., Kaczmarek, L., et al. (2004). Time-dependent changes in alcohol-seeking behavior during abstinence. *Eur Neuropsychopharmacol, 14*, 355-360.

Bilbao, A., Parkitna, J. R., Engblom, D., Perreau-Lenz, S., Sanchis-Segura, C., Schneider, M., et al. (2008). Loss of the ca2+/calmodulin-dependent protein kinase type iv in dopaminoceptive neurons enhances behavioral effects of cocaine. *Proc Natl Acad Sci U S A, 105*, 17549-17554.

Blanchard, D. C., & Blanchard, R. J. (1972). Innate and conditioned reactions to threat in rats with amygdaloid lesions. *J Comp Physiol Psychol, 81*, 281-290.

Bliss, T. V., & Collingridge, G. L. (1993). A synaptic model of memory: Long-term potentiation in the hippocampus. *Nature, 361*, 31-39.

Boudreau, A. C., Reimers, J. M., Milovanovic, M., & Wolf, M. E. (2007). Cell surface ampa receptors in the rat nucleus accumbens increase during cocaine withdrawal but internalize after cocaine challenge in association with altered activation of mitogen-activated protein kinases. *J Neurosci, 27*, 10621-10635.

Boulis, N. M., & Davis, M. (1989). Footshock-induced sensitization of electrically elicited startle reflexes. *Behav Neurosci, 103*, 504-508.

Bouton, M. E. (2002). Context, ambiguity, and unlearning: Sources of relapse after behavioral extinction. *Biol Psychiatry, 52*, 976-986.

Bouton, M. E., Mineka, S., & Barlow, D. H. (2001). A modern learning theory perspective on the etiology of panic disorder. *Psychol Rev, 108*, 4-32.

Bouton, M. E., Westbrook, R. F., Corcoran, K. A., & Maren, S. (2006). Contextual and temporal modulation of extinction: Behavioral and biological mechanisms. *Biol Psychiatry, 60*, 352-360.

Brown, R. M., & Lawrence, A. J. (2009). Neurochemistry underlying relapse to opiate seeking behavior. *Neurochem Res, 34*, 1876-1887.

Brown, R. M., Short, J. L., Cowen, M. S., Ledent, C., & Lawrence, A. J. (2009). A differential role for the adenosine a_{2a} receptor in opiate reinforcement vs opiate-seeking behavior. *Neuropsychopharmacology, 34*, 844-856.

Brown, S. A., Vik, P. W., Patterson, T. L., Grant, I., & Schuckit, M. A. (1995). Stress, vulnerability and adult alcohol relapse. *J Stud Alcohol, 56*, 538-545.

Calabresi, P., Pisani, A., Mercuri, N. B., & Bernardi, G. (1992). Long-term potentiation in the striatum is unmasked by removing the voltage-dependent magnesium block of nmda receptor channels. *Eur J Neurosci, 4*, 929-935.

Campbell, B. A., & Jaynes, J. (1966). Reinstatement. *Psychological review, 73*, 478-480.

Campbell, B. A., & Spear, N. E. (1972). Ontogeny of memory. *Psychological review, 79*, 215-236.

Campeau, S., Hayward, M. D., Hope, B. T., Rosen, J. B., Nestler, E. J., & Davis, M. (1991). Induction of the c-fos proto-oncogene in rat amygdala during unconditioned and conditioned fear. *Brain Res, 565*, 349-352.

Carr, D. B., & Sesack, S. R. (2000). Projections from the rat prefrontal cortex to the ventral tegmental area: Target specificity in the synaptic associations with mesoaccumbens and mesocortical neurons. *J Neurosci, 20*, 3864-3873.

Carroll, M. E. (1985). The role of food deprivation in the maintenance and reinstatement of cocaine-seeking behavior in rats. *Drug Alcohol Depend, 16*, 95-109.

Chandler, L. J., & Sutton, G. (2005). Acute ethanol inhibits extracellular signal-regulated kinase, protein kinase b, and adenosine 3':5'-cyclic monophosphate response element binding protein activity in an age- and brain region-specific manner. *Alcohol Clin Exp Res, 29*, 672-682.

Chen, L., & Xu, M. (2010). Dopamine d1 and d3 receptors are differentially involved in cue-elicited cocaine seeking. *J Neurochem, 114*, 530-541.

Chhatwal, J. P., Myers, K. M., Ressler, K. J., & Davis, M. (2005). Regulation of gephyrin and gabaa receptor binding within the amygdala after fear acquisition and extinction. *J Neurosci, 25*, 502-506.

Childress, A., Ehrman, R., McLellan, A. T., & O'Brien, C. (1988). Conditioned craving and arousal in cocaine addiction: A preliminary report. *NIDA Res Monogr, 81*, 74-80.

Childress, A. R., McLellan, A. T., & O'Brien, C. P. (1986). Role of conditioning factors in the development of drug dependence. *Psychiatr Clin North Am, 9*, 413-425.

Choe, E. S., Chung, K. T., Mao, L., & Wang, J. Q. (2002). Amphetamine increases phosphorylation of extracellular signal-regulated kinase and transcription factors in the rat striatum via group i metabotropic glutamate receptors. *Neuropsychopharmacology, 27*, 565-575.

Choe, E. S., & McGinty, J. F. (2001). Cyclic amp and mitogen-activated protein kinases are required for glutamate-dependent cyclic amp response element binding protein and elk-1 phosphorylation in the dorsal striatum in vivo. *J Neurochem, 76*, 401-412.

Chotiner, J. K., Nielson, J., Farris, S., Lewandowski, G., Huang, F., Banos, K., et al. (2010). Assessment of the role of map kinase in mediating activity-dependent transcriptional activation of the immediate early gene arc/arg3.1 in the dentate gyrus in vivo. *Learn Mem, 17*, 117-129.

Collins, R. J., Weeks, J. R., Cooper, M. M., Good, P. I., & Russell, R. R. (1984). Prediction of abuse liability of drugs using iv self-administration by rats. *Psychopharmacology (Berl), 82*, 6-13.

Comer, S. D., Sullivan, M. A., Yu, E., Rothenberg, J. L., Kleber, H. D., Kampman, K., et al. (2006). Injectable, sustained-release naltrexone for the treatment of opioid dependence: A randomized, placebo-controlled trial. *Arch Gen Psychiatry, 63*, 210-218.

Corcoran, K. A., & Maren, S. (2001). Hippocampal inactivation disrupts contextual retrieval of fear memory after extinction. *J Neurosci, 21*, 1720-1726.

Crombag, H. S., Bossert, J. M., Koya, E., & Shaham, Y. (2008). Review. Context-induced relapse to drug seeking: A review. *Philos Trans R Soc Lond B Biol Sci, 363*, 3233-3243.

Cui, S. Z., Wang, S. J., Li, J., Xie, G. Q., Zhou, R., Chen, L., et al. (2011). Alteration of synaptic plasticity in rat dorsal striatum induced by chronic ethanol intake and withdrawal via erk pathway. *Acta Pharmacol Sin, 32*, 175-181.

Dackis, C. A., & O'Brien, C. P. (2001). Cocaine dependence: A disease of the brain's reward centers. *J Subst Abuse Treat, 21*, 111-117.

Davis, M. (2000). The role of the amygdala in conditioned and unconditioned fear and anxiety. In J. P. Aggleton (Ed.), *The amygdala* (Vol. 2, pp. 213-287). Oxford, UK: Oxford University Press.

Davis, M., Falls, W. A., Campeau, S., & Kim, M. (1993). Fear-potentiated startle: A neural and pharmacological analysis. *Behav Brain Res, 58*, 175-198.

Davis, M., Myers, K. M., Chhatwal, J., & Ressler, K. J. (2006). Pharmacological treatments that facilitate extinction of fear: Relevance to psychotherapy. *NeuroRx, 3*, 82-96.

Davis, M., & Whalen, P. J. (2001). The amygdala: Vigilance and emotion. *Molecular psychiatry, 6*, 13-34.

Davis, S., Salin, H., Helme-Guizon, A., Dumas, S., Stephan, A., Corbex, M., et al. (2000). Dysfunctional regulation of alphacamkii and syntaxin 1b transcription after induction of ltp in the aged rat. *Eur J Neurosci, 12*, 3276-3282.

Davis, S., Vanhoutte, P., Pages, C., Caboche, J., & Laroche, S. (2000). The mapk/erk cascade targets both elk-1 and camp response element-binding protein to control long-term potentiation-dependent gene expression in the dentate gyrus in vivo. *J Neurosci, 20*, 4563-4572.

Davis, W. M., & Smith, S. G. (1976). Role of conditioned reinforcers in the initiation, maintenance and extinction of drug-seeking behavior. *Pavlov J Biol Sci, 11*, 222-236.

DeJong, W. (1994). Relapse prevention: An emerging technology for promoting long-term drug abstinence. *Int J Addict, 29*, 681-705.

Deroche-Gamonet, V., Belin, D., & Piazza, P. V. (2004). Evidence for addiction-like behavior in the rat. *Science, 305*, 1014-1017.

Di Benedetto, B., Kallnik, M., Weisenhorn, D. M., Falls, W. A., Wurst, W., & Holter, S. M. (2009). Activation of erk/mapk in the lateral amygdala of the mouse is required for acquisition of a fear-potentiated startle response. *Neuropsychopharmacology, 34*, 356-366.

Di Chiara, G. (2002). Nucleus accumbens shell and core dopamine: Differential role in behavior and addiction. *Behav Brain Res, 137*, 75-114.

Di Chiara, G., & Imperato, A. (1988). Drugs abused by humans preferentially increase synaptic dopamine concentrations in the mesolimbic system of freely moving rats. *Proc Natl Acad Sci U S A, 85*, 5274-5278.

Di Filippo, M., Picconi, B., Tantucci, M., Ghiglieri, V., Bagetta, V., Sgobio, C., et al. (2009). Short-term and long-term plasticity at corticostriatal synapses: Implications for learning and memory. *Behav Brain Res, 199*, 108-118.

Edwards, S., Bachtell, R. K., Guzman, D., Whisler, K. N., & Self, D. W. (2011). Emergence of context-associated glur(1) and erk phosphorylation in the nucleus accumbens core during withdrawal from cocaine self-administration. *Addict Biol, 16*, 450-457.

English, J. D., & Sweatt, J. D. (1997). A requirement for the mitogen-activated protein kinase cascade in hippocampal long term potentiation. *The Journal of biological chemistry, 272*, 19103-19106.

Farb, C., Aoki, C., Milner, T., Kaneko, T., & LeDoux, J. (1992). Glutamate immunoreactive terminals in the lateral amygdaloid nucleus: A possible substrate for emotional memory. *Brain Res, 593*, 145-158.

Fasano, S., D'Antoni, A., Orban, P. C., Valjent, E., Putignano, E., Vara, H., et al. (2009). Ras-guanine nucleotide-releasing factor 1 (ras-grf1) controls activation of extracellular signal-regulated kinase (erk) signaling in the striatum and long-term behavioral responses to cocaine. *Biol Psychiatry, 66*, 758-768.

Favata, M. F., Horiuchi, K. Y., Manos, E. J., Daulerio, A. J., Stradley, D. A., Feeser, W. S., et al. (1998). Identification of a novel inhibitor of mitogen-activated protein kinase kinase. *J Biol Chem, 273*, 18623-18632.

Feltenstein, M. W., & See, R. E. (2008). The neurocircuitry of addiction: An overview. *Br J Pharmacol, 154*, 261-274.

Ferguson, S. M., Fasano, S., Yang, P., Brambilla, R., & Robinson, T. E. (2006). Knockout of erk1 enhances cocaine-evoked immediate early gene expression and behavioral plasticity. *Neuropsychopharmacology, 31*, 2660-2668.

Fischer, A., Radulovic, M., Schrick, C., Sananbenesi, F., Godovac-Zimmermann, J., & Radulovic, J. (2007). Hippocampal mek/erk signaling mediates extinction of contextual freezing behavior. *Neurobiol Learn Mem, 87*, 149-158.

Fletcher, B. R., Calhoun, M. E., Rapp, P. R., & Shapiro, M. L. (2006). Fornix lesions decouple the induction of hippocampal arc transcription from behavior but not plasticity. *J Neurosci, 26*, 1507-1515.

Frey, U., Huang, Y. Y., & Kandel, E. R. (1993). Effects of camp simulate a late stage of ltp in hippocampal ca1 neurons. *Science, 260*, 1661-1664.

Frey, U., Krug, M., Reymann, K. G., & Matthies, H. (1988). Anisomycin, an inhibitor of protein synthesis, blocks late phases of ltp phenomena in the hippocampal ca1 region in vitro. *Brain Res, 452*, 57-65.

Fuchs, R. A., Eaddy, J. L., Su, Z. I., & Bell, G. H. (2007). Interactions of the basolateral amygdala with the dorsal hippocampus and dorsomedial prefrontal cortex regulate drug context-induced reinstatement of cocaine-seeking in rats. *Eur J Neurosci, 26*, 487-498.

Fuchs, R. A., Evans, K. A., Ledford, C. C., Parker, M. P., Case, J. M., Mehta, R. H., et al. (2005). The role of the dorsomedial prefrontal cortex, basolateral amygdala, and dorsal hippocampus in contextual reinstatement of cocaine seeking in rats. *Neuropsychopharmacology, 30*, 296-309.

Gale, G. D., Anagnostaras, S. G., Godsil, B. P., Mitchell, S., Nozawa, T., Sage, J. R., et al. (2004). Role of the basolateral amygdala in the storage of fear memories across the adult lifetime of rats. *J Neurosci, 24*, 3810-3815.

Gawin, F. H., & Kleber, H. D. (1986). Abstinence symptomatology and psychiatric diagnosis in cocaine abusers. Clinical observations. *Arch Gen Psychiatry, 43*, 107-113.

Gerdjikov, T. V., Ross, G. M., & Beninger, R. J. (2004). Place preference induced by nucleus accumbens amphetamine is impaired by antagonists of erk or p38 map kinases in rats. *Behav Neurosci, 118*, 740-750.

Giovannini, M. G., Efoudebe, M., Passani, M. B., Baldi, E., Bucherelli, C., Giachi, F., et al. (2003). Improvement in fear memory by histamine-elicited erk2 activation in hippocampal ca3 cells. *J Neurosci, 23*, 9016-9023.

Girault, J. A., Valjent, E., Caboche, J., & Herve, D. (2007). Erk2: A logical and gate critical for drug-induced plasticity? *Curr Opin Pharmacol, 7*, 77-85.

Gossop, M., Green, L., Phillips, G., & Bradley, B. (1989). Lapse, relapse and survival among opiate addicts after treatment. A prospective follow-up study. *Br J Psychiatry, 154*, 348-353.

Grant, S., London, E. D., Newlin, D. B., Villemagne, V. L., Liu, X., Contoreggi, C., et al. (1996). Activation of memory circuits during cue-elicited cocaine craving. *Proc Natl Acad Sci U S A, 93*, 12040-12045.

Greenberg, P. E., Sisitsky, T., Kessler, R. C., Finkelstein, S. N., Berndt, E. R., Davidson, J. R., et al. (1999). The economic burden of anxiety disorders in the 1990s. *J Clin Psychiatry, 60*, 427-435.

Greengard, P., Allen, P. B., & Nairn, A. C. (1999). Beyond the dopamine receptor: The darpp-32/protein phosphatase-1 cascade. *Neuron, 23*, 435-447.

Gresack, J. E., Schafe, G. E., Orr, P. T., & Frick, K. M. (2009). Sex differences in contextual fear conditioning are associated with differential ventral hippocampal extracellular signal-regulated kinase activation. *Neuroscience, 159*, 451-467.

Grillon, C., Baas, J. P., Lissek, S., Smith, K., & Milstein, J. (2004). Anxious responses to predictable and unpredictable aversive events. *Behav Neurosci, 118*, 916-924.

Grimm, J. W., Hope, B. T., Wise, R. A., & Shaham, Y. (2001). Neuroadaptation. Incubation of cocaine craving after withdrawal. *Nature, 412*, 141-142.

Grimm, J. W., Shaham, Y., & Hope, B. T. (2002). Effect of cocaine and sucrose withdrawal period on extinction behavior, cue-induced reinstatement, and protein levels of the dopamine transporter and tyrosine hydroxylase in limbic and cortical areas in rats. *Behav Pharmacol, 13*, 379-388.

Groblewski, P. A., Franken, F. H., & Cunningham, C. L. (2011). Inhibition of extracellular signal-regulated kinase (erk) activity with sl327 does not prevent acquisition, expression, and extinction of ethanol-seeking behavior in mice. *Behav Brain Res, 217*, 399-407.

Groblewski, P. A., Ryabinin, A. E., & Cunningham, C. L. (2011). Activation and role of the medial prefrontal cortex (mpfc) in extinction of ethanol-induced associative learning in mice. *Neurobiol Learn Mem.*

Guzowski, J. F., Lyford, G. L., Stevenson, G. D., Houston, F. P., McGaugh, J. L., Worley, P. F., et al. (2000). Inhibition of activity-dependent arc protein expression in the rat hippocampus impairs the maintenance of long-term potentiation and the consolidation of long-term memory. *J Neurosci, 20*, 3993-4001.

Harwood, H. J., Fountain, D., & Livermore, G. (1998). Economic costs of alcohol abuse and alcoholism. *Recent Dev Alcohol, 14*, 307-330.

Herry, C., & Garcia, R. (2002). Prefrontal cortex long-term potentiation, but not long-term depression, is associated with the maintenance of extinction of learned fear in mice. *J Neurosci, 22*, 577-583.

Herry, C., Trifilieff, P., Micheau, J., Luthi, A., & Mons, N. (2006). Extinction of auditory fear conditioning requires mapk/erk activation in the basolateral amygdala. *The European journal of neuroscience, 24*, 261-269.

Hofmann, S. G. (2007). Enhancing exposure-based therapy from a translational research perspective. *Behav Res Ther, 45*, 1987-2001.

Horvitz, J. C. (2002). Dopamine gating of glutamatergic sensorimotor and incentive motivational input signals to the striatum. *Behav Brain Res, 137*, 65-74.

Hoshi, M., Nishida, E., & Sakai, H. (1988). Activation of a ca2+-inhibitable protein kinase that phosphorylates microtubule-associated protein 2 in vitro by growth factors, phorbol esters, and serum in quiescent cultured human fibroblasts. *J Biol Chem, 263*, 5396-5401.

Huang, Y. Y., & Kandel, E. R. (1994). Recruitment of long-lasting and protein kinase a-dependent long-term potentiation in the ca1 region of hippocampus requires repeated tetanization. *Learn Mem, 1*, 74-82.

Huang, Y. Y., Martin, K. C., & Kandel, E. R. (2000). Both protein kinase a and mitogen-activated protein kinase are required in the amygdala for the macromolecular synthesis-dependent late phase of long-term potentiation. *J Neurosci, 20*, 6317-6325.

Hugues, S., Chessel, A., Lena, I., Marsault, R., & Garcia, R. (2006). Prefrontal infusion of pd098059 immediately after fear extinction training blocks extinction-associated prefrontal synaptic plasticity and decreases prefrontal erk2 phosphorylation. *Synapse, 60*, 280-287.

Hugues, S., Deschaux, O., & Garcia, R. (2004). Postextinction infusion of a mitogen-activated protein kinase inhibitor into the medial prefrontal cortex impairs memory of the extinction of conditioned fear. *Learn Mem, 11*, 540-543.

Hurd, Y. L., & Herkenham, M. (1993). Molecular alterations in the neostriatum of human cocaine addicts. *Synapse, 13*, 357-369.

Hyman, S. E. (2005). Addiction: A disease of learning and memory. *Am J Psychiatry, 162*, 1414-1422.

Impey, S., Obrietan, K., & Storm, D. R. (1999). Making new connections: Role of erk/map kinase signaling in neuronal plasticity. *Neuron, 23*, 11-14.

Izquierdo, I., & Medina, J. H. (1993). Role of the amygdala, hippocampus and entorhinal cortex in memory consolidation and expression. *Braz J Med Biol Res, 26*, 573-589.

Izquierdo, L. A., Barros, D. M., Ardenghi, P. G., Pereira, P., Rodrigues, C., Choi, H., et al. (2000). Different hippocampal molecular requirements for short- and long-term retrieval of one-trial avoidance learning. *Behav Brain Res, 111*, 93-98.

Jenab, S., Festa, E. D., Nazarian, A., Wu, H. B., Sun, W. L., Hazim, R., et al. (2005). Cocaine induction of erk proteins in dorsal striatum of fischer rats. *Brain Res Mol Brain Res, 142*, 134-138.

Jentsch, J. D., & Taylor, J. R. (1999). Impulsivity resulting from frontostriatal dysfunction in drug abuse: Implications for the control of behavior by reward-related stimuli. *Psychopharmacology (Berl), 146*, 373-390.

Jupp, B., Krstew, E., Dezsi, G., & Lawrence, A. (2011). Discrete cue-conditioned alcohol-seeking after protracted abstinence: Pattern of neural activation and involvement of orexin(1) receptors. *Br J Pharmacol, 162*, 880-889.

Jupp, B., & Lawrence, A. J. (2010). New horizons for therapeutics in drug and alcohol abuse. *Pharmacol Ther, 125*, 138-168.

Juraska, J. M. (1998). Neural plasticity and the development of sex differences. *Annu Rev Sex Res, 9*, 20-38.

Kalivas, P. W. (2009). The glutamate homeostasis hypothesis of addiction. *Nat Rev Neurosci, 10*, 561-572.

Kalivas, P. W., & O'Brien, C. (2008). Drug addiction as a pathology of staged neuroplasticity. *Neuropsychopharmacology, 33*, 166-180.

Kalivas, P. W., & Volkow, N. D. (2005). The neural basis of addiction: A pathology of motivation and choice. *Am J Psychiatry, 162*, 1403-1413.

Kandel, E. R. (2001). The molecular biology of memory storage: A dialogue between genes and synapses. *Science, 294*, 1030-1038.

Kandel, E. R. (2009). The biology of memory: A forty-year perspective. *J Neurosci, 29*, 12748-12756.

Kasanetz, F., Deroche-Gamonet, V., Berson, N., Balado, E., Lafourcade, M., Manzoni, O., et al. Transition to addiction is associated with a persistent impairment in synaptic plasticity. *Science, 328*, 1709-1712.

Kauer, J. A., & Malenka, R. C. (2007). Synaptic plasticity and addiction. *Nat Rev Neurosci, 8*, 844-858.

Kelleher, R. J., 3rd, Govindarajan, A., Jung, H. Y., Kang, H., & Tonegawa, S. (2004). Translational control by mapk signaling in long-term synaptic plasticity and memory. *Cell, 116*, 467-479.

Kelley, A. E. (2004). Memory and addiction: Shared neural circuitry and molecular mechanisms. *Neuron, 44*, 161-179.

Kerr, J. N., & Wickens, J. R. (2001). Dopamine d-1/d-5 receptor activation is required for long-term potentiation in the rat neostriatum in vitro. *J Neurophysiol, 85*, 117-124.

Kessler, R. C., Chiu, W. T., Demler, O., Merikangas, K. R., & Walters, E. E. (2005). Prevalence, severity, and comorbidity of 12-month dsm-iv disorders in the national comorbidity survey replication. *Arch Gen Psychiatry, 62*, 617-627.

Kim, J. H., Hamlin, A. S., & Richardson, R. (2009). Fear extinction across development: The involvement of the medial prefrontal cortex as assessed by temporary inactivation and immunohistochemistry. *J Neurosci, 29*, 10802-10808.

Kim, J. H., Li, S., Hamlin, A. S., McNally, G. P., & Richardson, R. (2012). Phosphorylation of mitogen-activated protein kinase in the medial prefrontal cortex and the amygdala following memory retrieval or forgetting in developing rats. *Neurobiol Learn Mem, doi:10.1016/j.nlm.2011.09.005*.

Kim, J. H., Li, S., & Richardson, R. (2011). Immunohistochemical analyses of long-term extinction of conditioned fear in adolescent rats. *Cereb Cortex, 21*, 530-538.

Kim, J. H., & Richardson, R. (2007a). A developmental dissociation in reinstatement of an extinguished fear response in rats. *Neurobiol Learn Mem, 88*, 48-57.

Kim, J. H., & Richardson, R. (2007b). Immediate post-reminder injection of gamma-amino butyric acid (gaba) agonist midazolam attenuates reactivation of forgotten fear in the infant rat. *Behav Neurosci, 121*, 1328-1332.

Kim, J. H., & Richardson, R. (2010). New findings on extinction of conditioned fear early in development: Theoretical and clinical implications. *Biol Psychiatry, 67*, 297-303.

Knackstedt, L. A., Moussawi, K., Lalumiere, R., Schwendt, M., Klugmann, M., & Kalivas, P. W. (2010). Extinction training after cocaine self-administration induces glutamatergic plasticity to inhibit cocaine seeking. *J Neurosci, 30*, 7984-7992.

Konorski, J. (1948). *Conditioned reflexes and neuron organization*. Cambridge: University Press.

Kornhauser, J. M., & Greenberg, M. E. (1997). A kinase to remember: Dual roles for map kinase in long-term memory. *Neuron, 18*, 839-842.

Koya, E., Uejima, J. L., Wihbey, K. A., Bossert, J. M., Hope, B. T., & Shaham, Y. (2008). Role of ventral medial prefrontal cortex in incubation of cocaine craving. *Neuropharmacology*.

Landauer, T. K. (1969). Reinforcement as consolidation. *Psychol Rev, 76*, 82-96.

Lattal, K. M., Radulovic, J., & Lukowiak, K. (2006). Extinction: [corrected] does it or doesn't it? The requirement of altered gene activity and new protein synthesis. *Biol Psychiatry, 60*, 344-351.

Le Moal, M. & Simon, H. (1991). Mesocorticolimbic dopaminergic network: Functional and regulatory roles. *Physiological Review, 71*, 155-234.

Ledgerwood, L., Richardson, R., & Cranney, J. (2003). Effects of d-cycloserine on extinction of conditioned freezing. *Behav Neurosci, 117*, 341-349.

LeDoux, J. E. (1993). Emotional memory systems in the brain. *Behav Brain Res, 58*, 69-79.

LeDoux, J. E. (2000). Emotion circuits in the brain. *Annual review of neuroscience, 23*, 155-184.

LeDoux, J. E., Farb, C., & Ruggiero, D. A. (1990). Topographic organization of neurons in the acoustic thalamus that project to the amygdala. *J Neurosci, 10*, 1043-1054.

Lee, H., & Kim, J. J. (1998). Amygdalar nmda receptors are critical for new fear learning in previously fear-conditioned rats. *J Neurosci, 18*, 8444-8454.

Li, Y. Q., Li, F. Q., Wang, X. Y., Wu, P., Zhao, M., Xu, C. M., et al. (2008). Central amygdala extracellular signal-regulated kinase signaling pathway is critical to incubation of opiate craving. *J Neurosci, 28*, 13248-13257.

Lin, C. H., Lee, C. C., & Gean, P. W. (2003). Involvement of a calcineurin cascade in amygdala depotentiation and quenching of fear memory. *Mol Pharmacol, 63*, 44-52.

Lin, C. H., Yeh, S. H., Leu, T. H., Chang, W. C., Wang, S. T., & Gean, P. W. (2003). Identification of calcineurin as a key signal in the extinction of fear memory. *J Neurosci, 23*, 1574-1579.

Lin, C. H., Yeh, S. H., Lu, H. Y., & Gean, P. W. (2003). The similarities and diversities of signal pathways leading to consolidation of conditioning and consolidation of extinction of fear memory. *J Neurosci, 23*, 8310-8317.

Lin, C. H., Yeh, S. H., Lu, K. T., Leu, T. H., Chang, W. C., & Gean, P. W. (2001). A role for the pi-3 kinase signaling pathway in fear conditioning and synaptic plasticity in the amygdala. *Neuron, 31*, 841-851.

Lin, H. C., Mao, S. C., Su, C. L., & Gean, P. W. (2010). Alterations of excitatory transmission in the lateral amygdala during expression and extinction of fear memory. *Int J Neuropsychopharmacol, 13*, 335-345.

Lu, K. T., Walker, D. L., & Davis, M. (2001). Mitogen-activated protein kinase cascade in the basolateral nucleus of amygdala is involved in extinction of fear-potentiated startle. *J Neurosci, 21*, RC162.

Lu, L., Grimm, J. W., Hope, B. T., & Shaham, Y. (2004). Incubation of cocaine craving after withdrawal: A review of preclinical data. *Neuropharmacology, 47 Suppl 1*, 214-226.

Lu, L., Hope, B. T., Dempsey, J., Liu, S. Y., Bossert, J. M., & Shaham, Y. (2005). Central amygdala erk signaling pathway is critical to incubation of cocaine craving. *Nat Neurosci, 8*, 212-219.

Lu, L., Koya, E., Zhai, H., Hope, B. T., & Shaham, Y. (2006). Role of erk in cocaine addiction. *Trends Neurosci, 29*, 695-703.

Lu, L., Zhang, B., Liu, Z., & Zhang, Z. (2002). Reactivation of cocaine conditioned place preference induced by stress is reversed by cholecystokinin-b receptors antagonist in rats. *Brain Res, 954*, 132-140.

Lynch, M. A. (2004). Long-term potentiation and memory. *Physiol Rev, 84*, 87-136.

Madeira, M. D., Sousa, N., & Paula-Barbosa, M. M. (1991). Sexual dimorphism in the mossy fiber synapses of the rat hippocampus. *Exp Brain Res, 87*, 537-545.

Malenka, R. C., & Bear, M. F. (2004). Ltp and ltd: An embarrassment of riches. *Neuron, 44*, 5-21.

Mao, L., Yang, L., Tang, Q., Samdani, S., Zhang, G., & Wang, J. Q. (2005). The scaffold protein homer1b/c links metabotropic glutamate receptor 5 to extracellular signal-regulated protein kinase cascades in neurons. *J Neurosci, 25*, 2741-2752.

Mao, S. C., Hsiao, Y. H., & Gean, P. W. (2006). Extinction training in conjunction with a partial agonist of the glycine site on the nmda receptor erases memory trace. *J Neurosci, 26*, 8892-8899.

Maren, S., Ferrario, C. R., Corcoran, K. A., Desmond, T. J., & Frey, K. A. (2003). Protein synthesis in the amygdala, but not the auditory thalamus, is required for consolidation of pavlovian fear conditioning in rats. *Eur J Neurosci, 18*, 3080-3088.

Markou, A., Weiss, F., Gold, L. H., Caine, S. B., Schulteis, G., & Koob, G. F. (1993). Animal models of drug craving. *Psychopharmacology (Berl), 112*, 163-182.

Marsicano, G., Wotjak, C. T., Azad, S. C., Bisogno, T., Rammes, G., Cascio, M. G., et al. (2002). The endogenous cannabinoid system controls extinction of aversive memories. *Nature, 418*, 530-534.

Martin, M., Chen, B. T., Hopf, F. W., Bowers, M. S., & Bonci, A. (2006). Cocaine self-administration selectively abolishes ltd in the core of the nucleus accumbens. *Nat Neurosci, 9*, 868-869.

Matsuda, S., Matsuzawa, D., Nakazawa, K., Sutoh, C., Ohtsuka, H., Ishii, D., et al. (2010). D-serine enhances extinction of auditory cued fear conditioning via erk1/2 phosphorylation in mice. *Prog Neuropsychopharmacol Biol Psychiatry, 34*, 895-902.

Mazzucchelli, C., Vantaggiato, C., Ciamei, A., Fasano, S., Pakhotin, P., Krezel, W., et al. (2002). Knockout of erk1 map kinase enhances synaptic plasticity in the striatum and facilitates striatal-mediated learning and memory. *Neuron, 34*, 807-820.

McClung, C. A., & Nestler, E. J. (2003). Regulation of gene expression and cocaine reward by creb and deltafosb. *Nat Neurosci, 6*, 1208-1215.

McDonald, A. J., Mascagni, F., & Guo, L. (1996). Projections of the medial and lateral prefrontal cortices to the amygdala: A phaseolus vulgaris leucoagglutinin study in the rat. *Neuroscience, 71*, 55-75.

McPherson, C. S., Mantamadiotis, T., Tan, S. S., & Lawrence, A. J. (2010). Deletion of creb1 from the dorsal telencephalon reduces motivational properties of cocaine. *Cereb Cortex, 20*, 941-952.

Milad, M. R., & Quirk, G. J. (2002). Neurons in medial prefrontal cortex signal memory for fear extinction. *Nature, 420*, 70-74.

Miller, C. A., & Marshall, J. F. (2005). Molecular substrates for retrieval and reconsolidation of cocaine-associated contextual memory. *Neuron, 47*, 873-884.

Mineka, S., & Zinbarg, R. (1996). Conditioning and ethological models of anxiety disorders: Stress-in-dynamic-context anxiety models. *Nebr Symp Motiv, 43*, 135-210.

Mizoguchi, H., Yamada, K., Mizuno, M., Mizuno, T., Nitta, A., Noda, Y., et al. (2004). Regulations of methamphetamine reward by extracellular signal-regulated kinase 1/2/ets-like gene-1 signaling pathway via the activation of dopamine receptors. *Mol Pharmacol, 65,* 1293-1301.

Morgan, M. A., Romanski, L. M., & LeDoux, J. E. (1993). Extinction of emotional learning: Contribution of medial prefrontal cortex. *Neurosci Lett, 163,* 109-113.

Morris, R. G., Anderson, E., Lynch, G. S., & Baudry, M. (1986). Selective impairment of learning and blockade of long-term potentiation by an n-methyl-d-aspartate receptor antagonist, ap5. *Nature, 319,* 774-776.

Moussawi, K., Pacchioni, A., Moran, M., Olive, M. F., Gass, J. T., Lavin, A., et al. (2009). N-acetylcysteine reverses cocaine-induced metaplasticity. *Nat Neurosci, 12,* 182-189.

Myers, K. M., Carlezon, W. A., Jr., & Davis, M. (2011). Glutamate receptors in extinction and extinction-based therapies for psychiatric illness. *Neuropsychopharmacology, 36,* 274-293.

Myers, K. M., & Davis, M. (2007). Mechanisms of fear extinction. *Mol Psychiatry, 12,* 120-150.

Nader, K., Schafe, G. E., & Le Doux, J. E. (2000). Fear memories require protein synthesis in the amygdala for reconsolidation after retrieval. *Nature, 406,* 722-726.

Nathaniel, Thomas I., Panksepp, Jaak, & Huber, Robert. (2009). Drug-seeking behavior in an invertebrate system: Evidence of morphine-induced reward, extinction and reinstatement in crayfish. *Behavioral Brain Research, 197,* 331-338.

Nestler, E. J. (2001). Molecular basis of long-term plasticity underlying addiction. *Nat Rev Neurosci, 2,* 119-128.

Newman, D. L., Moffitt, T. E., Caspi, A., Magdol, L., Silva, P. A., & Stanton, W. R. (1996). Psychiatric disorder in a birth cohort of young adults: Prevalence, comorbidity, clinical significance, and new case incidence from ages 11 to 21. *J Consult Clin Psychol, 64,* 552-562.

O'Brien, C., Childress, A. R., Ehrman, R., Robbins, S., & McLellan, A. T. (1992). Conditioning mechanisms in drug dependence. *Clin Neuropharmacol, 15 Suppl 1 Pt A,* 66A-67A.

O'Brien, C. P. (1997a). Progress in the science of addiction. *Am J Psychiatry, 154,* 1195-1197.

O'Brien, C. P. (1997b). A range of research-based pharmacotherapies for addiction. *Science, 278,* 66-70.

O'Brien, M. S., Wu, L. T., & Anthony, J. C. (2005). Cocaine use and the occurrence of panic attacks in the community: A case-crossover approach. *Subst Use Misuse, 40,* 285-297.

Omelchenko, N.Sesack, S. R. (2007). Glutamate synaptic inputs to ventral tegmental area neurons in the rat derive primarily from subcortical sources. *Neuroscience, 146,* 1259-1274.

Ortiz, J., Harris, H. W., Guitart, X., Terwilliger, R. Z., Haycock, J. W., & Nestler, E. J. (1995). Extracellular signal-regulated protein kinases (erks) and erk kinase (mek) in brain: Regional distribution and regulation by chronic morphine. *J Neurosci, 15,* 1285-1297.

Ozaki, S., Narita, M., Ozaki, M., Khotib, J., & Suzuki, T. (2004). Role of extracellular signal-regulated kinase in the ventral tegmental area in the suppression of the morphine-induced rewarding effect in mice with sciatic nerve ligation. *J Neurochem, 88,* 1389-1397.

Parker, L. A., & McDonald, R. V. (2000). Reinstatement of both a conditioned place preference and a conditioned place aversion with drug primes. *Pharmacol Biochem Behav, 66*, 559-561.

Pascoli, V., Besnard, A., Herve, D., Pages, C., Heck, N., Girault, J. A., et al. (2011). Cyclic adenosine monophosphate-independent tyrosine phosphorylation of nr2b mediates cocaine-induced extracellular signal-regulated kinase activation. *Biol Psychiatry, 69*, 218-227.

Pavlov, I. P. (1927). *Conditioned reflexes: An investigation of the physiological activity of the cerebral cortex.* London: Oxford University Press.

Pierce, R. C., Pierce-Bancroft, A. F., & Prasad, B. M. (1999). Neurotrophin-3 contributes to the initiation of behavioral sensitization to cocaine by activating the ras/mitogen-activated protein kinase signal transduction cascade. *J Neurosci, 19*, 8685-8695.

Price, J. L. (2003). Comparative aspects of amygdala connectivity. *Ann N Y Acad Sci, 985*, 50-58.

Quirk, G. J., Armony, J. L., & LeDoux, J. E. (1997). Fear conditioning enhances different temporal components of tone-evoked spike trains in auditory cortex and lateral amygdala. *Neuron, 19*, 613-624.

Quirk, G. J., Repa, C., & LeDoux, J. E. (1995). Fear conditioning enhances short-latency auditory responses of lateral amygdala neurons: Parallel recordings in the freely behaving rat. *Neuron, 15*, 1029-1039.

Quirk, G. J., Russo, G. K., Barron, J. L., & Lebron, K. (2000). The role of ventromedial prefrontal cortex in the recovery of extinguished fear. *J Neurosci, 20*, 6225-6231.

Rachman, S. (1989). The return of fear: Review and prospect. *Clin Psychol Rev, 9*, 147-168.

Radwanska, K., Caboche, J., & Kaczmarek, L. (2005). Extracellular signal-regulated kinases (erks) modulate cocaine-induced gene expression in the mouse amygdala. *Eur J Neurosci, 22*, 939-948.

Radwanska, K., Nikolaev, E., Knapska, E., & Kaczmarek, L. (2002). Differential response of two subdivisions of lateral amygdala to aversive conditioning as revealed by c-fos and p-erk mapping. *Neuroreport, 13*, 2241-2246.

Radwanska, K., Wrobel, E., Korkosz, A., Rogowski, A., Kostowski, W., Bienkowski, P., et al. (2008). Alcohol relapse induced by discrete cues activates components of ap-1 transcription factor and erk pathway in the rat basolateral and central amygdala. *Neuropsychopharmacology, 33*, 1835-1846.

Rasmusson, A. M., & Charney, D. S. (1997). Animal models of relevance to ptsd. *Ann N Y Acad Sci, 821*, 332-351.

Reichel, C. M., & Bevins, R. A. (2009). Forced abstinence model of relapse to study pharmacological treatments of substance use disorder. *Curr Drug Abuse Rev, 2*, 184-194.

Reinblatt, S. P., & Riddle, M. A. (2007). The pharmacological management of childhood anxiety disorders: A review. *Psychopharmacology (Berl), 191*, 67-86.

Repa, J. C., Muller, J., Apergis, J., Desrochers, T. M., Zhou, Y., & LeDoux, J. E. (2001). Two different lateral amygdala cell populations contribute to the initiation and storage of memory. *Nat Neurosci, 4*, 724-731.

Rescorla, R. A. (1988). Pavlovian conditioning. It's not what you think it is. *Am Psychol, 43*, 151-160.

Rescorla, R. A., & Wagner, A. R. (1972). A theory of pavlovian conditioning: Variations in the effectiveness of reinforcement and non-reinforcement. . In Prokasy A. H. (Ed.), *Classical conditioning ii: Current research and theory.* (pp. 64-99). New York: Appleton-Century-Croft.

Reynolds, J. N., Hyland, B. I., & Wickens, J. R. (2001). A cellular mechanism of reward-related learning. *Nature, 413,* 67-70.

Roberson, E. D., English, J. D., & Sweatt, J. D. (1996). A biochemist's view of long-term potentiation. *Learn Mem, 3,* 1-24.

Robinson, T. E., & Berridge, K. C. (1993). The neural basis of drug craving: An incentive-sensitization theory of addiction. *Brain Res Brain Res Rev, 18,* 247-291.

Robinson, T. E., & Berridge, K. C. (2001). Incentive-sensitization and addiction. *Addiction, 96,* 103-114.

Rogers, J. L., & See, R. E. (2007). Selective inactivation of the ventral hippocampus attenuates cue-induced and cocaine-primed reinstatement of drug-seeking in rats. *Neurobiol Learn Mem, 87,* 688-692.

Rosen, J. B., & Schulkin, J. (1998). From normal fear to pathological anxiety. *Psychol Rev, 105,* 325-350.

Rosenkranz, J. A., & Grace, A. A. (2002). Cellular mechanisms of infralimbic and prelimbic prefrontal cortical inhibition and dopaminergic modulation of basolateral amygdala neurons in vivo. *J Neurosci, 22,* 324-337.

Rosenkranz, J. A., Moore, H., & Grace, A. A. (2003). The prefrontal cortex regulates lateral amygdala neuronal plasticity and responses to previously conditioned stimuli. *J Neurosci, 23,* 11054-11064.

Rossato, J. I., Bevilaqua, L. R., Lima, R. H., Medina, J. H., Izquierdo, I., & Cammarota, M. (2006). On the participation of hippocampal p38 mitogen-activated protein kinase in extinction and reacquisition of inhibitory avoidance memory. *Neuroscience, 143,* 15-23.

Rothbaum, B. O., & Davis, M. (2003). Applying learning principles to the treatment of post-trauma reactions. *Ann N Y Acad Sci, 1008,* 112-121.

Royer, S., & Pare, D. (2002). Bidirectional synaptic plasticity in intercalated amygdala neurons and the extinction of conditioned fear responses. *Neuroscience, 115,* 455-462.

Rumpel, S., LeDoux, J., Zador, A., & Malinow, R. (2005). Postsynaptic receptor trafficking underlying a form of associative learning. *Science, 308,* 83-88.

Salzmann, J., Marie-Claire, C., Le Guen, S., Roques, B. P., & Noble, F. (2003). Importance of erk activation in behavioral and biochemical effects induced by mdma in mice. *Br J Pharmacol, 140,* 831-838.

SAMHSA. (2010). *Results from the 2009 national survey on drug use and health: Volume i. Summary of national findings.* Rockville, MD: Office of Applied Studies.

Santini, E., Muller, R. U., & Quirk, G. J. (2001). Consolidation of extinction learning involves transfer from nmda-independent to nmda-dependent memory. *J Neurosci, 21,* 9009-9017.

Satoh, Y., Endo, S., Ikeda, T., Yamada, K., Ito, M., Kuroki, M., et al. (2007). Extracellular signal-regulated kinase 2 (erk2) knockdown mice show deficits in long-term memory; erk2 has a specific function in learning and memory. *J Neurosci, 27*, 10765-10776.

Schafe, G. E., Atkins, C. M., Swank, M. W., Bauer, E. P., Sweatt, J. D., & LeDoux, J. E. (2000). Activation of erk/map kinase in the amygdala is required for memory consolidation of pavlovian fear conditioning. *J Neurosci, 20*, 8177-8187.

Schafe, G. E., & LeDoux, J. E. (2000). Memory consolidation of auditory pavlovian fear conditioning requires protein synthesis and protein kinase a in the amygdala. *J Neurosci, 20*, RC96.

Schafe, G. E., Nadel, N. V., Sullivan, G. M., Harris, A., & LeDoux, J. E. (1999). Memory consolidation for contextual and auditory fear conditioning is dependent on protein synthesis, pka, and map kinase. *Learn Mem, 6*, 97-110.

Schafe, G. E., Swank, M. W., Rodrigues, S. M., Debiec, J., & Doyere, V. (2008). Phosphorylation of erk/map kinase is required for long-term potentiation in anatomically restricted regions of the lateral amygdala in vivo. *Learn Mem, 15*, 55-62.

Schroeder, J. P., Spanos, M., Stevenson, J. R., Besheer, J., Salling, M., & Hodge, C. W. (2008). Cue-induced reinstatement of alcohol-seeking behavior is associated with increased erk1/2 phosphorylation in specific limbic brain regions: Blockade by the mglur5 antagonist mpep. *Neuropharmacology, 55*, 546-554.

Schumann, J., & Yaka, R. (2009). Prolonged withdrawal from repeated noncontingent cocaine exposure increases nmda receptor expression and erk activity in the nucleus accumbens. *J Neurosci, 29*, 6955-6963.

See, R. E. (2005). Neural substrates of cocaine-cue associations that trigger relapse. *Eur J Pharmacol, 526*, 140-146.

Seger, R., & Krebs, E. G. (1995). The mapk signaling cascade. *FASEB J, 9*, 726-735.

Selcher, J. C., Atkins, C. M., Trzaskos, J. M., Paylor, R., & Sweatt, J. D. (1999). A necessity for map kinase activation in mammalian spatial learning. *Learn Mem, 6*, 478-490.

Selcher, J. C., Nekrasova, T., Paylor, R., Landreth, G. E., & Sweatt, J. D. (2001). Mice lacking the erk1 isoform of map kinase are unimpaired in emotional learning. *Learn Mem, 8*, 11-19.

Sesack, S. R., & Pickel, V. M. (1992). Prefrontal cortical efferents in the rat synapse on unlabeled neuronal targets of catecholamine terminals in the nucleus accumbens septi and on dopamine neurons in the ventral tegmental area. *J Comp Neurol, 320*, 145-160.

Sestan, N., Artavanis-Tsakonas, S., & Rakic, P. (1999). Contact-dependent inhibition of cortical neurite growth mediated by notch signaling. *Science, 286*, 741-746.

Sgambato, V., Pages, C., Rogard, M., Besson, M. J., & Caboche, J. (1998). Extracellular signal-regulated kinase (erk) controls immediate early gene induction on corticostriatal stimulation. *J Neurosci, 18*, 8814-8825.

Shaham, Y., & Hope, B. T. (2005). The role of neuroadaptations in relapse to drug seeking. *Nat Neurosci, 8*, 1437-1439.

Shaham, Y., Shalev, U., Lu, L., De Wit, H., & Stewart, J. (2003). The reinstatement model of drug relapse: History, methodology and major findings. *Psychopharmacology (Berl), 168*, 3-20.

Shaham, Y., & Stewart, J. (1996). Effects of opioid and dopamine receptor antagonists on relapse induced by stress and re-exposure to heroin in rats. *Psychopharmacology (Berl), 125,* 385-391.

Shalev, U., Morales, M., Hope, B., Yap, J., & Shaham, Y. (2001). Time-dependent changes in extinction behavior and stress-induced reinstatement of drug seeking following withdrawal from heroin in rats. *Psychopharmacology (Berl), 156,* 98-107.

Shalin, S. C., Zirrgiebel, U., Honsa, K. J., Julien, J. P., Miller, F. D., Kaplan, D. R., et al. (2004). Neuronal mek is important for normal fear conditioning in mice. *J Neurosci Res, 75,* 760-770.

Shen, W., Flajolet, M., Greengard, P., & Surmeier, D. J. (2008). Dichotomous dopaminergic control of striatal synaptic plasticity. *Science, 321,* 848-851.

Shepard, J. D., Bossert, J. M., Liu, S. Y., & Shaham, Y. (2004). The anxiogenic drug yohimbine reinstates methamphetamine seeking in a rat model of drug relapse. *Biol Psychiatry, 55,* 1082-1089.

Shi, C., & Davis, M. (2001). Visual pathways involved in fear conditioning measured with fear-potentiated startle: Behavioral and anatomic studies. *J Neurosci, 21,* 9844-9855.

Shi, X., & McGinty, J. F. (2006). Extracellular signal-regulated mitogen-activated protein kinase inhibitors decrease amphetamine-induced behavior and neuropeptide gene expression in the striatum. *Neuroscience, 138,* 1289-1298.

Shiflett, M. W., & Balleine, B. W. (2011). Contributions of erk signaling in the striatum to instrumental learning and performance. *Behav Brain Res, 218,* 240-247.

Shiflett, M. W., Brown, R. A., & Balleine, B. W. (2010). Acquisition and performance of goal-directed instrumental actions depends on erk signaling in distinct regions of dorsal striatum in rats. *J Neurosci, 30,* 2951-2959.

Sindreu, C. B., Scheiner, Z. S., & Storm, D. R. (2007). Ca2+ -stimulated adenylyl cyclases regulate erk-dependent activation of msk1 during fear conditioning. *Neuron, 53,* 79-89.

Skinner, B.F. (1938). *The behavior of organisms : An experimental analysis* Appleton-Century, New York.

Smith, A. D., & Bolam, J. P. (1990). The neural network of the basal ganglia as revealed by the study of synaptic connections of identified neurones. *Trends Neurosci, 13,* 259-265.

Smith, M. D., Jones, L. S., & Wilson, M. A. (2002). Sex differences in hippocampal slice excitability: Role of testosterone. *Neuroscience, 109,* 517-530.

Sotres-Bayon, F., Cain, C. K., & LeDoux, J. E. (2006). Brain mechanisms of fear extinction: Historical perspectives on the contribution of prefrontal cortex. *Biol Psychiatry, 60,* 329-336.

Sotres-Bayon, F., Diaz-Mataix, L., Bush, D. E., & LeDoux, J. E. (2009). Dissociable roles for the ventromedial prefrontal cortex and amygdala in fear extinction: Nr2b contribution. *Cereb Cortex, 19,* 474-482.

Stanton, P. K., & Sarvey, J. M. (1984). Blockade of long-term potentiation in rat hippocampal ca1 region by inhibitors of protein synthesis. *J Neurosci, 4,* 3080-3088.

Stretch, R., Gerber, G. J., & Wood, S. M. (1971). Factors affecting behavior maintained by response-contingent intravenous infusions of amphetamine in squirrel monkeys. *Can J Physiol Pharmacol, 49,* 581-589.

Sullivan, R. M., Landers, M., Yeaman, B., & Wilson, D. A. (2000). Good memories of bad events in infancy. *Nature, 407*, 38-39.

Sutton, M. A., Schmidt, E. F., Choi, K. H., Schad, C. A., Whisler, K., Simmons, D., et al. (2003). Extinction-induced upregulation in ampa receptors reduces cocaine-seeking behavior. *Nature, 421*, 70-75.

Suzuki, A., Josselyn, S. A., Frankland, P. W., Masushige, S., Silva, A. J., & Kida, S. (2004). Memory reconsolidation and extinction have distinct temporal and biochemical signatures. *J Neurosci, 24*, 4787-4795.

Svenningsson, P., Nishi, A., Fisone, G., Girault, J. A., Nairn, A. C., & Greengard, P. (2004). Darpp-32: An integrator of neurotransmission. *Annu Rev Pharmacol Toxicol, 44*, 269-296.

Sweatt, J. D. (1999). Toward a molecular explanation for long-term potentiation. *Learn Mem, 6*, 399-416.

Sweatt, J. D. (2001). The neuronal map kinase cascade: A biochemical signal integration system subserving synaptic plasticity and memory. *J Neurochem, 76*, 1-10.

Szapiro, G., Vianna, M. R., McGaugh, J. L., Medina, J. H., & Izquierdo, I. (2003). The role of nmda glutamate receptors, pka, mapk, and camkii in the hippocampus in extinction of conditioned fear. *Hippocampus, 13*, 53-58.

Thiels, E., Kanterewicz, B. I., Norman, E. D., Trzaskos, J. M., & Klann, E. (2002). Long-term depression in the adult hippocampus in vivo involves activation of extracellular signal-regulated kinase and phosphorylation of elk-1. *J Neurosci, 22*, 2054-2062.

Thomas, G. M., & Huganir, R. L. (2004). Mapk cascade signalling and synaptic plasticity. *Nat Rev Neurosci, 5*, 173-183.

Thomas, M. J., Kalivas, P. W., & Shaham, Y. (2008). Neuroplasticity in the mesolimbic dopamine system and cocaine addiction. *Br J Pharmacol, 154*, 327-342.

Thorndike, E. L. (1898). *Animal intelligence; [electronic resource] an experimental study of the associative processes in animals*: Macmillan, New York.

Totterdell, S., & Smith, A. D. (1989). Convergence of hippocampal and dopaminergic input onto identified neurons in the nucleus accumbens of the rat. *J Chem Neuroanat, 2*, 285-298.

Trifilieff, P., Herry, C., Vanhoutte, P., Caboche, J., Desmedt, A., Riedel, G., et al. (2006). Foreground contextual fear memory consolidation requires two independent phases of hippocampal erk/creb activation. *Learning & memory, 13*, 349-358.

Tropea, T. F., Kosofsky, B. E., & Rajadhyaksha, A. M. (2008). Enhanced creb and darpp-32 phosphorylation in the nucleus accumbens and creb, erk, and glur1 phosphorylation in the dorsal hippocampus is associated with cocaine-conditioned place preference behavior. *J Neurochem, 106*, 1780-1790.

Tzschentke, T. M. (1998). Measuring reward with the conditioned place preference paradigm: A comprehensive review of drug effects, recent progress and new issues. *Prog Neurobiol, 56*, 613-672.

Valjent, E., Corbille, A. G., Bertran-Gonzalez, J., Herve, D., & Girault, J. A. (2006). Inhibition of erk pathway or protein synthesis during reexposure to drugs of abuse erases previously learned place preference. *Proc Natl Acad Sci U S A, 103*, 2932-2937.

Valjent, E., Corvol, J. C., Pages, C., Besson, M. J., Maldonado, R., & Caboche, J. (2000). Involvement of the extracellular signal-regulated kinase cascade for cocaine-rewarding properties. *J Neurosci, 20*, 8701-8709.

Valjent, E., Corvol, J. C., Trzaskos, J. M., Girault, J. A., & Herve, D. (2006). Role of the erk pathway in psychostimulant-induced locomotor sensitization. *BMC Neurosci, 7*, 20.

Valjent, E., Pages, C., Rogard, M., Besson, M. J., Maldonado, R., & Caboche, J. (2001). Delta 9-tetrahydrocannabinol-induced mapk/erk and elk-1 activation in vivo depends on dopaminergic transmission. *Eur J Neurosci, 14*, 342-352.

Valjent, E., Pascoli, V., Svenningsson, P., Paul, S., Enslen, H., Corvol, J. C., et al. (2005). Regulation of a protein phosphatase cascade allows convergent dopamine and glutamate signals to activate erk in the striatum. *Proc Natl Acad Sci U S A, 102*, 491-496.

Volkow, N. D., Fowler, J. S., & Wang, G. J. (2003). The addicted human brain: Insights from imaging studies. *J Clin Invest, 111*, 1444-1451.

Wagner, F. A., & Anthony, J. C. (2002). Into the world of illegal drug use: Exposure opportunity and other mechanisms linking the use of alcohol, tobacco, marijuana, and cocaine. *Am J Epidemiol, 155*, 918-925.

Walker, D. L., Ressler, K. J., Lu, K. T., & Davis, M. (2002). Facilitation of conditioned fear extinction by systemic administration or intra-amygdala infusions of d-cycloserine as assessed with fear-potentiated startle in rats. *J Neurosci, 22*, 2343-2351.

Walz, R., Roesler, R., Barros, D. M., de Souza, M. M., Rodrigues, C., Sant'Anna, M. K., et al. (1999). Effects of post-training infusions of a mitogen-activated protein kinase kinase inhibitor into the hippocampus or entorhinal cortex on short- and long-term retention of inhibitory avoidance. *Behav Pharmacol, 10*, 723-730.

Walz, R., Roesler, R., Quevedo, J., Rockenbach, I. C., Amaral, O. B., Vianna, M. R., et al. (1999). Dose-dependent impairment of inhibitory avoidance retention in rats by immediate post-training infusion of a mitogen-activated protein kinase kinase inhibitor into cortical structures. *Behav Brain Res, 105*, 219-223.

Walz, R., Roesler, R., Quevedo, J., Sant'Anna, M. K., Madruga, M., Rodrigues, C., et al. (2000). Time-dependent impairment of inhibitory avoidance retention in rats by posttraining infusion of a mitogen-activated protein kinase kinase inhibitor into cortical and limbic structures. *Neurobiol Learn Mem, 73*, 11-20.

Wasserman, D. A., Havassy, B. E., & Boles, S. M. (1997). Traumatic events and post-traumatic stress disorder in cocaine users entering private treatment. *Drug Alcohol Depend, 46*, 1-8.

Watson, J.B., & Rayner, R. (1920). Conditioned emotional reactions. *Journal of Experimental Psychology, 3*, 1-14.

Whitfield, T. W., Jr., Shi, X., Sun, W. L., & McGinty, J. F. (2011). The suppressive effect of an intra-prefrontal cortical infusion of bdnf on cocaine-seeking is trk receptor and extracellular signal-regulated protein kinase mitogen-activated protein kinase dependent. *J Neurosci, 31*, 834-842.

WHO. (2001). Mental health: New understanding, new hope. *The world health report*: World Health Organisation.

Wise, R. A. (1996). Neurobiology of addiction. *Curr Opin Neurobiol, 6*, 243-251.

Wu, S. P., Lu, K. T., Chang, W. C., & Gean, P. W. (1999). Involvement of mitogen-activated protein kinase in hippocampal long-term potentiation. *J Biomed Sci, 6*, 409-417.

Xia, Z., Dudek, H., Miranti, C. K., & Greenberg, M. E. (1996). Calcium influx via the nmda receptor induces immediate early gene transcription by a map kinase/erk-dependent mechanism. *J Neurosci, 16*, 5425-5436.

Xie, G. Q., Wang, S. J., Li, J., Cui, S. Z., Zhou, R., Chen, L., et al. (2009). Ethanol attenuates the hfs-induced, erk-mediated ltp in a dose-dependent manner in rat striatum. *Alcohol Clin Exp Res, 33*, 121-128.

Yang, Y. L.Lu, K. T. (2005). Facilitation of conditioned fear extinction by d-cycloserine is mediated by mitogen-activated protein kinase and phosphatidylinositol 3-kinase cascades and requires de novo protein synthesis in basolateral nucleus of amygdala. *Neuroscience, 134*, 247-260.

Zhai, H., Li, Y., Wang, X., & Lu, L. (2008). Drug-induced alterations in the extracellular signal-regulated kinase (erk) signalling pathway: Implications for reinforcement and reinstatement. *Cell Mol Neurobiol, 28*, 157-172.